Nuclear Theory in the Age of Multimessenger Astronomy

Over the last decade, astrophysical observations of neutron stars—both as isolated and binary sources—have paved the way for a deeper understanding of the structure and dynamics of matter beyond nuclear saturation density. The mapping between astrophysical observations and models of dense matter based on microscopic dynamics has been poorly investigated so far. However, the increased accuracy of present and forthcoming observations may be instrumental in resolving the degeneracy between the predictions of different equations of state. Astrophysical and laboratory probes have the potential to paint to a new coherent picture of nuclear matter—and, more generally, strong interactions—over the widest range of densities occurring in the Universe.

This book provides a self-contained account of neutron star properties, microscopic nuclear dynamics and the recent observational developments in multimessenger astronomy. It also discusses the unprecedented possibilities to shed light on long standing and fundamental issues, such as the validity of the description of matter in terms of point-like baryons and leptons and the appearance of deconfined quarks in the high-density regime.

It will be of interest to researchers and advanced PhD students working in the fields of Astrophysics, Gravitational Physics, Nuclear Physics, and Particle Physics.

Key Features:

- Reviews state-of-the-art theoretical and experimental developments
- Self-contained and cross-disciplinary
- While being devoted to a very lively and fast developing field, the book fundamentally addresses methodological issues. Therefore, it will not be subject to fast obsolescence.

Nuclear Theory in the Age of Multimessenger Astronomy

Edited by
Omar Benhar
Alessandro Lovato
Andrea Maselli
Francesco Pannarale

CRC Press is an imprint of the
Taylor & Francis Group, an informa business

Designed cover image: T. Dietrich, T. Hinderer, and A. Samajdar: published in General Relativity and Gravitation (2021) 53:27 https://doi.org/10.1007/s10714-020-02751-6

First edition published 2024
by CRC Press
2385 NW Executive Center Drive, Suite 320, Boca Raton FL 33431

and by CRC Press
4 Park Square, Milton Park, Abingdon, Oxon, OX14 4RN

CRC Press is an imprint of Taylor & Francis Group, LLC

ISBN: 978-1-032-30775-6 (hbk)
ISBN: 978-1-032-30776-3 (pbk)
ISBN: 978-1-003-30658-0 (ebk)

DOI: 10.1201/9781003306580

Typeset in Nimbus font
by KnowledgeWorks Global Ltd.

Publisher's note: This book has been prepared from camera-ready copy provided by the authors.

Contents

Preface

The first detection of a gravitational-wave signal consistent with emission from a coalescing binary neutron star system, and the complementary observation of electromagnetic radiation by space- and ground-based telescopes, arguably heralded the dawn of a new age for both astrophysics and nuclear physics research.

In years to come, multimessenger observations are expected to provide unprecedented insight into neutron star structure and dynamics, which will allow to shed light on fundamental issues, such as the nature of nuclear interactions and the transition to new forms of matter—featuring the appearance of strange baryons and deconfined quarks—in the high-density regime typical of the neutron star core.

Gravitational and and electromagnetic observations have already provided valuable information on average properties of neutron star matter, such as the compressibility and the symmetry energy, useful to parametrise the equation of state determining the star mass and radius.

The data collected by the Neutron Star Interior Composition Explorer (NICER) telescope, installed on the International Space Station, have allowed to obtain reliable estimates of the radius of a neutron star of 1.4 solar masses which fall in the range $11.5 \lesssim R \lesssim 13$ km. The measured values of M and R turn out to be compatible with the results of most theoretical calculations based on nuclear many-body theory, predicting that the central density of the star does not exceed $\sim\ 3\ \varrho_0$, ϱ_0 being the central density of atomic nuclei. The emerging picture strongly suggests that in this density regime the description of nuclear matter in terms of point-like neutrons and protons interacting through non relativistic two- and three-nucleon potentials may still be applicable.

The radius of the neutron star J0740+6620 of mass $M = 2.072^{+0.067}_{-0.066}\ M_\odot$, with $M_\odot$ being the solar mass, has been also estimated to be $R = 12.39^{+1.30}_{-0.98}$ km, corresponding to a central density $\sim\ 4\ \varrho_0$. The small difference between the radii of stars of mass 1.4 and 2.072 $M_\odot$, implying that the equation of state is still stiff at $\varrho > 3\varrho_0$, appears to rule out a strong first-order transition to a phase comprising particles other than nucleons in the density range $3\varrho_0 \lesssim \varrho \lesssim 4\varrho_0$, although the occurrence of a continuous crossover cannot be excluded.

The long anticipated detection of the gravitational-wave event GW170817 performed by the LIGO/Virgo Collaboration, providing a measure of the tidal deformability of coalescing neutron stars of mass $\sim 1.2 - 1.6\ M_\odot$, also supports the validity of the model of neutron star matter based on nucleon degrees of freedom, and appears to rule out some of the stiffest equations of state.

Astronomical data are complemented by the information coming from terrestrial experiments, such as heavy-ion collisions and the recent measurement of the neutron skin thickness of lead, performed at Jefferson Lab by the PREX-II collaboration.

Electron-nucleus scattering experiments also provide independent insight into the limits of applicability of the description of nuclear systems in terms of nucleons. The observation of y-scaling, showing that the beam particles interact with high-momentum nucleons belonging to strongly correlated pairs, points to the occurrence of large fluctuations of the nuclear density, that can locally reach values comparable to the central density of massive neutron stars.

The development of a theoretical framework suitable for a unified treatment of nuclear systems, from the deuteron to neutron stars, involves the description of interactions among nucleons embedded in the nuclear medium over a broad density range. The nuclear Hamiltonians commonly employed to perform calculations of neutron star properties, however, are mainly based on a set of phenomenological inputs, including the properties of the exactly solvable two- and three- nucleon bound states, the large data base of precise nucleon-nucleon scattering data, and the equilibrium density of isospin-symmetric matter extrapolated from nuclear systematic. The use of these Hamiltonians to describe the properties of nuclear matter at densities up to and above five times nuclear density obviously entails a largely arbitrary extrapolation, which hampers a reliable assessment of the accuracy of theoretical predictions.

The potential of multimessenger observations performed by space- and ground-based facilities is presently being exploited in a variety of studies, mainly aimed at constraining the existing models of the equation of state of neutron star matter. Recently, it has been also suggested that a conceptually different strategy, pushing the investigation based on multimessenger astrophysical data to a deeper level, may offer an unprecedented opportunity to constrain not just the equation of state, but also the underlying microscopic model of nuclear dynamics in dense matter. The results of pioneering work along this direction, based on Bayesian inference, suggest that multimessenger observations have the potential to pin down the strength of repulsive three-nucleon interactions, which are known to play a dominant role in determining the stiffness of the equation of state at high densities.

In view of the amount and quality of the forthcoming astrophysical data, the prospects for the achievement of a reliable description of neutron star properties and, more generally, nuclear dynamics in the high-density regime appear to be bright. The contributions to this volume provide an overview of the state-of-the-art of theoretical approaches for the description of neutron star structure and dynamics, as well as a survey of the groundbreaking results of recent measurements of neutron star properties.

Omar Benhar, Alessandro Lovato, Andrea Maselli, Francesco Pannarale

Roma, November 2023

About the Authors

Omar Benhar is an INFN Emeritus Research Director, and has been teaching Relativistic Quantum Mechanics, Quantum Electrodynamics and Structure of Compact Stars at "Sapienza" University of Rome for over twenty years. He has worked extensively in the United States, and since 2013 has served as an adjunct professor at the Center for Neutrino Physics of Virginia Polytechnic Institute and State University. Prof. Benhar has authored or co-authored three textbooks on Relativistic Quantum Mechanics, Gauge Theories, and Structure and Dynamics of Compact Stars, and published more than one hundred scientific papers on the theory of many-particle systems, the structure of compact stars and the electroweak interactions of nuclei.

Alessandro Lovato is a physicist at Argonne National Laboratory and an INFN researcher in Trento. His research in theoretical nuclear physics focuses on consistently modeling the self-emerging properties of atomic nuclei and neutron-star matter in terms of the microscopic interactions among the constituent protons and neutrons. He has co-authored more than eighty scientific publications on the theory of many-particle systems, the structure of compact stars, and the electroweak interactions of nuclei. He is at the forefront of high-performance computing applied to solving the quantum many-body problem.

Andrea Maselli is an Associate Professor at the Gran Sasso Science Institute, in L'Aquila, where he teaches Gravitation and Cosmology and Physics of Black Holes. His research focuses on strong gravity, which plays a crucial role in many astrophysical phenomena involving black hole and neutron stars, representing natural laboratories to test fundamental physics. Prof. Maselli has co-authored more than eighty scientific papers on the modelling of black holes and neutron stars in General Relativity and extension thereof, their gravitational wave emission, and on tests of gravity in the strong-field regime. He is active in various collaborations aimed at developing next generation of gravitational wave detectors, such as the LISA satellite, the Einstein Telescope, and the Lunar Gravitational Wave Antenna.

Francesco Pannarale is an Associate Professor at "Sapienza" University of Rome, where he teaches Gravitational Waves, Compact Objects and Black Holes, Computing Methods for Physics, and Electromagnetism. His research interests are in gravitational-wave physics and multimessenger astronomy, and they range from modeling compact binary sources to data analysis. He has co-authored over one hundred and eighty scientific publications and was at the forefront of the joint observation of GW170817 and GRB 170817A. He is currently serving as co-chair of the LIGO-Virgo-KAGRA Data Analysis Council.

Contributors

Nils Andersson
University of Southampton
UK

Milena Bastos Albino
University of Coimbra
Portugal

Omar Benhar
INFN
Roma, Italy

Fiorella Burgio
INFN
Catania, Italy

Márcio Ferreira
University of Coimbra
Portugal

Steven P. Harris
Institute for Nuclear Theory
Seattle, WA, USA

Brynmor Haskell
Nicolaus Copernicus Astronomical Center
Warsaw, Poland

Toru Kojo
Tohoku University
Sendai, Japan

Alessandro Lovato
Argonne National Laboratory
Lemont, IL, USA

Andrea Maselli
GSSI
L'Aquila. Italy

Domenico Logoteta
University of Pisa
Italy

Tuhin Malik
University of Coimbra
Portugal

Albino Perego
University of Trento
Italy

Francesco Pannarale
Sapienza University
Roma, Italy

Constança Providência
University of Coimbra
Portugal

Rahul Somasundaram
Los Alamos National Laboratory
Los Alamos, NM, USA

Hans-Joseph Schulze
INFN
Catania, Italy

Ingo Tews
Los Alamos National Laboratory
Los Alamos, NM, USA

Isaac Vidaña
INFN
Catania, Italy

Anna Watts
University of Amsterdam
The Netherlands

Jin-Biao Wei
University of Geosciences
Wuhan, P.R. of China

CHAPTER 1

Astronomical Constraints on Composition and Dynamics of Neutron-Star Matter

Omar Benhar, Alessandro Lovato

PRECISE measurements of masses and radii of compact stars, combined with the detection of gravitational waves emitted from both coalescing systems and isolated stars, have the potential to shed new light on the nature of matter at density exceeding by many orders of magnitude that typical of terrestrial macroscopic systems. Of utmost importance, in this context, will be studies aimed at determining the limits of applicability of the paradigm underlying nuclear theory—according to which nuclear systems can be treated as collections of point like nucleons, whose dynamics are described by a nonrelativistic Hamiltonian—and establishing the appearance of new forms of matter, such as strange baryonic matter or quark matter. In this chapter, we review the empirical information on the density regime in which nucleons are believed to be the relevant degrees of freedom, as well as the minimal requirements that must be met by theoretical models of nuclear structure and dynamics suitable for the interpretation of multimessenger observations of neutron star properties.

1.1 PROLOGUE

The development of a theoretical description of matter in the interior of neutron stars, the density of which, ϱ, encompasses many orders of magnitude, is a challenging—in fact, even daunting—endeavour. The properties of the crustal region, corresponding to densities in the range $4 \times 10^7 \leq \varrho \leq 4 \times 10^{11}$ g cm^{-3}, can be largely modelled resorting to the empirical information inferred from nuclear properties. On the other hand, the description of the inner region, in which ϱ is believed to be as high as $(5 - 10) \times \varrho_0$, with $\varrho_0 \approx$

DOI: 10.1201/9781003306580-1

2.6×10^{14} g cm^{-3} being the central density of atomic nuclei[1], involves a large degree of extrapolation and speculation.

The basic assumption underlying neutron star modelling based on nuclear theory is that, as long as the the range of nuclear forces is negligible compared to the radius of spacetime curvature, the microscopic dynamics of matter in the star interior are totally unaffected from gravitational effects. It follows that under this condition, nuclear interactions can be treated as if they were taking place in flat spacetime. In the classic book of Harrison *et al.*, the density at which the validity of this approximation breaks down is estimated to be $\varrho \sim 10^{49}$ g cm^{-3} [1].

All theoretical models predict that at densities $\varrho_0 \lesssim \varrho \lesssim 2\varrho_0$, the ground state of matter is a charge-neutral uniform fluid consisting mostly of neutrons, with a small admixture of protons and leptons in equilibrium with respect to neutron beta decay and lepton capture by protons, referred to as $npe\mu$ matter. On the other hand, a firm determination of the upper limit of the density region in which the description in terms of nucleons and leptons is applicable involves nontrivial difficulties. At densities beyond those typical of the $npe\mu$ regime, new forms of matter are expected to appear as a result of weak interaction processes turning nucleons into strange baryons, such as the hyperons Λ_0, and Σ^-. Moreover, at even larger densities the baryons—which are known to be composite finite-size systems—are bound to overlap, giving way to the appearance of a new phase consisting of deconfined quark matter, predicted by the fundamental theory of strong interactions.

From the above discussion, it follows that the quantitative understanding of nuclear matter properties is a crucial requisite for the achievement of a comprehensive description of neutron-star structure and dynamics. Owing to the complexity and nonperturbative nature of nuclear forces, however, the fulfilment of this goal entails severe difficulties. In fact, even the determination of the equilibrium properties of isospin symmetric matter, involving the calculation of the ground state energy as a function of density, requires the development of highly sophisticated dynamical models and computational techniques.

Advanced theoretical studies of dense nuclear matter are mainly based on either Relativistic Mean-Field (RMF) theory or nonrelativistic Nuclear Many-Body Theory (NMBT); see contributions of C. Providência *et al.* and G.F. Burgio *et al.* to this volume, respectively. Both RMF theory and NMBT are founded on the paradigm according to which all nuclear systems can be pictured as collections of interacting point like nucleons.

In RMF theory, nuclear matter is described using the formalism of quantum field theory, with the nucleons being treated as Dirac fermions interacting through exchange of scalar, vector and vector-isovector mesons. The equations of motion are solved within the mean-field approximation, which amounts to replacing the meson fields with classical fields. This scheme has the important advantage of being inherently consistent with relativity and manifestly covariant. On the other hand, the mean field approximation entails the use of an oversimplified model of nuclear dynamics. For example, within RMF theory the contribution of the one-pion exchange (OPE) interaction—famously proposed by Yukawa in 1935 [2] and associated with the exchange of the pseudo-scalar π-meson—turns out to be vanishing.

[1]According to the liquid drop model of the nucleus, ϱ_0 is the equilibrium density of isospin-symmetric matter, also referred to as nuclear matter saturation density.

In the nuclear Hamiltonians employed in studies based on NMBT, Yukawa's OPE potential, which provides an accurate description of the nucleon–nucleon (NN) force at large distances, is supplemented with largely phenomenological contributions meant to describe intermediate- and short-range interactions. In addition to the potential describing two-nucleon interactions, realistic Hamiltonians also comprise a potential taking into account the occurrence of irreducible three-nucleon (NNN) interactions. The emergence of these forces is a manifestation of the internal structure of the nucleon, which is disregarded altogether in the NMBT picture; see, e.g., Ref. [3].

Both the Lagrangian density of RMF theory and the Hamiltonian of NMBT are derived using a phenomenological approach, in which a set of parameters—determining the strength and range of the interactions—are adjusted in such a way as to reproduce the data obtained from experimental studies of nuclear structure and reactions. The inherent limitation of this procedure lies in the fact that it can mainly be used to constrain the description of interactions occurring in the density regime typical of atomic nuclei, corresponding to $\varrho \approx \varrho_0$. Extrapolation of the resulting dynamical models to densities relevant to the neutron-star core involves a great deal of uncertainty and requires careful consideration.

The combined analysis of multimessenger observations of neutron stars has the unprecedented potential to further constrain the available theoretical models of dense matter and address critical outstanding issues, such as the nature of its constituents and their interactions. Of foremost importance in this context will be analyses aimed at constraining the available models of NNN interactions, the contribution of which becomes large, or even dominant, at high densities. In this chapter, we analyse a nuclear Hamiltonian model extensively employed to carry out calculations of nuclear matter properties within the framework of NMBT and discuss its capability to describe nuclear dynamics at high density. The ongoing efforts aimed at improving upon the paradigm of Nuclear Theory, as well as the results of pioneering studies in which multimessenger astronomical data have been exploited to shed light on the details of microscopic nuclear dynamics, will be also discussed.

1.2 NUCLEON DYNAMICS IN DENSE MATTER

As pointed out above, NMBT is based on the hypothesis that all nucleon systems—from the deuteron, having mass number $A = 2$, to neutron stars, wherein $A \approx 10^{57}$—can be described in terms of point like protons and neutrons, the interactions of which are driven by the Hamiltonian

$$H = \sum_i \frac{\mathbf{p}_i^2}{2m} + \sum_{i<j} v_{ij} + \sum_{i<j<k} V_{ijk} \,, \tag{1.1}$$

where m and $\mathbf{p}_i$ denote the mass and momentum of the i-th particle.[2]

1.2.1 The nucleon-nucleon interaction

NN potentials that are local or semi-local in coordinate space are usually written in the form

$$v_{ij} = \sum_p v^p(r_{ij}) O^p_{ij} \,, \tag{1.2}$$

[2]In this Chapter, we adopt the system of natural units, in which $\hbar = c = k_B = 1$. Unless otherwise specified, we also neglect the small proton-neutron mass difference.

where $r_{ij} = |\mathbf{r}_i - \mathbf{r}_j|$ is the distance between the interacting particles. They are designed to reproduce the measured properties of the two-nucleon system, in both bound and scattering states, and reduce to Yukawa's OPE potential at large distances. The sum in Eq. (1.2) includes up to 18 terms, with the corresponding operators, O^p, being needed to describe the strong spin-isospin dependence and non central nature of nuclear forces ($i = 1, \ldots, 6$), as well as the occurrence of spin-orbit and other angular-momentum-dependent interactions ($i = 7, \ldots 14$). State-of-the-art phenomenological potentials, such as the Argonne v_{18} (AV18) model, also feature additional terms accounting for small violations of charge symmetry and charge independence ($i = 15, \ldots, 18$) [4].

The definition of the AV18 potential involves 40 parameters, the values of which are adjusted in such a way as to reproduce the properties of the only observed two-nucleon bound state—the nucleus of deuterium, or deuteron—as well as the measured NN cross sections. In general, phenomenological potential models are capable of providing an accurate description of NN scattering in vacuum by construction. However, their use to describe collisions in dense matter deserves thorough consideration.

It has to be kept in mind that in strongly degenerate fermion systems, such as cold nuclear matter, only nucleons with momentum close to the Fermi momentum k_F can participate in scattering processes. It follows because k_F is trivially related to matter density through $k_F = (6\pi^2\varrho/\nu)^{1/3}$—where ν denotes the degeneracy of the momentum eigenstates—so is the typical center-of-mass energy of NN scattering. For example, in the case of head-on collisions in pure neutron matter (PNM) at density ϱ, one finds [5]

$$E_{\rm cm} = \frac{1}{m}(3\pi^2\varrho)^{2/3} . \tag{1.3}$$

The above equation suggests that phenomenological potentials can be reliably used to describe nuclear matter up to a density determined by the upper limit of the energy range in which they accurately reproduce NN scattering data. For example, the AV18 model, which provides an excellent fit of the phase shifts at energy as high as 600 MeV, is expected to be adequate up to densities $\varrho \gtrsim 4\varrho_0$.

Over the past two decades, a more elegant and fundamental approach originally proposed by S. Weinberg [6], in which the NN potential is obtained within the framework of chiral effective field theory (χEFT), has enjoyed an enormous amount of popularity within the Nuclear Theory community. Within this framework, long- and intermediate-range nuclear forces, originating from pion exchange processes, are fully determined by pion-nucleon observables, whereas short-range interactions are described by contact terms involving a set of additional parameters, adjusted in such a way as to reproduce the NN scattering phase shifts. However, being derived from a low-momentum expansion, potentials based on χEFT are inherently limited when it comes to describe NN scattering at high energy. For this reason, they are generally regarded as adequate to provide reliable predictions of nuclear matter properties only for $\varrho \lesssim 2\varrho_0$ [5, 7].

1.2.2 Irreducible three-nucleon interactions

The NN potentials discussed in the previous section, designed to explain the properties of the two-nucleon system, fail to reproduce the binding energies of ^{3}He and ^{4}He obtained from highly accurate solutions of the Schrödinger equation. This problem is taken care of by including in the nuclear Hamiltonian a potential describing irreducible NNN interactions.

Phenomenological NNN potential such as the Urbana IX model [8]—which has been extensively used in conjunction with the AV18 NN potential to study neutron star properties [9, 10]—are written in the form

$$V_{ijk} = V_{ijk}^{2\pi} + V_{ijk}^{R} \, , \tag{1.4}$$

where $V_{ijk}^{2\pi}$ is an attractive contribution arising from two-pion exchange processes in which one of the participating nucleons is excited to the spin $3/2$ state of mass 1232 MeV referred to as Δ resonance. The second term on the right-hand side of Eq. (1.4) is purely phenomenological, repulsive and independent of isospin. The strengths of the two contributions to V_{ijk} are determined in such a way as to reproduce the binding energies of ^{3}He and ^{4}He, as well as the empirical equilibrium density of isospin symmetric nuclear matter (SNM), whose value can be inferred from nuclear systematics.

Potentials describing NNN interactions can also be consistently derived within the formalism of χEFT. A detailed comparison between the results of this approach and those obtained from the purely phenomenological approach outlined above can be found in Ref. [11]. Note that unlike the NN potentials, both the χEFT and the phenomenological NNN potentials are constrained at densities around ϱ_0 only.

1.2.3 Effective nuclear Hamiltonians

A prominent feature of the NN potential is the presence of a strongly repulsive core, which prevents from using v_{ij} to perform perturbative calculations in the basis of eigenstates of the non interacting system. Historically, this problem has been circumvented using well-behaved effective potentials designed to explain the bulk properties of nuclear matter, such as the Skyrme interaction [12, 13]. While being remarkably successful in a number of applications; however, this approach, being inherently inadequate to describe NN scattering and the few-nucleon bound states, lacks a direct connection with nuclear dynamics at microscopic level.

Early attempts to derive an effective interaction from a microscopic NN potential are described in Refs. [14, 15]. They were based either on the replacement of the bare potential with the G-matrix, describing NN scattering in the nuclear medium, or on a modification of the two-nucleon wave function meant to suppress the probability of finding two nucleons at distances $r \lesssim r_c$, with r_c being the radius of the repulsive core. The G-matrix formalism is thoroughly described in the contribution of G.F. Burgio *et al.* to this volume.

Following the groundbreaking work of Ref. [16], the authors of Refs. [17] have developed a procedure—based on the same conceptual framework underlying the work of Ref. [14]—to determine the effective interaction using the formalism of Correlated Basis Functions (CBF) and the cluster expansion technique, described in, e.g., Ref.. [18]. The CBF effective Hamiltonian is defined by the equation

$$\langle H \rangle = \langle H_{\rm eff} \rangle_{FG} = \langle T \rangle_{FG} + \langle V_{\rm eff} \rangle_{FG} \, , \tag{1.5}$$

where $\langle \ldots \rangle$ and $\langle \ldots \rangle_{FG}$ denote expectation values in the ground state of the interacting and noninteracting systems, respectively, and T is the kinetic energy operator. The definition of $V_{\rm eff}$ involves a set of correlation functions, meant to account for screening effects arising from the repulsive core of the NN potential. Their range is adjusted in such a way as to reproduce the value of $\langle H \rangle$ obtained from the bare Hamiltonian using advanced computational techniques, such as the Auxiliary Field Diffusion Monte Carlo (AFDMC) method or the variational approach known as Fermi Hyper-Netted Chain/Single Operator Chain (FHNC/SOC); see Ref. [17].

The effective interaction of Ref. [17], which consistently includes the contributions of both NN and NNN interactions, has been derived from the nuclear Hamiltonian consisting of the Argonne v_6' NN potential (AV6P)—obtained by projecting the full AV18 potential into the operator basis including the O^p_{ij} of Eq.(1.2) with $p \leq 6$—and the UIX NNN potential. The results of accurate calculations of the energy of PNM, reported in Ref. [19], show that the predictions of the somewhat simplified AV6P+UIX Hamiltonian turn out to be in close agreement with those obtained using the full AV18+UIX model. The main advantage of using the CBF effective Hamiltonian lies in the fact that it is well behaved and can be employed to perform perturbative calculations of any properties of nuclear matter at both zero and finite temperatures [20].

1.2.4 Relativistic corrections to the nuclear Hamiltonian

The approach based on NMBT, while allowing a good description of nuclear-bound states and NN scattering data, fails to fulfill the constraint of causality, because it leads to predict a speed of sound in matter that exceeds the speed of light at large density.

Possible modifications of nuclear dynamics due to relativistic effects can be treated as corrections by using an extension of the formalism of NMBT. They have been analysed in the few-nucleon systems, ^{3}He and ^{4}He, using a Hamiltonian defined as in Eq. (1.1) and advanced Quantum Monte Carlo computational techniques [21]. The authors of this study considered effects arising from the replacement

$$\sum_{i=1}^{A} \frac{\mathbf{p}_i^2}{2m} \to \sum_{i=1}^{A} \sqrt{m^2 + \mathbf{p}_i^2} - m \, , \tag{1.6}$$

as well as corrections associated with the relative and center-of-mass momenta of the interacting nucleon pair

$$\mathbf{p}_{ij} = \frac{1}{2}(\mathbf{p}_i - \mathbf{p}_j) \quad , \quad \mathbf{P}_{ij} = \mathbf{p}_i + \mathbf{p}_j \, . \tag{1.7}$$

Note that within the phenomenological approach, in which v_{ij} is determined by fitting two-nucleon data, the effects of replacing the nonrelativistic kinetic energy with its relativistic counterpart are buried in the NN potential.

The results of Ref. [21] do not show significant relativistic effects originating from large relative momentum. On the other hand, corrections involving the momentum $\mathbf{P}_{ij}$—taking into account the fact that the NN potential is determined in the center-of-mass frame of the interacting nucleon pair, corresponding to $|\mathbf{P}_{ij}| = 0$—turn out to be important.

In order to describe NN interactions in the nuclear medium, v_{ij} must be boosted to a frame in which $|\mathbf{P}_{ij}| \neq 0$. The resulting expression is

$$v_{ij}(\mathbf{r}) \to v_{ij}(\mathbf{r}) + \delta v_{\mathbf{ij}}(\mathbf{P}, \mathbf{r}) , \tag{1.8}$$

with

$$\delta v_{ij}(\boldsymbol{P}, \boldsymbol{r}) = -\frac{P^2}{8m^2} v^s_{ij}(\boldsymbol{r}) + \frac{(\boldsymbol{P} \cdot \boldsymbol{r})}{8m^2} \boldsymbol{P} \cdot \boldsymbol{\nabla} v^s_{ij}(\boldsymbol{r}) , \tag{1.9}$$

where $\mathbf{r}$ is the NN distance and $v^s_{ij}(\boldsymbol{r})$ denotes the static part of the potential [22, 23].

In order to explain the data, the nuclear Hamiltonian employed in the fit must be modified to include both the boost-corrected NN potential and a *modified* NNN potential. This amounts to substituting

$$H \to H_R = \sum_i \frac{\mathbf{p_i}^2}{2m} + \sum_{i<j} [v_{ij} + \delta v_{ij}] + \sum_{i<j<k} V^*_{ijk} \; . \tag{1.10}$$

The inclusion of δv entails a significant softening of the repulsive part of the NNN potential, V^R_{ijk}, while $V^{2\pi}_{ijk}$ is left unchanged. The full correction to the energy $\langle H_R \rangle$ can be accurately described in the form

$$\delta\langle H_R \rangle = \langle \delta v \rangle - \gamma \langle V^R_{ijk} \rangle , \tag{1.11}$$

with $\gamma \approx 0.3$. From the above equations it follows that $V^*_{ijk} = (1-\gamma)V_{ijk}$.

1.2.5 The equation of state of $npe\mu$ matter

The equation of state (EOS) is a nontrivial relation between the thermodynamic functions specifying the state of a physical system. The simplest example is Boyle's law of ideal gases, stating that the pressure of a collection of non interacting classical particles at temperature T depends on density according to $P = \varrho T$. In the case of neutron star matter in the density regime in which nucleons and leptons are the dominant degrees of freedom, the EOS provides the relation between pressure and density—or, equivalently, energy density—of matter, the temperature of which can be safely assumed to be zero.[3]

Nucleons and leptons are in equilibrium with respect to the weak interaction processes

$$n \to p + \ell + \bar{\nu}_\ell \quad , \quad p + \ell \to n + \nu_\ell , \tag{1.12}$$

where $\ell = e, \mu$ denotes the lepton flavour. It follows that for any given matter density ϱ, the neutron, proton and lepton densities are uniquely determined by the requirements of charge neutrality and chemical equilibrium

$$\varrho_p = \sum_\ell \varrho_\ell , \tag{1.13}$$

$$\mu_n = \mu_p + \mu_\ell , \tag{1.14}$$

[3] The reference neutron star temperature, $T \sim 10^9$ K, corresponds to a thermal energy $\sim$ 100 keV, to be compared to Fermi energies of tens of MeV.

where ϱ_α and μ_α denote, respectively, density and chemical potential of the particles of species α. Note that Eq. (1.14) is derived under the assumption that neutrinos and antineutrinos do not interact appreciably with matter; therefore, they have vanishing densities and chemical potentials.

The EOS of $npe\mu$ matter is generally computed treating the charged leptons as non interacting particles, and assuming that the energy per nucleon at matter density ϱ and proton fraction $Y_p = \varrho_p/\varrho$ depend quadratically on the neutron excess $\delta = 1 - 2Y_p$. This assumption, the validity of which is confirmed by the results of detailed microscopic calculations [17], allows to determine the energy at arbitrary Y_p from the energies of PNM and SNM, corresponding to $Y_p = 0$ and 1, respectively. The pressure is then obtained from the thermodynamic definition

$$P(\varrho) = \varrho^2 \frac{\partial(E/N)}{\partial \varrho} , \tag{1.15}$$

where E and N denote the energy and the number of nucleons.

Figure 1.1 illustrates the effect of irreducible NNN interactions and relativistic boost corrections on the EOS of $npe\mu$ matter, obtained from results of the accurate variational calculations performed by Akmal and Pandharipande and Ravenhall using the FHNC/SOC scheme [9, 10]. For reference, the shaded area shows the region consistent with the flow data extracted from the analysis of relativistic heavy-ion collisions, reported in Ref. [24].

The dashed line represents the EOS corresponding to the AV18+UIX nuclear Hamiltonian—comprising the AV18 NN potential and the UIX NNN potential—whereas the dotted line has been obtained neglecting the contribution of the NNN potential. It is apparent that the occurrence of interactions involving (NNN) plays a critical role, and leads to a significant increase in the *stiffness* of the EOS, measured by the matter incompressibility modulus $K = 9(\partial P/\partial \varrho)_{\varrho=\varrho_0}$. On the other hand, the inclusion of the relativistic boost correction of Eq. (1.11), which entails a significant suppression of the isoscalar repulsive contribution to the NNN potential, leads to an appreciable softening EOS. The contribution of the NNN potential to the energy per particle of PNM is illustrated in Fig. 1.2, showing the density dependence on the expectation value $\langle V_{ijk} \rangle / N$ in PNM corresponding to the UIX and UIX* models.

The incompressibility modules obtained from the AV18, AV18+UIX and AV18 + δv + UIX* EOSs turn out to be all compatible with the results obtained from analyses of the available experimental data, pointing to a value $K_{\rm expt} = 240 \pm 20$ MeV [25, 26]. It has to be kept in mind, however, that stiffness is a *local*—that is, density-dependent—property of the EOS, while the compressibility module, which is often used to parametrise the ϱ-dependence of the energy of nuclear matter, only provides information relevant to the region $\varrho \approx \varrho_0$. In this region the contribution of NNN forces to the total interaction energy is comparatively small; for example, in PNM at $\varrho = \varrho_0$, $\langle V_{ijk} \rangle$ turns out to account for $\sim$ 20% of the total potential energy [27]. In this chapter, we will refer to the dynamical models based on the AV18+UIX and AV18 + δv + UIX* Hamiltonians as APR1 and APR2 model, respectively.

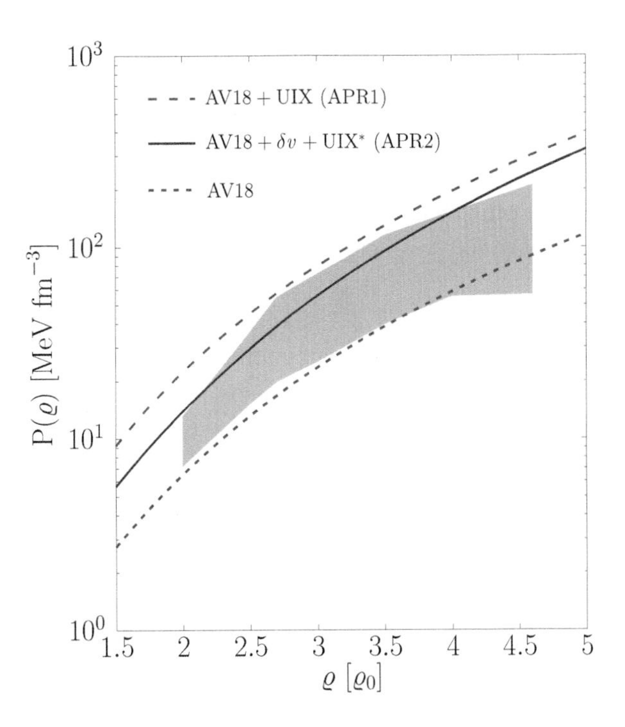

Figure 1.1 Density dependence of the pressure of $npe\mu$ matter, obtained from the results reported by Akmal, Pandharipande and Ravenhall in Refs. [9, 10]. The dotted and dashed lines correspond to the AV18, and AV18+UIX nuclear Hamiltonians, respectively, while the solid line represents the results of calculations performed using the Hamiltonian H_R of Eq. (1.10), which includes relativistic boost corrections. The shaded area corresponds to the region consistent with the flow data extracted from the analysis of relativistic heavy-ion collisions [24]. The density is given in units of the equilibrium density of SNM.

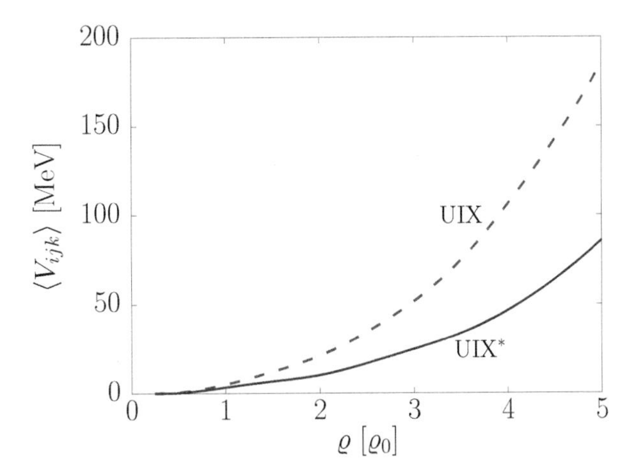

Figure 1.2 Density dependence of the expectation value $\langle V_{ijk} \rangle / N$ in PNM corresponding to the UX and UIX* models, represented by the dashed and solid lines, respectively.

1.3 NEUTRON STARS PROPERTIES

The zero temperature approximation discussed previously, while being capable to provide an accurate description of matter in neutron stars, is long known to be inadequate when it comes to studying the properties of supernovae and protoneutron stars, the temperature of which can reach several tens of MeV. Thermal modifications of nuclear matter properties are expected to play an even more important role in the merger and postmerger phases of binary neutron star coalescence, in which the temperature can be as high as $\sim$100 MeV [28–33].

The results discussed in Section 1.2.5 suggest that the present development of Nuclear Theory allows to perform accurate calculations of the EOS of cold nuclear matter using the Hamiltonian of Eq. (1.1) and reliable computational techniques. On the other hand, effective Hamiltonians appear to be best suited to perform thermodynamically consistent calculations of the properties of matter at non-zero temperature [20, 34], as well as to describe properties other than the EOS. Notable examples are the neutrino mean free path in matter [35–38] and the bulk viscosity leading to the appearance of dissipative processes [39]. Detailed discussions of the bulk viscosity of nuclear matter can be found in the contributions of S.P. Harris and N. Andersson & B. Haskel to this volume.

1.3.1 Zero-temperature regime

Mass and radius of stable neutron star

The equilibrium properties of neutron stars are obtained by combining the equations of hydrostatic equilibrium with Einstein's equations of General Relativity (GR). This procedure leads to the determination of a system of two first-order differential equations, known as Tolman–Oppenheimer–Volkoff (TOV) equations. Given a model of EOS, for any value of the energy density at the center of the star, ϵ_c, the solution of TOV equations provides the values of mass and radius of the stable equilibrium configurations of the star.

Panels (A) and (B) of Fig. 1.3 illustrate the ϵ_c-dependence of the mass, $M(\epsilon_c)$, and the mass-radius relation, $M(R)$, obtained using the AV18, AV18 + UIX and AV18 + δv + UIX* nuclear Hamiltonians, displayed by the dotted, dashed and solid lines, respectively. Note that *stable* equilibrium configurations lie in the region of $dM/d\epsilon_c \geq 0$, with the maximum mass corresponding to $dM/d\epsilon_c = 0$.

The increased stiffness of the EOS resulting from the inclusion of NNN interactions is associated with a larger maximum mass and a larger radius for stars of equal mass. Relativistic boost corrections, the main effect of which in the density regime relevant to neutron stars is the softening of the repulsive NNN potential, also turn out to have a significant impact on both the maximum mass and the mass-radius relation.

Tidal deformation in coalescing binary systems

A tide is the deformation of a body produced by the gravitational pull of another nearby body. Because the deformation depends on the body's internal structure, observations of tidal effects in binary neutron star systems have the potential to provide valuable information on the EOS of neutron star matter.

The orbital motion of two stars gives rise to the emission of gravitational waves (GW), that carry away energy and angular momentum. This process leads to a decrease in the

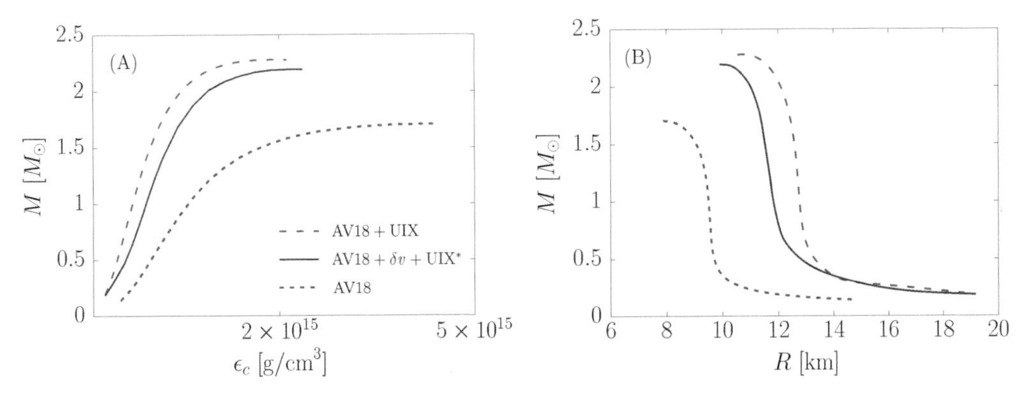

Figure 1.3 (A) Central density dependence of the mass of stable neutron stars, in units of the solar mass, obtained from the solution of TOV with the EOS of $npe\mu$ matter discussed in the text. (B) Mass-radius relation of stable neutron stars obtained from the solution of TOV equations. The meaning of the curves is the same as in panel (A).

orbital radius and, conversely, to an increase in the orbital frequency. In the early stage of the inspiral, characterised by large orbital separation and low frequency, the two stars—having mass M_1 and M_2, with $M_1 \geq M_2$—behave as point-like bodies, and the evolution of the frequency is primarily determined by the chirp mass $\mathcal{M}$, defined as

$$\mathcal{M} = \frac{(M_1 M_2)^{3/5}}{(M_1 + M_2)^{1/5}} . \tag{1.16}$$

The details of the internal structure become important as the orbital separation approaches the size of the stars. The tidal field associated with one of the bodies induces a mass-quadrupole moment on the companion, which in turn generates the same effect on the first one, thus accelerating coalescence. This effect is quantified by the tidal deformability, defined as

$$\Lambda = \frac{2}{3} k_2 \left(\frac{R}{GM} \right)^5 , \tag{1.17}$$

where M and R are the star mass and radius, respectively, and k_2 is called second tidal *Love number* [40]. For any given stellar mass, the radius and the tidal Love number are uniquely determined by the EOS of neutron star matter, and so is the tidal deformability [41]. A detailed discussion of tidal deformations of neutron stars is provided in the contribution of A. Maselli and F. Pannarale to this volume.

The sensitivity of the tidal deformability to the dynamics of neutron star matter is illustrated in Fig. 1.4, showing the stellar mass dependence of Λ obtained from the AV18+UIX and AV18 + δv + UIX* Hamiltonians. For a star of mass $M = 1.4\ M_\odot$, the predictions of the two models turn out to be $\sim$ 50% apart, and appreciable differences are visible for mass values up to $\sim$ 1.6 M$_\odot$. The circles of Fig. 1.4, represent the tidal deformability obtained using the CBF effective potential of Ref. [17], derived from the AV18+UIX Hamiltonian. The close agreement with the results displayed by the full line suggests that the CBF formalism provides a reliable framework, suitable to renormalise the NN potential and consistently take into account (NN) and NNN interactions.

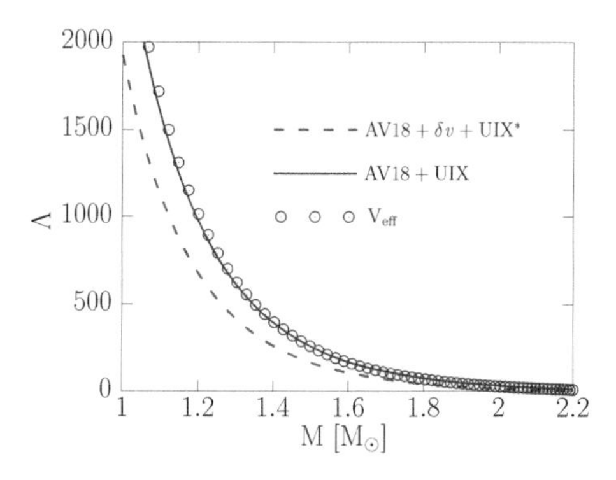

Figure 1.4 Stellar mass dependence of the tidal deformability Λ, defined by Eq. (1.16), obtained using different nuclear Hamiltonians. See text for details.

Neutron star oscillations

When a neutron star is perturbed by an external or internal event, it can be set into non-radial oscillations, leading to the emission of GW at the characteristic *complex* frequencies of its quasi-normal—that is, *damped*—modes (QNM). The frequencies and damping times of the QNMs carry valuable information on both the structure of the star and the properties of matter in its interior. They can be obtained from the analysis of the adiabatic perturbations of an equilibrium configuration corresponding to a given EOS, which involves the solution of the linearised Einstein equations, coupled to the equations of hydrodynamics, with suitable boundary conditions; see, e.g., Ref. [42].

If the unperturbed star is assumed to be static and non-rotating, it is convenient to expand the perturbation in tensorial spherical harmonics. This procedure allows to pin down even- and odd-parity solutions, referred to as polar and axial modes, respectively.

The polar oscillation modes are relativistic generalisation of the tidal perturbations of Newton's theory and couple the perturbations of the gravitational field to the perturbations of matter in the star interior. They can be classified according to the source of the restoring force which dominates in bringing a perturbed fluid element back to equilibrium. This classification scheme, dating back to the 1940s, comprises g, f and p modes. In the case of g-modes, or gravity modes, the restoring force originates from a change of density, that is, from buoyancy, while for p-modes is due to a change of pressure. The *fundamental* mode, or f-mode, is the relativistic generalisation of the only possible oscillation mode of an incompressible homogeneous sphere.

In general relativity, there exists an additional family of modes, called w-modes, associated with pure space-time oscillations, the corresponding motion of the fluid being negligible. The w-modes are characterised by frequencies typically higher than those of the g, f and p modes, and much smaller damping times, implying that these modes are highly damped.

The frequency of the f-mode has been determined using several EOSs, including some obtained from highly realistic dynamical models. In addition, it has been shown that it is possible to infer empirical relations between the mode frequency and the macroscopic parameters specifying the star configuration, that is, mass and radius [43, 44].

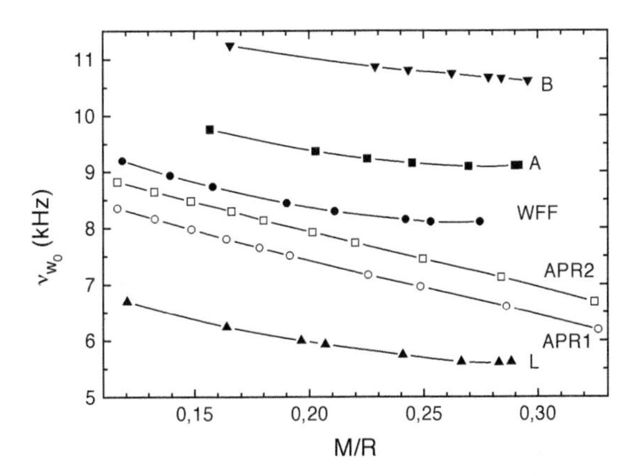

Figure 1.5 Frequency of the first axial w-mode obtained from different models of the nuclear matter EOS—ordered according to stiffness—displayed as a function of the star compactness. The models labelled B and L correspond to the softest and stiffest EOSs, respectively. Reprinted from Ref. [45] with permission. © RAS (1999).

The imprint of the nuclear matter EOS of on the properties of QNMs is clearly illustrated by the results of the study of Ref. [45], whose authors have computed the frequency of the first w-mode of neutron star configurations spanning a broad range of masses using a set of EOSs derived from different models of neutron star matter. Figure 1.5 shows that for any given value of the star compactness—that is, of the ratio M/R—the pulsation frequencies turn out to be ordered according to the stiffness of the EOS. Higher frequencies always correspond to softer EOSs, regardless of the value of M/R, and the pattern observed in the figure reflects the monotonic increase in compressibility moving from model B to model L. In addition, because the curves corresponding to different models do not cross each other within the range displayed in the figure, the results of Fig. 1.5 suggest that a measurement of the frequency of the first w-mode may allow to pin down the underlying EOS independent of the star compactness.

1.3.2 Finite-temperature regime

The description of hot nuclear matter using the formalism of NMBT is based on the assumption that at temperatures $T \ll m_\pi$, $m_\pi \approx 135$ MeV being the mass of the π-meson, thermal effects do not significantly affect strong-interaction dynamics [20, 34]. In this temperature regime, thermal modifications of nuclear matter properties arise primarily from the Fermi distribution

$$n_\alpha(k,T) = \Big\{1 + \exp\left[\beta(e_{\alpha k} - \mu_\alpha)\right]\Big\} \ , \tag{1.18}$$

where $\beta = 1/T$, $e_{\alpha k}$ denotes the energy of a particle of species α carrying momentum k, and μ_α is the corresponding chemical potential. Comparison to the $T \to 0$ limit

$$n_\alpha(k,0) = \theta(\mu_\alpha - e_{\alpha k}) \ , \tag{1.19}$$

where $\theta(x)$ is the Heaviside theta-function, shows that the probability distribution $n_\alpha(k, T > 0)$ is reduced from unity in the region corresponding to $\mu_\alpha - T \lesssim e_{\alpha k} \lesssim \mu_\alpha$,

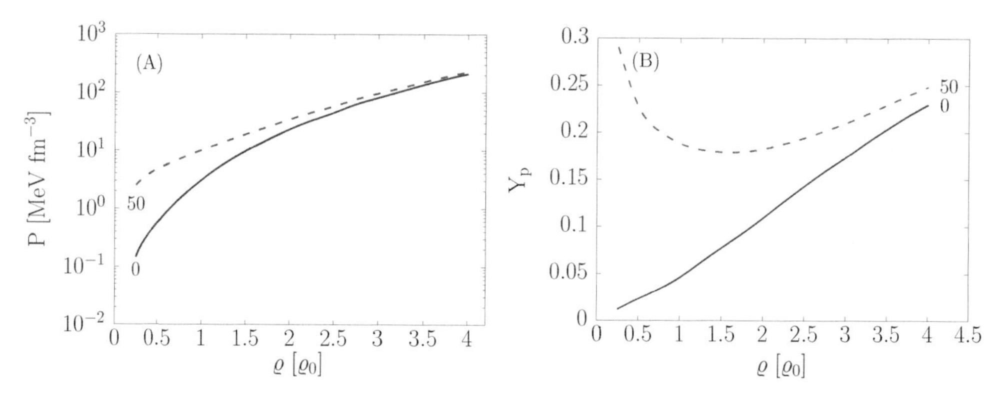

Figure 1.6 (A) Density dependence of the pressure of $npe\mu$ matter computed by the authors of Ref. [20] using an effective nuclear Hamiltonian derived from the AV18 + UIX model [17]. The solid and dashed lines correspond to temperature $T = 0$ and 50 MeV, respsctively. (B) Density dependence of the proton fraction in charge-neutral β-stable matter. The line code is the same as in panel (A).

while acquiring a non-vanishing positive value for $\mu_\alpha \lesssim e_{\alpha k} \lesssim \mu_\alpha + T$. It follows that for any given temperature T, the extent of thermal modifications to the Fermi distribution is driven by the ratio $2T/\mu_{\alpha k}$. This observation in turn implies that because the chemical potential is a monotonically increasing function of the particle density ϱ_α over a broad range of temperatures, for any given T thermal effects are more pronounced at lower ϱ_α. On the other hand, they become vanishingly small in the high-density regime, in which degeneracy dominates.

This feature clearly emerges from the results displayed in panel (A) of Fig. 1.6, showing a comparison between the EOSs of $npe\mu$ matter at temperature $T = 0$ and 50 MeV. It is apparent that at densities above $\sim 2.5\ \varrho_0$ the deviation from the zero-temperature EOS becomes hardly visible, and degeneracy pressure dominates.

Density-dependent thermal effects entail significant modifications of the nucleon effective masses and chemical potentials and play an important role in determining the properties of multicomponent systems, such as charge-neutral β-stable matter, in which different particles have very different densities; see Ref. [34]. A remarkable departure from the zero-temperature results is illustrated in panel (B) of Fig. 1.6, showing the density dependence of the proton fraction in $npe\mu$ matter at 0 an 50 MeV. At $T = 50$ MeV and low density, the dominance of thermal effects leads to a breakdown of the monotonic behaviour of the proton fraction typical of cold matter.

The strong thermal modifications of the proton fraction in $npe\mu$ matter are mirrored by the density dependence of the bulk viscosity—the appearance of which is a consequence of an instantaneous departure from β-equilibrium induced by some perturbation—displayed in Fig. 1.7 [39]. The peculiar behaviour of the solid line corresponding to the highest temperature, $T =$ 50 MeV, results from the occurrence of the minimum in the proton fraction $Y_p(\varrho)$—clearly visible in panel (B) of Fig. 1.6—and from the fact that the value of Y_p exceeds the threshold for neutrino emission through direct Urca processes over the whole density range. On the other hand, for $T = 10$ and 30 MeV $Y_p(\varrho)$ turns out to be below threshold for $\varrho/\varrho_0 \lesssim 2.5$ and $0.25 \lesssim \varrho/\varrho_0 \lesssim 2.25$, respectively. The results of Figs. 1.6

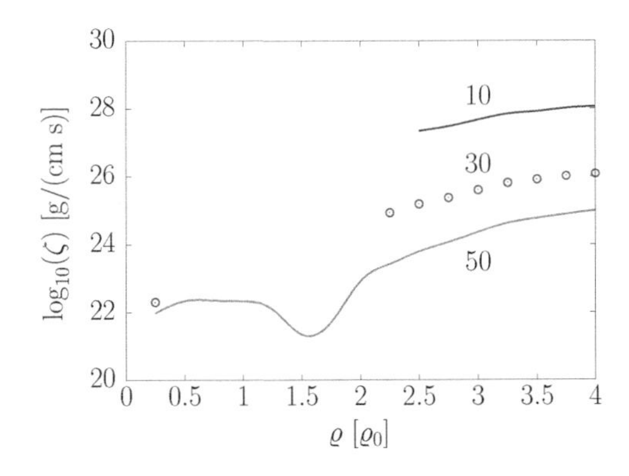

Figure 1.7 Density dependence of the bulk viscosity coefficient of β-stable matter associated with a density fluctuation of frequency $\omega = 2\pi \times 1$ kHz.The labels specify the temperature in units of MeV.

and 1.7 have been obtained by the authors of Refs. [34, 39], using the effective nuclear Hamiltonian derived from the AV18+UIX model of nuclear dynamics [17]. The calculation of the viscosity has been performed setting the frequency of the density oscillation leading to the departure from equilibrium to $\omega = 2\pi \times 1$ kHz, a value typical of neutron star pulsation; see contribution of S. Harris to this volume.

1.4 THE GOLDEN AGE OF NEUTRON STAR ASTRONOMY

On August 17, 2017, the Advanced LIGO–Virgo detector network made the first observation of the GW event GW170817, consistent with emission from a coalescing binary neutron star system [46]. The detection of this signal, and the later observation of electromagnetic radiation by space- and ground-based telescopes [47] arguably marked the dawning of the long-anticipated age of multimessenger astronomy. A prominent role in this effort has been played the Neutron Star Internal Composition Explorer (NICER), deployed on the International Space Station; see A.Watts contribution to this volume. In view of the wealth of precise data that are expected to flow from experimental facilities in the coming years, it has been claimed that we are entering the *golden age* of neutron star physics, in which many outstanding and fundamental issues may eventually be resolved [48].

Most ongoing studies are primarily aimed at exploiting the available and forthcoming empirical information to determine the occurrence of phase transitions, leading to the appearance of new forms of matter comprising strange baryons or deconfined quarks—possibly in a colour-superconducting phase—in the inner core of the star. However, here we focus on the regime in which nucleons are believed to be the relevant degrees of freedom, and NMBT is expected to be applicable. Special emphasis will be paid to the possibility to shed light on repulsive NNN interactions, which are known to play a critical role in determining the nuclear matter EOS at densities typical of the neutron star core, but—unlike the NN potential—are totally unconstrained at $\varrho \gtrsim \varrho_0$.

1.4.1 Measurements of mass and radius

The connection between the EOS of neutron star matter and the mass of the star clearly emerges from the results of the study carried out by Oppenheimer and Volkoff in the 1930s, showing that the mass of a star consisting of non interacting neutrons cannot exceed $\sim 0.8\ M_\odot$ [49]. This finding rules out the description of nuclear matter based on the Fermi gas model, and, more generally, on any models predicting an EOS too soft to support stable stars with masses compatible with observations.

Historically, the masses of neutron stars belonging to binary systems have been determined to remarkable accuracy using a procedure based on Kepler's third law. More recently, the results of the observations of binary systems of radio pulsars, allowing for accurate timing measurements, have provided information on post-keplerian orbital parameters describing effects of general relativity—such as the longitudinal advance of the periastron of the neutron star orbit, the time derivative of the period, and the gravitational redshift—discussed in, e.g., Ref. [50].

A measurement of post-keplerian parameters famously led to the precise determination of the masses of the neutron stars in the binary system PSR1913+16, discovered in 1974 by Hulse and Taylor [51]. Their analysis—that provided the first striking, although indirect, evidence of GW emission—yielded the results $M_1 = 1.39 \pm 0.15$ and $M_2 = 1.44 \pm 0.15\ M_\odot$ [52].

From the discussion of stellar equilibrium in Section 1.3.1, it follows that each EOS determines a specific value for the maximum mass of stable neutron star configurations, $M_{\rm max}$. Therefore, the most straightforward test of an EOS and the underlying microscopic model is a direct comparison between the corresponding $M_{\rm max}$ and the measured neutron star masses, the distribution of which is peaked around the canonical mass 1.4 $M_\odot$. The most interesting information, however, is provided by the precise determination of the masses of pulsars J1614-2230, J0348+0432 and J0740+6620, which turned out to be as high as $M = 1.97 \pm 0.04$, 2.01 ± 0.04 and $2.08 \pm 0.07\ M_\odot$, respectively [53–55].

Neutron star masses around and above 2 $M_\odot$ tend to rule out dynamical models predicting a soft EOS, notably those featuring the onset of a high-density phase including hyperons. It should be kept in mind, however, that little is known about hyperon interactions in dense matter, and these models often imply rather crude assumptions; see contributions of G.F. Burgio and C. Providência to this volume. On the other hand, it is a fact that many models involving only nucleon degrees of freedom and based on NMBT—notably those using the nuclear Hamiltonians AV18+UIX, AV18 + δv + UIX*, as well as the effective Hamiltonian $H_{\rm eff}$ defined in Section 1.2.3—turn out to be compatible with the recent measurements.

The importance of NNN interactions in this context can be gauged considering the modification of $M_{\rm max}$ associated with the inclusion of relativistic boost corrections, the main effect of which is a $\approx 30\%$ suppression of the repulsive component of the NNN potential; see panel (A) of Fig 1.3. A detailed study of the dependence of the maximum mass on the strength of V^R_{ijk} has been carried out by the authors of Ref. [56]. The results of these calculations, reported in Fig. 1.8, have been obtained by rescaling V^R_{ijk} according to

$$V^R_{ijk} \to \tilde{V}^R_{ijk} = \alpha \times \gamma V^R_{ijk} \, . \tag{1.20}$$

with $\gamma = 0.3$, and varying the value of α within the range $0.8 \leq \alpha \leq 1.8$. Note that $\alpha = 1$ corresponds to the NNN potential modified by relativistic boost corrections, that is, UIX* of Eq. (1.11). It is apparent that while there is a distinct sensitivity to the value of α, all Hamiltonians with α between 1.0 and 1.4 predict the occurrence of stable neutron stars with masses compatible with the recent observations in the region of central energy

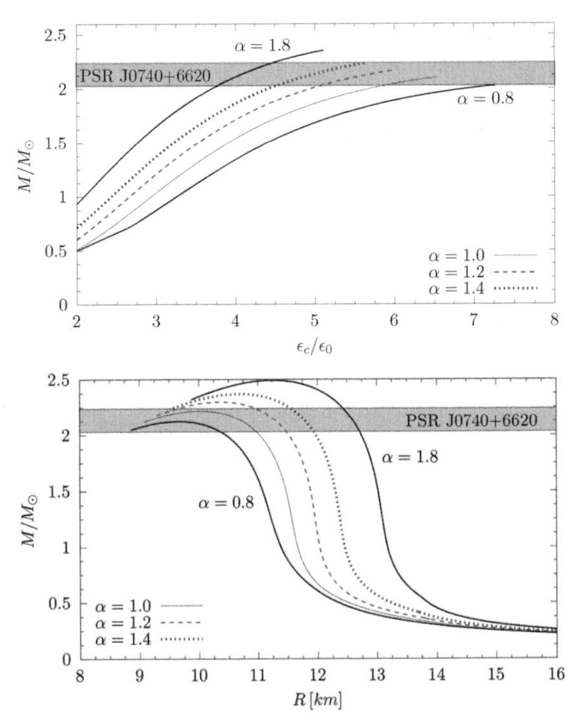

Figure 1.8 Top panel: dependence of the neutron star mass on the central energy density, obtained from the UIX* NNN potential modified as in Eq. (1.20), with $0.8 \leq \alpha \leq 1.8$. All curves extend to the maximum value of density for which the speed of sound in matter satisfies the causality constraint $v_s < 1$; Bottom panel: mass-radius relations of the stable neutron star configurations obtained using the UIX* NNN potential modified as in Eq. (1.20), with $0.8 \leq \alpha \leq 1.8$. Reprinted from Ref. [56] with permission. © APS (2021).

density in which the causality constraint is satisfied. Also note that, as shown in Ref. [10], the inclusion of boost corrections appears to alleviate the violation of causality, pushing its occurrence towards higher values of energy density.

The overall picture emerging from the analyses of the measured neutron star masses—a summary of which can be found in, e.g., Ref. [57]—shows a significant degree of degeneracy. In fact, most EOSs derived from models of neutron star matter including only nucleons turn out to support a stable configuration of mass around two solar masses. Precise measurements of the star radius, providing an additional constraint on the EOS and the underlying dynamical model, are needed to resolve this issue.

The radius of an ordinary star is obtained under the assumptions that: i) the radiation emitted by the star can be described by a blackbody spectrum, which allows to determine the surface temperature T_s; and ii) the star radiates uniformly and isotropically. Within this scheme, a measurement of flux and distance, denoted F and D, respectvely, allows to determine the radius from

$$R = \sqrt{\frac{FD^2}{\sigma T_s^4}}, \tag{1.21}$$

where σ is the constant of Stephan-Boltzmann.

In the case of neutron stars, however, the models of spectrum and flux employed in data analysis involve uncertainties arising from astrophysical effects, that need to be carefully taken into account. Even more critical is the assumption that the whole surface of the star radiates uniformly. The emitted X-ray bursts actually originate from *hot spots*, first observed in the early 2000 by the X-ray Multi-Mirror Mission (XMM-Newton) [58]. More recently, the unprecedented capabilities of NICER's X-ray Timing Instrument (XTI), allowed to detect the relativistic effects affecting the motion of the hot spots. These observations have provided accurate information on both the mass and radius of the emitting star.

Recent analyses of NICER data allowed to infer the radius of a neutron star of mass $M = 1.4\ M_\odot$, the central density of which does not exceed $\sim 3\varrho_0$. The reported values—$R = 12.35 \pm 0.65$ km [59], $12.33^{+0.76}_{-0.81}$ km [60] and $12.18^{+0.56}_{-0.79}$ km [61]—turn out to be generally compatible with the predictions of theoretical calculations performed using EOSs of purely nucleonic matter. The equatorial radius of the neutron star PSR J0740+6620, having mass $M = 2.072^{+0.067}_{-0.066}\ M_\odot$ and central density $\sim 4\varrho_0$, has been also evaluated to be $R = 12.39^{+1.30}_{-0.98}$ km [60]. The small difference between the radii of stars of mass 1.4 and $\sim 2.1\ M_\odot$, implying that the EOS remains still quite stiff at $\varrho > 3\varrho_0$, appears to rule out the occurrence of a strong first order phase transition in the density range $3\varrho_0 \lesssim \varrho_B \lesssim 4\varrho_0$. However, it should be kept in mind that—as pointed out by the authors of Ref [62]—a continuous crossover associated with the formation of a mixed phase cannot be excluded. This issue is discussed in detail by T. Kojo in its contribution to this volume.

Figure 1.9 shows a comparison between the mass-radius relations predicted by the APR1 and APR2 models and the results of recent neutron star observations, including the NICER data discussed above and the estimated mass and radius of the millisecond pulsar PSR J0030+0451 [63]. In addition, the box labelled GW170817 represents the 90%-confidence-level estimate of the neutron star mass and radius reported by the LIGO-Virgo Collaboration, the determination of which depends on the properties of the EOS employed in the analysis [64].

The data appear to favour the APR1 model, corresponding to a stiffer EOS in the high density region relevant to the more massive star. It follows that the pattern emerging from Fig. 1.9 implies a distinct sensitivity on the strength of repulsive NNN interactions, which, in turn, is largely driven by relativistic boost corrections to the nuclear Hamiltonian. A comparison with the results displayed in the bottom panel of Fig 1.8 suggests that, using the parametrisation of Eq.(1.20), a better agreement between the data and the results of the APR2 model—based on the AV18 + δv + UIX* nuclear Hamiltonian—may be achieved by increasing α to a value larger than unity.

1.4.2 Measurements of tidal deformability

The detection of the gravitational wave signal of event GW170817, reported by the LIGO-Virgo Collaboration in 2017 [46], allowed to determine the chirp mass of the coalescing binary neutron star system, defined by Eq. (1.16), with extraordinary accuracy. The measured value turned out to be $\mathcal{M} = 1.188^{+0.004}_{-0.002}\ M_\odot$. On the other hand, the determination of the masses of the coalescing stars and their ratio, $q = M_2/M_1$ depends on the assumptions made on their spins, and involves a larger uncertainty. The data of Ref. [46] have been

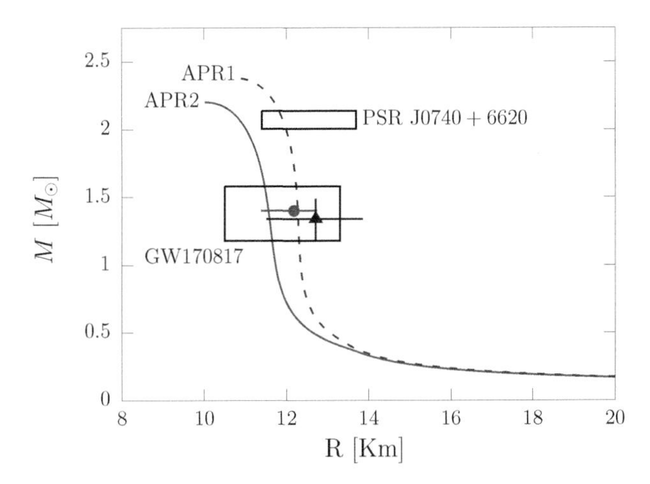

Figure 1.9 Comparison between the mass-radius relations predicted by the APR1 and APR2 models of the nuclear matter EOS and the data obtained from neutron star observations. The boxes correspond to the NICER results for PSR J0740+6620 [60] and the estimate obtained from the analysis of the GW170817 event [64]. The data points marked with a circle and a triangle correspond to NICER results for a neutron star of 1.4 solar masses [60] and the millisecond pulsar PSR J0030+0451, respectively [63].

analysed considering two different scenarios, corresponding to high and low spin. The results, yielding the range of both masses at 90% confidence level, are $1.36 \leq M_1 \leq 2.26$ $M_\odot$ and $0.86 \leq M_2 \leq 1.36\ M_\odot$ for the high-spin scenario, and $1.36 \leq M_1 \leq 1.60\ M_\odot$ and $1.17 \leq M_2 \leq 1.36\ M_\odot$ for the low-spin scenario.

Figure 1.10 shows the two-dimensional probability density of the tidal deformabilities of the coalescing stars, Λ_1 and Λ_2, obtained from the analysis of the observation of the GW170817 event in the high spin scenario [46]. The predictions obtained from the APR1 and APR2 EOSs discussed above, displayed by the solid lines, are compared to the boundaries of the regions enclosing 50% and 90% of the probability density, represented by the dashed lines.The emerging pattern shows that the data favour EOSs predicting more compact stars. To see this, consider that the compactness of a star of 1.4 $M_\odot$ predicted by the APR2 and APR1 models turn out to be $M/R = 0.108$ and $0.100\ M_\odot/\mathrm{Km}$, respectively.

1.4.3 Towards multimessenger astronomy

Besides marking the beginning of the new era of GW astronomy, the landmark observation of event GW170817 contributed to highlight the potential of combining gravitational and electromagnetic observations. The association of the GW detection with that of the γ-ray burst GRB 170817A —carried out by the Fermi Gamma-ray Burst Monitor (GBM) [65] and the International Gamma-ray Astrophysics Laboratory 1.7 s after the coalescence (INTEGRAL) [66]—has been critical to confirm the hypothesis of neutron star merger, and provided the first direct evidence of the connection between these processes.

As pointed out above, the observations of gravitational and electromagnetic signals can be exploited to constrain the EOS, or specific equilibrium properties, of neutron star matter. Work along this line has been done to reconstruct the EOS within both phenomenological

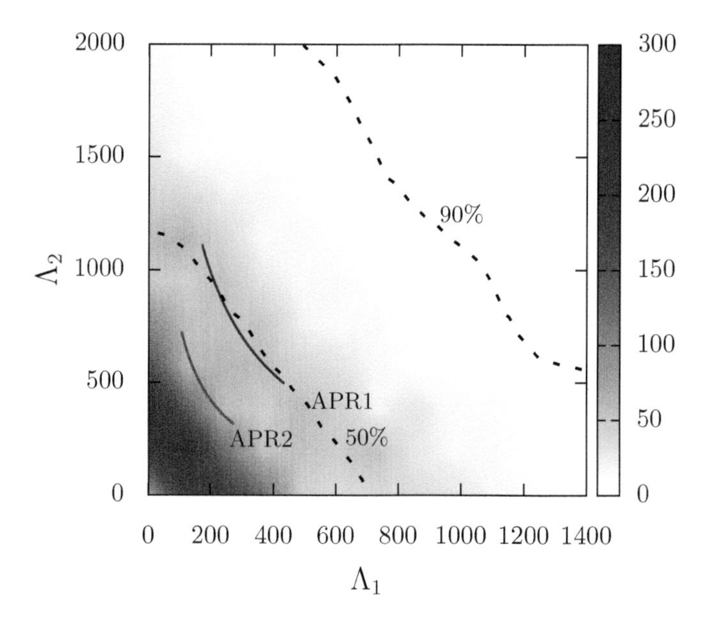

Figure 1.10 Probability density of the tidal deformability parameters, Λ_1 and Λ_2, obtained from the analysis of the observation of the GW170817 event [46]. The thick solid lines represent the results of calculations carried out using the EOSs described in the text. The dashed lines show the boundaries of the regions enclosing 50% and 90% of the probability density.

and nonparametric frameworks, determine the occurrence of phase transitions, or pin down the behaviour of the symmetry energy above nuclear saturation density [41, 61, 67–91].

A conceptually different approach, pioneered by the groundbreaking studies discussed in Refs. [92, 93], aims at pushing the analyses based on multimessenger astronomical data to a deeper level. The proposed strategy rests on the premise that the data currently available—as well as those to be collected by existing facilities operating at design sensitivity and next-generation detectors—may offer an unprecedented opportunity to constrain the microscopic models of nuclear dynamics at supranuclear density. The authors of Refs. [92, 93] focused on the repulsive NNN interaction, which, as pointed out above, is the dominant factor determining the stiffness of the nuclear matter EOS at large densities.

In Ref. [92], the strength of the repulsive NNN potential, parametrised according to Eq. (1.20), was inferred from a Bayesian analysis based on the dataset including (i) the tidal deformabilities obtained from the observation of the GW170817 event performed by the LIGO-Virgo Collaboration [46]; (ii) the mass and radius of the millisecond pulsar PSR J0030+0451, measured by the NICER Collaboration [60]; and (iii) the mass of PSR J0740+6620, which is the most massive pulsar discovered so far [54]. The resulting probability distribution is displayed in Fig. 1.11.

While being rather broad, $\mathcal{P}(\alpha)$ exhibits a clearly visible maximum, showing the sensitivity of multimessenger neutron star observations to the strength of the repulsive component of the NNN potential. Note also that, among the currently available priors employed in the analysis, the constraint on $M_{\rm Max}$ is dominant. The position of the maximum, located at $\alpha > 1$, suggests that the data favour a potential more repulsive than the UIX* model,

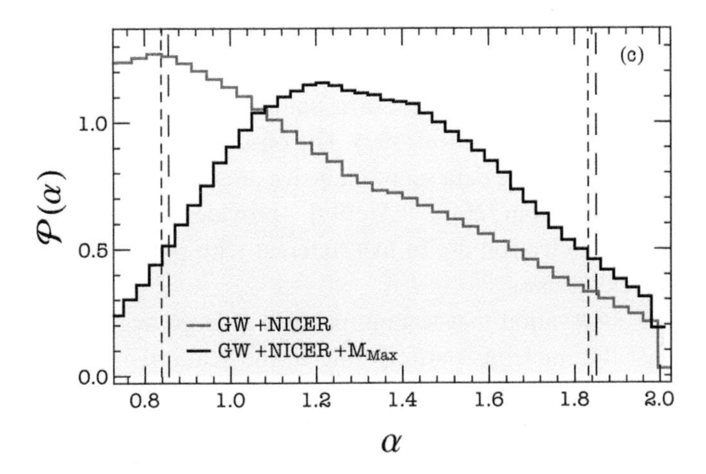

Figure 1.11 The grey area represents the posterior probability distribution of the parameter α of Eq. (1.20)—determining the strength of the repulsive three-nucleon force—inferred combining (i) the GW observation of the binary system GW170817; (ii) the mass and radius obtained by the NICER analysis of the millisecond pulsar PSR J0030+0451; and (iii) the bound on the maximum neutron star mass, $M_{\rm Max}$, imposed by the observations of PSR J0740+6620. Reprinted from Ref. [92] with permission. © APS (2021).

which is recovered setting $\alpha = 1$. The corresponding EOS would therefore be stiffer than the APR2 EOS at large density.

The analysis of Ref. [92] has been extended by the authors of Ref. [93], who investigated the potential to achieve an accurate determination of α exploiting the capability of the proposed Einstein Telescope [94]. More recently, s similar study has been performed by the authors of Ref. [95].

1.5 OUTLOOK

For many years to come, exploiting the full potential of multimessenger astronomical observations of neutron stars will require a high degree of synergy between experimentalists, theorists with a broad spectrum of competences, and developers of simulation codes. Efficient numerical implementations of the available theoretical models will, in fact, be required for the interpretation of the detected signals, which will in turn provide information indispensable to test the validity of the underlying models—unavoidably involving a number of assumptions—and shed light on outstanding nuclear theory issues.

According to the current consensus—largely based on the extrapolation of nuclear data to higher densities—matter in the interior of stars of mass around the canonical vaule of 1.4 $M_\odot$—the density of which is expected not to exceed $\sim 3\varrho_0$—consists mainly of nucleons and leptons in β-equilibrium. The possibility that this form of matter may be dominant even in the most massive stars, having $M \gtrsim 2M_\odot$, is suggested by both astronomical data [60] and theoretical studies [62].

It has to be pointed out that valuable complementary information on the properties of dense nuclear matter has been also extracted from electron-nucleus scattering cross sections measured in the kinematical regime corresponding to momentum transfer around and above 1 GeV; for a review, see, e.g. Ref. [97]. The onset of y-scaling [98]—clearly emerging from the analysis of the large data set of inclusive cross sections collected using a broad range of targets, extending from ^{2}H to ^{197}Au [99]—provides ample *model-independent* evidence that the incoming electron do, in fact, interact with point like spin 1/2 particles of mass equal to the nucleon mass.

In addition, the observation that scaling persists at large negative values of the variable y indicates that the nucleons participating in the scattering process carry momenta as high as 0.7 GeV. The appearance of high-momentum nucleons belonging to strongly correlated pairs gives rise to fluctuations of the matter density, the amplitude of which can be estimated using the simple relation linking density and Fermi momentum in degenerate fermion systems. For example, according to this relation, in SNM a nucleon Fermi momentum of 0.5 GeV corresponds to $\varrho \sim 7\varrho_0$,

The fluctuations of nuclear density induced by short-range NN correlations have been further investigated using exclusive electron-nucleus scattering cross sections measured at the Thomas Jefferson National Accelerator Facility (Jefferson Lab) [100, 101]. The results of these studies, suggesting that the density of correlated pairs inferred from the data can be as high as $\sim\ 5\ \varrho_0$, turn out to be consistent with those obtained from the y-scaling analysis [102].

In view of the above considerations, it is apparent that a quantitative description of the properties of stars of mass up to $\sim 2\ M_\odot$ requires a reliable model of nuclear matter, based on a Hamiltonian strongly constrained by data at both nuclear and supranuclear densities.

In this chapter, we have focused on the determination of the potential describing repulsive NNN interactions, which are known to provide the dominant contribution to the EOS of dense matter. The results discussed in Section 1.4 show that the results of multimessenger astronomical observations of neutron star properties such as mass, radius and tidal deformability exhibit a distinct sensitivity to the softening of the repulsive NNN potential associated with the inclusion of relativistic boost corrections in the NN potential. Moreover, recent analyses carried out within the Bayesian inference framework suggest that astronomical data can be used to pin down the size of these corrections—whose theoretical treatment involves significant uncertainties—at fully quantitative level.

These studies demonstrate the potential for using multimessenger astronomy to improve the present understanding of microscopic nuclear dynamics. The development of advanced theoretical models, capable to provide an accurate description of neutron star properties, will in turn improve the ability to perform realistic simulations of the astrophysical processes relevant to the observed signals.

Thanks to the unprecedented availability of multimessenger astronomical data, the ultimate goal of nuclear theory, that is, the achievement of a unified description of all nucleon systems, from the deuteron to neutron stars, appears to be eventually reacheable.

Bibliography

[1] Harrison, B.K., Thorne, K.S., Wakano, M., and Wheeler, J.A. (1965). *Gravitation theory and gravitational collapse*. University of Chicago Press. Chap. 9.

[2] Yukawa, H. (1935). On the interaction of elementary particles. *Proc. Phys. Math. Soc. Jpn.*, 17: 48–57.

[3] Benhar, O. (2023). *Structure and Dynamics of Compact Stars*. Springer. Chap. 3.

[4] Wiringa, R.B., Stoks, V.G.J., and Schiavilla, R. (1995). Accurate nucleon-nucleon potential with charge-independence breaking. *Phys. Rev. C*, 51: 38–51.

[5] Benhar, O. (2021). Scale dependence of the nucleon-nucleon potential. *Int. J. Mod. Phys. E*, 30: 213009 1–11.

[6] Weinberg, S. (1990). Nuclear forces from chiral lagrangians. *Phys. Lett. B*, 251: 288–292.

[7] Tews, I., Carlson, J., Gandolfi, S., and and Reddy, S. (2018). Constraining the speed of sound inside neutron stars with chiral effective field theory interactions and observations. *Astrophys. J.*, 860: 149 1–14.

[8] Pudliner, B.S., Pandharipande, V.R., Carlson, J., and Wiringa, R.B. (1995). Quantum Monte Carlo calculations of $A \leq 6$ nuclei. *Phys. Rev. Lett.*, 74: 4396–4399.

[9] Akmal, A. and Pandharipande, V.R. (1997). Spin-isospin structure and pion condensation in nucleon matter. *Phys. Rev. C*, 56: 2261–2279.

[10] Akmal, A., Pandharipande, V.R., and Ravenhall, D.G. (1998). Equation of state of nucleon matter and neutron star structure. *Phys. Rev. C*, 58: 1804–1828.

[11] Lovato, A., Benhar, O., Fantoni, S., and Schmidt, K.E. (2012). Comparative study of three-nucleon potentials in nuclear matter. *Phys. Rev. C*, 85: 024003 1–13.

[12] Rikovska Stone, J., Miller, J.C., Koncewicz, R., Stevenson, P.D., and Strayer, M.R. (2003). *Phys. Rev. C*, 68: 034324 1–16.

[13] Chabanat, E., Bonche, P., Haensel, P., Meyer, J, and Schaeffer, R. (1997). *Nucl. Phys. A*, 627: 710–746.

[14] Amundsen, L. and Østgaard, E. (1985). *Nucl. Phys. A*, 437: 487–508.

[15] Wambach, J., Ainsworth, T.L., and Pines, D. (1993). *Nucl. Phys. A*, 555: 128–150.

[16] Lovato, A., Benhar, O., Fantoni, S., Illarionov, A.Y., and Schmidt, K.E. (2011). Density-dependent nucleon-nucleon interaction from three-nucleon forces. *Phys. Rev. C*, 83: 054003 1–16.

[17] Benhar, O. and Lovato, A. (2017). Perturbation theory of nuclear matter with a microscopic effective interaction. *Phys. Rev. C*, 96: 054301 1–10.

[18] Clark, J.W. (1979). Variational treatment of nuclear matter. *Prog. Part. Nucl. Phys.*, 2: 89–199.

[19] Lovato, A., Bombaci, I., Logoteta, D., Piarulli, M., and Wiringa, R.B. (2022), Benchmark calculations of infinite neutron matter with realistic two- and three-nucleon potentials. *Phys. Rev. C*, 105: 055808 1–14.

[20] Benhar, O., Lovato, A., and Camelio, G. (2022). Modelling neutron star matter in the age of multimessenger astrophysics. *Astrophys. J.*, 939: 52 1–14.

[21] Forest, J.L, Pandharipande, V.R., and Arriaga, A. (1999). Quantum Monte Carlo studies of relativistic effects in light nuclei. *Phys. Rev. C*, 60: 0140012 1–16.

[22] Friar, J.L. (1975). Relativistic effects on the wave function of a moving system. *Phys. Rev. C*, 12: 695–698.

[23] Forest, J.L., Pandharipande, V.R., and Friar, J.L. (1995). Relativistic nuclear Hamiltonians. *Phys. Rev. C*, 52: 568–575.

[24] Danielewicz, P., Lacey, R., and Lynch, W.G. (2002), Determination of the equation of state of dense matter. *Science*, 298: 1592–1596.

[25] Shlomo, S., Kolomietz, V.M., and Colò, G. (2006), Deducing the nuclear-matter incompressibility coefficient from data on isoscalar compression modes. *Eur. Phys. J. A*, 30: 23–30.

[26] Colò, G. (2006).The compression modes in atomic nuclei and their relevance for the nuclear equation of state. *Phys. Part. Nucl.*, 39: 286–305.

[27] Akmal, A. (1998). *Variational studies of nucleon matter with realistic potential.* PhD Thesis, University of Illinois at Urbana-Champaign.

[28] Baiotti, L. and Rezzolla, L. (2017), Binary neutron star mergers: a review of Einstein's richest laboratory. *Rep. Prog. Phys.* 80: 096801 1–65.

[29] Raithel, C.A.;, Paschalidis, V., and Özel, F. (2021), Realistic finite-temperature effects in neutron star merger simulations. *Phys. Rev. D*, 104: 063016 1–19.

[30] Figura, A., Lu, J.J., Burgio, G.F., Li, Z.H., and Schulze, H.J. (2020), Hybrid equation of state approach in binary neutron-star merger simulations. *Phys. Rev. D*, 102: 043006 1–16.

[31] Figura, A., Lu, J.J., Burgio, G.F., Li, Z.H., and Schulze, H.J. (2021), Binary neutron star merger simulations with hot microscopic equations of state. *Phys. Rev. D*, 103: 083012 1–14.

[32] Hammond, P., Hawke, I., and Andersson, N. (2021), Thermal aspects of neutron star mergers. *Phys. Rev. D*, 104: 103006 1–22.

[33] Camelio, G., Lovato, A., Gualtieri, L., Benhar, O., Pons, J.A., and Ferrari, V. (2017), Evolution of a proto-neutron star with a nuclear many-body equation of state: Neutrino luminosity and gravitational wave frequencies. *Phys. Rev. D*, 96: 043015 1–24.

[34] Tonetto, L. and Benhar, O. (2022). Thermal effects on nuclear matter properties. *Phys. Rev. D*, 106: 10320 1–11.

[35] Cowell, S. and Pandharipande, V.R. (2003). Quenching of weak interactions in nucleon matter. *Phys. Rev. C*, 67: 035504 1–19.

[36] Cowell, S. and Pandharipande, V.R. (2004). Neutrino mean free paths in cold symmetric nuclear matter. *Phys. Rev. C*, 70: 035801 1–15.

[37] Lovato, A., Losa, C., and Benhar, O. (2013). Weak response of cold symmetric nuclear matter at three-body cluster level. *Nucl. Phys. A*, 901: 22–50.

[38] Lovato, A., Benhar, O., Gandolfi, S., and Losa, C. (2014). Neutral-current interactions of low-energy neutrinos in dense neutron matter. *Phys. Rev. C*, 89: 025804 1–8.

[39] Benhar, O., Lovato, A., and Tonetto, L. (2023). Properties of hot nuclear matter. *Universe*, 9: 345 1–18.

[40] Hinderer, T. (2008). Tidal Love Numbers of Neutron Stars. *Astrophys. J.*, 677: 1216–1220.

[41] Sabatucci, A. and Benhar, O. (2020). Tidal deformation of neutron stars from microscopic models of nuclear dynamics. *Phys. Rev. C*, 101: 045807 1–7.

[42] Ferrari, V., Gualtieri, L., and Pani, P. (2021). *General relativity and its Applications*, CRC Press, Chap. 15.

[43] Andersson, N. and Kokkotas, K.D. (1988), Towards gravitational wave asteroseismology. *Mon. Not. R. Astron. Soc.*, 299: 1059–1068.

[44] Benhar, O., Gualtieri L., and Ferrari, V. (2004). Gravitational wave asteroseismology reexamined. *Phys. Rev. D*, 70: 124015 1–9.

[45] Benhar, O., Berti, E., and Ferrari, V. (1999). The imprint of the equation of state on the axial w-modes of oscillating neutron stars. *Mon. Not. R. Astron. Soc.*, 310: 797–803.

[46] Abbott, B.P. *et al.* (LIGO Scientific Collaboration and Virgo Collaboration) (2017). GW170817: Observation of Gravitational Waves from a Binary Neutron Star Inspiral. *Phys. Rev. Lett.* 119: 161101 1–18.

[47] Abbott, B.P. *et al.* (LIGO Scientific Collaboration and Virgo Collaboration) (2017). Multi-messenger Observations of a Binary Neutron Star Merger. *Astrophys. J. Lett.*, 848: L12 1–59.

[48] Mann, A. (2020). The strange hearts of neutron stars. *Nature* 579: 20–22.

[49] Oppenheimer, J.R. and Volkoff, G.M. (1939). On massive neutron cores. *Phys. Rev.* 55: 374–381.

[50] Damour, T. and Deruelle, N. (1986). General relativistic celestial mechanics. II The post-Newtonian timing formula. *Ann. Inst. Henri Poincaré, Phys. Theor.* 4: 263–292.

[51] Hulse, R.A. and Taylor, J.H. (1975). Discovery of a pulsar in a binary system. *Astrophys. J. Lett.*, 195: L51–L53.

[52] Taylor, J.H., Flower, L.A. and McCulloch P.M. (1979), Measurement of general relativistic effects in the binary pulsar PSR1913+16. *N*ature 277: 437–440.

[53] Antoniadis, J. *et al.* (2013). A Massive Pulsar in a Compact Relativistic Binary. *Science* 340: 1233232 1–9.

[54] Fonseca, E. *el al* (2021). Refined mass and geometric measurements of the high-mass PSR J0740+6620. *Astrophys. J. Lett.*, 915: L12 1–15.

[55] Demorest, P., Pennucci, T., Ransom, S., Roberts, M., and Hessels, J. (2010). A two-solar-mass neutron star measured using Shapiro delay. *Nature* 467: 1081–1083.

[56] Tonetto, L., Sabatucci, A., and Benhar, O. (2021). Impact of three-nucleon forces on gravitational wave emission from neutron stars. *Phys. Rev. D*, 104: 083034 1–9.

[57] Lattimer, J.M. (2019). Neutron star mass and radius measurements, *Universe*, 5: 159 1–20.

[58] De Luca, A. *et al.* (2005). On the polar caps of the Three Musketeers. *Astrophys. J.*, 623: 1051–1069.

[59] Miller, M.C. *et al.* (2021). The radius of the PSR J7040+6620 from NICER and XMM Newton data. *Astrophys. J. Lett.*, 918, L28 1–31.

[60] Riley, T.E. *et al.* (2021). A NICER view of the massive pulsar PSR J0740+6620 informed by radio timing and XMM-Newton spectroscopy. *Astrophys. J. Lett.*, 918: L27 1–30.

[61] Raaijmakers, G. *et al.* (2021). Constraints on the dense matter equation of state and neutron star properties from NICER's mass-radius estimate of PSR J0740+6620 and multimessenger observations. *Astrophys. J. Lett.*, 918: L29 1–13.

[62] Brandes, L., Weise, W., and Kaiser, N. (2023), Inference of the sound speed and related properties of neutron stars. *Phys. Rev. D*, 107: 014011 1–25.

[63] Riley, T.E. *et al.* (2019). A NICER view of PSR J0030+0451: millisecond pulsar parameter estimation. *Astrophys. J. Lett.*, 887: L29 1–60.

[64] Abbott, B.P. *et al.* (LIGO Scientific Collaboration and Virgo Collaboration) (2018). GW170817: measurements of neutron star radii and equation of state. *Phys. Rev. Lett.*, 121: 161101 1–16.

[65] Goldstein, A. *et al.* (2017). An ordinary short gamma-ray burst with extraordinary implications: Fermi-GBM detection of GRB 170817A. *Astrophys. J. Lett.*, 848, L14 1–14.

[66] Savchenko, V. *et al.* (2017). INTEGRAL detection of the first prompt gamma-ray signal coincident with the gravitational-wave event GW170817. *Astrophys. J. Lett.*, 848: L15 1–8.

[67] Annala, E., Gorda, T., Kurkela, A., and Vuorinen, A. (2018).Gravitational-wave constraints on the neutron-star-matter equation of state. *Phys. Rev. Lett.*, 120: 172703 1–5.

[68] Margalit, B. and Metzger, B.D. (2017). Constraining the maximum mass of neutron stars from multi-messenger observations of GW170817. *Astrophys. J. Lett.*, 850: L19 1–8.

[69] Radice, D., Perego, A., Zappa, F., and Bernuzzi, S. (2018). GW170817: Joint constraint on the neutron star equation of state from multimessenger observations. *Astrophys. J. Lett.*, 852: L29 1–5.

[70] Bauswein, A., Just, O., Janka, H.Th., and Stergioulas, N. (2017). Neutron-star radius constraints from GW170817 and future detections. *Astrophys. J. Lett.*, 850: L34 1–5.

[71] Lim, Y. and Holt, J.W. (2018). Neutron star tidal deformabilities constrained by nuclear theory and experiment. *Phys. Rev. Lett.*, 121: 062701 1–6.

[72] Lim, Y., Bhattacharya, A., Holt, J.W., and Pati, D. (2021). Radius and equation of state constraints from massive neutron stars and GW190814. *Phys. Rev. C*, 104: L032802 (2021) 1–6.

[73] Most, E.R., Weih, L.R., Rezzolla, L., and Schaffner-Bielich, J. (2018). New constraints on radii and tidal deformabilities of neutron stars from GW170817. *Phys. Rev. Lett.*,120: 261103 1–6.

[74] De, S., Finstad, D., Lattimer, J.M. Brown, D.A., Berger, E., and Biwer, C.M. (2018). Tidal deformabilities and radii of neutron stars from the observation of GW170817. *Phys. Rev. Lett.* 121: 091102 1–6.

[75] Annala, E., Gorda, T., Kurkela, A., Nättilä, J., and Vuorinen, A., (2020). Evidence of quark-matter cores in massive neutron stars. *Nature Phys.*, 16: 907–910.

[76] Raaijmakers, G. *et al.* (2020). Constraining the dense matter equation of state with joint analysis of NICER and LIGO/Virgo measurements. *Astrophys. J. Lett.*, 893: L21 1–13.

[77] Miller, M.C., Chirenti, C., and Lamb, F.K. (2020). Constraining the equation of state of high-density cold matter using nuclear and astronomical measurements. *Astrophys. J.*, 888: 12 1–13.

[78] Kumar, B. and Landry, P. (2019). Inferring neutron star properties from GW170817 with universal relations. *Phys. Rev. D*, 99: 123026 1–16.

[79] Fasano, M., Abdelsalhin, T., Maselli, A., and Ferrari, V. (2019). Constraining the neutron star equation of state using multiband independent measurements of radii and tidal deformabilities. *Phys. Rev. Lett.*, 123: 141101 1–6.

[80] Landry, P., Essick, R., and Chatziioannou, K. (2020). Nonparametric constraints on neutron star matter with existing and upcoming gravitational wave and pulsar observations. *Phys. Rev. D*, 101: 123007 1–21.

[81] Güven, H., Bozkurt, K., Khan, E., and Margueron, J. (2020). Multimessenger and multiphysics Bayesian inference for the GW170817 binary neutron star merger. *Phys. Rev. C*, 102: 015805.

[82] Traversi, S., Char, P., and Pagliara, G. (2020). Bayesian inference of dense matter equation of state within relativistic mean field models using astrophysical measurements. *Astrophys. J.* **897**: 165 1–24.

[83] Zimmerman, J, Carson, Z., Schumacher, K., Steiner, A.W., and Yagi, K. (2020). Measuring nuclear matter parameters with NICER and LIGO/Virgo. arXiv:2002.03210 [astro-ph.HE].

[84] Silva, H.O., Holgado, A.M., Cárdenas-Avendaño, A., and Yunes, N. (2021). Astrophysical and theoretical physics implications from multimessenger neutron star observations. *Phys. Rev. Lett.*, 126: 181101 1–7.

[85] Blaschke, D., Ayriyan, A., Alvarez-Castillo, D.E., and Grigorian, H. (2020). Was GW170817 a canonical neutron star merger? Bayesian analysis with a third family of compact stars. *Universe*, 6: 81 1–22.

[86] Tang, S.-P., Jiang, J.-L., Gao, W.-H., Fan, Y.-Z., and Wei, D.-M. (2021). Constraint on phase transition with the multimessenger data of neutron stars. *Phys. Rev. D*, **103**, 063026 1–9.

[87] Biswas, B., Char, P., Nandi R., and Bose, S. (2021). Constraint on phase transition with the multimessenger data of neutron stars. *Phys. Rev. D*, 103: 103015 1–9.

[88] Pacilio C., Maselli A., Fasano, M., and Pani P. (2022). Ranking Love numbers for the neutron star equation of state: the need for third-generation detectors. *Phys. Rev. Lett.*, 128: 101101 1–8.

[89] Malik, T. and Providência, C. (2022). Bayesian inference of signatures of hyperons inside neutron stars. *Phys. Rev. D*, 106, 063024 1–14.

[90] Altiparmak, S., Ecker, C., and Rezzolla, L. (2022). On the Sound Speed in Neutron Stars. *Astrophys. J. Lett.*, 939: L34 1–9.

[91] Gupta, P.K. *et al.* (2022). Determining the equation of state of neutron stars with Einstein Telescope using tidal effects and r-mode excitations from a population of binary inspirals. arXiv:2205.01182 [gr-qc] (2022).

[92] Maselli, A., Sabatucci, A., and Benhar, O. (2021). Constraining three-nucleon forces with multimessenger data. *Phys. Rev. C*, 103: 065804 1–6.

[93] Sabatucci, A., Benhar, O., Maselli, A., and Pacilio, C. (2022). Sensitivity of neutron star observations to three-nucleon forces. *Phys. Rrev. D*, 106: 083010 1–13.

[94] Maggiore, M. *et al.* (2020). Science case for the Einstein telescope. *Journal of Cosmology and Astroparticle Physics*: 2020 050 1–47.

[95] Rose, H. *et al.* (2023). Revealing the strength of three-nucleon interactions with the proposed Einstein Telescope. *Phys. Rev. C*, 108: 025811 1–13.

[96] Malik, T., Ferreira, M., Bastos Albino, M., and Providência, C. (2023). Spanning the full range of neutron star properties within a microscopic description. *Phys. Rev. D*, 107: 103018 1–18.

[97] Benhar, O., Day, D, and Sick, I. (2008). Inclusive quasi elastic electron-nucleus scattering. *Rev. Mod. Phys.*, 80: 189–224.

[98] Day, D. *et al.* (1987). y-scaling in electron-nucleus scattering. *Phys. Rev. Lett*, 59 427–430.

[99] Arrington, J. *et al.* (1999). Inclusive electron-nucleus scattering at large momentum transfer, *Phys. Rev. Lett.* 82: 2056–2059.

[100] Subedi, R. *et al.* (2008). Probing cold dense nuclear matter, *Science*, 320: 1476–1478.

[101] Schmidt, A. *et al.* (2020). Probing the core of the strong nuclear interaction, *Nature*, 578: 840–844.

[102] Benhar, O. (2023). Testing the paradigm of nuclear many-body theory. *Particles*, 6: 611–621.

CHAPTER 2

Measuring Neutron Star Mass and Radius Using Pulse Profile Modelling – NICER and Beyond

Anna Watts

NICER, NASA's Neutron Star Interior Composition Explorer (or ExploreR, if you want to be picky!), is an X-ray telescope mounted on the International Space Station. Its primary science goal – as the name suggests – is to explore the nature of the supranuclear density matter in the cores of neutron stars. The high-quality data sets that NICER collects are used for Pulse Profile Modelling (PPM), a technique that exploits relativistic effects which imprint information about the star's mass and radius in the radiation emitted from rotation-powered millisecond pulsars. Mass and radius, in turn, constrain the dense matter equation of state. PPM also enables us to make surface maps of the X-ray emitting hot spots - no mean feat for stars 20–30 km in diameter, thousands of lightyears from Earth. This chapter provides an overview of the PPM process, from data to inferred parameters. It summarizes NICER's results to date, and looks ahead to what we can expect over the course of the mission lifetime. We will also look at how the PPM technique is being applied to new source classes, ready for the next generation of X-ray telescopes.

2.1 THE NEUTRON STAR INTERIOR

Since NICER's main science goal is to explore the neutron star interior, this is where we will start. Theory predicts that neutron stars have a complex, layered structure (Figure 2.1). Underneath an atmosphere (perhaps even an ocean), neutron stars have a solid crust, an ionic lattice of nuclei that become gradually more neutron-rich as one goes deeper into the star. In the outer crust, the dominant pressure contribution is from degenerate electrons. At a certain density, the neutron drip point, it becomes energetically favourable for some

DOI: 10.1201/9781003306580-2

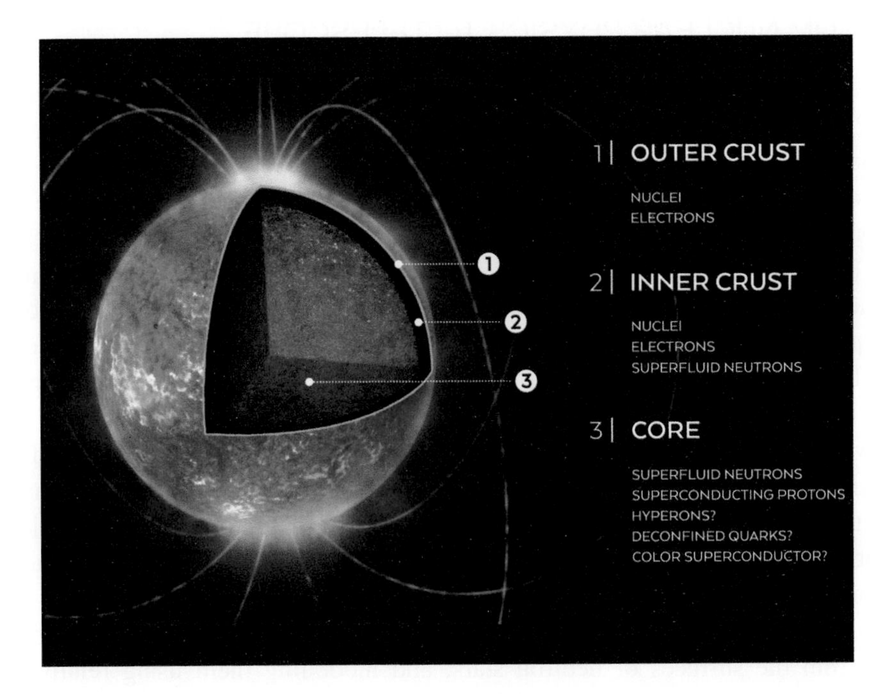

Figure 2.1 A schematic cut-through of the neutron star interior, showing crust, core and possible composition. Reprinted from Ref. [52] with permission. © APS (2016).

neutrons to exist outside of the nuclei, and this neutron (super)fluid starts to contribute to the pressure as well. At the base of the crust, where the nuclei come into contact, we pass through the pasta phase (where the nuclei are highly deformed) and finally the nuclei merge together to form the nuclear density fluid of the core. Here the composition becomes more uncertain. The core could consist of a neutron-rich nucleonic fluid (perhaps 95% neutrons, 5% protons, and some electrons and muons). But there are other possibilities, such as the formation of stable states of strange matter – with the strange quarks either locked up in hyperons or in the form of deconfined quark matter (an up, down and strange quark mix).

Putting this into the broader context, neutron stars form part of efforts to understand the strong force across the full range of temperature and density. Although we have a theoretical framework – quantum chromodynamics – it is difficult to calculate properties of even regular nuclei (those making up your office furniture or your cat, for example) from first principles. To understand what happens as we go to extremes of pressure and temperature, for now, theorists develop phenomenological models that are then tested experimentally and observationally. Heavy ion collision experiments explore the high temperature, low to intermediate density regime – whereas neutron stars sit in the high-density low-temperature regime (with newly-born and newly-merged neutron stars at higher temperatures where thermal effects on the structure become important). Neutron stars access the highest densities, and are unique in letting us study extreme states of neutron-richness and (potentially) stable states of strange matter.

2.2 FROM NUCLEAR PHYSICS TO TELESCOPE

So how do we, as astronomers, go about determining the properties of the neutron star interior? We need something that we can measure with our telescopes, that links back to the nuclear physics. Fortunately, such a chain exists.

All of the unknowns about the microphysics of the neutron star core – the types of particles present, and the forces between them – set, on the macroscopic scale, the Equation of State (EOS). This is the relation between pressure and density[1]. The EOS is an ingredient in the stellar structure equations (for neutron stars, the Tolman-Oppenheimer-Volkoff, or TOV, equations for relativistic stellar structure). For a given EOS model, and a choice of central density, these equations can be solved to give mass M and radius R. Change central density, repeat and you can trace out an M-R relation for that EOS.

Now M and R, together with the spin of the neutron star (which can be very fast), sets the gravitational field and hence the exterior space-time of the neutron star. Any photon leaving the surface of the star propagates towards us through this environment, picking up the imprint of M and R. Thanks to decades of work on rotating neutron star space-times, we can model this process extremely well (see [8] and references therein).

This is the physics that NICER relies on – by measuring the properties of the X-rays emitted from the surfaces of neutron stars, and modelling them using relativistic ray-tracing, we are able to extract M and R. We can then loop back to the EOS, using the structure equations. This we then hand off to our nuclear physics colleagues, to figure out what it means for the nature of dense matter.

2.3 INTRODUCING NICER

NICER [18] is a soft X-ray telescope, sensitive in the 0.2 - 12 keV band, with X-ray concentrator optics focusing the incoming photons onto 56 Silicon Drift Detectors[2]. It operates in the same band and with similar energy resolution as XMM-Newton, but with twice the effective area and much better time-tagging resolution. This combination of spectral and timing capability is key to NICER's science.

NICER was launched on a SpaceX Falcon 9 rocket on June 3, 2017 as part of Crew Resupply Mission 11. It was extracted from the Dragon capsule on June 11 and installed onto the International Space Station (ISS, Figure 2.2), beginning science observations in July after a brief period of commissioning and calibration. Being on the ISS is very different from operating as a free-flying telescope. NICER operates in a complex environment with all of the other moving parts of the ISS, which imposes some restrictions. NICER must occasionally be stowed, for example, during the arrival of visiting spacecraft or during spacewalks (although it is designed to withstand an 'astronaut kick load' [18]). However, there are also benefits. For a start, we have access to camera and film footage of the telescope in operation! NICER takes advantage of the ISS's power and data transmission capabilities, and it can make repairs more feasible.

[1] In theory, temperature should be included here as well but for the old, cold neutron stars that NICER studies, thermal effects on the structure are unimportant.

[2] Four of these have been inactive since launch.

Figure 2.2 NICER on the International Space Station (Image credit: NASA, with label added by author).

2.4 PULSE PROFILE MODELLING

As explained in Section 2.2, the technique that we use to extract M and R from NICER data relies on relativistic effects on X-ray emission from the neutron star surface. However NICER specifically targets stars whose X-ray emission varies – in both brightness and spectrum – as the star spins. This variation with rotational phase means that the emission is *pulsed* at the spin period of the neutron star. It is this phase-resolved pulsed emission that we model, taking advantage of NICER's high time resolution.

The specific class of neutron stars that NICER focuses on are rotation-powered millisecond X-ray pulsars (RMPs). These are old neutron stars, that have been recycled and spun-up to millisecond periods by accretion from a companion star. The accretion phase has now ended (in some cases the companion is still present, in others it has evaporated). The magnetic poles of the star emit thermal X-rays as they are heated by magnetospheric return currents. Many are also radio (and gamma-ray) pulsars, which means that we know the spin ephemeris very precisely. And if the pulsar is in a binary, radio pulsar timing can monitor orbital variations that provide important prior information about the mass of the neutron star and its inclination and distance.

PPM analysis for NICER is done using Bayesian inference (Figure 2.3). This means that we need the following: pulse profile data; a lightcurve and noise model that lets us simulate pulse profile data (this is the basis of the likelihood evaluation), with priors on the model parameters; and a framework for statistical inference that delivers posteriors on

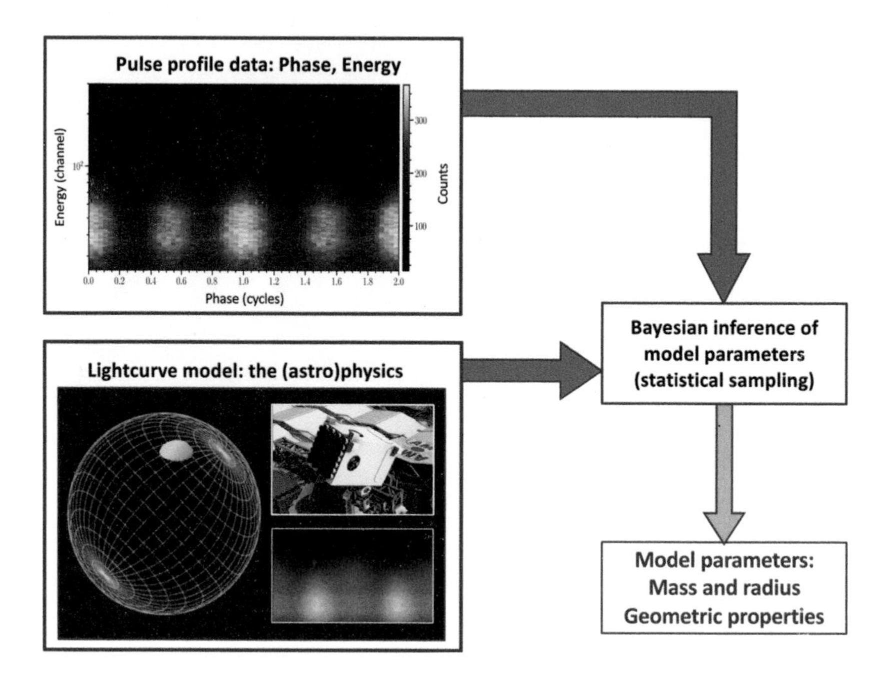

Figure 2.3 Pulse Profile Modelling is a Bayesian inference process that involves several different elements: pulse profile data, a lightcurve model, that generates synthetic pulse profiles for a given choice of model parameters, and statistical sampling. PPM results in posterior distributions on the model parameters: not only mass and radius but also the other parameters in the model such as the properties of the surface hot spots. Figure includes elements from Ref. [38] and NASA.

those model parameters. We will now step through the most important elements of the PPM process.

2.4.1 Data

The top left panel of Figure 2.3 shows an example pulse profile data set (for PSR J0030+0451, [38]). For every photon (or 'count') that NICER records we know the energy channel it was recorded in and its arrival time. Using the spin ephemeris of the pulsar, the arrival time is mapped to the rotational phase of the pulsar, and the photon is assigned to a phase bin. Over time the counts build up until we have a full profile (in the figure, two rotational cycles are shown for clarity). The brighter colours indicate more counts, and it is clear that this source has a strong double-peaked pulse profile. Pulse profiles for RMPs are extremely stable, with no significant variation on timescales of interest for NICER analysis.

For the data set to deliver useful constraints on mass and radius we need a certain minimum number of photons [24, 29]. This requires large data sets: the pulse profile for

PSR J0030+0451 for which NICER released its first results involved 1.9 Ms (22 days) of data, built up over nearly a year and a half [7].

2.4.2 Lightcurve model

The lightcurve model is where we find the physics and the astrophysics, including all of our understanding of relativistic ray-tracing of neutron star surface emission as it propagates towards the observer (see [8] and references therein). It allows us to simulate a synthetic pulse profile, for a given set of model parameters. These include the parameters setting the exterior space-time (mass, radius, spin rate), the surface emission pattern (the size, shape, location and temperature of the hot spots, and the atmospheric beaming model), observer inclination and distance, and interstellar hydrogen column density (which determines interstellar absorption as the radiation propagates from source to telescope). We also of course need to set priors on the parameters, which may or may not be well-constrained. For example, we may know mass to within a few percent if the target is in a binary (see above). If it is not, the mass prior is constrained only by neutron star formation models and current EOS models (which determine the maximum possible mass).

One particularly complex aspect is how we model the hot spots, since from pulsar theory we do not have a single clear answer as to the expected location, shape, size, or temperature distribution. Our models have to allow a wide range of possibilities, but with as few parameters as possible, to keep computational cost to manageable levels. At present, the NICER team uses simple parameterized hot spot shape and temperature distribution models using (potentially overlapping) circles or ovals, that can nevertheless capture the key features of the options currently being discussed by pulsar theorists. The hot spots generate pulsed emission. However, they may also be able to generate an unpulsed contribution that is phase-invariant, for example, due to strong gravitational lightbending or a fortuitous geometry in which at least part of a spot is always in view (for example because it is close to the rotational poles).

The lightcurve model is the basis of the likelihood evaluation in the Bayesian inference process. For a given choice of model parameters, how likely is it that we would obtain our pulse profile data set (the exact number of photons that NICER has recorded in every channel and phase bin)? We assume Poisson counting statistics apply to the predicted incident signal, but we also have to model the instrument response – how NICER actually records an incident signal. For this purpose, we use the NICER instrument response matrix files. These describe the effective area of the instrument and how incident photons of a given energy are redistributed to energy channels. And although NICER is of course perfect, our understanding of it is not! So we also add some parameterized calibration uncertainty to the overall instrument response in our modelling.

The final element to take into account when comparing data to a synthetic pulse profile is background: by which we mean anything that contributes to the observed pulse profile but which is not generated by the neutron star surface (the hot spots and the region exterior to the hot spots[3]). The background is by definition unpulsed and will include contributions from instrumental noise, particles, the cosmic X-ray background and other astrophysical

[3]We can assign this a finite emitting temperature in our modelling, but for NICER's analyses to date the rest of the stellar surface has not been found to contribute at an appreciable level in the NICER waveband.

sources in the field of view. To 'match' the data as closely as possible, the lightcurve model assumes that it can 'add on' an unpulsed background component to the pulsed (and possibly unpulsed) emission generated by the hot spots. In the early stages of the NICER mission, we worked without prior constraints on the background. More recently, we have been able to place priors on the expected range of instrumental and astrophysical background, and this can act as a strong constraint on the parameter space.

2.4.3 Statistical inference

The final stage of the Bayesian inference process is statistical sampling – an exploration of the entire (prior-bounded) parameter space for the assumed model, and which results in posterior distributions on the model parameters. Statistical sampling is done using samplers such as Multinest [9, 15] or pt-emcee [50], with the computational costs of the sampling process being sufficiently high for our current models that we require High Performance Computing facilities. Different samplers perform in different ways – Multinest, for example uses a specific implementation of a nested sampling technique [45], and aims to maximise evidence – these things set how the sampler proceeds and when it terminates. Sampler performance is tested using parameter recovery simulations where the entire PPM process is run using synthetic data with known input parameters (see, for example [6, 48]).

The process also lets us assess how well the given model performs – is it capable of producing something that resembles the data, or are there discrepancies that point to shortcomings in the model? For example, we can compare the pulse profile generated by the 'best-performing' parameter vectors with the data. By looking at the differences between them (the residuals) we can determine whether the overall distribution is consistent with what we would expect due to Poisson statistics, or whether the deviations are anomalously large. In the NICER analysis to date, for example, models where the hot spots are restricted to be antipodal (as they would be for a textbook dipole magnetic field configuration) often lead to prominent structures in the pulse profile residuals – antipodal hot spot models are simply not capable of explaining the data satisfactorily. We can also compute measures that we can use to compare models, such as the Bayesian evidence. In our NICER analysis, we have used this type of measure to compare more complex hot spot models, to determine a preferred geometry.

2.5 NICER'S FIRST MAJOR RESULTS RELEASES

Following its launch and installation in 2017, NICER set to work building up pulse profile data sets on its priority RMP targets [7] (in addition to carrying out observations of a range of other sources from magnetars to accreting black holes). While this was happening the PPM analysis teams started to work on preliminary data sets, uncovering issues such as the distorting effects of optical loading from solar photons, which resulted in us setting a minimum sun angle filter for RMP data collection. Tests were carried out on the relativistic ray-tracing packages in use within the collaboration [8], and on the overall parameter recovery performance of the lightcurve modelling and inference pipelines [6]. Slowly and surely, definitive data sets were acquired and analysed, and in late 2019 the team were finally able to release their first results.

2.5.1 PSR J0030+0451

The source that was chosen for NICER's first release is the second brightest source in its target list: the isolated RMP PSR J0030+0451 (hereafter J0030)[4]. J0030 has a spin frequency of 205 Hz and lies at a distance of 329^{+6}_{-5} pc [14]. The initial data set, 1.93 Ms in total, was taken between July 2017 and December 2018 [7].

Results released by the NICER collaboration to date come from two independent PPM analysis teams, using different simulation and inference pipelines: an Amsterdam-led team using the X-ray Pulse Simulation and Inference (X-PSI, [39]) software package and a team led by the University of Maryland. This allows us to cross-test our results and assess how sensitive we are to differences in – for example – modelling assumptions and prior or sampler choices. Although in what follows I will show primarily figures using the X-PSI results, I will summarize the key results from both sets of papers and would encourage anyone using NICER results to study both analyses.

The teams were able to identify a preferred geometric configuration for the hot spots, and an associated mass and radius. Figure 2.4 shows the posterior distributions for the mass, radius and compactness for the preferred 'ST+PST' model obtained by the X-PSI team [38]: a mass and radius of $1.34^{+0.15}_{-0.16}$ $M_\odot$ and $12.71^{+1.14}_{-1.19}$ km, respectively (68% credible intervals). The model quality is good, as can be seen from the residuals. The hot spot geometry, however, was very unexpected. The inference suggests that both hot spots occupy the opposite hemisphere to the observer. One hot spot is small, while the other is a long, thin crescent. This points to a complex magnetic field geometry (Figure 2.5), certainly not a centered dipole configuration, which raises interesting follow-on questions for both pulsar emission mechanisms and neutron star evolution [5].

The results obtained by the Maryland team [25] were similar: an inferred mass and radius of $1.44^{+0.15}_{-0.14}$ $M_\odot$ and $13.02^{+1.24}_{-1.06}$ km, and a similar hot spot configuration with a small spot and a long thin oval (reflecting differences in how the hot spot shapes are parameterized), again on the opposite hemisphere from the observer.

The fact that PPM could indeed deliver constraints on mass and radius for a star with no prior information on either was a very positive sign. With uncertainties $\sim 10\%$ on mass and radius, derived constraints on the EOS were not particularly strong, but some improvement could already be seen [25, 31, 33].

2.5.2 PSR J0740+6620

The NICER collaboration had originally intended, once we had published our first results on J0030, to prioritise analysis of the brightest source in the target list, PSR J0437-4715. But in the mean time, our radio astronomy colleagues had thrown us a curveball when they measured a mass of $2.14^{+0.10}_{-0.09}$ $M_\odot$ for PSR J0740+6620 (hereafter J0740) [12]), making it the heaviest pulsar known at the time. A heavy pulsar means a high central density, making it a very interesting source for dense matter constraints. J0740 therefore leapfrogged up our priority list.

[4]Why not the brightest source, PSR J0437-4715, a source with a tight mass prior? Two main reasons - firstly, a wish to confirm that PPM could constrain both mass and radius in the absence of informative priors on either. And secondly, because PSR J0437-4715 has to be observed off-axis, complicating the instrument response modelling (see Section 2.7.1)

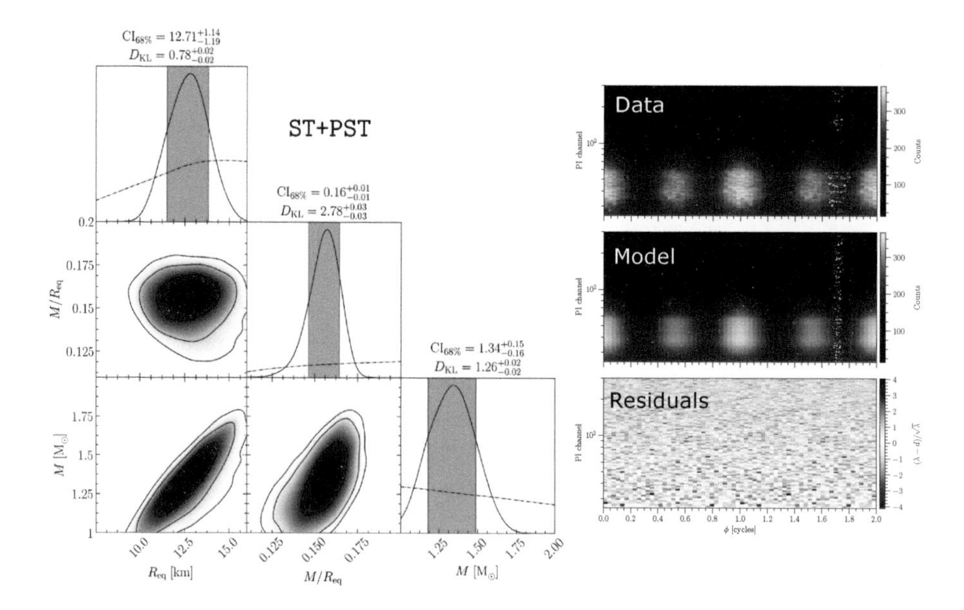

Figure 2.4 X-PSI team results for PSR J0030+0451 (figures from Ref. [38]). Left: posterior distributions for mass, radius and compactness. Contour levels for the 2D posterior distribution plots indicate the 68.3%, 95.4% and 99.7% credible regions. On the 1D posterior plots, the red band indicates the 68.3% credible interval, and the dashed line shows the prior. Right: A residual plot showing the differences between data and model (for details of the scaling used see Ref. [38]). No obvious clusters are seen in the residuals, and the overall distribution is as expected.

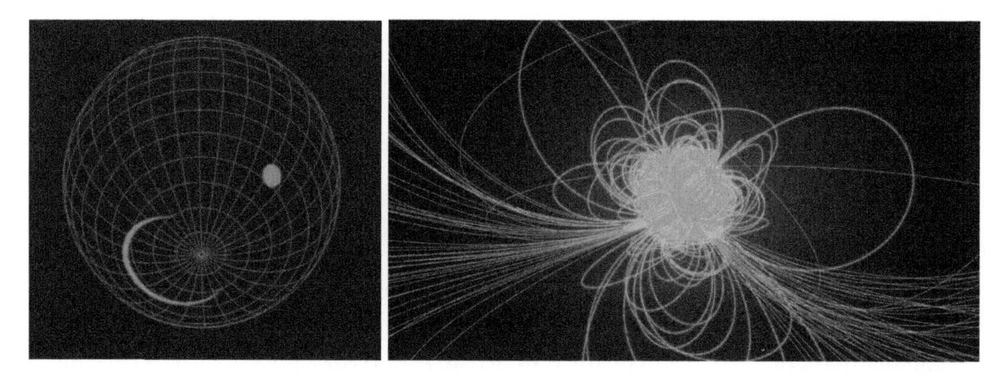

Figure 2.5 Left: X-PSI team inferred hot spot geometry for PSR J0030+0451 [38] (Image credit: NASA). From Earth we view the star from the opposite hemisphere. Right: one possible magnetic field configuration that might lead to such hot spot locations and shapes (Image credit: NASA).

J0740 is a 346 Hz RMP in a binary system and has a near edge-on inclination which permits a measurement of Shapiro delay: this delivers an informative prior on not only the mass but also the inclination and distance. Updated values for all of these parameters (with mass shifting to 2.08 ± 0.07 $M_\odot$) were provided to the NICER collaboration for us to use in our PPM analysis [16]. At a distance ~ 1.1 kpc J0740 is faint and a real stretch goal for NICER in terms of signal to noise, but from September 2018 to April 2020 the team were able to build up a 1.6 Ms dataset to be used for PPM [54]. Collaboration results were reported by both the X-PSI team [40] and the Maryland team [26].

Hot spot maps for J0740 appear to be simpler than for J0030, with plain circular hot spots being sufficient to explain the data well, and further complexity in hot spot properties not appearing to offer any improvement. However, the spots are not antipodal, again pointing to a magnetic field configuration more complex than a simple centered dipole. We view J0740 from a near-equatorial inclination (as already known from the radio observations), with one hot spot near the equator and the other at a latitude closer to either north or south pole (we cannot determine which).

In our analysis, we also took into account XMM-Newton data of the source. Although the XMM-Newton data set was much smaller, and consisted of just a spectrum rather than a phase-resolved pulse-profile, it turned out to be very important because for the XMM-Newton data set we are able to get a much better estimate of the background. So now, instead of just using the NICER data set in the inference process, we did a joint fit for both data sets; the full NICER pulse profile and (by phase-averaging the output of the simulated pulse profiles and folding them through the XMM-Newton instrument response) the XMM-Newton data set. For the latter, we imposed the informative priors on the background, thereby constraining the number of photons that are allowed to come from the source (i.e. the hot spots). This then constrains indirectly the NICER background by placing limits on the spot contribution to the unpulsed emission in the NICER data set.

The results, and the impact of having the background constraints, can be seen in Figure 2.6. The inferred mass was essentially unchanged from the radio prior (and indeed without the radio information we would not have been able to place tight constraints using this data set). Using NICER data alone, the X-PSI team analysis inferred a radius of $11.29^{+1.20}_{-0.81}$ km. Including the XMM-Newton data in the joint fit, however, increases the inferred radius to $12.39^{+1.30}_{-0.98}$ km. So what is happening? The answer can be seen by looking at the compactness M/R. When XMM-Newton data are included, high compactness solutions - in which spots can make more of a contribution to the unpulsed component of the pulse profile due to more extreme lightbending - are removed from the solution space. The high compactness solutions result in too many source counts for the XMM-Newton data set, where the source to background ratio is well-constrained. With mass essentially fixed, this reduces low radius solutions.

The Maryland team inferred a similar hot spot geometry, and a radius from the combined NICER and XMM-Newton fits of $13.7^{+2.6}_{-1.5}$ km [26]. The reasons for the differences between the values reported by the two teams are discussed in more depth in the papers (see Section 4.4 of [40] and Section 4.6 of [26]), but the main factors are differences in priors, assumptions about NICER/XMM-Newton cross-calibration uncertainty, treatment of the XMM-Newton background information, and sampler choice. The two analyses also use slightly different channel selections for the NICER data. Despite the differences in

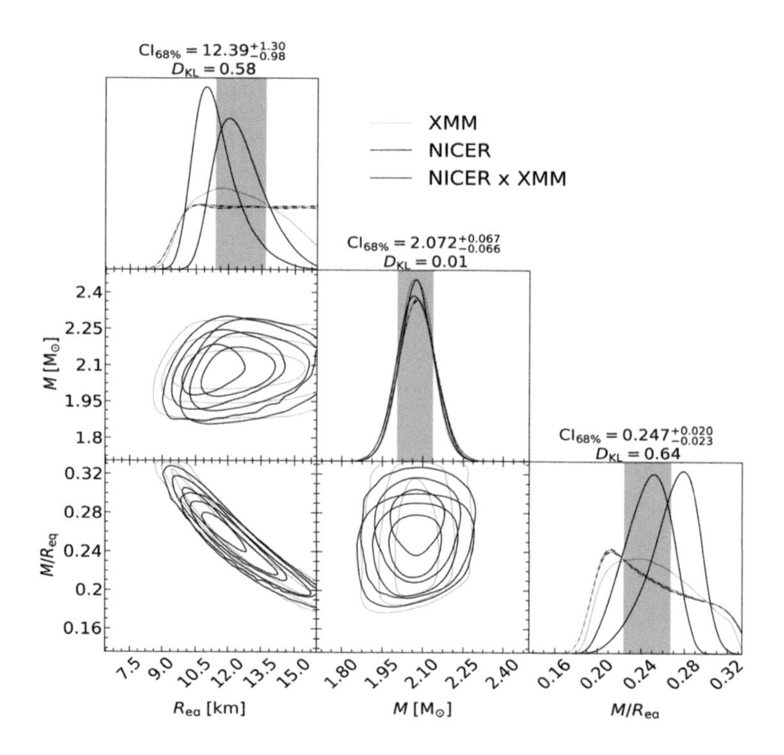

Figure 2.6 Posterior distributions from the X-PSI team analysis of PSR J0740+6620 for mass, radius and compactness (figure adapted from Ref. [40]). Contour levels for the 2D posterior distribution plots indicate the 68.3%, 95.4% and 99.7% credible regions. On the 1D posterior plots, the red band indicates the 68.3% credible interval, and the dashed line shows the prior. Results are shown for three different data sets: NICER only; XMM-Newton only; and NICER and XMM-Newton combined.

credible intervals, however, both analyses find median radii similar to those found previously for J0030, and appear to disfavour very low radii. This already starts to place interesting constraints on the EOS [26, 30].

2.6 NICER'S LATEST RESULTS

2.6.1 PSR J0740+6620

Our initial work on the heavy pulsar J0740 illustrated the importance of good constraints on the background. And while indirect constraints such as those imposed by joint-fitting with XMM-Newton are useful, they introduce another source of uncertainty due to instrumental cross-calibration. Our previous modelling took into account energy-independent calibration uncertainty, but did not take into account possible energy-dependent effects. Meanwhile, however, NICER background characterization has advanced to the point where we can include it directly in our analysis.

In [43] we revisited our analysis of J0740 and included prior information from both the NICER space weather model [17] and the 3C50 background model [36]. For space weather, we were able to use the same data set used previously [54]. For 3C50 we require a specially filtered data set, and used NICER data taken over a slightly longer period, from September 2018 to December 2021. The inferred mass and radius were compatible with the already published results, with a slight dependence on how tight we assumed the background constraints to be. This was both reassuring in terms of the robustness of our results, and encouraging for future analysis where we can now include such background constraints.

These data sets have also been used to explore a number of aspects in our models in more detail. In addition to looking at the effects of background models, [43] also considered a broader range of antipodal spot models than [40]. In particular, we expanded our analysis to consider the option of the two hot spots being antipodal but differing in size and temperature (the previous antipodal model analysis assumed the two spots to be identical). Antipodal models are still strongly disfavoured, strengthening the case for a complex magnetic field.

Meanwhile, [44] explored the uncertainties associated with our atmosphere models, using X-PSI. The initial analysis for J0740 by [40] considered atmospheres composed of fully and partially ionized hydrogen and helium. [44] conducted a broader analysis including partially ionized carbon composition, externally heated hydrogen and an empirical atmospheric beaming parametrization to explore deviations in the expected anisotropy of the emitted radiation. None of the studied atmosphere cases were found to have any significant influence on the inferred radius of J0740, possibly due to its X-ray faintness, tighter external constraints and/or viewing geometry.

2.6.2 PSR J0030+0451

While J0740 has proven to be robust to changes in the modelling of background and atmosphere, the picture for J0030 has become more complex. Some of this complexity was already apparent in the initial analysis. The X-PSI analysis of [38] showed that simpler hot spot models, while not preferred by measures such as the evidence, were acceptable in terms of residuals. If at some point pulsar theorists were to conclude that our model space was too broad, or our preferred configuration with the extended crescent was inconsistent with emission models for radio or gamma-rays, we could discard this model and return to a simpler one without a problem. This would, however, change the inferred mass and radius.

Our analysis of J0740 indicated that testing higher resolution sampler settings for X-PSI would be desirable, as would formally including background constraints. In the 2019 analysis no constraints were placed on the J0030 background, but the inferred background was compared a posteriori with predictions from XMM-Newton observations. Although they were close and the uncertainties were large, it was noted that the inferred background was somewhat too high and the spots should perhaps contribute more to the unpulsed component [25, 38].

We have now started to look at some of these issues in more detail, in anticipation of a new larger data set for this source. [48] has carried out detailed parameter recovery simulations with X-PSI using hot spot configurations inspired by those found in our previous analysis [38] for J0030. In addition to confirming that X-PSI is capable of recovering mass and radius within reasonable credible intervals, the simulations show that the posterior surface could be highly multimodal, setting stringent minimum requirements for sampler settings.

The X-PSI team have also re-analysed the 2017-2018 J0030 data set, with higher sampler resolution, a larger range of hot spot models, and taking into account both background constraints and updates to the instrument response matrix that were implemented in our analysis of J0740 [49]. If we analyse only the NICER data set, with the same hot spot models used in the 2019 analysis, we obtain results that are fully consistent with those reported previously (in terms of mass, radius and geometry). However, the posterior surface is indeed multimodal, and this becomes important once we carry out joint fitting with the XMM-Newton data set in order to impose indirect constraints on the NICER background. Including more complex models and the background constraints changes the preferred configurations. Subject to some caveats (discussed in detail in [49]), our analysis finds inferred masses and radii $\sim$ [1.4 $M_{\odot}$, 11.5 km] and $\sim$ [1.7 $M_{\odot}$, 14.5 km], depending on the assumed model (with the latter being preferred by the evidence). The two hot spots need no longer be in the same hemisphere, and may have temperature gradients rather than being elongated. J0030 also appears to be more sensitive than J0740, in terms of the inferred parameters, to different choices of atmosphere model [44].

2.7 THE FUTURE OF PPM

2.7.1 RMPs: Upcoming NICER results and beyond

The NICER collaboration is currently analysing pulse profile data sets for five more RMPs [7, 54]: PSR J0437-4715 (with known mass $\sim$ 1.4 $M_{\odot}$ [35]), PSR J1614-2230 (with known mass $\sim$ 1.9 $M_{\odot}$ [4]), PSR J0614-3329, PSR J1231-1411 and PSR J2124-3358. All future analyses are likely to include background constraints, from either XMM-Newton observations or using NICER's own background models.

For PSR J0437-4715, NICER's brightest source, analysis is complicated by the presence of a bright Active Galactic Nucleus (AGN) in the field of view. The source must be viewed off-axis to reduce the contribution from the AGN, which makes characterising the instrument response more difficult. We also need to consider how to address potential variability of the AGN in our models, since it is not possible to monitor the AGN independently and simultaneously every single time that NICER points at J0437. How to deal with these issues has now been resolved, and results are (dare I say it?!) imminent.

NICER's mission has now been extended to at least 2025, and it continues to collect more data on all of our sources, including J0030 and J0740. Larger data sets – for which NICER background estimates are available – are already being analysed for both sources. Constraints on mass and radius from PPM are predicted to get tighter as data sets improve [24, 29] and preliminary analysis of the updated pulse profiles for both J0030 and J0740 indicates that this is indeed the case. It will be particularly interesting to see how the results for J0030 evolve when we analyse the 3C50-filtered data set with background estimates, given indications from the joint NICER and XMM-Newton analysis of the older data set in [49]. Other aspects of our modelling will also continue to improve, for example with new NICER background models such as SCORPEON being incorporated into the analysis pipelines.

There are many more RMPs that could be interesting targets for PPM but which are simply too faint for NICER to build up data sets in a reasonable time, including some with known masses from radio pulsar timing (such as PSR J2222-0137, PSR J0751+1807 and PSR J1909-3744). They could be important targets for the next generation of X-ray telescopes, however. Proposed missions such as ESA's Advanced Telescope for High Energy Astrophysics (Athena), the joint Chinese/European Enhanced X-ray Timing and Polarimetry (eXTP) mission [53], or the NASA Probe concept the Spectroscopic Time-Resolving Observatory for Broadband X-rays (STROBE-X) [34] could all build up high quality PPM data sets on both the NICER RMPs and these fainter sources, with the additional advantage of having lower background.

2.7.2 PPM for accreting and bursting neutron stars

RMPs are not the only type of rapidly-rotating[5] neutron stars with X-ray emitting hot spots. There are accretion-powered millisecond X-ray pulsars (AMXPs, [28]), where hot spots form as accreting material is channeled by the magnetic field onto the magnetic poles of the star; and Thermonuclear Burst Oscillation (TBO) sources [51], where hot regions form during thermonuclear-powered Type I X-ray bursts in the accreted ocean of the neutron star. AMXPs and TBO sources are attractive targets for PPM: they are numerous; many of them spin faster than the RMPs; and they offer the prospect for cross-checks. By this I mean not only cross-checks against results from the RMP population, but also cross-checks for an individual source. Some neutron stars, for example, have both accretion-powered pulsations and burst oscillations; by modelling both types of pulsation using PPM, we can check for systematic uncertainties in our models (for further discussion of this see [53]). Building up pulse profiles for these stars ideally requires broader band X-ray coverage than NICER (since pulsed emission is also found at higher photon energies) and they are major targets for broadband missions like STROBE-X and eXTP.

Efforts are now underway to apply PPM to both AMXPs and TBO sources [21, 22, 24, 27, 41, 42, 46]. Just as for the RMPs, the surface emission pattern is not well constrained, so again we need to use broad families of (theory-informed) parameterized models of the surface temperature distribution. For TBOs, this is particularly difficult since the mechanism that generates the patterns is not yet understood [51]. For AMXPs, the accretion column can also affect the beamed emission, and this needs to be accounted for in the atmosphere modelling [41]. And for both types of source the pulse profile is variable – for TBO sources, where the thermonuclear burst evolves on timescales of a second, dramatically so. For AMXPs the variability timescales (of order a few days) are such that future large-area telescopes should be able to collect enough data in a short-enough observing time to mitigate this. However, for TBOs variability is unavoidable, so we are developing and testing analysis methods that can take this into account [21, 22, 24].

[5] Rapid rotation is important to break the degeneracy between mass and radius, [24, 29].

2.8 NICER'S IMPACT

2.8.1 Dense matter EOS

The primary purpose of NICER's mass and radius inference is to constrain the dense matter EOS. One can of course simply check whether a specific EOS model produces mass-radius curves that are consistent with NICER's results. However, one can also use the mass radius posteriors as input to a second stage of Bayesian inference using a parameterized EOS model, and determine posterior distributions on those model parameters (see for example [3, 23, 26, 30, 47]. All NICER collaboration PPM papers have linked Zenodo repositories where users who wish to do follow-on EOS analysis can find both credible region mass-radius contour data (for simple checks) and the full posterior samples necessary for subsequent EOS inference. Note that careful attention must be paid to priors, both on mass and radius in the original set-up of the PPM (to ensure that set-up is consistent with what is done subsequently [32, 37]) and for the EOS model parameters when doing the EOS inference itself [19].

Figure 2.7 shows an example from one such follow-on EOS inference. The mass-radius relation shown in the figure results from EOS analysis using a model that couples constraints from Chiral Effective Field Theory at low densities with a piecewise-polytropic parameterization at higher densities [30]. The analysis uses the X-PSI team's mass-radius posteriors for J0030 and J0740, and mass and tidal deformability posteriors from gravitational wave observations of two binary neutron star binary mergers. The derived mass-radius relation as these measurements are combined is noticeably narrower than the individual NICER mass-radius posteriors, with the radius of a 1.4 $M_\odot$ neutron star $R_{1.4} = 12.30^{+0.72}_{-0.76}$ km [30]. This provides tight constraints on the pressure of neutron star matter at around twice saturation density. A similar narrowing can also be seen in Maryland team's EOS analysis [26], which uses different EOS models and their own NICER mass-radius posteriors. And as NICER's measurements improve, these constraints will tighten still further.

2.8.2 Surface maps and magnetic fields

One assumption underpinning our PPM analysis of NICER RMP data is that the hot spots are the magnetic poles of the star, heated by return currents. This means that their properties reveal (indirectly) the magnetic field structure of the star at the stellar surface. NICER's results to date show that the properties of both J0030 and J0740 are very different from the textbook simple dipole picture of pulsar magnetic fields, for which the magnetic poles would be antipodal and identical. This has broad consequences for our understanding of pulsar magnetospheres, emission mechanisms and evolution (from birth through the binary evolution and accretion phase that caused the NICER pulsars to spin up to millisecond spin periods) (see, e.g. [10, 11, 13, 20]).

2.9 NICER AS A PIONEER

The results that we are generating using NICER data are some of the first measurements made using PPM. This is a huge step forward for a technique that has been many years in development, and which forms the basis for the science case for several future large-area

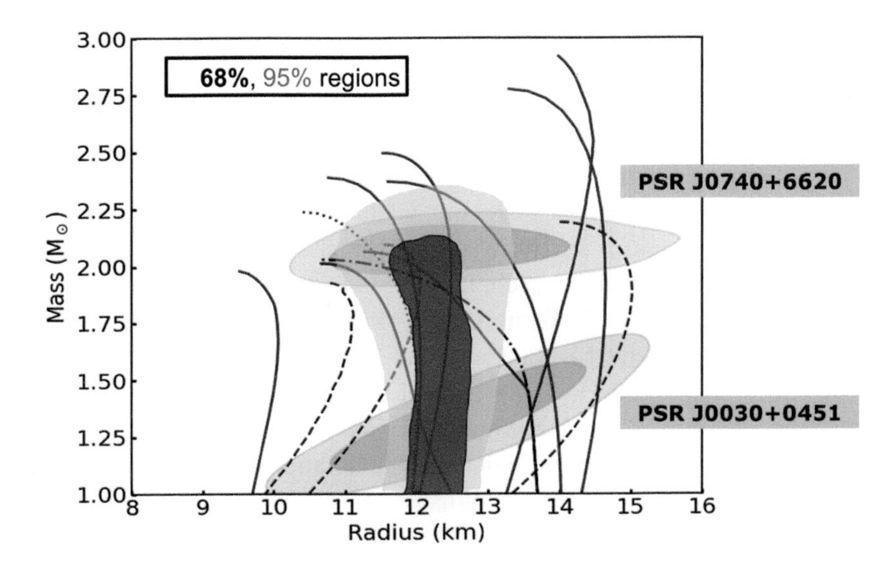

Figure 2.7 EOS inference. The blue contours show the inferred mass-radius relation associated with one of the parameterized EOS models considered in Ref. [30]. The analysis used the mass-radius posteriors from the X-PSI team PPM analysis (shown, Refs. [38, 40]) and mass-tidal deformability constraints from two binary neutron star mergers detected in gravitational waves, GW170817 [1] and GW190425 [2]. The individual lines show, for comparison, mass-radius relations for a number of individual EOS models from the literature (see Figure 3 of Ref. [52] for details).

X-ray telescopes. It is a vital proof of principle and a demonstration that the technique is viable and reliable. So far we have measured the size of two neutron stars, including the heaviest pulsar known to date, with more results to come over the anticipated mission lifetime. And as a by-product we are making surface maps of these tiny stars, thousands of light years from Earth. Not bad for a small X-ray telescope on a 20-year-old space station!

ACKNOWLEDGEMENTS

The author thanks Tuomo Salmi and Serena Vinciguerra for comments, and acknowledges support from ERC Consolidator grant No. 865768 AEONS.

Bibliography

[1] B. P. Abbott et al. Properties of the Binary Neutron Star Merger GW170817. *Physical Review X*, 9(1):011001, January 2019.

[2] B. P. Abbott et al. GW190425: Observation of a Compact Binary Coalescence with Total Mass $\sim$ 3.4 $M_\odot$. *Astrophysical Journal Letters*, 892(1):L3, March 2020.

[3] Eemeli Annala, Tyler Gorda, Joonas Hirvonen, Oleg Komoltsev, Aleksi Kurkela, Joonas Nättilä, and Aleksi Vuorinen. Strongly interacting matter exhibits deconfined behavior in massive neutron stars. *arXiv e-prints*, page arXiv:2303.11356, March 2023.

[4] Zaven Arzoumanian, Adam Brazier, Sarah Burke-Spolaor, Sydney Chamberlin, Shami Chatterjee, Brian Christy, James M. Cordes, Neil J. Cornish, Fronefield Crawford, H. Thankful Cromartie, Kathryn Crowter, Megan E. DeCesar, Paul B. Demorest, Timothy Dolch, Justin A. Ellis, Robert D. Ferdman, Elizabeth C. Ferrara, Emmanuel Fonseca, Nathan Garver-Daniels, Peter A. Gentile, Daniel Halmrast, E. A. Huerta, Fredrick A. Jenet, Cody Jessup, Glenn Jones, Megan L. Jones, David L. Kaplan, Michael T. Lam, T. Joseph W. Lazio, Lina Levin, Andrea Lommen, Duncan R. Lorimer, Jing Luo, Ryan S. Lynch, Dustin Madison, Allison M. Matthews, Maura A. McLaughlin, Sean T. McWilliams, Chiara Mingarelli, Cherry Ng, David J. Nice, Timothy T. Pennucci, Scott M. Ransom, Paul S. Ray, Xavier Siemens, Joseph Simon, Renée Spiewak, Ingrid H. Stairs, Daniel R. Stinebring, Kevin Stovall, Joseph K. Swiggum, Stephen R. Taylor, Michele Vallisneri, Rutger van Haasteren, Sarah J. Vigeland, Weiwei Zhu, and NANOGrav Collaboration. The NANOGrav 11-year Data Set: High-precision Timing of 45 Millisecond Pulsars. *Astrophysical Journal Supplement Series*, 235(2):37, April 2018.

[5] A. V. Bilous, A. L. Watts, A. K. Harding, T. E. Riley, Z. Arzoumanian, S. Bogdanov, K. C. Gendreau, P. S. Ray, S. Guillot, W. C. G. Ho, and D. Chakrabarty. A NICER View of PSR J0030+0451: Evidence for a Global-scale Multipolar Magnetic Field. *Astrophysical Journal Letters*, 887(1):L23, December 2019.

[6] Slavko Bogdanov, Alexander J. Dittmann, Wynn C. G. Ho, Frederick K. Lamb, Simin Mahmoodifar, M. Coleman Miller, Sharon M. Morsink, Thomas E. Riley, Tod E. Strohmayer, Anna L. Watts, Devarshi Choudhury, Sebastien Guillot, Alice K. Harding, Paul S. Ray, Zorawar Wadiasingh, Michael T. Wolff, Craig B. Markwardt, Zaven Arzoumanian, and Keith C. Gendreau. Constraining the Neutron Star Mass-Radius Relation and Dense Matter Equation of State with NICER. III. Model Description and Verification of Parameter Estimation Codes. *Astrophysical Journal Letters*, 914(1):L15, June 2021.

[7] Slavko Bogdanov, Sebastien Guillot, Paul S. Ray, Michael T. Wolff, Deepto Chakrabarty, Wynn C. G. Ho, Matthew Kerr, Frederick K. Lamb, Andrea Lommen, Renee M. Ludlam, Reilly Milburn, Sergio Montano, M. Coleman Miller, Michi Bauböck, Feryal Özel, Dimitrios Psaltis, Ronald A. Remillard, Thomas E. Riley, James F. Steiner, Tod E. Strohmayer, Anna L. Watts, Kent S. Wood, Jesse Zeldes, Teruaki Enoto, Takashi Okajima, James W. Kellogg, Charles Baker, Craig B. Markwardt, Zaven Arzoumanian, and Keith C. Gendreau. Constraining the Neutron Star Mass-Radius Relation and Dense Matter Equation of State with NICER. I. The Millisecond Pulsar X-Ray Data Set. *Astrophysical Journal Letters*, 887(1):L25, December 2019.

[8] Slavko Bogdanov, Frederick K. Lamb, Simin Mahmoodifar, M. Coleman Miller, Sharon M. Morsink, Thomas E. Riley, Tod E. Strohmayer, Albert K. Tung, Anna L. Watts, Alexander J. Dittmann, Deepto Chakrabarty, Sebastien Guillot, Zaven Arzoumanian, and Keith C. Gendreau. Constraining the Neutron Star Mass-Radius Relation and Dense Matter Equation of State with NICER. II. Emission from Hot Spots on a

Rapidly Rotating Neutron Star. *Astrophysical Journal Letters*, 887(1):L26, December 2019.

[9] J. Buchner, A. Georgakakis, K. Nandra, L. Hsu, C. Rangel, M. Brightman, A. Merloni, M. Salvato, J. Donley, and D. Kocevski. X-ray spectral modelling of the AGN obscuring region in the CDFS: Bayesian model selection and catalogue. *Astronomy & Astrophysics*, 564:A125, April 2014.

[10] F. Carrasco, J. Pelle, O. Reula, D. Viganò, and C. Palenzuela. Relativistic force-free models of the thermal X-ray emission in millisecond pulsars observed by NICER. *Monthly Notices of the Royal Astronomical Society*, 520(2):3151–3163, April 2023.

[11] Alexander Y. Chen, Yajie Yuan, and Georgios Vasilopoulos. A Numerical Model for the Multiwavelength Lightcurves of PSR J0030+0451. *Astrophysical Journal Letters*, 893(2):L38, April 2020.

[12] H. T. Cromartie, E. Fonseca, S. M. Ransom, P. B. Demorest, Z. Arzoumanian, H. Blumer, P. R. Brook, M. E. DeCesar, T. Dolch, J. A. Ellis, R. D. Ferdman, E. C. Ferrara, N. Garver-Daniels, P. A. Gentile, M. L. Jones, M. T. Lam, D. R. Lorimer, R. S. Lynch, M. A. McLaughlin, C. Ng, D. J. Nice, T. T. Pennucci, R. Spiewak, I. H. Stairs, K. Stovall, J. K. Swiggum, and W. W. Zhu. Relativistic Shapiro delay measurements of an extremely massive millisecond pulsar. *Nature Astronomy*, 4:72–76, January 2020.

[13] Pushpita Das, Oliver Porth, and Anna L. Watts. GRMHD simulations of accreting neutron stars with non-dipole fields. *Monthly Notices of the Royal Astronomical Society*, 515(3):3144–3161, September 2022.

[14] H. Ding, A. T. Deller, B. W. Stappers, T. J. W. Lazio, D. Kaplan, S. Chatterjee, W. Brisken, J. Cordes, P. C. C. Freire, E. Fonseca, I. Stairs, L. Guillemot, A. Lyne, I. Cognard, D. J. Reardon, and G. Theureau. The MSPSRπ catalogue: VLBA astrometry of 18 millisecond pulsars. *Monthly Notices of the Royal Astronomical Society*, 519(4):4982–5007, March 2023.

[15] F. Feroz, M. P. Hobson, and M. Bridges. MULTINEST: an efficient and robust Bayesian inference tool for cosmology and particle physics. *Monthly Notices of the Royal Astronomical Society*, 398(4):1601–1614, October 2009.

[16] E. Fonseca, H. T. Cromartie, T. T. Pennucci, P. S. Ray, A. Yu. Kirichenko, S. M. Ransom, P. B. Demorest, I. H. Stairs, Z. Arzoumanian, L. Guillemot, A. Parthasarathy, M. Kerr, I. Cognard, P. T. Baker, H. Blumer, P. R. Brook, M. DeCesar, T. Dolch, F. A. Dong, E. C. Ferrara, W. Fiore, N. Garver-Daniels, D. C. Good, R. Jennings, M. L. Jones, V. M. Kaspi, M. T. Lam, D. R. Lorimer, J. Luo, A. McEwen, J. W. McKee, M. A. McLaughlin, N. McMann, B. W. Meyers, A. Naidu, C. Ng, D. J. Nice, N. Pol, H. A. Radovan, B. Shapiro-Albert, C. M. Tan, S. P. Tendulkar, J. K. Swiggum, H. M. Wahl, and W. W. Zhu. Refined Mass and Geometric Measurements of the High-mass PSR J0740+6620. *Astrophysical Journal Letters*, 915(1):L12, July 2021.

[17] K. Gendreau. Nicer background estimator tools, 2020.

[18] Keith C. Gendreau, Zaven Arzoumanian, Phillip W. Adkins, Cheryl L. Albert, John F. Anders, Andrew T. Aylward, Charles L. Baker, Erin R. Balsamo, William A. Bamford, Suyog S. Benegalrao, Daniel L. Berry, Shiraz Bhalwani, J. Kevin Black, Carl Blaurock, Ginger M. Bronke, Gary L. Brown, Jason G. Budinoff, Jeffrey D. Cantwell, Thoniel Cazeau, Philip T. Chen, Thomas G. Clement, Andrew T. Colangelo, Jerry S. Coleman, Jonathan D. Coopersmith, William E. Dehaven, John P. Doty, Mark D. Egan, Teruaki Enoto, Terry W. Fan, Deneen M. Ferro, Richard Foster, Nicholas M. Galassi, Luis D. Gallo, Chris M. Green, Dave Grosh, Kong Q. Ha, Monther A. Hasouneh, Kristofer B. Heefner, Phyllis Hestnes, Lisa J. Hoge, Tawanda M. Jacobs, John L. Jørgensen, Michael A. Kaiser, James W. Kellogg, Steven J. Kenyon, Richard G. Koenecke, Robert P. Kozon, Beverly LaMarr, Mike D. Lambertson, Anne M. Larson, Steven Lentine, Jesse H. Lewis, Michael G. Lilly, Kuochia Alice Liu, Andrew Malonis, Sridhar S. Manthripragada, Craig B. Markwardt, Bryan D. Matonak, Isaac E. Mcginnis, Roger L. Miller, Alissa L. Mitchell, Jason W. Mitchell, Jelila S. Mohammed, Charles A. Monroe, Kristina M. Montt de Garcia, Peter D. Mulé, Louis T. Nagao, Son N. Ngo, Eric D. Norris, Dwight A. Norwood, Joseph Novotka, Takashi Okajima, Lawrence G. Olsen, Chimaobi O. Onyeachu, Henry Y. Orosco, Jacqualine R. Peterson, Kristina N. Pevear, Karen K. Pham, Sue E. Pollard, John S. Pope, Daniel F. Powers, Charles E. Powers, Samuel R. Price, Gregory Y. Prigozhin, Julian B. Ramirez, Winston J. Reid, Ronald A. Remillard, Eric M. Rogstad, Glenn P. Rosecrans, John N. Rowe, Jennifer A. Sager, Claude A. Sanders, Bruce Savadkin, Maxine R. Saylor, Alexander F. Schaeffer, Nancy S. Schweiss, Sean R. Semper, Peter J. Serlemitsos, Larry V. Shackelford, Yang Soong, Jonathan Struebel, Michael L. Vezie, Joel S. Villasenor, Luke B. Winternitz, George I. Wofford, Michael R. Wright, Mike Y. Yang, and Wayne H. Yu. The Neutron star Interior Composition Explorer (NICER): design and development. In Jan-Willem A. den Herder, Tadayuki Takahashi, and Marshall Bautz, editors, *Space Telescopes and Instrumentation 2016: Ultraviolet to Gamma Ray*, volume 9905 of *Society of Photo-Optical Instrumentation Engineers (SPIE) Conference Series*, page 99051H, July 2016.

[19] S. K. Greif, G. Raaijmakers, K. Hebeler, A. Schwenk, and A. L. Watts. Equation of state sensitivities when inferring neutron star and dense matter properties. *MNRAS*, 485:5363–5376, 2019.

[20] Constantinos Kalapotharakos, Zorawar Wadiasingh, Alice K. Harding, and Demosthenes Kazanas. The Multipolar Magnetic Field of the Millisecond Pulsar PSR J0030+0451. *Astrophysical Journal*, 907(2):63, February 2021.

[21] Yves Kini, Tuomo Salmi, Serena Vinciguerra, Anna L. Watts, Devarshi Choudhury, Slavko Bogdanov, Johannes Buchner, Zach Meisel, and Valery Suleimanov. Pulse Profile Modelling of Thermonuclear Burst Oscillations II: Handling variability. *arXiv e-prints*, page arXiv:2308.12895, August 2023.

[22] Yves Kini, Tuomo Salmi, Anna L. Watts, Serena Vinciguerra, Devarshi Choudhury, Siem Fenne, Slavko Bogdanov, Zach Meisel, and Valery Suleimanov. Pulse profile

modelling of thermonuclear burst oscillations - I. The effect of neglecting variability. *Monthly Notices of the Royal Astronomical Society*, 522(3):3389–3404, July 2023.

[23] Isaac Legred, Katerina Chatziioannou, Reed Essick, Sophia Han, and Philippe Landry. Impact of the PSR J 0740 +6620 radius constraint on the properties of high-density matter. *Physical Review D*, 104(6):063003, September 2021.

[24] Ka Ho Lo, M. Coleman Miller, Sudip Bhattacharyya, and Frederick K. Lamb. Determining Neutron Star Masses and Radii Using Energy-resolved Waveforms of X-Ray Burst Oscillations. *Astrophysical Journal*, 776(1):19, October 2013.

[25] M. C. Miller, F. K. Lamb, A. J. Dittmann, S. Bogdanov, Z. Arzoumanian, K. C. Gendreau, S. Guillot, A. K. Harding, W. C. G. Ho, J. M. Lattimer, R. M. Ludlam, S. Mahmoodifar, S. M. Morsink, P. S. Ray, T. E. Strohmayer, K. S. Wood, T. Enoto, R. Foster, T. Okajima, G. Prigozhin, and Y. Soong. PSR J0030+0451 Mass and Radius from NICER Data and Implications for the Properties of Neutron Star Matter. *Astrophysical Journal Letters*, 887(1):L24, December 2019.

[26] M. C. Miller, F. K. Lamb, A. J. Dittmann, S. Bogdanov, Z. Arzoumanian, K. C. Gendreau, S. Guillot, W. C. G. Ho, J. M. Lattimer, M. Loewenstein, S. M. Morsink, P. S. Ray, M. T. Wolff, C. L. Baker, T. Cazeau, S. Manthripragada, C. B. Markwardt, T. Okajima, S. Pollard, I. Cognard, H. T. Cromartie, E. Fonseca, L. Guillemot, M. Kerr, A. Parthasarathy, T. T. Pennucci, S. Ransom, and I. Stairs. The Radius of PSR J0740+6620 from NICER and XMM-Newton Data. *Astrophysical Journal Letters*, 918(2):L28, September 2021.

[27] M. Coleman Miller and Frederick K. Lamb. Determining Neutron Star Properties by Fitting Oblate-star Waveform Models to X-Ray Burst Oscillations. *Astrophysical Journal*, 808(1):31, July 2015.

[28] A. Patruno and A. L. Watts. Accreting Millisecond X-Ray Pulsars. *ArXiv e-prints*, 2012.

[29] Dimitrios Psaltis, Feryal Özel, and Deepto Chakrabarty. Prospects for Measuring Neutron-star Masses and Radii with X-Ray Pulse Profile Modeling. *Astrophysical Journal*, 787(2):136, June 2014.

[30] G. Raaijmakers, S. K. Greif, K. Hebeler, T. Hinderer, S. Nissanke, A. Schwenk, T. E. Riley, A. L. Watts, J. M. Lattimer, and W. C. G. Ho. Constraints on the Dense Matter Equation of State and Neutron Star Properties from NICER's Mass-Radius Estimate of PSR J0740+6620 and Multimessenger Observations. *Astrophysical Journal Letters*, 918(2):L29, September 2021.

[31] G. Raaijmakers, S. K. Greif, T. E. Riley, T. Hinderer, K. Hebeler, A. Schwenk, A. L. Watts, S. Nissanke, S. Guillot, J. M. Lattimer, and R. M. Ludlam. Constraining the Dense Matter Equation of State with Joint Analysis of NICER and LIGO/Virgo Measurements. *Astrophysical Journal Letters*, 893(1):L21, April 2020.

[32] G. Raaijmakers, T. E. Riley, and A. L. Watts. A pitfall of piecewise-polytropic equation of state inference. *Monthly Notices of the Royal Astronomical Society*, 478:2177–2192, 2018.

[33] G. Raaijmakers, T. E. Riley, A. L. Watts, S. K. Greif, S. M. Morsink, K. Hebeler, A. Schwenk, T. Hinderer, S. Nissanke, S. Guillot, Z. Arzoumanian, S. Bogdanov, D. Chakrabarty, K. C. Gendreau, W. C. G. Ho, J. M. Lattimer, R. M. Ludlam, and M. T. Wolff. A Nicer View of PSR J0030+0451: Implications for the Dense Matter Equation of State. *Astrophysical Journal Letters*, 887(1):L22, December 2019.

[34] P. S. Ray et al. STROBE-X: X-ray Timing and Spectroscopy on Dynamical Timescales from Microseconds to Years. *arXiv e-prints*, 2019.

[35] D. J. Reardon, G. Hobbs, W. Coles, Y. Levin, M. J. Keith, M. Bailes, N. D. R. Bhat, S. Burke-Spolaor, S. Dai, M. Kerr, P. D. Lasky, R. N. Manchester, S. Osłowski, V. Ravi, R. M. Shannon, W. van Straten, L. Toomey, J. Wang, L. Wen, X. P. You, and X. J. Zhu. Timing analysis for 20 millisecond pulsars in the Parkes Pulsar Timing Array. *Monthly Notices of the Royal Astronomical Society*, 455(2):1751–1769, January 2016.

[36] Ronald A. Remillard, Michael Loewenstein, James F. Steiner, Gregory Y. Prigozhin, Beverly LaMarr, Teruaki Enoto, Keith C. Gendreau, Zaven Arzoumanian, Craig Markwardt, Arkadip Basak, Abigail L. Stevens, Paul S. Ray, Diego Altamirano, and Douglas J. K. Buisson. An Empirical Background Model for the NICER X-Ray Timing Instrument. *Astronomical Journal*, 163(3):130, March 2022.

[37] T. E. Riley, G. Raaijmakers, and A. L. Watts. On parametrized cold dense matter equation-of-state inference. *Monthly Notices of the Royal Astronomical Society*, 478:1093–1131, 2018.

[38] T. E. Riley, A. L. Watts, S. Bogdanov, P. S. Ray, R. M. Ludlam, S. Guillot, Z. Arzoumanian, C. L. Baker, A. V. Bilous, D. Chakrabarty, K. C. Gendreau, A. K. Harding, W. C. G. Ho, J. M. Lattimer, S. M. Morsink, and T. E. Strohmayer. A NICER View of PSR J0030+0451: Millisecond Pulsar Parameter Estimation. *Astrophysical Journal Letters*, 887(1):L21, December 2019.

[39] Thomas E. Riley, Devarshi Choudhury, Tuomo Salmi, Serena Vinciguerra, Yves Kini, Bas Dorsman, Anna L. Watts, Daniela Huppenkothen, and Sebastien Guillot. X-PSI: A Python package for neutron star X-ray pulse simulation and inference. *The Journal of Open Source Software*, 8(82):4977, February 2023.

[40] Thomas E. Riley, Anna L. Watts, Paul S. Ray, Slavko Bogdanov, Sebastien Guillot, Sharon M. Morsink, Anna V. Bilous, Zaven Arzoumanian, Devarshi Choudhury, Julia S. Deneva, Keith C. Gendreau, Alice K. Harding, Wynn C. G. Ho, James M. Lattimer, Michael Loewenstein, Renee M. Ludlam, Craig B. Markwardt, Takashi Okajima, Chanda Prescod-Weinstein, Ronald A. Remillard, Michael T. Wolff, Emmanuel Fonseca, H. Thankful Cromartie, Matthew Kerr, Timothy T. Pennucci, Aditya Parthasarathy, Scott Ransom, Ingrid Stairs, Lucas Guillemot, and Ismael Cognard. A NICER View of the Massive Pulsar PSR J0740+6620 Informed by Radio

Timing and XMM-Newton Spectroscopy. *Astrophysical Journal Letters*, 918(2):L27, September 2021.

[41] T. Salmi, J. Nättilä, and J. Poutanen. Bayesian parameter constraints for neutron star masses and radii using X-ray timing observations of accretion-powered millisecond pulsars. *Astronomy & Astrophysics*, 618:A161, October 2018.

[42] Tuomo Salmi, Vladislav Loktev, Karri Korsman, Luca Baldini, Sergey S. Tsygankov, and Juri Poutanen. Neutron star parameter constraints for accretion-powered millisecond pulsars from the simulated IXPE data. *Astronomy & Astrophysics*, 646:A23, February 2021.

[43] Tuomo Salmi, Serena Vinciguerra, Devarshi Choudhury, Thomas E. Riley, Anna L. Watts, Ronald A. Remillard, Paul S. Ray, Slavko Bogdanov, Sebastien Guillot, Zaven Arzoumanian, Cecilia Chirenti, Alexander J. Dittmann, Keith C. Gendreau, Wynn C. G. Ho, M. Coleman Miller, Sharon M. Morsink, Zorawar Wadiasingh, and Michael T. Wolff. The Radius of PSR J0740+6620 from NICER with NICER Background Estimates. *Astrophysical Journal*, 941(2):150, December 2022.

[44] Tuomo Salmi, Serena Vinciguerra, Devarshi Choudhury, Anna L. Watts, Wynn C. G. Ho, Sebastien Guillot, Yves Kini, Bas Dorsman, Sharon M. Morsink, and Slavko Bogdanov. Atmospheric Effects on Neutron Star Parameter Constraints with NICER. *Astrophysical Journal in press*, August 2023.

[45] John Skilling. Nested Sampling. In Rainer Fischer, Roland Preuss, and Udo Von Toussaint, editors, *Bayesian Inference and Maximum Entropy Methods in Science and Engineering: 24th International Workshop on Bayesian Inference and Maximum Entropy Methods in Science and Engineering*, volume 735 of *American Institute of Physics Conference Series*, pages 395–405, November 2004.

[46] A. L. Stevens, J. D. Fiege, D. A. Leahy, and S. M. Morsink. Neutron Star Mass-Radius Constraints Using Evolutionary Optimization. *Astrophysical Journal*, 833(2):244, December 2016.

[47] János Takátsy, Péter Kovács, György Wolf, and Jürgen Schaffner-Bielich. What neutron stars tell about the hadron-quark phase transition: a Bayesian study. *arXiv e-prints*, page arXiv:2303.00013, February 2023.

[48] Serena Vinciguerra, Tuomo Salmi, Anna L. Watts, Devarshi Choudhury, Yves Kini, and Thomas E. Riley. X-PSI Parameter Recovery for Temperature Map Configurations Inspired by PSR J0030+0451. *arXiv e-prints*, page arXiv:2308.08409, August 2023.

[49] Serena Vinciguerra, Tuomo Salmi, L. Anna Watts, Devarshi Choudhury, E. Thomas Riley, Paul S. Ray, Slavko Bogdanov, Yves Kini, Sebastien Guillot, Deepto Chakrabarty, Wynn C.H. Ho, Daniela Huppenkothen, Sharon M. Morsink, Zorawar Wadiasingh, and Michael T. Wolff. An updated mass-radius analysis of the 2017-2018 NICER data set of PSR J0030+0451. *Astrophysical Journal submitted*, August 2023.

[50] W. D. Vousden, W. M. Farr, and I. Mandel. Dynamic temperature selection for parallel tempering in Markov chain Monte Carlo simulations. *Monthly Notices of the Royal Astronomical Society*, 455(2):1919–1937, January 2016.

[51] A. L. Watts. Thermonuclear Burst Oscillations. *Ann. Rev. Astron. Astrophys.*, 50:609–640, 2012.

[52] A. L. Watts, N. Andersson, D. Chakrabarty, M. Feroci, K. Hebeler, G. Israel, F. K. Lamb, M. C. Miller, S. Morsink, F. Özel, A. Patruno, J. Poutanen, D. Psaltis, A. Schwenk, A. W. Steiner, L. Stella, L. Tolos, and M. van der Klis. Colloquium: Measuring the neutron star equation of state using x-ray timing. *Reviews of Modern Physics*, 88(2):021001, April 2016.

[53] A. L. Watts, W. Yu, J. Poutanen, S. Zhang, et al. Dense matter with eXTP. *Science China Physics, Mechanics, and Astronomy*, 62:29503, 2019.

[54] Michael T. Wolff, Sebastien Guillot, Slavko Bogdanov, Paul S. Ray, Matthew Kerr, Zaven Arzoumanian, Keith C. Gendreau, M. Coleman Miller, Alexander J. Dittmann, Wynn C. G. Ho, Lucas Guillemot, Ismael Cognard, Gilles Theureau, and Kent S. Wood. NICER Detection of Thermal X-Ray Pulsations from the Massive Millisecond Pulsars PSR J0740+6620 and PSR J1614-2230. *Astrophysical Journal Letters*, 918(2):L26, September 2021.

CHAPTER 3

Multimessenger Constraints on the Neutron Star Equation of State Microphysics

Andrea Maselli, Francesco Pannarale

RECENT gravitational-wave observations by ground-based interferometers have proved to be effective in constraining the nuclear equation of state, exploring sources in a mass range complementary to the one sampled in the electromagnetic band. The loudness and number of events expected to be observed by current detectors at design sensitivity, as well as by next-generation facilities, pave the way for turning gravitational waves into a magnifying glass that connects the properties of macroscopic objects and the fundamental properties of nuclear interactions.

The scope of this chapter is to introduce the basic matter-dependent ingredients that characterise gravitational-wave signals emitted by neutron star coalescences. We discuss the impact of such contributions on the waveform models, and their relevance on constraining the equation of state of dense matter, based on current observations. We will focus specifically on a recently proposed pioneering approach to directly constrain the strength of three-body nuclear forces with gravitational waves [82, 108, 111].

3.1 INTRODUCTION

Observations of neutron stars (NSs) evolving in astrophysical environments with a large variety of dynamical setups have proved to be unique laboratories for fundamental physics ever since the discovery of the first radio-pulsar [58]. In particular, NSs are of primary interest for our understanding of the nuclear equation of state (EoS). Current models of dense matter feature large uncertainties due to the complexity in describing strong interactions above the saturation density, $\rho_0 \simeq 2.14\,\mathrm{g\,cm^{-3}}$, in regimes where constituents other than nucleons may appear, and that are ubiquitous within the inner core of typical NSs [76]. With laboratory constraints being limited by the density range that can

DOI: 10.1201/9781003306580-3

be achieved by Earth-based experiments [37, 77, 109, 120, 125], major contributions are expected from astrophysical observations [91]. In this regard, mass-radius measurements of spinning NSs in the electromagnetic (EM) band have been the predominant source of information so far, recently boosted by pulsar observations of the NICER satellite [17, 34, 39, 46, 56, 83, 84, 106, 107, 114].

The breakthrough discovery of GW170817 by LIGO and Virgo, which was followed by other observations of NS binary coalescences, established yet a new channel of investigation, immediately providing novel constraints on the EoS [2, 5, 6, 19]. The latter leaves an imprint on the emitted GW signals due to tidal interactions, which in the last phase of the coalescence are strong enough to affect the dynamical evolution of the source. Binary NS mergers are also expected to be rich multimessenger factories of photons through gamma-ray burst and kilonovae events. The observation of such EM counterparts associated with GW170817 has indeed driven a new quest for joint GW-EM constraints on nuclear matter, e.g., [16, 33, 38, 40, 44, 63, 64, 66, 67, 71, 80, 101, 102].

GW-based analyses have exploited a variety of approaches and methodologies which include constraints inferred from both microscopic and phenomenological representations of the EoS, aimed at studying the occurrence of new physical phenomena, such as phase transitions to strange matter and quark deconfinement, or at determining specific properties of dense matter, such as the behavior of the symmetry energy above ρ_0. While the constraining power of current interferometers is limited, GW observations and their synergy with EM probes have already been incredibly fruitful in bounding *global* features, such as the EoS stiffness, which translates into the stellar compactness.

Future detections by the LIGO-Virgo-KAGRA network , including LIGO-India [112], and more crucially by the 3G interferometers Einstein Telescope [59] and Cosmic Explorer [105] will dramatically enhance the discovery potential of binary NS mergers. Moreover, the large number of expected events will allow to stack thousands of observations leading to population-based constraints [24]. This richness will eventually turn GWs into a new probe for the EoS *microphysical* properties, and for the dynamics of strong interactions beyond the nuclear saturation density.

Hereafter, unless otherwise specified, we will use geometric units in which $G = c = 1$.

3.2 TIDAL DEFORMATIONS OF COMPACT OBJECTS

3.2.1 Newtonian tides

We generically refer to tidal effects as *finite size* effects, as they arise on extended bodies embedded within an external gravitational field . An intuitive picture of tidal deformations can be understood by considering a spherical object and a companion point-like body which acts as a source of the external field. The side of the object closer to (farther from) the source experiences a larger (smaller) gravitational attraction compared to that felt by the center of mass. As a result, the spherical object deforms from its equilibrium configuration stretching in the direction of the point-like source. Classically such effect can be described in terms of a tidal potential which drives the deformations. To determine its expression let's consider two particles, with masses $m_1 \gg m_2$, and let $X_{i=1,2,3}$ be the Cartesian coordinates of m_2 with respect to an inertial observer $\mathcal{O}_{m_1}$ fixed in m_1. Moreover, let $x_{i=1,2,3}$ be the Cartesian coordinates of the displacement vector that identifies the position of a third massive particle relative to m_2, such that with respect to $\mathcal{O}_{m_1}$ the position of this body is given by $r_i = X_i + x_i$. The Newtonian gravitational potential at $\vec{r}$ is then given by:

$$\Phi_{\rm N}(\vec{r}) = \Phi_{\rm N}(\vec{r}=\vec{X}) + \left.\frac{\partial \Phi_{\rm N}(r)}{\partial r_i}\right|_{\vec{r}=\vec{X}} x_i + \frac{1}{2}\left.\frac{\partial^2 \Phi_{\rm N}(r)}{\partial r_i \partial r_j}\right|_{\vec{r}=\vec{X}} x_i x_j + \mathcal{O}(x^3)\,, \tag{3.1}$$

where $\Phi_{\rm N}(\vec{r}) = -\frac{Gm_1}{r}$ is the Newtonian potential. With respect to $\mathcal{O}_{m_2}$, the third particle experiences an acceleration

$$\ddot{x}_k = \ddot{r}_k - \ddot{X}_k = -\frac{\partial \Phi_{\rm N}(\vec{r})}{\partial r_k} + \frac{\partial \Phi_{\rm N}(\vec{X})}{\partial X_k} = -C^{\rm N}_{ki} x_i \tag{3.2}$$

where we defined the Newtonian tidal tensor

$$C^{\rm N}_{kj} = \left.\frac{\partial^2 \Phi_{\rm N}(\vec{r})}{\partial r_k \partial r_j}\right|_{\vec{r}=\vec{X}}\,. \tag{3.3}$$

For equatorial orbits in the XY, plane, the latter takes the simple form

$$C^{\rm N}_{ij} = \frac{Gm_1}{r^3}\begin{pmatrix} 1-3\frac{X^2}{r^2} & -3\frac{XY}{r^2} & 0 \\ -3\frac{XY}{r^2} & 1-3\frac{Y^2}{r^2} & 0 \\ 0 & 0 & 1 \end{pmatrix}\,. \tag{3.4}$$

Using Eqs. (3.1)–(3.3), we can write the Newtonian tidal potential as

$$\Phi^{\rm T}_{\rm N}(\vec{r}) = \frac{1}{2} C^{\rm N}_{ij} x_i x_j\,. \tag{3.5}$$

While the classical theory of tides dates back to Newton, and in particular to Laplace's description in the 18th century, it was the seminal work of A. Love in 1911 that provided a fully rigorous treatment of tidal phenomena, developing a consistent approach to study the multipolar response of an object immersed in an external tidal field [79]. Such framework builds on the definition of Love numbers, which encode the deformability properties of the body. We shall now discuss their relativistic formulation.

3.2.2 Relativistic tides

In this section, we shall discuss the key ingredients needed for a relativistic description of tidal effects . We follow the formalism of [21, 36, 60, 118], which is valid for a static and spherically symmetric compact object embedded within an external stationary tidal field acting as a perturbation of the equilibrium configuration of the central body. While this is a typical scenario encountered during the inspiral phase of a binary coalescence, we do not expect tides developing during the evolution to be stationary. In this setup we can identify two timescales: (i) the time on which the tidal field varies, $T_{\rm tid}$, which is dictated by the orbital dynamics and (ii) the time on which the object rearranges its structure after being deformed by the tidal field, $T_{\rm def}$, related to the internal fluid dynamics. When the binary components are sufficiently far apart $T_{\rm tid} \gg T_{\rm def}$ and the tidal field can be considered effectively as stationary. Such *adiabatic* evolution breaks down as the system approaches the late stage of the merger, when dynamical tides become relevant [81, 115].

In this framework, we assume that the multipole moments tidally induced on the compact objects are linearly proportional to the external field. In General Relativity, this allows to define two sets of Love numbers, which are introduced as constants of proportionality between such moments and the tidal field [21, 36, 60]:

$$Q_L = \lambda_\ell G_L \quad , \quad S_L = \sigma_\ell H_L \qquad \text{for} \qquad \ell \geq 2\,, \tag{3.6}$$

where λ_ℓ and σ_ℓ are the even (electric) and odd (magnetic) tidal deformabilities of order ℓ, respectively. The corresponding mass Q_L and current S_L multipole moments are sourced by the electric G_L and magnetic H_L tidal moments.[1]

Electric and magnetic deformabilities are associated with tidal deformations of mass and current distributions. In the Newtonian limit, the even sector translates to the classical theory of Love numbers, such that

$$Q_L = \int_{\mathcal{V}} \rho(\vec{x}) x_{\langle L \rangle} d^3x \quad , \quad G_L = -[\partial_{\langle L \rangle}\phi(\vec{x})]_{\vec{x}=0}\,, \tag{3.7}$$

where $\rho(\vec{x})$ is the mass density of the compact object with volume $\mathcal{V}$, $\phi(\vec{x})$ is the external gravitational field, and angular brackets identify trace-free symmetrization over the indices L. The magnetic sector does not have any correspondence in Newtonian gravity, as current distributions do not gravitate and therefore they do not source any deformation.

The tidal deformabilities defined in Eqs. (3.6) can be recast in terms of the *dimensionless Love numbers*:

$$k_\ell^{\mathrm{E}} = \frac{(2\ell-1)!!}{2}\frac{\lambda_\ell}{R^{(2\ell+1)}} \quad , \quad k_\ell^{\mathrm{B}} = \frac{4(\ell+2)(2\ell-1)!!}{\ell-1}\frac{\sigma_\ell}{R^{(2\ell+1)}}\,, \tag{3.8}$$

where R is the radius of the compact object . The Love numbers, or equivalently the tidal deformabilities, carry information on the internal composition of the compact object. In the case of NSs, for a given EoS, they uniquely depend on the stellar compactness $\mathcal{C} = M/R$, where M is the NS gravitational mass. As we will discuss in the next section, λ_ℓ and σ_ℓ enter the gravitational signal as *finite size* corrections to point-particle waveforms, and can potentially be extracted by astrophysical observations of coalescing binaries. The $\ell = 2$ quadrupolar component provides the dominant contribution, with the electric component k_2^{E} yielding the leading order correction to the GW phase.

The multipole moments defined in Eq. (3.6) can be extracted from the asymptotic expansion at spatial infinity of the spacetime metric of a stationary object perturbed by an external tidal field. Two common choices used to define the multipole moments are given by the Geroch-Hansen [49, 50] and the Thorne [118] approaches, which have been shown to be fully equivalent [55]. Here we follow the Thorne prescription, such that in an Asymptotically Cartesian and Mass Centered frame (ACMC) the g_{tt} and g_{ti} metric components, from which mass and current multipole can be extracted, read:

$$g_{tt}^{\mathrm{ACMC}} = -1 + \frac{2M}{r} + \frac{3Q_{ij}n^i n^j}{r^3} + r^2 G_{ij}n^i n^j + \sum_{\ell>2} \ldots\,, \tag{3.9}$$

$$g_{ti}^{\mathrm{ACMC}} = -\frac{2\epsilon_{ijk}J^j n^k}{r^2} - \frac{4\epsilon_{ijk}S^{js}n_s n^k}{r^3} + r^3\frac{1}{3}\epsilon_{ijk}H^{js}n_s n^k + \sum_{\ell>2} \ldots\,, \tag{3.10}$$

[1] In our notation, capital Latin letters are used as a short-hand notation for indices $a_1, \ldots, a_\ell$.

where $r = \delta_{ij}x^i x^j$ is the radial coordinate, and $n^i = x^i/r$ is the unit radial vector, and the reader should refer to [118] for the definitions of the various terms. The mass M and the angular momentum J^i of the central object correspond to the mass monopole and the current dipole term of the expansion.[2] As discussed previously, we focus here on non-rotating objects, such that $J^i = 0$. Equations (3.9) and (3.10) show a clear separation between the decaying (r^{-3}) and growing (r^2) solution of the metric, which correspond to the multipole moments induced on the central object and to the tidal moments, respectively. Such distinction becomes ambiguous for more complex configurations, as in the case of spinning objects [53, 95].

3.2.3 Relativistic perturbations

In this section, we sketch the basic steps needed to numerically compute the Love numbers (or equivalently the tidal deformabilities) . For the sake of simplicity, we will focus on the leading (even parity) quadrupolar contribution λ_2, although calculations of other coefficients follow the same strategy.

We consider the linear perturbations of a static (non-rotating), spherically symmetric star, specified by the following line element

$$ds^2 = g^{(0)}_{\alpha\beta}dx^\alpha dx^\beta = -e^\nu dt^2 + e^f dr^2 + r^2(d\theta^2 + \sin^2\theta d\phi^2) \quad x^\mu = (t, r, \theta, \phi) \ , \tag{3.11}$$

where the metric functions ν and f depend on the radial coordinate only. The stellar interior is modelled in terms of a perfect fluid with a stress-energy tensor

$$T^{(0)}_{\mu\nu} = (\epsilon + p)u_\mu u_\nu + p g_{\mu\nu} \ , \tag{3.12}$$

where $u^\mu = (e^{-\nu/2}, 0, 0, 0)$, ϵ and p are the fluid 4-velocity, energy density and pressure, respectively. For a given barotropic EoS, $\epsilon = \epsilon(p)$, $g^{(0)}_{\mu\nu}$ are solutions of the Einstein field equations sourced by the matter components:

$$G^{(0)}_{\mu\nu}[g^{(0)}_{\alpha\beta}] = 8\pi T^{(0)}_{\mu\nu} \ . \tag{3.13}$$

The latter yield the Tolman–Oppenheimer–Volkoff equations which describe the relativistic stellar structure at equilibrium :

$$\frac{dm}{dr} = 4\pi\epsilon r^2 \quad , \quad \frac{dp}{dr} = -\frac{(\epsilon + p)(m + 4\pi r^3 p)}{r(r - 2m)} \quad , \quad \frac{d\nu}{dr} = -\frac{2}{\epsilon + p}\frac{dp}{dr} \ , \tag{3.14}$$

where $m(r) = r/2(1 - e^{-f})$ is the mass enclosed in a sphere of radius r [85].

We now assume that the star is embedded within an external tidal field, and introduce the metric linear perturbations $h_{\mu\nu}$, such that $|h_{\mu\nu}| \ll |g^{(0)}_{\mu\nu}|$. Similarly, we also introduce the stress-energy tensor linear fluctuations $|\delta T_{\mu\nu}| \ll |T^{(0)}_{\mu\nu}|$, which include the (Eulerian) perturbations of pressure, energy and fluid velocity:

$$\delta T_{\mu\nu} = (\delta\epsilon + \delta p)u_\mu u_\nu + (\epsilon + p)(\delta u_\mu u_\nu + \delta u_\nu u_\mu) + \delta p g^{(0)}_{\mu\nu} + p h_{\mu\nu} \ , \tag{3.15}$$

[2]The mass dipole vanishes as the coordinates are mass centered.

where $\delta u^k = d\xi^k/d\tau$ and the vector ξ^k represents the spatial displacement of the fluid element due to the perturbation, with τ proper time. With all terms in hand, we can solve the Einstein equations for the first-order perturbations:

$$G^{(1)}[h_{\mu\nu}] = 8\pi\delta T_{\mu\nu} \,. \tag{3.16}$$

We note that, given the assumption of stationary tidal field, the metric and the fluid perturbations are independent of time, and are functions of the (r, θ, ϕ) coordinates. In order to solve Eqs. (3.16), we exploit the symmetry properties of the background and expand the metric and fluid perturbations in tensor, vector and spherical harmonics [119]. In doing so, we adopt the so-called Regge-Wheeler gauge [104], which allows to set to zero four components in $h_{\mu\nu}$. For the actual computation of the quadrupolar Love number, the relevant[3] metric functions are given by

$$h_{\mu\nu} = \sum_{\ell=2}^{\infty}\sum_{m=-\ell}^{\ell} \begin{pmatrix} e^{\nu}H_{0,\ell m}(r) & H_{1,\ell m}(r) & 0 & 0 \\ H_{1,\ell m}(r) & e^{f}H_{2,\ell m}(r) & 0 & 0 \\ 0 & 0 & r^2 K_{\ell m}(r) & 0 \\ 0 & 0 & 0 & r^2\sin^2\theta K_{\ell m}(r) \end{pmatrix} Y_{\ell m}(\theta,\phi) \,, \tag{3.17}$$

where the $Y_{\ell m}(\theta, \phi)$ functions denotes the standard spherical harmonics, the pressure and energy density perturbations are expanded as

$$\delta\epsilon = \sum_{\ell=2}^{\infty}\sum_{m=-\ell}^{\ell} \delta\epsilon_{\ell m} Y_{\ell m}(\theta,\phi) \,, \quad \delta p = \sum_{\ell=2}^{\infty}\sum_{m=-\ell}^{\ell} \delta p_{\ell m} Y_{\ell m}(\theta,\phi) \,, \tag{3.18}$$

and finally the even-parity coefficients of the velocity components are given by

$$\delta u^r = \sum_{\ell=2}^{\infty}\sum_{m=-\ell}^{\ell} W_{\ell m}(r) Y_{\ell m}(\theta,\phi) \,, \tag{3.19}$$

$$\delta u^\theta = \frac{\delta u^\phi}{\sin^2\theta} = \sum_{\ell=2}^{\infty}\sum_{m=-\ell}^{\ell} V_{\ell m}(r)\partial_\theta Y_{\ell m}(\theta,\phi) \,. \tag{3.20}$$

Given the spherically symmetric background, the solution is independent of the azimuthal number m. Moreover, perturbations with different values of ℓ do not couple to each other and can be solved independently. For the sake of clarity, hereafter we will drop both multipolar indices.

Assuming a barotropic EoS fixes the energy and pressure perturbations,

$$\delta\epsilon(r) = \frac{d\epsilon(r)}{dp(r)}\delta p(r) = \frac{1}{c_s^2}\delta p(r) \,, \tag{3.21}$$

[3]In general, tensor and vector perturbations decompose in two families of modes, on the basis of their transformation properties under parity: *even* (or polar) and *odd* (axial) modes transform as $(-1)^\ell$ and $(-1)^{\ell+1}$, respectively. Given the spherical background, the two families decouple and can be treated separately. Only the polar sector is necessary in the case of the electric Love number.

where c_s is the of speed of sound. We can further reduce the number of independent variables through Einstein's equations since

$$\delta p(r) = \frac{\epsilon(r) + p(r)}{2} H_0(r) \quad , \quad H_2(r) = H_0(r) \, . \tag{3.22}$$

This procedure allows to find[4] a second-order ordinary differential equation for H_0 which can be integrated along with the Tolman–Oppenheimer–Volkoff equations (3.14),

$$\begin{aligned} H_0''(r) &+ \left[4\pi e^f \left(5\epsilon + 9p + \frac{\epsilon + p}{dp/d\epsilon}\right) - \nu'^2 - \frac{6e^f}{r^2}\right] H_0(r) \\ &+ \left[\frac{2}{r} + e^f \left(\frac{2m(r)}{r^2} + 4\pi r(p - \epsilon)\right)\right] H_0'(r) = 0 \, , \end{aligned} \tag{3.23}$$

where a prime denotes a derivative with respect to r.

To numerically compute the Love number, we first find a solution for $H_0(r)$ *inside* the star, with boundary condition at the center

$$H_0(r \to 0) = a_0 r^2 [1 + \mathcal{O}(r^2)] \, , \tag{3.24}$$

a_0 being an integration constant that can be chosen freely as it scales away from the final calculation of λ_2. We match the numerical value found at the stellar radius, $r = R$, where Eq. (3.23) has an analytical solution in terms of the associated Legendre polynomials

$$H_0(r \geq R) = c_{\mathcal{P}} \mathcal{P}_{22} \left(r/M - 1\right) + c_{\mathcal{Q}} \mathcal{Q}_{22} \left(r/M - 1\right) \, , \tag{3.25}$$

where $(c_{\mathcal{P}}, c_{\mathcal{Q}})$ are constants found through the match, and M is the mass of the star. Finally, we plug Eq. (3.25) into the metric, and expand the g_{tt} component at spatial infinity,

$$g_{tt}(r \to \infty) = -1 + \frac{2M}{r} + \left(\frac{8M^3}{5r^3} C_{\mathcal{Q},20} + \frac{3r^2}{M^2} C_{\mathcal{P},20}\right) Y_{20} + \sum_{\ell > 2} \cdots \, , \tag{3.26}$$

which can be matched with the ACMC expression (3.9):

$$g_{tt}^{\mathrm{ACMC}} = -1 + \frac{2M}{r} + \left(\frac{3}{r^3} Q_{20} + r^2 G_{20}\right) Y_{20} + \sum_{\ell > 2} \cdots \, . \tag{3.27}$$

In the last two expressions, we have decomposed the tidal and source multipole moments in trace-free tensors.

$$Q_L = \sum_{m=-\ell}^{\ell} Q_{\ell m} \mathcal{Y}_L^{\ell m} \quad , \quad G_L = \sum_{m=-\ell}^{\ell} G_{\ell m} \mathcal{Y}_L^{\ell m} \, , \tag{3.28}$$

with $Y_{\ell m} = \mathcal{Y}_L^{\ell m} n_L$ [119]. Moreover, as discussed above the solution is independent of m, which we have then fixed for sake of simplicity to $m = 0$.

[4] An algebraic equation for $K(r)$ can be found in terms of $H_0(r)$ and its first derivative, while $H_1(r)$ vanishes.

Comparing Eqs. (3.26) and (3.27), we identify the growing (tidal field) and decaying (quadrupole deformation) solutions, which can be related to the two integration constants

$$C_{\mathcal{Q},20} = \frac{15}{8}\frac{Q_{20}}{M^3} = \frac{15}{8}\lambda_2 \frac{G_{20}}{M^3} \quad , \quad C_{\mathcal{P},20} = \frac{1}{3}M^2 G_{20} \, , \tag{3.29}$$

where in the first expression of Eq. (3.29), we used the adiabatic approximation (3.6). We can finally express k_2^{E} in a closed form in terms of $H_0(R)$ and $H_0'(R)$

$$\begin{aligned} k_2^{\mathrm{E}} =& \frac{8\mathcal{C}^5}{5}(1-2\mathcal{C})^2 \left[2+2\mathcal{C}(y-1)-y\right] \Big\{ 2\mathcal{C}[6-3y+3\mathcal{C}(5y-8)] \\ &+ 4\mathcal{C}^3[13-11y+\mathcal{C}(3y-2)+2\mathcal{C}^2(1+y)] \\ &+ 3(1-2\mathcal{C})^2[2-y+2\mathcal{C}(y-1)]\log(1-2\mathcal{C}) \Big\}^{-1} \, , \end{aligned} \tag{3.30}$$

where $\mathcal{C} = M/R$ is the stellar compactness introduced in Sec. 3.2.2 and $y = RH_0'(R)/H(R)$.

In Fig. 3.1, we show mass-radius and mass-tidal deformability profiles for a variety of EoSs. We normalise the tidal deformability to the NS mass, $\Lambda = \lambda_2/M^5$, as it represents the relevant parameter entering the GW phase (see discussion in Sec. 3.3). For a given EoS, the Love number decreases as the compactness increase, and it vanishes in the Schwarzschild limit, that is, as $\mathcal{C} \to 1/2$.

3.3 TIDAL EFFECTS IN COMPACT BINARY MERGERS

Gravitational wave signals emitted by compact binary mergers are sensitive to matter effects induced by the NS internal structure, the imprint on the waveform phase[5] being encoded by the Love numbers (see [31, 54], for recent reviews) . Their contribution can be modelled through the post-Newtonian (pN) approach, used to describe the orbital dynamics and the GW emission during the inspiral stage of the coalescence [22]. In the pN framework, Einstein's field equations are systematically expanded in terms of a small parameter

$$\varepsilon \sim \frac{v}{c} \sim \frac{M}{d} \, , \tag{3.31}$$

where v and d are the characteristic velocity and scale of the system, with total mass M. This procedure leads to a semi-analytical expression for the waveform, in which both amplitude and phase are described by a Taylor series in ε, in which a factor of order $\mathcal{O}(\varepsilon^{2n}) = \mathcal{O}\left((v/c)^{2n}\right)$ counts as an n-pN order.

Tidal effects add linearly to the point-particle terms of the GW frequency phase , which depend on the masses and the spins of the binary components:

$$\varphi = \varphi_{\text{p-p}} + \varphi_{\text{tidal}} \, , \tag{3.32}$$

[5]Tidal interactions modify the amplitude of the signal as well, although the effect is smaller.

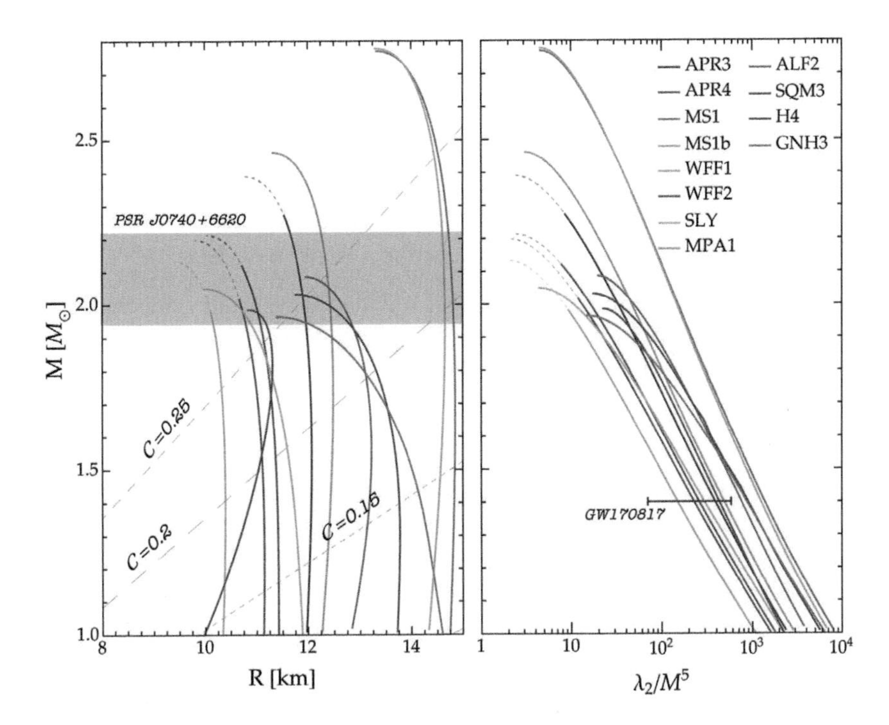

Figure 3.1 Mass-radius and mass-(dimensionless) tidal deformability profiles for a set of EoSs which include plain $npe\mu$ nuclear matter (APR3, APR4, MPA1, MS1, MS1b, SLY, WFF1, WFF2) [15, 41, 86, 87, 121], hyperons (GNH3, H4) [51, 68], and hybrid mixtures of nucleonic and quark matter (ALF2, SQM3) [75, 99] . The tidal deformability is normalised to the star mass as λ_2/M^5. The shaded region in the left panel brackets the most massive pulsar observed so far through EM observations with its measurement uncertainty [45]. The horizontal interval in the right panel identifies the dimensioneless tidal deformability for a $M = 1.4\, M_\odot$ NS, as inferred from GW170817 [4]. Dashed lines in the left panel correspond to stellar configurations with fixed compactness $\mathcal{C}$. Solid and dashed curves identify stars with the speed of sound at the center smaller and larger than the speed of light, respectively. Adapted from Ref. [92].

where, at its lowest order, φ_{tidal} provides an $\varepsilon^5 = v^5/c^5$ correction to $\varphi_{\text{p-p}}$. Matter effects enter[6] at their leading order through the effective tidal parameter $\tilde{\Lambda}$:

$$\tilde{\Lambda} = \frac{16}{13}\left[\left(\frac{13}{\eta_1} - 11\right)\eta_1^5\frac{\lambda_2^{(1)}}{m_1^5} + \left(\frac{13}{\eta_2} - 11\right)\eta_2^5\frac{\lambda_2^{(2)}}{m_2^5}\right], \tag{3.33}$$

[6]Along with tidal effects, the imprint of the stellar EoS is also encoded in the spin-induced quadrupole moments, which modify the GW phase at 2pN (ε^4) order. However, such terms are expected to be small and highly correlated with other particle contributions, making their detectability hard with current generation of detectors. Nonetheless expressing the spin-induced quadrupoles in terms of the tidal deformabilities through semi-analytic universal relations discovered in [123], can improve the measurements accuracy of the binary spins [124].

where $M = m_1 + m_2$, $\eta_{1,2} = m_{1,2}/M$ and $\lambda_2^{(1,2)}$ are the tidal deformabilities of the two compact objects in the binary. In the frequency domain, the GW waveform phase can be schematically written as

$$\varphi = \frac{3}{128\nu\varepsilon^5}\left[\underbrace{1 + \ldots + \mathcal{O}(\varepsilon^8)}_{\text{point-particle}} \underbrace{- \frac{39}{2}\tilde{\Lambda}\varepsilon^{10} + \ldots}_{\text{tidal}}\right], \tag{3.34}$$

where $\nu = m_1 m_2/M^2$ is the symmetric mass ratio. The point-particle contribution includes terms up to ε^8, namely 4-pN order [18, 23, 25, 98]. Higher order terms in φ_{tidal} contain different combinations of $\lambda_2^{(1,2)}$, as well as other families of tidal parameters, such as (i) magnetic tidal Love numbers [12, 20, 57, 72, 73], (ii) rotational tidal Love numbers [29, 30, 48, 65, 69, 74, 94] given by the coupling between the tidal field and the NS angular momentum, and (iii) time-dependent tidal Love numbers [61, 96, 97, 115, 116].

The series expansion (3.34) shows that $\tilde{\Lambda}$ leads the largest contribution to φ_{tidal}, and it is indeed the parameter most tightly constrained by GW observations. However, neglecting other tidal Love numbers may introduce a bias in the EoS reconstruction for very loud GW observations, as those expected from the next generation of ground-based interferometers, e.g., [42]. Finally, we note that for a typical NS $\tilde{\Lambda}$ can be as large as $\sim 10^3 - 10^4$ [cfr. Fig. (3.1)], amplifying the overall contribution of the tidal phase and making it comparable to the lower point-particle terms, unlike a simple power counting in terms of ε may suggest.

3.3.1 Gravitational-wave observations

In the context of scrutinizing the behaviour of matter in the interior of NSs, eight GW events are of potential interest.

- The sources of six GW events uncovered so far are *compatible* with the coalescence of an NS and a black hole, as the mass of the secondary compact object inferred from the GW data is *compatible* with that of a NS, i.e., its posterior distribution has support below $3\,M_\odot$, with a strong likelihood of being an NS as opposed to a black hole. These six events are GW190814 ($m_2 = 2.59^{+0.08}_{-0.09}\,M_\odot$, where we quote medians and 90% symmetric credible intervals) [8], GW190917_114630 ($m_2 = 2.1^{+1.1}_{-0.4}\,M_\odot$) [9], GW191219_163120 [10], GW200115_042309 ($m_2 = 1.44^{+0.85}_{-0.29}\,M_\odot$), GW200105_162426 ($m_2 = 1.91^{+0.33}_{-0.24}\,M_\odot$) [11], and GW200210_092254 ($m_2 = 2.83^{+0.47}_{-0.42}\,M_\odot$) [10]. However, the GW data do not allow to pinpoint the nature of the secondary compact object in the sources of these six GW events, as tidal deformation and tidal disruption effects are either absent or suppressed [8, 11] and hence too weak to be measurable. Further, no counterparts to these signals were observed.
- Two GW signals — GW170817 [2] and GW190425 [6] — are instead *consistent* with the coalescence of a binary neutron star system and will be the focus of this section.

The masses estimated for GW170817 [2] and GW190425 [6] source components are in the $1 - 2\,M_\odot$ range, and therefore consistent with galactic NS masses [117]. In the case of GW170817, the presence of at least one NS in the source is ensured by the EM counterpart to the GW signal [1, 3] (see the next section); nevertheless, one cannot rule out the possibility of it originating from a neutron star-black hole binary [32, 62]. GW190425, on the other hand, lacks any counterpart and the GW signal remains compatible also with emission from a low-mass binary black hole, as well as with a neutron star-black hole binary. The definitive proof of the nature of the constituents of the compact binaries that sparked GW170817 and GW190425 would require the distinct measurement of tidal effects from the two binary components, individually, which is not a trivial task for current GW detectors, under general circumstances [7, 42].

The most straightforward approach to analyse these signals makes no assumptions about the individual dimensionless tidal deformabilities when the data is compared to waveform models generated after sampling the source physical parameters. Specifically, the priors for the two dimensionless tidal deformabilities are independent of one another and of the mass of the compact objects — $\pi(m_1, m_2, \Lambda_1, \Lambda_2) = \pi(m_1, m_2)\pi(\Lambda_1)\pi(\Lambda_2)$ — while these physical quantities are mass-dependent and are determined by an underlying EoS. In the case of GW170817, these minimal assumptions result in a constraint for the mass-reweighted dimensionless deformability parameter $\tilde{\Lambda}$ which is found to be in the range $0 - 630$ when considering dimensionless spins up to 0.89 for the binary components; if, instead, the uniform prior on the dimensionless spins is restricted to $[0, 0.05]$, $\tilde{\Lambda} = 300^{+420}_{-230}$ (90% highest posterior density interval), a result that rules out several EoS models at the 90% credible level [5] . The corresponding constraints for GW190425 are $\tilde{\Lambda} \leq 1100$ and $\tilde{\Lambda} \leq 600$, respectively [6] (see also, e.g., [88, 89] for similar results). These looser constraints on the mass-reweighted tidal deformability indicate that this specific kind of analysis of GW190425 did not shed further insight on the EoS of NS matter: this is a consequence of the measurability of tidal effects being suppressed by the overall lower signal-to-noise ratio of GW190425, and by its higher mass[7] [see Eqs. (3.33) and (3.34)]. Indeed [6] shows that the results of GW190425 for $\tilde{\Lambda}$ are not competitive with those given by GW170817. Nevertheless, GW190425 demonstrated that the NS EoS must support masses heavier than those seen generally for NS in the EM spectrum: future GW observations may strengthen this constraint for the high-density regime of the NS EoS .

A more elaborate analysis may be carried out on the GW data by assuming that both bodies in the source of the signal are NSs and that they are governed by the same, unknown, EoS. This yields more stringent constraints as parameter space regions compatible with the data, but not with the NS coalescence hypothesis, are dropped. This approach is used to process GW170817 in [4], where a spectral parametrization of the EoS function $p(\rho)$ [78] is adopted to directly handle the unknown EoS (taken to be barotropic): the parameters of this representation are variables sampled and constrained within a Bayesian inference setup. Further, EoS-insensitive relations connecting macroscopic properties of NSs [123, 124] are then used to measure the radii of the NSs from the GW data. These are found to be

[7]The chirp mass $\mathcal{M} = (m_1 m_2)^{3/5}/(m_1 + m_2)^{1/5}$ measurements at the 90% credible level are $1.186^{+0.001}_{-0.001}\,M_\odot$ and $1.44^{+0.02}_{-0.02}\,M_\odot$ for GW170817 and GW190425, respectively. The inferred total mass depends instead on assumptions made about the component spins, as spin and mass ratio are correlated [35, 90]. A conservative prior with dimensionless spins up to 0.89 yields $2.77^{+0.22}_{-0.05}\,M_\odot$ and $3.4^{+0.3}_{-0.1}\,M_\odot$, respectively.

$10.8^{+2.0}_{-1.7}$ km and $10.7^{+2.1}_{-1.5}$ km for the more massive and the less massive NS, respectively, at the 90% credible level.

The scaling of the tidal deformability with the inverse of the fifth power of the NS mass can be used to map $\tilde{\Lambda}$ constraints onto constraints on the dimensionless tidal deformability of a fiducial $1.4\,M_\odot$ NS. The original analysis of GW170817 reports $\Lambda_{1.4} < 800$ [2], while the more complex investigation in [4] finds $\Lambda_{1.4} = 190^{+390}_{-120}$, at the 90% level. The latter measurement favours 'soft' EoSs, i.e., ones predicting more compressible NS matter and smaller values of the tidal deformability parameter, over 'stiff' ones.

3.4 RETRIEVING MULTIMESSENGER CONSTRAINTS

Gravitational-wave observations can be combined with data obtained with other messengers and/or from other experimental setups to produce tighter constraints on the EoS of NS matter. One can either focus on counterparts to the *same* GW event, or fold into the analysis additional data/information coming from EM observations of other NS sources, for example, and/or from nuclear physics experiments; naturally, doing so assumes that all the data considered pertains to the same class of sources, that is, NSs and ultimately the matter in their interior. This section will summarize a few key examples of strategies of this kind.

The discovery of GW170817 represented the first GW detection of a binary neutron star system and also marked the beginning of the multimessenger era. The event was accompanied by the observation of two different EM counterparts: the prompt gamma-ray emission GRB 170817A [1] and the EM transient kilonova/macronova AT2017gfo [3]. GRB 170817A was detected independently of the GW170817 observation by the Fermi Gamma-ray Burst Monitor [52] and by the Anti-Coincidence Shield for the Spectrometer for the International Gamma-Ray Astrophysics Laboratory (INTEGRAL) [113] (1.74 ± 0.05) s after the merger epoch . AT2017gfo was instead discovered ~ 11 h after the merger as a result of the extensive follow-up observing campaign sparked by the initial LIGO-Virgo sky-localization of GW170817 which spanned an area of about $30\,\mathrm{deg}^2$. These distinct observations of the same event can be used together to address various open questions in physics; here we will concentrate solely on the NS EoS. We note that multiple multimessenger events could also be combined, but to date GW170817 and its counterparts constitute the only GW multimessenger event.

One of the earliest multimessenger constraints on the NS EoS from GW170817 was provided in [102], where the UV/optical/infrared counterpart was compared to kilonova models and numerical-relativity results consistent with GW170817 to place a *lower bound* on $\tilde{\Lambda}$ complementary to the upper bound reported in the first analysis of GW170817 [2]. The simulations were performed with four mean-field theory EoS, and the requirement set to power the kilonova is that the system that ejects $0.05\,M_\odot$. The authors were able to place the lower bound on $\tilde{\Lambda}$ at 400 so that the combination of this result and the one from [2] (tentatively) excluded extremely stiff, extremely soft, and mildly soft NS EoS models. This lower bound was loosened in several follow-up studies as new numerical-relativity simulations were performed and other EoS models were adopted, e.g. [33, 101]. Notably in [66] it was driven down to 242.

A different approach to establish multimessenger constraints with GW170817 was reported in [38], where Bayesian inference is applied to GW170817 data with the source location and distance informed by EM observations, and assuming that the two NSs share the same equation of state. This is achieved by setting up a uniform prior for $\tilde{\Lambda}$ and imposing $\Lambda_1/\Lambda_2 = (m_2/m_1)^6$, so that the individual tidal deformabilities are correlated. The 90% credible intervals obtained for $\tilde{\Lambda}$ are 222^{+420}_{-138}, 245^{453}_{-151} and 233^{+448}_{-144} for a uniform component mass prior, a component mass prior informed by radio observations of Galactic double NSs, and for a component mass prior informed by radio pulsars, respectively.

Shifting to approaches that are not based on multiple observation of a unique event, indeed [4] carries out a second analysis on GW170817 that factors in the additional constraint that the NS EoS[8] must support masses larger than $1.97\,M_\odot$, as prescribed by the observation of PSR J0348 + 0432 with mass $(2.01 \pm 0.04)\,M_\odot$ [17]. This method tightens the NS radius measurements quoted in the previous section to $11.9^{+1.4}_{-1.4}$ km and $11.9^{+1.4}_{-1.4}$ km, again at the 90% credible level. The method also leads to a pressure at twice ρ_0 of $3.5^{+2.7}_{-1.4} \times 10^{34}$ dyn cm^{-2}, at the 90% credible level.

Along the lines of combining constraints from diverse sources and messengers, further improvements in terms of posterior widths were obtained notably in [44, 67], albeit with an exposure to potential additional systematic errors. [67] used EoS-insensitive relations among macroscopic NS observables [123, 124] to combine radii measurements for a set of sources targeted by radio or X-ray observatories and the tidal deformability measurements from GW170817 reported in [4]. This yielded the 90% credible level results $\Lambda_{1.4} = 196^{+92}_{-63}$ and $R_{1.4} = 10.9^{+1.9}_{-1.5}$ km for the tidal deformability and radius of a fiducial $1.4\,M_\odot$ NS, respectively. Similarly, [44] used a spectral EoS parametrization [78] and processed with a Bayesian approach the masses and tidal deformabilities of GW170817, and the masses and stellar radii measured for nuclear bursts in accreting low-mass X-ray binaries. In turn, the posteriors on the parameters of the EoS spectral representation were used to determine $R = 12.4^{+0.5}_{-0.4}$ km for the radius of the more massive component of the GW170817 source.

Mass-radius constraints set by NICER on isolated pulsars over the last few years (see the contribution to this book by A.Watts) have also led to studies aimed at establishing joint constraints with GW events, observations of X-ray binaries, nuclear data, and also the GW170817 counterpart. GW170817 data and the data of 6 low mass X-ray binaries with thermonuclear burst are used in [63], alongside a parametrized EoS description; the same reference also considers GW170817 data and data for the symmetry energy of the nuclear interaction. The posteriors on the EoS parameters and EoS-independent relations on NS macroscopic quantities are then utilized to determine the results $\Lambda_{1.4} = 220^{+90}_{-90}$ and $R_{1.4} = 11.1^{+0.7}_{-0.6}$ km, and $\Lambda_{1.4} = 390^{+280}_{-210}$ and $R_{1.4} = 11.8^{+1.2}_{-0.7}$ km, for the two sets of combined data, respectively (90% highest posterior density intervals). This study was then revisited in [64], which combined data from nuclear experiments, from the LIGO-Virgo Collaboration for GW170817, and from NICER for PSR J0030+0451. The authors adopted two distinct generic representations of the EoS [78, 103] which gave consistent results, namely: $p(2\rho_0) = 3.38^{+2.43}_{-1.50} \times 10^{34}$ dyn cm^{-2}, $\Lambda_{1.4} = 390^{+320}_{-140}$ and $R_{1.4} = 12.2^{+1.0}_{-0.9}$ km, at the 90% credible level, which are tighter than those reported in [4], as they factor in more, heterogenous information. In summary, GW170817 and NICER results, while being mutually consistent, tend to favour slightly softer and stiffer EoS models, respectively.

---

[8] Recall that the EoS function $p(\rho)$ is parametrized in [4].

Finally, [40] and [71] are examples of multimessenger analysis applied to more than one GW signal. [40] looked jointly at GW170817, its EM counterparts GRB 170817A and AT2017gfo, and at the GW190425 signal, under the hypothesis that both GW signals originated from the merger of two NSs. Results from this multimessenger, multi-event data set were combined with existing measurements of pulsars using X-ray and radio observations, and nuclear-theory computations adopting chiral effective field theory. The resulting reference NS radius was found to be $R_{1.4} = 11.75^{+0.86}_{-0.81}$ km at 90% confidence. This framework was later generalized in [93] to simultaneously analyse for the first time GW170817, GRB 170817A, and AT2017gfo data while incorporating nuclear-physics constraints at low densities and constraints from X-ray and radio observations of isolated neutron stars. This resulted in the estimate $R_{1.4} = 11.98^{+0.35}_{-0.40}$ km at 90% confidence. [71] used a non-parametric EoS representation based on a Gaussian process conditioned on a set of nuclear models [43, 70], with data from GW170817, GW190425, heavy pulsars and J0030+0451, finding $R_{1.4} = 12.32^{+1.09}_{-1.47}$ km and $p(2\rho_0) = 3.8^{+2.9}_{-2.7} \times 10^{34}$ dyn cm^{-2} at the 90% credible level, with lower and upper driven by EM and GW observations, respectively.

3.5 THREE-BODY NUCLEON FORCES

In this section, we will focus on a recent effort aimed at exploiting binary NS mergers to directly constrain three-body nucleon forces beyond ρ_0. We focus on the investigation carried out in Refs. [82, 111], which worked within a Nuclear Many-Body Theory approach, promoting the strength of (repulsive) three-body forces as a free parameter to be constrained by observations. A different analysis was performed by Rose and collaborators in [108] (for further details, we refer the reader to the original manuscript, as well as to Chapter 6, and in particular to Section 6.4 of the present book). These two approaches build around different methodologies and EoS description, yet showing complementarity as well as demonstrating the potential of GW observations to infer the properties of nucleon dynamics.

3.5.1 A Nuclear Many-Body Theory approach

The authors of [82] model the EoS within a Nuclear Many-Body Theory approach, with nucleons dynamics described by the non-relativistic Hamiltonian

$$\mathcal{H} = \sum_i \frac{\mathbf{p}_i^2}{2m} + \sum_{i<j} v_{ij} + \sum_{i<j<k} V_{ijk} \,, \tag{3.35}$$

where $m \sim 939$ MeV is the mass of the i-th nucleon with momentum $\mathbf{p}_i$. In this framework, two-body (NN) interactions are encoded within v_{ij} , while three-body (NNN) forces are determined by the potential V_{ijk}. The inclusion of NNN interactions within the Hamiltonian is necessary to reproduce certain properties of 3-nucleon systems, such as ^{3}H and ^{4}He, which are not captured by the NN term alone. Three-body forces play the physical analogue of tidal interactions in gravity, and can be understood with the need to overcome the point-particle description of nucleons, which have rather an internal structure .

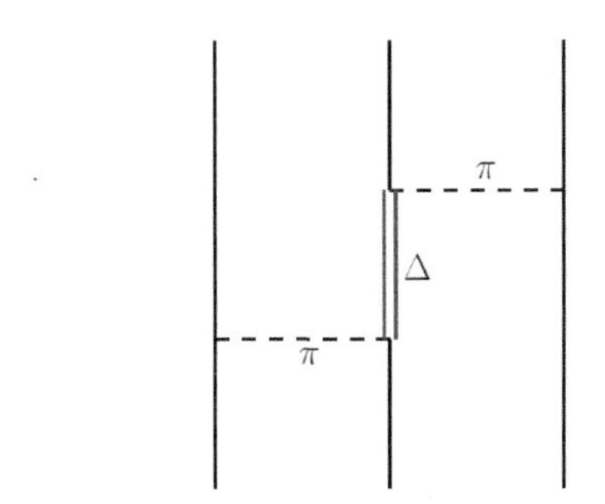

Figure 3.2 Diagrammatic representation of the three-body interaction described by the potential $V_{ijk}^{2\pi}$. Dashed and solid lines represent the π-exchange and the Δ resonance, respectively.

Commonly used phenomenological potentials can be split as sum of two contributions [100]:

$$V_{ijk} = V_{ijk}^{2\pi} + V_{ijk}^{\mathrm{R}} \,. \tag{3.36}$$

The attractive potential $V_{ijk}^{2\pi}$, introduced by the pioneering work of Fujita-Miyiazawa [47], describes the dominant NNN contribution in terms of a two-pion exchange process with the excitation of a Δ resonance (schematically drawn in Fig. 3.2). The second term, V_{ijk}^{R}, identifies a purely phenomenological repulsive potential.

The expectation value of the ground state ψ_0 of V_{ijk} is in general much smaller than the NN contribution, with $\langle\psi_0|V_{ijk}|\psi_0\rangle/\langle\psi_0|v_{ij}|\psi_0\rangle \sim 3\%$. Nonetheless, the overall contribution of three-body interactions can be larger, due to cancellations between the kinetic and the NN term, leading to a correction of $\sim 15\%$ for the ground energy state of ^{3}H and ^{4}He [26, 27]. The Hamiltonian formulation which includes the three-body potential is able to accurately reproduce the properties of atomic nuclei with more than three nucleons, as shown by calculations of the ground and of the low-lying excited states for atomic nuclei with $A \leq 12$ [26]. These results support the idea that while NNN are necessary to match experimental data, higher order corrections can be safely neglected.

The amplitude of the phenomenological potential V_{ijk}^{R} affects the stiffness of the EoS, and it is poorly constrained above the nuclear saturation density, where it is also expected to provide a major contribution. This gap opens the possibility to infer such amplitude from astrophysical data, as demonstrated by the recent investigations carried out in [82, 111]. In such studies the authors use the Hamiltonian formalism in Eq. (3.35), exploiting the Argonne $v18$ NN potential (AV18) [122] and a modified Urbana IX NNN potential (UIX) [28, 100], which suitably takes into account relativistic boost corrections to the AV18 contribution (see also discussion in [110]). This *baseline* model is used to construct the APR EoS [13, 14], which provides one of the most reliable description of NS dense matter, still compatible with observations.

To investigate the detectability of the three-body forces, [82, 110] introduce a new parameter, the three-body *strength*, such that

$$\langle V_{ijk}^{\mathrm{R}}\rangle \to \alpha \langle V_{ijk}^{\mathrm{R}}\rangle \,, \tag{3.37}$$

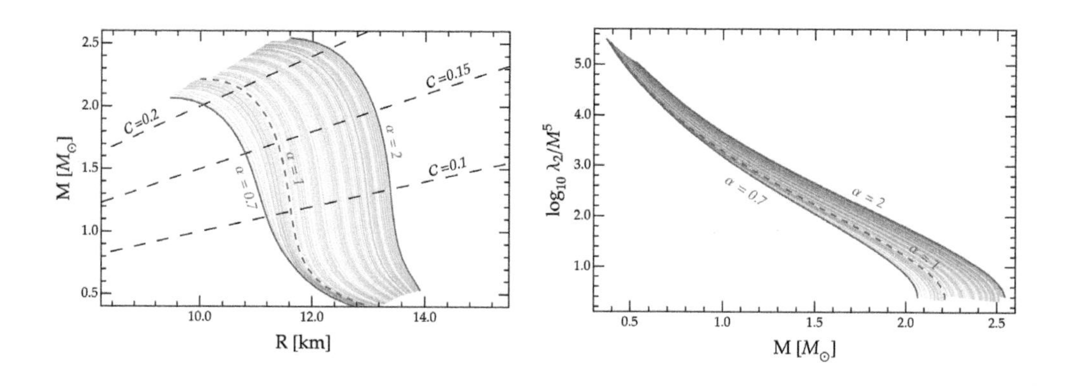

Figure 3.3 Mass-radius (left panel) and mass-tidal deformability (right panel) profiles for the EoS described in Sec. 3.5.1, as a function of the three-body amplitude α. Dashed lines in the left plot identify configurations with fixed compactness. Adapted from Ref. [111].

where $\alpha = 1$ corresponds to the baseline model discussed above. Larger values of α lead to stronger repulsive three-nucleon forces, namely, to stiffer EoSs, which in turn affect the NS structure. Figure 3.3 shows indeed that macroscopic observables are sensitive to the value of α, leading to large variations in the stellar profiles sampled by both GW and EM probes. The three-body strength can hence be promoted to a free parameter to be constrained by data, for a given set of EoS-dependent astrophysical observables, as masses, radii and tidal deformabilities.

The authors of [82, 110] followed this approach, building a Bayesian pipeline based on Monte Carlo Markov Chains simulations to infer the posterior probability distribution of α, exploiting a multimessenger set that includes: (i) measurements of masses and tidal deformabilities from GW170817 [2, 5], (ii) mass-radius constraints obtained by the NICER satellite for the millisecond pulsars PSR J0030+0451 [83, 106] and PSR J0740+6620 [45, 84, 107] . Note that varying the amplitude of $V^{\rm R}_{ijk}$ also affects the value of the nuclear saturation density. For this reason α was allowed to vary within $[0.7, 2.0]$, in a regime where ρ_0 changes at most by $\sim 15\%$, and the energy per particle would never exceed 3%. The left column of Fig. 3.4 shows the posterior densities of α obtained by this study for different observations. The distribution inferred from GW170817 alone has the smallest constraining power: the posterior rallies against the lower prior[9]. Conversely, posteriors obtained from EM pulsar data are informative and show a remarkable agreement, with a median $\alpha \sim 1.4$. For J0740+6620, which is the most massive NS analysed, the inference suggests a higher value of the strength, since a large α tends to support more massive configurations. Given these independent results, the multimessenger analysis is dominated by the pulsar measurements, leading to values peaking around $\alpha \sim 1.5$. Interestingly, while the constraints are still broad and support the baseline model $\alpha = 1$ within the 90% of the distributions, EM data seem to consistently favour larger values of the 3-body strength, i.e. stronger repulsive NNN interactions.

[9]In [111], the parameter α is sampled from a uniform distribution within $[0.7, 2]$.

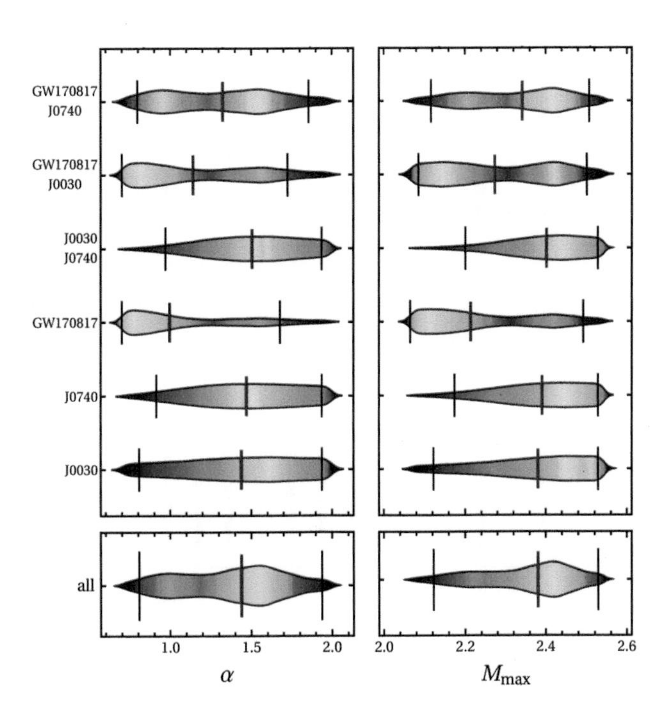

Figure 3.4 (Left) Posterior density distributions of the three-body strength α obtained from observations in the GW and in the EM band. (Right) Posterior densities for the maximum mass of the EoS corresponding to the inferred values of α. The bottom panels in each column show results obtained by combining all datasets. Red and black lines identify the median and the 90% credible intervals of the distribution, respectively. Reprinted from Ref. [111] with permission. © APS (2022).

Bounds obtained on α using the microscopic Hamiltonian in Eq. (3.35) can be used to forecast macroscopic observable properties, such as the maximum mass allowed for a stable stellar configuration. The right column of Fig. 3.4 shows the density distribution of $M_{\max}$ derived from the corresponding values of α drawn in the left panels. All observations yield $M_{\max} \sim 2.2 M_{\odot}$. The multimessenger inference in particular provides a large support for $M_{\max} \sim 2.5 M_{\odot}$.

With current GW data being almost uninformative on the strength of three-body forces, Ref. [111] simulated a campaign of GW observations for current (2G) detectors at design sensitivity, and for next generation (3G) ground-based interferometers, to assess their ability to infer the properties of nucleon dynamics. Figure 3.5 shows the results of this analysis for two sets of 30 detections made by (i) a newtwork (HLV) of 2G detectors with two LIGO and one Virgo sites and by (ii) the Einstein Telescope (ET) . The two sets of 30 GW events were built assuming two values for the strength, $\alpha = (1, 1.3)$, marked by the horizontal dashed lines within the panels. The top and bottom horizontal axes provide the signal-to-noise ratio and the chirp mass[10] of each binary.

[10]The chirp mass $\mathcal{M} = (m_1 m_2)^{3/5}/(m_1 + m_2)^{1/5}$ controls the leading order GW emission of a binary systems.

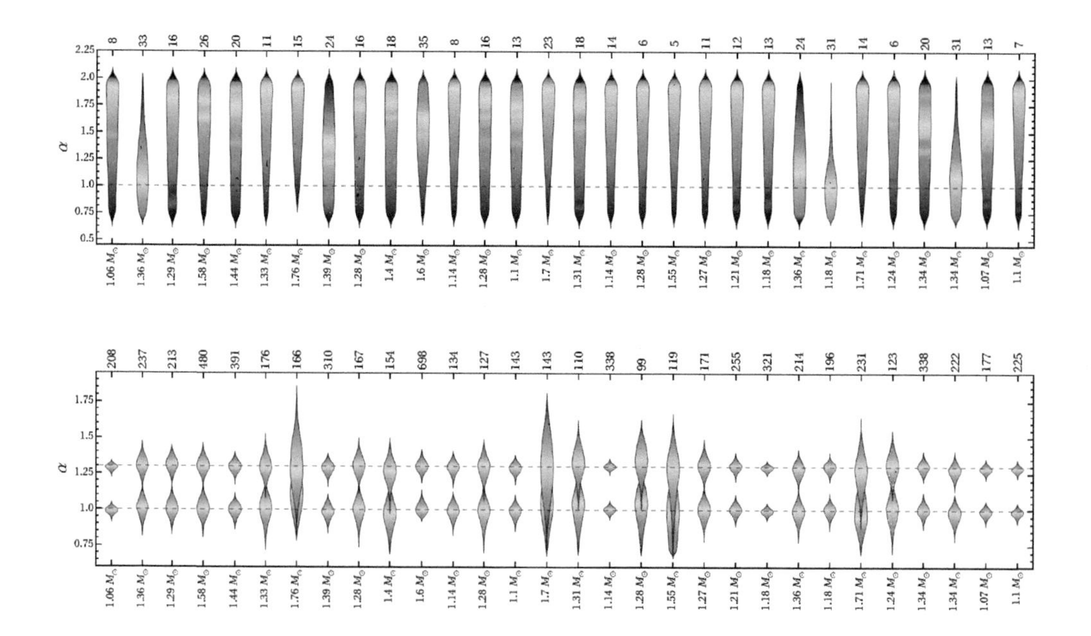

Figure 3.5 (Top panel) Posterior density distributions for the strength of three-body forces obtained from a set of 30 simulated binary NS mergers observed by a network of advanced GW detectors composed of two LIGO and one Virgo interferometers. (Bottom panel) Same as top figure, but assuming that the NS coalescences are observed by the Einstein Telescope. GW signals are simulated assuming two values of α, identified by the horizontal dashed lines. Yellow and green regions correspond to high and low probability values of the distributions. Finally, top and bottom horizontal axes provide the signal-to-noise ratio and the chirp mass of each coalescence. Reprinted from Ref. [111] with permission. © APS (2022).

The top panel of Fig. 3.5 shows the results for binary NSs with $\alpha = 1$, observed by HLV. In this setup the accuracy of 2G measurements strongly depends on *both* the signal-to-noise ratio and the NS masses. For signal-to-noise ratios $\lesssim 25$ or $\mathcal{M} \gtrsim 1.4 M_\odot$, α is unconstrained, with the posterior distribution leaving the injected (true) value outside the 90% confidence interval. Interestingly, accurate constraints on α require both loud signals, with signal-to-noise ratio larger than 30 and low-mass systems. All other configurations feature a strong bias in the reconstruction of α, which drives the posteriors to large values of the strength, close to the prior boundary. The authors of [111] also investigated how 2G constraints could change by stacking multiple events, finding a mild improvement in the resulting final posterior for α.

A significant improvement with respect to the 2G scenario can be achieved with ET observations, as the increase in detector sensitivity translates into a dramatic increase in the signal-to-noise ratio for a given physical source. The bottom panel of Fig. 3.5 shows the key output of this quality enhancement: (i) the accuracy of the Bayesian inference improves significantly, with posterior densities of all events peaking at the injected value of α, and constraints in the best (worst) case scenario of $\sim 2\%$ $(\sim 30\%)$ at 68% credible level, showing no support on the prior boundaries; (ii) posteriors for $\alpha = 1$ and $\alpha = 1.3$ are well separated for the majority of events, showing that ET could distinguish the two values with a single observation; (iii) when posteriors overlap, stacking of $2 - 3$ events would allow to distinguish between the two injected values of α. While low-mass systems still provide the best constraints, results for ET show a milder dependence on the NS masses, compared to HLV results.

3.6 CONCLUSIONS

The long quest to determine the laws of physics that regulate the interior of neutron stars, i.e. the nuclear matter equation of state in the many-body regime, has recently reached an exciting turning point. Over the last few years, this field has received outstanding observational results from the traditional electromagnetic band, from the newly born gravitational-wave band, and from the combination of the two. The impact of multimessenger astronomy on fundamental physics is expected to grow, as more data and more events are accumulated, and as new observing facilities are designed and constructed.

GW170817 has allowed us to scrutinize the tidal properties of neutron stars, while NICER is providing more stringent constraints on their radii, and the bar for their maximum mass is constantly monitored and updated via radio-observations of pulsars. These macroscopic measurables all depend upon the equation of state that governs neutron star interiors, so that neutron star observations are collectively advancing our understanding of the fundamental properties of nuclear interactions.

This chapter covered the basis of the formalism required to include matter-dependent effects in the modelling of gravitational-wave signals emitted by neutron star binaries during the last stages of their evolution. The dominant quantity driving this dependence is the (mass quadrupolar) tidal deformability, which was measured for the first time with GW170817. We summarized some of the measurements and constraints that were posed on the neutron star equation of state with GW170817 alone or within a multimessenger approach, where we distinguished between approaches that target the same event viewed in distinct bands and ones that target the collective population of sources containing neutron stars that is common to the gravitational-wave and the electromagnetic band.

Finally, we focus in greater detail on a recent approach that aims at directly constraining the strength of three-body nuclear forces with gravitational-wave observations. The simulations performed within this approach provide evidence that future observations from the upcoming third generation of gravitational-wave detectors, such as the Einstein Telescope and Cosmic Explorer, alongside information from other messengers will allow us to dive into probing fundamental physics at the Fermi scale. To date, this is the clearest avenue towards pinpointing the equation of state of neutron stars.

ACKNOWLEDGEMENTS

We would like to dedicate this contribution to our friend and colleague Stefania Marassi, an invaluable inspiration and a reference for us both when we first approached research and neutron stars as students.

Bibliography

[1] B. P. Abbott et al. Gravitational Waves and Gamma-rays from a Binary Neutron Star Merger: GW170817 and GRB 170817A. *Astrophys. J. Lett.*, 848(2):L13, 2017.

[2] B. P. Abbott et al. GW170817: Observation of Gravitational Waves from a Binary Neutron Star Inspiral. *Phys. Rev. Lett.*, 119(16):161101, 2017.

[3] B. P. Abbott et al. Multi-messenger Observations of a Binary Neutron Star Merger. *Astrophys. J. Lett.*, 848(2):L12, 2017.

[4] B. P. Abbott et al. GW170817: Measurements of neutron star radii and equation of state. *Phys. Rev. Lett.*, 121(16):161101, 2018.

[5] B. P. Abbott et al. Properties of the binary neutron star merger GW170817. *Phys. Rev. X*, 9(1):011001, 2019.

[6] B. P. Abbott et al. GW190425: Observation of a Compact Binary Coalescence with Total Mass $\sim 3.4 M_{\odot}$. *Astrophys. J. Lett.*, 892(1):L3, 2020.

[7] B. P. Abbott et al. Model comparison from LIGO–Virgo data on GW170817's binary components and consequences for the merger remnant. *Class. Quant. Grav.*, 37(4):045006, 2020.

[8] R. Abbott et al. GW190814: Gravitational Waves from the Coalescence of a 23 Solar Mass Black Hole with a 2.6 Solar Mass Compact Object. *Astrophys. J. Lett.*, 896(2):L44, 2020.

[9] R. Abbott et al. GWTC-2.1: Deep Extended Catalog of Compact Binary Coalescences Observed by LIGO and Virgo During the First Half of the Third Observing Run. 8 2021. *Phys. Rev. D*, 109(2):022001, 2024.

[10] R. Abbott et al. GWTC-3: Compact Binary Coalescences Observed by LIGO and Virgo During the Second Part of the Third Observing Run. *Phys. Rev. X*, 13(4):041039, 2023.

[11] R. Abbott et al. Observation of Gravitational Waves from Two Neutron Star–Black Hole Coalescences. *Astrophys. J. Lett.*, 915(1):L5, 2021.

[12] T. Abdelsalhin, Leonardo Gualtieri, and Paolo Pani. Post-Newtonian spin-tidal couplings for compact binaries. *Phys. Rev. D*, 98(10):104046, 2018.

[13] A. Akmal and V. R. Pandharipande. Spin-isospin structure and pion condensation in nucleon matter. *Phys. Rev. C*, 56:2261–2279, Oct 1997.

[14] A. Akmal, V. R. Pandharipande, and D. G. Ravenhall. Equation of state of nucleon matter and neutron star structure. *Phys. Rev. C*, 58:1804–1828, Sep 1998.

[15] A. Akmal, V. R. Pandharipande, and D. G. Ravenhall. The Equation of state of nucleon matter and neutron star structure. *Phys. Rev. C*, 58:1804–1828, 1998.

[16] Eemeli Annala, Tyler Gorda, Aleksi Kurkela, and Aleksi Vuorinen. Gravitational-wave constraints on the neutron-star-matter Equation of State. *Phys. Rev. Lett.*, 120(17):172703, 2018.

[17] John Antoniadis et al. A Massive Pulsar in a Compact Relativistic Binary. *Science*, 340:6131, 2013.

[18] K. G. Arun, Alessandra Buonanno, Guillaume Faye, and Evan Ochsner. Higher-order spin effects in the amplitude and phase of gravitational waveforms emitted by inspiraling compact binaries: Ready-to-use gravitational waveforms. *Phys. Rev. D*, 79:104023, 2009. [Erratum: Phys.Rev.D 84, 049901 (2011)].

[19] Luca Baiotti and Luciano Rezzolla. Binary neutron star mergers: a review of Einstein's richest laboratory. *Rept. Prog. Phys.*, 80(9):096901, 2017.

[20] Batoul Banihashemi and Justin Vines. Gravitomagnetic tidal effects in gravitational waves from neutron star binaries. *Phys. Rev. D*, 101(6):064003, 2020.

[21] Taylor Binnington and Eric Poisson. Relativistic theory of tidal Love numbers. *Phys. Rev. D*, 80:084018, 2009.

[22] Luc Blanchet. Gravitational Radiation from Post-Newtonian Sources and Inspiralling Compact Binaries. *Living Rev. Rel.*, 17:2, 2014.

[23] Luc Blanchet, Guillaume Faye, Quentin Henry, François Larrouturou, and David Trestini. Gravitational Wave Flux and Quadrupole Modes from Quasi-Circular Non-Spinning Compact Binaries to the Fourth Post-Newtonian Order. *Phys. Rev. D*, 108:064041, 2023.

[24] Marica Branchesi et al. Science with the Einstein Telescope: a comparison of different designs. *JCAP*, 07:068, 2023.

[25] Alessandra Buonanno, Bala Iyer, Evan Ochsner, Yi Pan, and B. S. Sathyaprakash. Comparison of post-Newtonian templates for compact binary inspiral signals in gravitational-wave detectors. *Phys. Rev. D*, 80:084043, 2009.

[26] J. Carlson, S. Gandolfi, F. Pederiva, Steven C. Pieper, R. Schiavilla, K. E. Schmidt, and R. B. Wiringa. Quantum Monte Carlo methods for nuclear physics. *Rev. Mod. Phys.*, 87:1067, 2015.

[27] J. Carlson, V. R. Pandharipande, and R. Schiavilla. Variational monte carlo calculations of ^{3}H and ^{4}He with a relativistic hamiltonian. *Phys. Rev. C*, 47:484–497, Feb 1993.

[28] J. Carlson, V. R. Pandharipande, and Robert B. Wiringa. Three-nucleon interaction in 3-body, 4-body, and infinite-body systems. *Nucl. Phys. A*, 401:59–85, 1983.

[29] Gonçalo Castro, Leonardo Gualtieri, Andrea Maselli, and Paolo Pani. Impact and detectability of spin-tidal couplings in neutron star inspirals. *Phys. Rev. D*, 106(2):024011, 2022.

[30] Gonçalo Castro, Leonardo Gualtieri, and Paolo Pani. Hidden symmetry between rotational tidal Love numbers of spinning neutron stars. *Phys. Rev. D*, 104(4):044052, 2021.

[31] Katerina Chatziioannou. Neutron star tidal deformability and equation of state constraints. *Gen. Rel. Grav.*, 52(11):109, 2020.

[32] Michael W. Coughlin and Tim Dietrich. Can a black hole–neutron star merger explain GW170817, AT2017gfo, and GRB170817A? *Phys. Rev. D*, 100(4):043011, 2019.

[33] Michael W. Coughlin et al. Constraints on the neutron star equation of state from AT2017gfo using radiative transfer simulations. *Mon. Not. Roy. Astron. Soc.*, 480(3):3871–3878, 2018.

[34] H. T. Cromartie et al. Relativistic Shapiro delay measurements of an extremely massive millisecond pulsar. *Nature Astron.*, 4(1):72–76, 2019.

[35] Curt Cutler and Éanna E. Flanagan. Gravitational waves from merging compact binaries: How accurately can one extract the binary's parameters from the inspiral waveform? *Phys. Rev. D*, 49:2658–2697, Mar 1994.

[36] Thibault Damour and Alessandro Nagar. Relativistic tidal properties of neutron stars. *Phys. Rev. D*, 80:084035, 2009.

[37] Pawel Danielewicz, Roy Lacey, and William G. Lynch. Determination of the equation of state of dense matter. *Science*, 298:1592–1596, 2002.

[38] Soumi De, Daniel Finstad, James M. Lattimer, Duncan A. Brown, Edo Berger, and Christopher M. Biwer. Tidal Deformabilities and Radii of Neutron Stars from the Observation of GW170817. *Phys. Rev. Lett.*, 121(9):091102, 2018. [Erratum: Phys.Rev.Lett. 121, 259902 (2018)].

[39] Paul Demorest, Tim Pennucci, Scott Ransom, Mallory Roberts, and Jason Hessels. Shapiro Delay Measurement of A Two Solar Mass Neutron Star. *Nature*, 467: 1081–1083, 2010.

[40] Tim Dietrich, Michael W. Coughlin, Peter T. H. Pang, Mattia Bulla, Jack Heinzel, Lina Issa, Ingo Tews, and Sarah Antier. Multimessenger constraints on the neutron-star equation of state and the Hubble constant. *Science*, 370(6523):1450–1453, 2020.

[41] F. Douchin and P. Haensel. A unified equation of state of dense matter and neutron star structure. *Astron. Astrophys.*, 380:151, 2001.

[42] Reetika Dudi, Francesco Pannarale, Tim Dietrich, Mark Hannam, Sebastiano Bernuzzi, Frank Ohme, and Bernd Brügmann. Relevance of tidal effects and post-merger dynamics for binary neutron star parameter estimation. *Phys. Rev. D*, 98(8):084061, 2018.

[43] Reed Essick, Philippe Landry, and Daniel E. Holz. Nonparametric Inference of Neutron Star Composition, Equation of State, and Maximum Mass with GW170817. *Phys. Rev. D*, 101(6):063007, 2020.

[44] Margherita Fasano, Tiziano Abdelsalhin, Andrea Maselli, and Valeria Ferrari. Constraining the Neutron Star Equation of State Using Multiband Independent Measurements of Radii and Tidal Deformabilities. *Phys. Rev. Lett.*, 123(14):141101, 2019.

[45] E. Fonseca et al. Refined Mass and Geometric Measurements of the High-mass PSR J0740+6620. *Astrophys. J. Lett.*, 915(1):L12, 2021.

[46] Emmanuel Fonseca et al. The NANOGrav Nine-year Data Set: Mass and Geometric Measurements of Binary Millisecond Pulsars. *Astrophys. J.*, 832(2):167, 2016.

[47] J. Fujita and H. Miyazawa. Pion Theory of Three-Body Forces. *Prog. Theor. Phys.*, 17:360–365, 1957.

[48] Jérémie Gagnon-Bischoff, Stephen R. Green, Philippe Landry, and Néstor Ortiz. Extended I-Love relations for slowly rotating neutron stars. *Phys. Rev. D*, 97(6):064042, 2018.

[49] Robert Geroch. Multipole Moments. I. Flat Space. *Journal of Mathematical Physics*, 11(6):1955–1961, June 1970.

[50] Robert Geroch. Multipole Moments. II. Curved Space. *Journal of Mathematical Physics*, 11(8):2580–2588, August 1970.

[51] N. K. Glendenning. Neutron stars are giant hypernuclei ? *Astrophys. J.*, 293:470–493, June 1985.

[52] A. Goldstein et al. An Ordinary Short Gamma-Ray Burst with Extraordinary Implications: Fermi-GBM Detection of GRB 170817A. *Astrophys. J. Lett.*, 848(2):L14, 2017.

[53] Samuel E. Gralla. On the Ambiguity in Relativistic Tidal Deformability. *Class. Quant. Grav.*, 35(8):085002, 2018.

[54] Andreas Guerra Chaves and Tanja Hinderer. Probing the equation of state of neutron star matter with gravitational waves from binary inspirals in light of GW170817: a brief review. *J. Phys. G*, 46(12):123002, 2019.

[55] Yekta Gürsel. Multipole moments for stationary systems: The equivalence of the Geroch-Hansen formulation and the Thorne formulation. *General Relativity and Gravitation*, 15(8):737–754, August 1983.

[56] Tolga Guver and Feryal Ozel. The mass and the radius of the neutron star in the transient low mass X-ray binary SAX J1748.9-2021. *Astrophys. J. Lett.*, 765:L1, 2013.

[57] Quentin Henry, Guillaume Faye, and Luc Blanchet. Tidal effects in the gravitational-wave phase evolution of compact binary systems to next-to-next-to-leading post-Newtonian order. *Phys. Rev. D*, 102(4):044033, 2020.

[58] A. Hewish, S. J. Bell, J. D. H. Pilkington, P. F. Scott, and R. A. Collins. Observation of a Rapidly Pulsating Radio Source. *Nature*, 217(5130):709–713, February 1968.

[59] S. Hild et al. Sensitivity Studies for Third-Generation Gravitational Wave Observatories. *Class. Quant. Grav.*, 28:094013, 2011.

[60] Tanja Hinderer. Tidal Love numbers of neutron stars. *Astrophys. J.*, 677:1216–1220, 2008.

[61] Tanja Hinderer et al. Effects of neutron-star dynamic tides on gravitational waveforms within the effective-one-body approach. *Phys. Rev. Lett.*, 116(18):181101, 2016.

[62] Tanja Hinderer et al. Distinguishing the nature of comparable-mass neutron star binary systems with multimessenger observations: GW170817 case study. *Phys. Rev. D*, 100(6):06321, 2019.

[63] Jin-Liang Jiang, Shao-Peng Tang, Dong-Sheng Shao, Ming-Zhe Han, Yin-Jie Li, Yuan-Zhu Wang, Zhi-Ping Jin, Yi-Zhong Fan, and Da-Ming Wei. The equation of state and some key parameters of neutron stars: constraints from GW170817, the nuclear data and the low mass X-ray binary data. *Astrophys. J.*, 885:39, 2019.

[64] Jin-Liang Jiang, Shao-Peng Tang, Yuan-Zhu Wang, Yi-Zhong Fan, and Da-Ming Wei. PSR J0030+0451, GW170817 and the nuclear data: joint constraints on equation of state and bulk properties of neutron stars. *Astrophys. J.*, 892:1, 2020.

[65] Xisco Jiménez Forteza, Tiziano Abdelsalhin, Paolo Pani, and Leonardo Gualtieri. Impact of high-order tidal terms on binary neutron-star waveforms. *Phys. Rev. D*, 98(12):124014, 2018.

[66] Kenta Kiuchi, Koutarou Kyutoku, Masaru Shibata, and Keisuke Taniguchi. Revisiting the lower bound on tidal deformability derived by AT 2017gfo. *Astrophys. J. Lett.*, 876(2):L31, 2019.

[67] Bharat Kumar and Philippe Landry. Inferring neutron star properties from GW170817 with universal relations. *Phys. Rev. D*, 99(12):123026, 2019.

[68] Benjamin D. Lackey, Mohit Nayyar, and Benjamin J. Owen. Observational constraints on hyperons in neutron stars. *Phys. Rev. D*, 73:024021, 2006.

[69] Philippe Landry. Tidal deformation of a slowly rotating material body: Interior metric and Love numbers. *Phys. Rev. D*, 95(12):124058, 2017.

[70] Philippe Landry and Reed Essick. Nonparametric inference of the neutron star equation of state from gravitational wave observations. *Phys. Rev. D*, 99(8):084049, 2019.

[71] Philippe Landry, Reed Essick, and Katerina Chatziioannou. Nonparametric constraints on neutron star matter with existing and upcoming gravitational wave and pulsar observations. *Phys. Rev. D*, 101(12):123007, 2020.

[72] Philippe Landry and Eric Poisson. Dynamical response to a stationary tidal field. *Phys. Rev. D*, 92(12):124041, 2015.

[73] Philippe Landry and Eric Poisson. Gravitomagnetic response of an irrotational body to an applied tidal field. *Phys. Rev. D*, 91(10):104026, 2015.

[74] Philippe Landry and Eric Poisson. Tidal deformation of a slowly rotating material body. External metric. *Phys. Rev. D*, 91:104018, 2015.

[75] J. M. Lattimer and M. Prakash. Neutron star structure and the equation of state. *Astrophys. J.*, 550:426, 2001.

[76] James M. Lattimer and Maddapa Prakash. Neutron Star Observations: Prognosis for Equation of State Constraints. *Phys. Rept.*, 442:109–165, 2007.

[77] Bao-An Li and Xiao Han. Constraining the neutron-proton effective mass splitting using empirical constraints on the density dependence of nuclear symmetry energy around normal density. *Phys. Lett. B*, 727:276–281, 2013.

[78] Lee Lindblom. Spectral representations of neutron-star equations of state. *Phys. Rev. D*, 82:103011, Nov 2010.

[79] A. E. H. Love. *Some Problems of Geodynamics*. 1911.

[80] Ben Margalit and Brian D. Metzger. Constraining the Maximum Mass of Neutron Stars From Multi-Messenger Observations of GW170817. *Astrophys. J. Lett.*, 850(2):L19, 2017.

[81] Andrea Maselli, Leonardo Gualtieri, Francesco Pannarale, and Valeria Ferrari. On the validity of the adiabatic approximation in compact binary inspirals. *Phys. Rev. D*, 86:044032, 2012.

[82] Andrea Maselli, Andrea Sabatucci, and Omar Benhar. Constraining three-nucleon forces with multimessenger data. *Phys. Rev. C*, 103(6):065804, 2021.

[83] M. C. Miller et al. PSR J0030+0451 Mass and Radius from $NICER$ Data and Implications for the Properties of Neutron Star Matter. *Astrophys. J. Lett.*, 887(1):L24, 2019.

[84] M. C. Miller et al. The Radius of PSR J0740+6620 from NICER and XMM-Newton Data. *Astrophys. J. Lett.*, 918(2):L28, 2021.

[85] Charles W. Misner, K. S. Thorne, and J. A. Wheeler. *Gravitation*. W. H. Freeman, San Francisco, 1973.

[86] Horst Mueller and Brian D. Serot. Relativistic mean field theory and the high density nuclear equation of state. *Nucl. Phys. A*, 606:508–537, 1996.

[87] H. Müther, M. Prakash, and T.L. Ainsworth. The nuclear symmetry energy in relativistic brueckner-hartree-fock calculations. *Physics Letters B*, 199(4):469–474, 1987.

[88] Tatsuya Narikawa, Nami Uchikata, Kyohei Kawaguchi, Kenta Kiuchi, Koutarou Kyutoku, Masaru Shibata, and Hideyuki Tagoshi. Discrepancy in tidal deformability of GW170817 between the Advanced LIGO twin detectors. *Phys. Rev. Research.*, 1:033055, 2019.

[89] Tatsuya Narikawa, Nami Uchikata, Kyohei Kawaguchi, Kenta Kiuchi, Koutarou Kyutoku, Masaru Shibata, and Hideyuki Tagoshi. Reanalysis of the binary neutron star mergers GW170817 and GW190425 using numerical-relativity calibrated waveform models. *Phys. Rev. Res.*, 2(4):043039, 2020.

[90] Frank Ohme, Alex B. Nielsen, Drew Keppel, and Andrew Lundgren. Statistical and systematic errors for gravitational-wave inspiral signals: A principal component analysis. *Phys. Rev. D*, 88(4):042002, 2013.

[91] Feryal Ozel, Dimitrios Psaltis, Tolga Guver, Gordon Baym, Craig Heinke, and Sebastien Guillot. The Dense Matter Equation of State from Neutron Star Radius and Mass Measurements. *Astrophys. J.*, 820(1):28, 2016.

[92] Costantino Pacilio, Andrea Maselli, Margherita Fasano, and Paolo Pani. Ranking Love Numbers for the Neutron Star Equation of State: The Need for Third-Generation Detectors. *Phys. Rev. Lett.*, 128(10):101101, 2022.

[93] P. T. H. Pang et al. An updated nuclear-physics and multi-messenger astrophysics framework for binary neutron star mergers. *Nature Commun*, 14:8352, 2023.

[94] Paolo Pani, Leonardo Gualtieri, and Valeria Ferrari. Tidal Love numbers of a slowly spinning neutron star. *Phys. Rev. D*, 92(12):124003, 2015.

[95] Paolo Pani, Leonardo Gualtieri, Andrea Maselli, and Valeria Ferrari. Tidal deformations of a spinning compact object. *Phys. Rev. D*, 92(2):024010, 2015.

[96] Eric Poisson. Gravitomagnetic Love tensor of a slowly rotating body: post-Newtonian theory. *Phys. Rev. D*, 102(6):064059, 2020.

[97] Eric Poisson. Compact body in a tidal environment: New types of relativistic Love numbers, and a post-Newtonian operational definition for tidally induced multipole moments. *Phys. Rev. D*, 103(6):064023, 2021.

[98] Eric Poisson and Jean Doucot. Gravitomagnetic tidal currents in rotating neutron stars. *Phys. Rev. D*, 95(4):044023, 2017.

[99] M. Prakash, J. R. Cooke, and J. M. Lattimer. Quark-hadron phase transition in protoneutron stars. *Phys. Rev. D*, 52:661–665, Jul 1995.

[100] B. S. Pudliner, V. R. Pandharipande, J. Carlson, and Robert B. Wiringa. Quantum Monte Carlo calculations of A <= 6 nuclei. *Phys. Rev. Lett.*, 74:4396–4399, 1995.

[101] David Radice and Liang Dai. Multimessenger Parameter Estimation of GW170817. *Eur. Phys. J. A*, 55(4):50, 2019.

[102] David Radice, Albino Perego, Francesco Zappa, and Sebastiano Bernuzzi. GW170817: Joint Constraint on the Neutron Star Equation of State from Multimessenger Observations. *Astrophys. J. Lett.*, 852(2):L29, 2018.

[103] Jocelyn S. Read, Benjamin D. Lackey, Benjamin J. Owen, and John L. Friedman. Constraints on a phenomenologically parameterized neutron-star equation of state. *Phys. Rev. D*, 79:124032, 2009.

[104] Tullio Regge and John A. Wheeler. Stability of a Schwarzschild singularity. *Phys. Rev.*, 108:1063–1069, 1957.

[105] David Reitze et al. Cosmic Explorer: The U.S. Contribution to Gravitational-Wave Astronomy beyond LIGO. *Bull. Am. Astron. Soc.*, 51(7):035, 2019.

[106] Thomas E. Riley et al. A $NICER$ View of PSR J0030+0451: Millisecond Pulsar Parameter Estimation. *Astrophys. J. Lett.*, 887(1):L21, 2019.

[107] Thomas E. Riley et al. A NICER View of the Massive Pulsar PSR J0740+6620 Informed by Radio Timing and XMM-Newton Spectroscopy. *Astrophys. J. Lett.*, 918(2):L27, 2021.

[108] Henrik Rose, Nina Kunert, Tim Dietrich, Peter T. H. Pang, Rory Smith, Chris Van Den Broeck, Stefano Gandolfi, and Ingo Tews. Revealing the strength of three-nucleon interactions with the proposed Einstein Telescope. *Phys. Rev. C*, 108(2):025811, 2023.

[109] P. Russotto et al. Results of the ASY-EOS experiment at GSI: The symmetry energy at suprasaturation density. *Phys. Rev. C*, 94(3):034608, 2016.

[110] Andrea Sabatucci and Omar Benhar. Tidal Deformation of Neutron Stars from Microscopic Models of Nuclear Dynamics. *Phys. Rev. C*, 101(4):045807, 2020.

[111] Andrea Sabatucci, Omar Benhar, Andrea Maselli, and Costantino Pacilio. Sensitivity of neutron star observations to three-nucleon forces. *Phys. Rev. D*, 106(8):083010, 2022.

[112] M. Saleem et al. The science case for LIGO-India. *Class. Quant. Grav.*, 39(2):025004, 2022.

[113] V. Savchenko et al. INTEGRAL IBIS, SPI, and JEM-X observations of LVT151012. *Astron. Astrophys.*, 603:A46, 2017.

[114] Andrew W. Steiner, James M. Lattimer, and Edward F. Brown. The Equation of State from Observed Masses and Radii of Neutron Stars. *Astrophys. J.*, 722:33–54, 2010.

[115] Jan Steinhoff, Tanja Hinderer, Alessandra Buonanno, and Andrea Taracchini. Dynamical Tides in General Relativity: Effective Action and Effective-One-Body Hamiltonian. *Phys. Rev. D*, 94(10):104028, 2016.

[116] Jan Steinhoff, Tanja Hinderer, Tim Dietrich, and Francois Foucart. Spin effects on neutron star fundamental-mode dynamical tides: Phenomenology and comparison to numerical simulations. *Phys. Rev. Res.*, 3(3):033129, 2021.

[117] T. M. Tauris et al. Formation of Double Neutron Star Systems. *Astrophys. J.*, 846(2):170, 2017.

[118] Kip S. Thorne. Multipole expansions of gravitational radiation. *Rev. Mod. Phys.*, 52:299–339, Apr 1980.

[119] Kip S. Thorne and Alfonso Campolattaro. Non-Radial Pulsation of General-Relativistic Stellar Models. I. Analytic Analysis for $L >= 2$. *Astrophys. J.*, 149: 591, September 1967.

[120] M. B. Tsang, Yingxun Zhang, P. Danielewicz, M. Famiano, Zhuxia Li, W. G. Lynch, and A. W. Steiner. Constraints on the density dependence of the symmetry energy. *Phys. Rev. Lett.*, 102:122701, 2009.

[121] R. B. Wiringa, V. Fiks, and A. Fabrocini. Equation of state for dense nucleon matter. *Phys. Rev. C*, 38:1010–1037, Aug 1988.

[122] Robert B. Wiringa, V. G. J. Stoks, and R. Schiavilla. An Accurate nucleon-nucleon potential with charge independence breaking. *Phys. Rev. C*, 51:38–51, 1995.

[123] Kent Yagi and Nicolas Yunes. I-Love-Q. *Science*, 341:365–368, 2013.

[124] Kent Yagi and Nicolas Yunes. I-Love-Q Relations in Neutron Stars and their Applications to Astrophysics, Gravitational Waves and Fundamental Physics. *Phys. Rev. D*, 88(2):023009, 2013.

[125] Zhen Zhang and Lie-Wen Chen. Constraining the symmetry energy at subsaturation densities using isotope binding energy difference and neutron skin thickness. *Phys. Lett. B*, 726:234–238, 2013.

CHAPTER 4

Neutron Stars and Multimessenger Observations: A Challenge for Nuclear Physics Theory

Fiorella Burgio, Hans-Joseph Schulze, Isaac Vidaña, Jin-Biao Wei

THE recent observations of neutron stars have shed some light on their structure properties and inner composition, thus helping in improving the knowledge of the equation of state of very dense matter. Nuclear many-body theories, both microscopic and phenomenological, have significantly contributed to the interpretation of the experimental data, either coming from terrestrial laboratories or from X-ray telescopes and gravitational-wave interferometers. The detection of more and more astrophysical events will hopefully help to improve the quality of the theoretical models, thus advancing our knowledge of the nuclear physics theory.

4.1 INTRODUCTION

The equation of state (EoS) of isospin-asymmetric nuclear matter is a fundamental ingredient in the description of the static and dynamical properties of neutron stars (NSs), core-collapse supernova (CCSNe) and compact-star mergers (BNSs) [1–3]. Its determination, however, is very challenging due to the wide range of densities, temperatures and isospin asymmetries found in these astrophysical scenarios, and it constitutes nowadays one of the key problems in nuclear astrophysics.

Current models of the nuclear EoS are based on reliable experimental data on atomic nuclei and nucleon scattering. However, terrestrial nuclear physics experiments cannot probe the physical state of matter under extreme conditions, and therefore, theoretical models that have been successfully applied to ordinary nuclear structure, need to be extrapolated to these unknown regions. In doing so, the main difficulties are associated with

DOI: 10.1201/9781003306580-4

our lack of knowledge of the in-medium nuclear interaction, and to the complex solution of the nuclear many-body problem [4]. Theoretical calculations of the nuclear EoS at such extreme conditions can be tested almost exclusively by astrophysical observations. The recent NS observations performed with the X-ray NICER telescope [5–10] along with the gravitational wave observations of the well-known GW170817 event [11–14] have strongly contributed to the advance of the nuclear physics theory for dense matter and improved our knowledge of the internal structure of those compact objects.

Our task is to illustrate the current state of the art of the nuclear many-body approaches, their limitations imposed by the experimental data and the possible improvements due to new observational data. For that the manuscript is organised in the following way. In Sect. 4.2, we illustrate and compare some of the mostly used theoretical approaches for the interpretation of the data. The experimental constraints on the nuclear EoS are presented in Sect. 4.3, whereas the astrophysical ones are discussed in Sect. 4.4. Sect. 4.5 is devoted to the discussion of the EoS at finite temperature, which is crucial for the interpretation of NS binary mergers data. An alternative and likely scenario besides the purely nucleonic one, i.e. hyperonic matter, is depicted in Sect. 4.6. Finally, a summary is given in Sec. 4.7.

4.2 THEORETICAL APPROACHES TO THE NUCLEAR EOS

The nuclear EoS is usually determined following two alternative many-body approaches, i.e. either a purely phenomenological or a microscopic one. Phenomenological approaches, nonrelativistic or relativistic, are based on effective interactions that are built upon the ground-state properties of atomic nuclei [15]; Skyrme interactions [15–20] and relativistic mean-field (RMF) models [21, 22] are among the most used ones. Microscopic approaches, on the other hand, are based on realistic two- and three-body forces (TBFs) that describe scattering data in free space and the properties of the deuteron. These interactions are realised as meson-exchange models [23–29] or, more recently, within chiral perturbation theory [30–33]. In those approaches, to obtain the EoS, one has to solve the complicated many-body problem [4, 34], whose main difficulty lies in the treatment of the short-range repulsive part of the nucleon-nucleon (NN) interaction. The mostly used microscopic many-body approaches include the (Dirac)-Brueckner-Bethe-Goldstone theories [4, 34–37], the variational method [38], the correlated basis function formalism [39], the self-consistent Green's function technique [40, 41], the $V_{\text{low-k}}$ approach [42] and Quantum Monte Carlo techniques [43–45].

In the subsections below, we give a brief review of these models.

4.2.1 Microscopic approaches

The Bethe-Brueckner-Goldstone many-body theory is well known among the diagrammatic techniques. It is essentially based on the re-summation of the perturbation expansion of the ground-state energy of nuclear matter [4, 34, 35, 46], the key point being the solution of the Bethe-Goldstone integral equation for the G-matrix, which can be written as

$$\langle 12|G(\omega)|34\rangle = \langle 12|V|34\rangle + \sum_{5,6} \langle 12|V|56\rangle \frac{[1-\Theta_F(k_5)]\,[1-\Theta_F(k_6)]}{\omega - e_5 + e_6 + i\eta} \langle 56|G(\omega)|34\rangle \,, \tag{4.1}$$

where V is the bare NN interaction, ω is the starting energy, the two Pauli-blocking factors $[1 - \Theta_F(k)]$ force the intermediate momenta to be above the Fermi momentum ("particle states"), the single-particle energy being $e_k = k^2/2m + U(k)$, with m the particle mass, and $U(k)$ the in-medium single-particle potential felt by each nucleon with momentum k. The multi-indices $1, 2, \ldots$ denote in general momentum, isospin and spin.

The crucial point of this technique is that the original bare NN interaction V is systematically replaced by the G-matrix, an effective interaction that describes the in-medium scattering processes, and that takes into account the effect of the Pauli principle on the scattered particles. The main advantage of the G-matrix is that it does not feature the hard core of the original bare NN interaction, and it is defined even for bare interactions with an infinite hardcore. The introduction and choice of $U(k)$ are essential to make the resummed expansion convergent, and it is calculated self-consistently with the G-matrix itself in order to incorporate as many higher-order correlations as possible. The resulting nuclear EoS can be calculated with good accuracy at the so-called Brueckner-Hartree-Fock (BHF) level, thus keeping the two-body correlations [47–49].

It is widely known that all non-relativistic many-body approaches fail to reproduce the correct saturation point of nuclear matter, which is an essential requirement; for that, TBFs have to be introduced in the calculations as effective density-dependent two-body forces. Unfortunately, nowadays a complete theory of TBFs does not exist, and we have to resort to phenomenological models. Among the mostly used ones, we mention the Urbana model [50, 51], which is built on the properties of ^{3}H and ^{3}He and produces at the saturation point a shift of about $+1$ MeV in the binding energy and of $-0.01\ \text{fm}^{-3}$ in density. The problem of such a procedure is that the TBF is dependent on the two-body force. The connection between two-body and three-body forces within the meson-nucleon theory of nuclear interaction is extensively discussed and developed in [52–54]. At present the theoretical status of microscopically derived TBFs is still quite rudimentary; however, a tentative approach has been proposed using the same meson-exchange parameters as the underlying NN potential. Results have been obtained [54, 55] with the Argonne V_{18} [56], the Bonn B [37], and the Nijmegen 93 potentials [57]. The BHF+TBF approach can reproduce nuclear-matter properties near the saturation density with a quite good accuracy [3, 58].

The relativistic BHF formalism, i.e. the Dirac-Brueckner approach [24], has been developed in analogy with the non-relativistic case, where the two-body correlations are described by introducing the in-medium relativistic G-matrix. In the DBHF method, the nuclear mean field is described in terms of a scalar and a vector component, whose combination provides an explanation for the binding of nucleons. In particular, the use of the spinor formalism has been shown [59] to be equivalent to introducing a particular TBF, the so-called Z-diagram, which is repulsive and consequently provides a better saturation point than the non-relativistic method. Therefore, the most striking feature of the DBHF theory is its ability to describe the saturation properties of nuclear matter, without introducing additional TBFs. It turns out that the inclusion in BHF of the Z-diagram allows to obtain results close to DBHF calculations [60], and that the EoS calculated within the DBHF method is stiffer above saturation than the one calculated from the BHF+TBF method. A further point of difference is that the relativistic DBHF produces a superluminal EoS at higher densities than the BHF approach. The reader is referred to Ref. [61] for a relatively recent review of the DBHF method and a variety of applications to both nuclear matter and nuclei.

A further many-body approach extensively used in nuclear astrophysics is the variational one [62]. This method is based on the Ritz-Raleigh variational principle, according to which the trial ground-state energy

$$E_{\text{trial}} = \frac{\langle\Psi|H|\Psi\rangle}{\langle\Psi|\Psi\rangle} , \tag{4.2}$$

calculated from the system's Hamiltonian H with a trial many-body wave function Ψ, provides an upper bound for the true ground-state energy of the system. It is usually assumed that the ground-state wave function Ψ can be written as

$$\Psi(r_1, r_2, ...) = \prod_{i<j} f(r_{ij})\Phi(r_1, r_2, ...) , \tag{4.3}$$

where Φ is the unperturbed ground-state wave function, properly antisymmetrised (the product runs over all possible distinct pairs of particles), and the correlation factors f are determined by the Ritz-Raleigh variational principle, i.e. by assuming that the expectation value of the Hamiltonian reaches a minimum:

$$\frac{\delta}{\delta f} \frac{\langle\Psi|H|\Psi\rangle}{\langle\Psi|\Psi\rangle} = 0 . \tag{4.4}$$

Therefore, the main task in the variational method is to find a suitable ansatz for the correlation factors f. Several different methods exist for the calculation of f, e.g. in the nuclear context the Fermi-Hyper-Netted-Chain (FHNC) calculations [62, 63] have been proven to be efficient.

For nuclear matter, it is necessary to introduce channel-dependent correlation factors, which is equivalent to assume that f is actually a two-body operator $\hat{F}_{ij}$, which can be expanded in the same spin-isospin, spin-orbit and tensor operators appearing in the NN interaction [39, 64]. Most variational calculations have been performed with the Argonne NN forces, because of their formal structure, and often supplemented by the Urbana TBFs. The best known and most used variational nuclear-matter EoS is the one of Akmal-Pandharipande-Ravenhall (APR) [38]. Many excellent review papers exist in the literature on the variational method for the determination of the nuclear matter EoS, e.g. [62].

A further EoS of the variational class that we include in our study, is the so-called CBF-EI ("Correlated Basis Functions–Effective Interaction") EoS [65, 66], which has been obtained within nonrelativistic many-body theory, using a realistic nuclear Hamiltonian with the Argonne V_6' [67] and the Urbana IX nuclear potentials as TBF. We remind that the V_6' potential is not simply a truncated version of the full V_{18} potential, but rather its projection on the basis of the six spin-isospin operators. As already mentioned for the BHF method, the TBF is included by making an average over the general coordinates of the third particle, and therefore an effective density-dependent interaction is again obtained. The one from the latest and most advanced implementation of this approach, has been applied to the determination of a variety of properties of nuclear matter, at both zero and in particular nonzero temperature in the scenario of protoneutron stars [65, 68].

4.2.2 Phenomenological approaches

Phenomenological approaches treat dense matter making use of effective interactions instead of bare ones, and mostly rely on the energy density functional (EDF) theory, which is able to recast the complex many-body problem of interacting particles into an effective independent-particle approach [69]. In the EDF theory, the total energy of the system is usually expressed as a functional of the nucleon number densities, the kinetic energy densities, and the spin-current densities, but the exact form of the functional itself is not known a priori. Therefore, one has to rely on phenomenological functionals, which depend on a certain number of parameters fitted to reproduce some properties of known nuclei and nuclear matter, as well as microscopic calculations of infinite nuclear matter.

Non-relativistic approaches usually start from an Hamiltonian $\hat{H} = \hat{T} + \hat{V}$ for the many-body system, where $\hat{T} = \sum_i \hat{p}^2/2m_i$ is the kinetic term ($\hat{p}$ being the momentum operator and m_i the mass of the species i) and $\hat{V}$ is the potential term, which accounts for the two-body potential and incorporates physical properties like effective masses, but TBFs can also be included explicitly [17]. A very popular scheme is based on the Skyrme-type effective interactions, which are zero-range density-dependent interactions and allow for fast numerical computations. Since the pioneering work of Skyrme [70], several extensions have been proposed [71–75], which accurately reproduce experimentally measured properties of finite nuclei and nuclear matter and are applied to NSs.

On the basis of the Skyrme nuclear effective force, a class of so-called "unified" EoSs has been generated, in which the whole range of NS density, from the outer crust to the inner core, covering about 14 orders of magnitude, is described within the same theoretical framework. Douchin and Haensel (DH) [76] formulated a unified EoS for NSs, where some parameters of the Skyrme interaction were adjusted to reproduce a microscopic calculation of neutron matter above saturation density [77]. Hence, the DH EoS contains certain microscopic input. In the DH model the inner crust, where non-homogeneous matter is present, was treated within the compressible liquid-drop model (CLDM) approach. More recently, unified EoSs for NSs have been derived by the Brussels-Montreal group [78–81]. They are based on the BSk family of Skyrme nuclear effective forces [74]. Each force is fitted to the known masses of nuclei and adjusted among other constraints to reproduce a microscopic EoS of neutron matter with different stiffness at high density. The inner crust is treated in the extended Thomas-Fermi (TF) approach. Analytical fits of these NS EoSs have been constructed in order to facilitate their inclusion in astrophysical simulations [81].

Besides the Skyrme effective interactions, new EDFs have been recently constructed on the basis of the Kohn-Sham density-functional theory [82, 83]. These Barcelona-Catania-Paris-Madrid (BCPM) EDFs have been derived by implementing the functional the results of microscopic nuclear and neutron-matter BHF calculations, thus yielding a very good description of properties of finite nuclei and NS masses with a reduced number of parameters. More recently, in [84] has been proposed a unified EoS from the outer crust to the core based on modern microscopic calculations using the Argonne V_{18} potential plus TBFs computed with the Urbana model. To deal with the inhomogeneous structures of matter in the NS crust, the authors used a nuclear energy-density functional that is directly based on the same microscopic calculations, and which is able to reproduce the ground-state properties of nuclei.

RMF models have been constructed on the basis of a field-theoretical formalism where nucleons are represented by four-component Dirac spinors, and the NN interaction is modeled by exchange of mesons [85]. These models have been successfully employed in nuclear structure, to describe both stable and exotic nuclei [86]. The common starting point in RMF models is an effective Lagrangian $\mathcal{L} = \mathcal{L}_{\rm nuc} + \mathcal{L}_{\rm mes} + \mathcal{L}_{\rm int}$, where the different terms account for the nucleon, the free mesons and the interaction contribution, respectively. The isoscalar-scalar σ meson and the isoscalar-vector ω meson mediate the long-range attraction and short-range repulsion of the nuclear interaction, respectively, in symmetric nuclear matter (SNM). Isovector mesons (like the isovector-vector ρ meson and the isovector-scalar δ meson) need to be included as well to treat neutron-proton asymmetric systems. The interaction term depends on the nucleon-meson coupling constants that are usually determined by fitting nuclei or nuclear-matter properties. From the Lagrangian above, the field equations for nucleons and mesons are derived, and they are solved self-consistently, usually in the RMF approximation.

However, the correct description of nuclear matter and finite nuclei requires the extension of that simple Lagrangian in order to include a medium-dependent effective interaction. This effect can either be introduced by including non-linear (NL) meson self-interaction terms in the Lagrangian, or by assuming an explicit density dependence (DD) for the meson-nucleon couplings, which gives rise to the so-called rearrangement contributions mostly important for the thermodynamic consistency of the model. The former approach has been employed in constructing several phenomenological RMF interactions, e.g. the NL3 [87] and FSUGold [88] models. In the second approach, the functional form of the density dependence of the coupling constants can be motivated by comparing the results with the microscopic DBHF calculations or it can be fully phenomenological, with parameters adjusted to experimental data (DD-RMF models [89, 90]).

Due to their simplicity and easy implementation, both Skyrme and RMF-based approaches turn out to be very popular in astrophysical applications, i.e. core-collapse supernovae and binary NS mergers simulations, where a wide range of densities, temperatures and charge fractions, describing both clustered and homogeneous matter, is needed. This range is covered by the so-called "general-purpose" EoSs. However, at present, only a few of them are available and directly applicable to simulations; in this work we discuss the currently most used ones in astrophysical applications.

The Lattimer and Swesty (LS) EoS [91] models matter as a mixture of heavy nuclei treated in the single-nucleus approximation (SNA), α particles, free neutrons and protons, immersed in a uniform gas of leptons and photons. Nuclei are described within a medium-dependent liquid-drop model, and a simplified NN interaction of Skyrme type is employed for nucleons. With increasing density, shape deformations of nuclei (non-spherical nuclei and bubble phases) are taken into account by modifying the Coulomb and surface energies, and the transition to uniform matter is described by a Maxwell construction. Three version of the LS EoS do exist, each characterized by a different value of the nuclear-matter compressibility at saturation, i.e. 180, 220 and 375 MeV. In the following we illustrate results for the LS220 EoS.

The Shen et al. (STOS) EoS [92] is also widely used. As the LS EoS, matter is described as a mixture of heavy nuclei (treated in SNA), α particles, and free neutrons and protons, immersed in a homogeneous lepton gas. For nucleons, a RMF model with the

TM1 interaction [93] is used; α particles are described as an ideal Boltzmann gas with excluded-volume corrections. The properties of the heavy nucleus are determined by WS-cell calculations within the TF approach employing parameterised density distributions of nucleons rand α particles. We employ here the latest version Shen20 [94].

The SFHo EoS [95] is based on the Hempel and Schaffner-Bielich [96] approach, inspired by the extended nuclear statistical equilibrium (NSE) network model, that takes into account an ensemble of nuclei and interacting nucleons. Nuclei are considered as classical Maxwell-Boltzmann particles, and nucleons are described within an RMF model employing new parametrizations fitted to some NS radius determinations. Binding energies are taken from experimental data or from theoretical nuclear mass tables. Excluded-volume effects are implemented in a thermodynamically consistent way so that it is possible to describe the transition to uniform matter. We remind that NSE models have been employed for typical conditions of core-collapse supernova [97–99]; for a comparison among the different methods, please refer to Ref. [100].

4.2.3 Comparing microscopic and phenomenological approaches

We now turn to discuss the main differences between microscopic and phenomenological EoS models. The results are summarised in Fig. 4.1. On the left side, we display the result for SNM, in the upper panel the binding energy per particle and in the lower panel the pressure, as functions of the nucleon density.

As far as microscopic approaches are concerned, we adopt two BHF EoSs based on the Argonne V_{18} potential, supplemented either by a microscopic TBF (labelled as V18) [52, 55, 58, 101–104] or a phenomenological UIX one (labelled as UIX) [84, 105], the latter being constructed from the BCPM energy density functional. Moreover, we compare these BHF EoSs with the often-used results of the APR [38] and the CBF-EI EoS [65, 66], both based on the variational method. Representative of phenomenological EoSs, we show results obtained for a couple of Skyrme forces, namely, DH EoS with the SLy4 effective force [106] and BSk26 EoS [107]. We also consider the general-purpose SFHo EoS [95], belonging to the RMF family, the Lattimer-Swesty EoS (LS220), and the Shen20 EoS. Constraints on the nuclear EoS for SNM from heavy ion collision (HIC) data of the KaoS experiment [108] (pink area) and flow data [109] (blue area) are shown in the lower panel. Any reliable model for the EoS should pass through the region defined by the experimental data. We notice that most of the EoSs considered are in general compatible with these data, with some exception in the case of the microscopic V18 one, which is only marginally compatible at large density, and the Shen20, which is generally too high. We should, however, point our that the constraints inferred from HICs are indirect and model dependent, in particular at large density, since the analysis of the measured data requires the use of transport models.

We see that all approaches yield similar results up to about twice the saturation density, and diverge at larger density. For the microscopic approaches of BHF type, this is certainly due to the different high-density behaviour of TBFs, where three-body correlations and interactions make a difference. For phenomenological models, a similar spread at high density can be observed. Indeed, these models are characterized by parameters which are fitted on experimental data approximately known around saturation density, therefore their

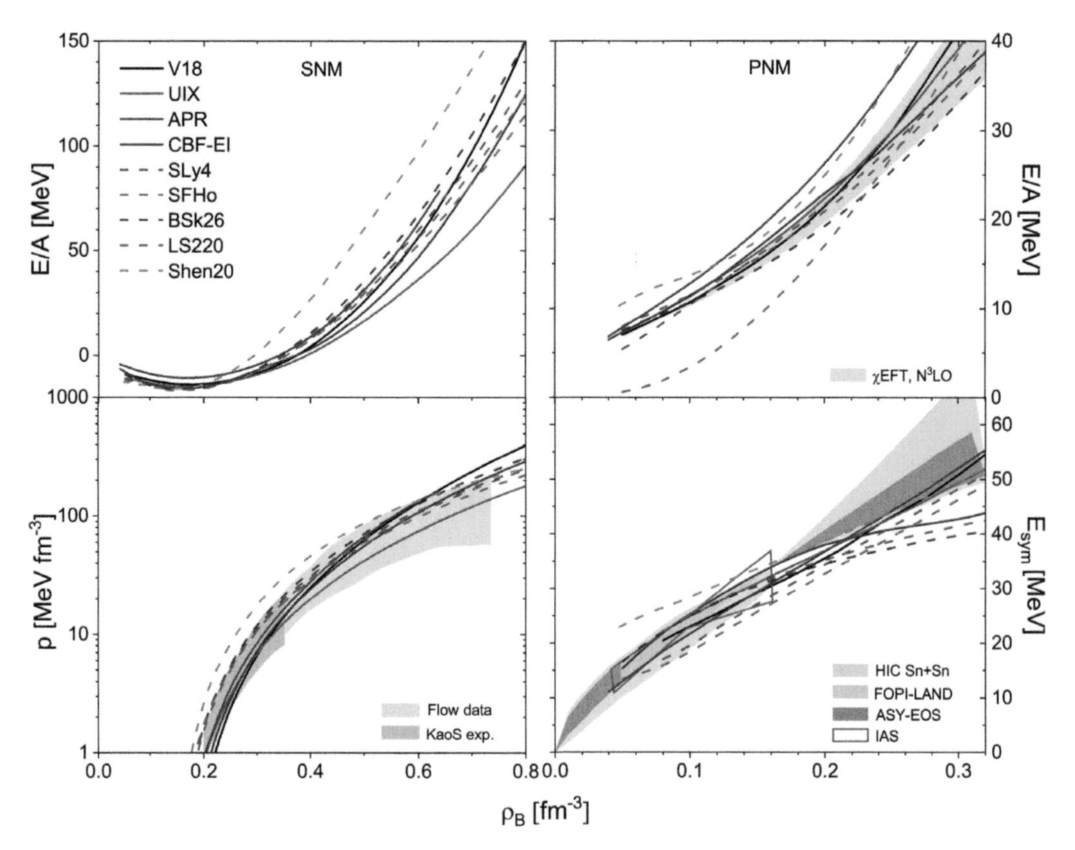

Figure 4.1 Left plot: Binding energy per nucleon of symmetric nuclear matter (upper panel) and the corresponding pressure (lower panel). Right plot: Binding energy per nucleon of pure neutron matter (upper panel) and the symmetry energy (lower panel). Results obtained with different EoSs are compared. Shaded areas represent experimental and theoretical constraints of various type. See text for details.

extrapolation to high density, where no experimental data are available, can produce large differences.

It is worthwhile to perform a comparison of the above-discussed EoSs with additional available constraints. In the upper right panel of Fig. 4.1, we compare with the results obtained for pure neutron matter (PNM) within the chiral EFT interactions up to N3LO order [110, 111], shown by the shaded blue band. Please note that this comparison is limited to a density range up to $\rho = 0.34\,\mathrm{fm}^{-3}$, where the accuracy estimate of the chiral expansion becomes prohibitively large. We stress that neutron matter is an ideal laboratory for nuclear interactions derived from chiral effective field theory since all contributions are predicted up to N3LO in the chiral expansion, including TBFs [112]. We notice that most of the microscopic and phenomenological calculations fall well inside the chiral band, but some inconsistency is evident for the Shen20 and LS220 EoSs in particular, being respectively too stiff and too soft in PNM. In general the phenomenological EoSs exhibit a larger variation in PNM, since in many cases no data have been used for calibration and the predictions are purely theoretical.

In the lower right panel, we display the symmetry energy over the same density range. We observe a monotonically increasing behaviour for all the considered EoSs, and strong divergences at large density, which depend on either the nucleonic interaction and/or the many-body scheme; those are important for determining structure and composition of NSs, and also their cooling mechanisms. Results are compared with available experimental data, which are displayed by the shaded areas. In particular, the grey area represents the diffusion data of HICs, the red contour shows the results of the isobaric analog states (IAS), obtained in Ref. [113], the light blue area includes the data obtained by the FOPI-LAND collaboration [114] on the collective flow, and the blue area is the experimental region checked by the ASY-EOS collaboration [115]. We see that most of the considered EoSs are compatible with experimental data up to around saturation density, whereas for larger densities some EoSs tend to predict smaller values of the symmetry energy, below the experimental areas. This is a clear sign of discrepancy, which results in a much larger difference at larger values of the baryon density, as the ones characterising the inner core of a NS. We notice that the inferred constraints are model dependent, since the data interpretation requires theoretical simulations. Those discrepancies give rise to different predictions for NS structure, as will be shown later.

4.3 EXPERIMENTAL CONSTRAINTS OF THE NUCLEAR EOS

Around saturation density ρ_0 and isospin asymmetry $\beta \equiv (\rho_n - \rho_p)/(\rho_n + \rho_p) = 0$, the nuclear EoS can be characterized by a set of a few isoscalar (E_0, K_0, Q_0) and isovector $(S, L, K_{\text{sym}}, Q_{\text{sym}})$ parameters. These parameters can be constrained by nuclear experiments and are related to the coefficients of a Taylor expansion of the energy per particle of asymmetric nuclear matter in density and isospin asymmetry:

$$\frac{E}{A}(\rho,\beta) = E_0 + \frac{1}{2}K_0 x^2 + \frac{1}{6}Q_0 x^3 + \left(S_0 + Lx + \frac{1}{2}K_{\text{sym}} x^2 + \frac{1}{6}Q_{\text{sym}} x^3\right)\beta^2 + \mathcal{O}(\beta^4)\,. \tag{4.5}$$

Here $x \equiv (\rho - \rho_0)/3\rho_0$, E_0 is the energy per particle of SNM at ρ_0, K_0 the incompressibility parameter, Q_0 the so-called skewness, S_0 the value of the nuclear symmetry energy at ρ_0, L the slope of the symmetry energy, K_{sym} the symmetry incompressibility, and Q_{sym} the third derivative of the symmetry energy with respect to the density.

Measurements of nuclear masses [116] and density distributions [117] yield $E_0 = -16 \pm 1$ MeV and $\rho_0 = 0.15 - 0.16\,\text{fm}^{-3}$, respectively. The value of K_0 can be extracted from the analysis of isoscalar giant monopole resonances in heavy nuclei. However, its extraction is complicated by the adopted analysis model. Results of Ref. [118] suggest $K_0 = 240 \pm 10$ MeV, whereas in Ref. [119] a value of $K_0 = 248 \pm 8$ MeV is reported, and Ref. [120] obtained $K_0 = 230 \pm 40$ MeV, based on the calculation of the third derivative of the energy per unit volume of SNM. HIC experiments would point to a rather “soft” EoS, i.e. a lower value of K_0 [121]. However, the constraints inferred from HICs are model dependent because the analysis of the measured data requires the use of transport models. The value of the skewness parameter Q_0 is more uncertain and is not very well constrained yet, being estimated in the range $-500 \leq Q_0 \leq 300$ MeV.

Experimental information on the isovector parameters of the nuclear EoS can be obtained from several sources such as the analysis of giant [132] and pygmy [133, 134] resonances, isospin diffusion measurements [135], isobaric analog states [113], isoscaling [136], measurements of the neutron skin thickness in heavy nuclei [137–142] or

TABLE 4.1 Saturation properties of the considered models.

Model	Source/Ref.	ρ_0 [fm^{-3}]	$-E_0$ [MeV]	K_0 [MeV]	S_0 [MeV]	L [MeV]
V18	[122]	0.178	13.9	207	32.3	67
UIX	[122]	0.171	14.9	171	33.5	61
CBF-EI	[65, 68]	0.160	10.9	240	30.0	68
APR	[123]	0.160	16.0	266	32.6	59
SLy4	[106]	0.160	15.2	232	35.2	50
BSk26	[107]	0.159	15.2	241	30.0	38
SFHo	[123]	0.157	16.2	244	32.8	53
LS220	[124]	0.155	16.0	220	29.3	74
Shen20	[94]	0.145	16.3	281	31.4	40
Exp.		$\sim$ 0.14–0.17	$\sim$ 15–16	220–260	28.5–34.9	30–87
Ref.		[125]	[125]	[126, 127]	[1, 128]	[1, 128]

meson production in HICs [143, 144]. However, whereas S_0 is more or less well established (≈ 30 MeV), the values of L, and specially those of $K_{\rm sym}$ and $Q_{\rm sym}$, are still uncertain and poorly constrained. For example, combining different data the authors of Ref. [145] give $29.0 < S_0 < 32.7$ MeV and $40.5 < L < 61.9$ MeV, while a more recent work [113] suggests $30.2 < S_0 < 33.7$ MeV and $35 < L < 70$ MeV. Why the isovector part of the nuclear EoS is so uncertain is still an open question whose answer is related to our limited knowledge of the nuclear force and, in particular, of its spin and isospin dependence. We report in Table 4.1 the values of the isoscalar and isovector parameters calculated in each model.

Another important correlation is the one between the slope L of the symmetry energy and its value at saturation S_0, which is shown in Fig. 4.2. The different symbols show the predictions of microscopic approaches (black circles) and phenomenological models (green squares). Currently available experimental data from neutron-skin thickness in Sn isotopes [146], isospin diffusion in HICs [147], electric dipole polarizability [148], isobaric-analog-state (IAS) phenomenology combined with the skin-width data [113] or the Bayesian analysis of mass and radius measurements in NSs [95] are also indicated in the figure. The dashed line shows the unitary-limit constraint determined in Ref. [129], which combined with the experimental and observational data indicates that only values of (S_0, L) to the right of this line are permitted. As can be seen, most of the models considered fulfil all these constraints. A notable exception is the values of both S_0 and L obtained in a recent analysis of the PREX-II measurement of the neutron skin of ^{208}Pb [130]. In particular $S_0 = 38.1 \pm 4.7$ is much larger than the otherwise well-constrained results of all other methods. However, that analysis relied exclusively on a few specific RMF functionals. This has been clearly exposed in [131], where instead $S_0 = 33 \pm 2$ MeV, $L = 53 \pm 15$ MeV were obtained in a more model-independent analysis of the same PREX-II data.

Astrophysical observations can also be used to constrain these parameters. It has been shown, for instance, that the slope parameter L of the symmetry energy is correlated with

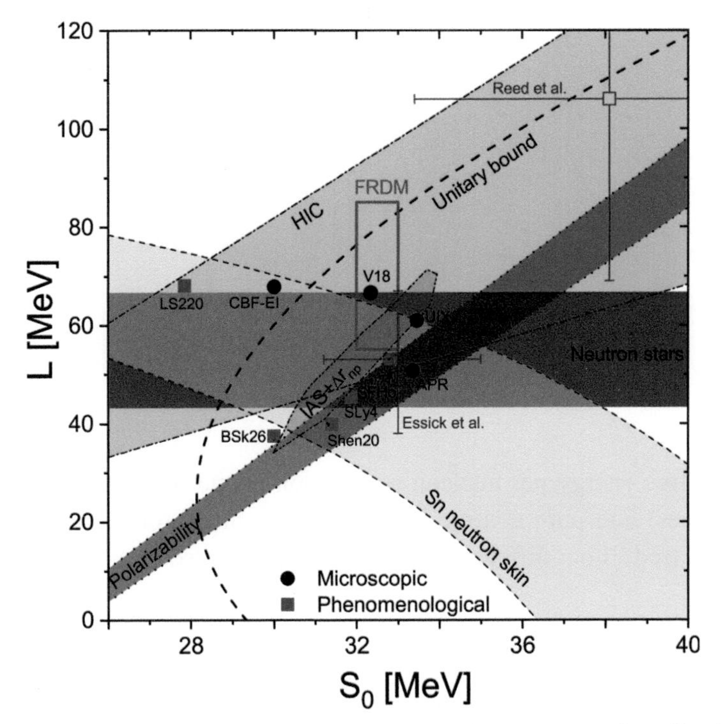

Figure 4.2 Slope L of the symmetry energy vs its value at saturation S_0. Markers show the predictions of microscopic (black circles) and phenomenological (green squares) approaches adopted in this work. The shaded areas represent experimental bands, whereas the dashed line shows the unitary-limit constraint determined in Ref. [129]. The results of two different theoretical analyses [130, 131] of the PREX-II experiment are also indicated (blue and red crosses).

the radius [149] and the tidal deformability [150, 151] of a $1.4\,M_\odot$ NS, and that precise and independent measurements of the radius and the tidal deformability from multiple observations of NSs can potentially pin down the correlation between $K_{\rm sym}$ and L and thus the high-density behaviour of the nuclear symmetry energy [152]. We come back to these features in Sect. 4.5.

4.4 FINITE-TEMPERATURE EOS

The EoS at finite temperature of asymmetric nuclear matter plays a major role in the dynamics of core-collapse supernovae [153–155], the evolution of proto-neutron stars (PNS) and the post-merger phase of binary neutron star mergers (BNS) [156, 157]. In fact, it determines the final evolution of a transitory state to either collapse to a black hole or to the formation of a NS, and this requires an EoS depending on three thermodynamic parameters, typically chosen as temperature T, baryon number density ρ_B and electron fraction Y_e. These need to cover wide domains: $10^{-14}\,\text{fm}^{-3} \lesssim \rho_B \lesssim 1.5\,\text{fm}^{-3}$, $0 \leq Y_e \lesssim 0.6$ and $0 \leq T \lesssim 100$ MeV.

As previously discussed, the recent NS observations have allowed to considerably narrow down the parameter space for the cold β-equilibrated EoS. On the contrary, the situation looks more complicated for the finite-temperature EoS because it is potentially out of

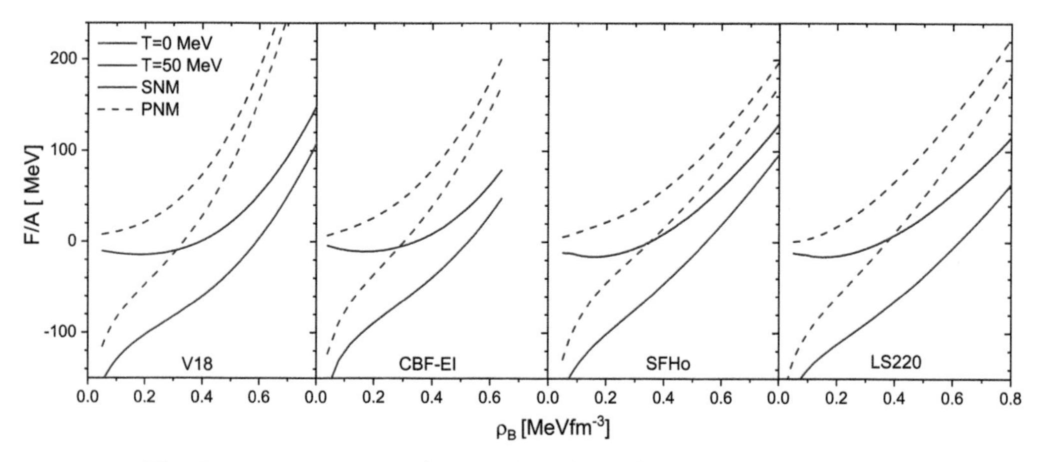

Figure 4.3 The free energy per nucleon as function of the baryon density for symmetric matter (solid lines) and pure neutron matter (dashed lines) for the cold EoS (blue) and T=50 MeV EoS (red). Four different EoSs have been selected. See text for details.

weak equilibrium, and no firm constraints exist so far. In many numerical approximations, thermal effects are approximately included by adding to the cold EoS thermal contributions in the form of a Γ law relating thermal pressure and thermal energy density [158–161]. Currently, a few finite-temperature nuclear EoSs for astrophysical simulations are available [91, 92, 94–96, 162–164], and the predictions for the thermal effects on stellar stability are strongly model dependent; in fact BHF results indicate a slight reduction of the maximum mass, whereas RMF models usually predict increasing stability (maximum mass) with temperature. For a recent review, the reader is referred to Ref. [165].

Regarding the theoretical approaches illustrated in this paper, we mention in particular the non-relativistic Brueckner-like calculations [4, 166], based on the formalism by Bloch and De Dominicis [167] at finite temperature and the CBF-EI variational scheme. Those calculations confirmed that the resulting EoS for SNM shows a typical Van der Waals behaviour, which entails a liquid-gas phase transition, and this is clearly shown in Fig. 4.3, where the free energy is displayed as function of the baryon density for SNM (solid lines) and PNM (dashed lines) at $T = 0$ (blue) and $T = 50$ MeV (red). A similar behaviour is present also for the phenomenological approaches SFHo and LS220. All the results show a quite similar behaviour up to about twice saturation density, but diverge at large density. We mention that the number of phenomenological EoSs available at finite temperature is also quite small. These are mostly based on density-functional theory (DFT), either in a relativistic version with various parametrizations or a non-relativistic one based on Skyrme functionals. SFHo and LS220 belong to the class of the so-called "general-purpose" EoSs, which are able to cover a wide range of densities, temperatures and charge fractions and describe both clustered and homogeneous matter. These EoSs are therefore suitable for applications to core-collapse supernovae and mergers. The interested reader can refer to the CompOSE repository [123] for their practical use.

It is important to mention that the main drawback of the microscopic approaches is the violation of the Hugenholtz-Van Hove (HVH) theorem. In other words, the pressure p

calculated from the thermodynamic relation $p = -f + \mu\rho$, being f the free energy density, μ the chemical potential and ρ the number density, does not coincide with the pressure calculated from $p = -\Omega/V$, Ω being the grand potential and V the volume. In order to overcome this problem, the procedure illustrated in Ref. [4] is usually adopted, and in this way, the HVH theorem is automatically satisfied. The same drawback is present also in the relativistic DBHF. On the contrary, the HVH theorem is strictly fulfilled within the SCGF method [168].

4.5 ASTROPHYSICAL CONSTRAINTS

The main astrophysical constraints on the nuclear EoS are those arising from the observation of NSs. After fifty years of observations we have collected an enormous amount of data on different NS observables, from which it is possible to infer valuable information on the internal structure of these objects and, therefore, also on the nuclear EoS, which is the only ingredient needed to solve the structure equations of NSs. In the following, we shortly review some of the observables, which are mostly important for the determination of the EoS, i.e. the NS mass, the radius and the tidal polarizability.

4.5.1 Masses and radii

NS masses can be inferred directly from observations of binary systems and likely also from supernova explosions. In any binary system, there exist five orbital parameters (usually known as Keplerian parameters), which can be precisely measured, and can be related to the masses of the NS (M_p) and its companion (M_c) through the so-called mass function

$$f(M_p, M_c, i) = \frac{(M_c \sin i)^3}{(M_p + M_c)^2} = \frac{P_b v_1^3}{2\pi G}, \qquad (4.6)$$

where $v_1 = 2\pi a_1 \sin i / P_b$ is the projection of the orbital velocity of the NS along the line of sight, P_b is the orbital period, a_1 is the semi-major axis and i is the orbit inclination.

One cannot proceed further than Eq. (4.6) if only one mass function can be measured for a binary system, unless additional assumptions are made. Deviations from the Keplerian orbit due to general-relativity effects can be detected fortunately, and this allows to measure the masses of the two components of the binary system. An example of a high-precision mass measurement is that of the Hulse–Taylor binary pulsar [169] with measured masses $M_p = 1.4408 \pm 0.0003\, M_\odot$ and $M_c = 1.3873 \pm 0.0003\, M_\odot$. Other examples are those of the millisecond pulsars PSR J1614-2230 [170], PSR J0348+0432 [171] and the most recently observed PSR J0740+6620 [172] with masses $M_p = 1.928 \pm 0.017\, M_\odot$, $M_p = 2.01 \pm 0.04\, M_\odot$ and $M_p = 2.14^{+0.10}_{-0.09}\, M_\odot$, respectively. These are binary systems formed by a NS and a white dwarf. The measurement of these unusually high NS masses constitutes nowadays one of the most stringent astrophysical constraints on the nuclear EoS.

We now turn to the discussion of the radius. NS radii are very difficult to measure, mainly because NSs are very small objects and are very far away from us. Direct measurements of radii do not exist. However, a possible way to determine them is to use the thermal emission of low-mass X-ray binaries. The observed X-ray flux F and temperature T, assumed to be originated from a uniform black body, together with a determination of the distance D of the star can be used to obtain the radius through the following implicit relation

$$R = \sqrt{\frac{FD^2}{\sigma T^4}\left(1 - \frac{2GM}{R}\right)}, \tag{4.7}$$

where σ is the Stefan-Boltzmann constant and M the mass of the star. The major uncertainties in the measurement of the radius through Eq. (4.7) originate from the determination of the temperature, which requires the assumption of an atmospheric model, and the estimation of the distance of the star. The analysis of present observations from quiescent low-mass X-ray binaries is still controversial. Whereas the analysis of Ref. [95, 173] indicates NS radii in the range of $10.4 - 12.9$ km, that of Ref. [174, 175] points towards smaller radii of ~ 10 km or less.

The simultaneous measurement of both mass and radius of the same NS would provide the most definite observational constraint on the nuclear EoS. The Neutron Star Interior Composition Explorer (NICER) mission has recently provided a Bayesian parameter estimation of the mass and equatorial radius for the millisecond pulsars PSR J0030+0451 with $R(1.44^{+0.15}_{-0.14}\,M_\odot) = 13.02^{+1.24}_{-1.06}$ km [6] and $R(1.34^{+0.15}_{-0.16}\,M_\odot) = 12.71^{+1.14}_{-1.19}$ km [5] and for J0740+6620 with $R(2.08 \pm 0.07\,M_\odot) = 13.7^{+2.6}_{-1.5}$ km [8] and $R(2.072^{+0.067}_{-0.066}\,M_\odot) = 12.39^{+1.30}_{-0.98}$ km [7]. The difference between these estimates reflects the model dependence of the experimental analyses. The combined (strongly model-dependent) analysis of both pulsars together with GW170817 event observations [11, 12] yields improved limits on $R_{2.08} = 12.35 \pm 0.75$ km [8], but in particular on the radius $R_{1.4}$, namely 12.45 ± 0.65 km [8], $11.94^{+0.76}_{-0.87}$ km [9], and $12.33^{+0.76}_{-0.81}$ km or $12.18^{+0.56}_{-0.79}$ km [10] We want to stress here that the measurement performed by the NICER mission does not make use of Eq. (4.7) and constitutes the first model-independent one, since only the geometry of the hot spots of the NS and general-relativity effects enter the determination of the star radius.

These various astrophysical constraints are shown in Fig. 4.4 in comparison with the theoretical mass-radius relations obtained with different microscopic (solid lines) and phenomenological (broken lines) EoSs. Most of the models considered, except the soft microscopic UIX, predict values of the maximum mass larger than $2\,M_\odot$, therefore being compatible with current observational limits. Constraints arising from the GW170817 event [11] will be further discussed in the next subsection. Some recent theoretical analyses of this event indicate also an upper limit of the maximum mass of $\sim 2.2 - 2.3\,M_\odot$ [176–178], with which several of the microscopic and phenomenological EoSs considered would be compatible as well. The theoretical values are also compiled in Table 4.2. One may conclude that in particular the NICER+GW170817 constraints (shown as black horizontal bars in the figure) pose a challenge for the EoS and might for example exclude the CBF-EI, LS220 and Bsk26 of the current sample. Taking the $R_{2.08}$ limits for granted, only the V18, CBF-EI and Shen20 EoSs would be permissible. Future refinements of these radii limits will allow to pin down the EoS very well.

4.5.2 Gravitational waves

Gravitational waves originate from the oscillation modes of NSs or during the coalescence of a NS with a black hole or another NS. On August 17th 2017, the gravitational-wave signal from a binary NS merger (now known as GW170817), was detected for the first

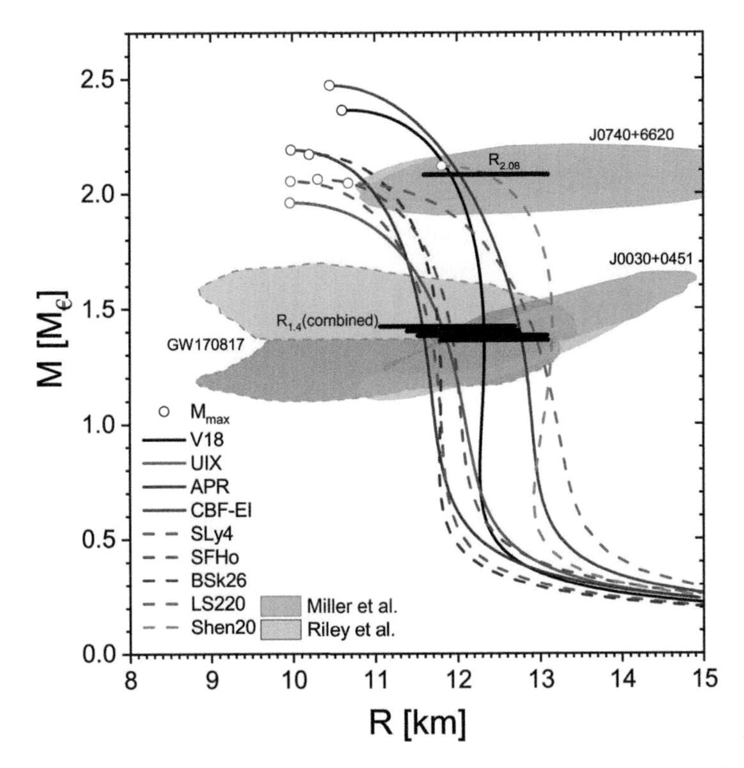

Figure 4.4 Mass-radius relations obtained with different microscopic (solid lines) and phenomenological (broken lines) EoSs. The mass of the most heavy pulsar PSR J0740+6620 [172] observed until now is also shown, together with the constraints from the GW170817 event [12], the mass-radius constraints of the NICER mission for J0030+0451 [5, 6] and J0740+6620 [7, 8] and their joint analysis with GW170817 [9, 10] (black bars).

time by the Advanced LIGO and VIRGO collaborations [11], inaugurating a new era in the observation of NSs. The analysis of GW170817 indicates that NS radii should be $R < 13$ km or even smaller than 12 km (some analysis suggest $R < 11$ km). That could put an additional stringent constraint on the nuclear EoS, since those predicting large radii would be excluded.

In the GW170817 event, the so-called tidal deformability Λ, or equivalently the tidal Love number $k_2 = \frac{3}{2}\beta^5\Lambda$ of a NS [179–181], was measured. This can provide valuable information and constraints on the related EoS, because it depends strongly on the compactness $\beta \equiv M/R$ of the object. In fact a universal relation of the individual tidal deformabilities of NSs as function of the stellar compactness was introduced in Ref. [182], and in Ref. [183] the following fit was proposed

$$\beta = 0.36 - 0.0355 \ln \Lambda + 0.000705 (\ln \Lambda)^2 \,, \tag{4.8}$$

which holds to within 6.5% for a large set of NS EoSs [183].

TABLE 4.2 Neutron stars astrophysical predictions of the considered EoS models.

Model	Ref.	$M_{\max}$ $[M_\odot]$	$\Lambda_{1.4}$	$R_{1.4}$ [km]	$R_{2.08}$ [km]
V18	[122]	2.36	440	12.3	11.9
UIX	[122]	1.96	309	11.8	-
CBF-EI	[65, 68]	2.47	501	12.8	12.0
APR	[123]	2.19	250	11.3	10.7
SLy4	[106]	2.05	287	11.7	-
BSk26	[107]	2.17	326	11.8	11.0
SFHo	[123]	2.06	334	11.9	-
LS220	[124]	2.04	540	12.9	-
Shen20	[94]	2.11	680	13.2	12.4
Exp.		$> 2.14^{+0.10}_{-0.09}$	70–580	11.1–12.7	11.6–13.1
Ref.		[172]	[12]	[9]	[7]

More specifically, the Love number can be computed in general relativity as

$$k_2 = \frac{8}{5}\frac{\beta^5 z}{F}\,, \qquad z \equiv (1-2\beta)^2[2 - y_R + 2\beta(y_R - 1)]\,,$$
$$F \equiv 6\beta(2-y_R) + 6\beta^2(5y_R - 8) + 4\beta^3(13 - 11y_R) + 4\beta^4(3y_R - 2) + 8\beta^5(1+y_R) + 3z\ln(1-2\beta)\,, \tag{4.9}$$

together with the TOV equations for pressure p and enclosed mass m of a static NS, along with the following first-order differential equation for y, being $y_R \equiv y(R)$, [184–186],

$$\frac{dp}{dr} = -\frac{m\varepsilon}{r^2}\frac{(1+p/\varepsilon)(1+4\pi r^3 p/m)}{1-2m/r}\,, \tag{4.10}$$

$$\frac{dm}{dr} = 4\pi r^2 \varepsilon\,, \tag{4.11}$$

$$\frac{dy}{dr} = -\frac{y^2}{r} - \frac{y-6}{r-2m} - rQ\,,$$
$$Q \equiv 4\pi\frac{(5-y)\varepsilon + (9+y)p + (\varepsilon+p)/c_s^2}{1-2m/r} - \left[\frac{2(m+4\pi r^3 p)}{r(r-2m)}\right]^2, \tag{4.12}$$

with the EoS $\varepsilon(p)$ as input, $c_s^2 = dp/d\varepsilon$ the squared speed of sound, and boundary conditions given by

$$[p, m, y](r=0) = [p_c, 0, 2]\,, \tag{4.13}$$

and the mass-radius relation $M(R)$ provided by the condition $p(R) = 0$ for varying p_c.

In the case of an asymmetric binary system, $(M,R)_1 + (M,R)_2$, with mass asymmetry $q = M_2/M_1$, and known chirp mass

$$M_c = \frac{(M_1 M_2)^{3/5}}{(M_1+M_2)^{1/5}}\,, \tag{4.14}$$

the effective tidal deformability is given by

$$\tilde{\Lambda} = \frac{16}{13} \frac{(1+12q)\Lambda_1 + (q+12)\Lambda_2}{(1+q)^5} \tag{4.15}$$

with

$$\frac{[M_1, M_2]}{M_c} = \frac{297}{250}(1+q)^{1/5}[q^{-3/5}, q^{2/5}] . \tag{4.16}$$

A value of $M_c = 1.186^{+0.001}_{-0.001}\, M_\odot$ was obtained from the analysis of the GW170817 event [11, 12, 187], corresponding to $M_1 = M_2 = 1.365\, M_\odot$ for a symmetric binary system, $q = M_2/M_1 = 1$, and a maximum asymmetry $(M_1, M_2) = (1.64, 1.15)\, M_\odot$ for $q = 0.73$. The limits $70 < \tilde{\Lambda} < 720$ were obtained from the phase-shift analysis of the observed signal. The GW170817 observation puts a strong constraint on the radius of a $1.4\, M_\odot$ NS. Requiring both NSs to have the same EoS leads to the constraints $70 < \Lambda_{1.4} < 580$ and $10.5 < R_{1.4} < 13.3\,\text{km}$ [12] for this radius. In Ref. [188], a general polytropic parametrization of the EoS compatible with perturbative QCD at very high density was used, and the constraint $\Lambda_{1.4} < 800$ yielded a similar upper limit $R_{1.4} < 13.4\,\text{km}$.

The high luminosity of the kilonova AT2017gfo following the NS merger event imposes also a *lower* limit on the average tidal deformability, $\tilde{\Lambda} > 400$, which was deduced in order to justify the amount of ejected material being heavier than $0.05\, M_\odot$ [189]. This lower limit, which was used in Refs. [190–193] in order to constrain the EoS, could indicate that $R_{1.4} \gtrsim 12\,\text{km}$. However, this constraint of $\tilde{\Lambda}$ has to be taken with great care and, in fact, it has been recently revised to $\tilde{\Lambda} \geq 300$ [194, 195], but considered of limited significance in Ref. [196]. We thus notice that the determination of the average tidal deformability of the binary NS system GW170817 has imposed constraints on the NS radii, to lie between about 12 and 13 kilometers [197]. This is complementary to the mass-radius measurement by NICER mentioned before and contributes to selecting the suitable EoS.

Fig. 4.5 shows Λ as a function of M or R for the various EoSs considered in this chapter. The shaded areas represent the observational limits derived in Refs. [12], and thus visualise the constraints on the radius $R_{1.4}$ discussed above. All selected EoSs apart from marginally Shen20 fulfil these constraints. The exact theoretical results agree fairly well with the universal relation for $\Lambda(M/R)$, Eq. (4.8).

4.6 HYPERONS, QUARKS AND OTHER EXOTIC COMPONENTS OF THE NEUTRON STAR EOS

The core of NSs has been traditionally modeled as a uniform fluid of neutron-rich nuclear matter in β equilibrium. However, due to the large value of the density, new hadronic degrees of freedom are expected to appear in addition to nucleons. Hyperons, baryons with a strangeness content, are an example of these new degrees of freedom [198]. Contrary to terrestrial conditions, where hyperons are unstable and decay into nucleons through weak interactions, matter in NSs maintains the weak equilibrium between the decays and their inverse capture processes. Hyperons may appear in the inner core of NSs at densities of about 2–3 times saturation density. At such densities, the nucleon chemical potential is large enough to make the conversion of nucleons into hyperons energetically favourable. This conversion relieves the Fermi pressure exerted by the baryons, and makes the EoS

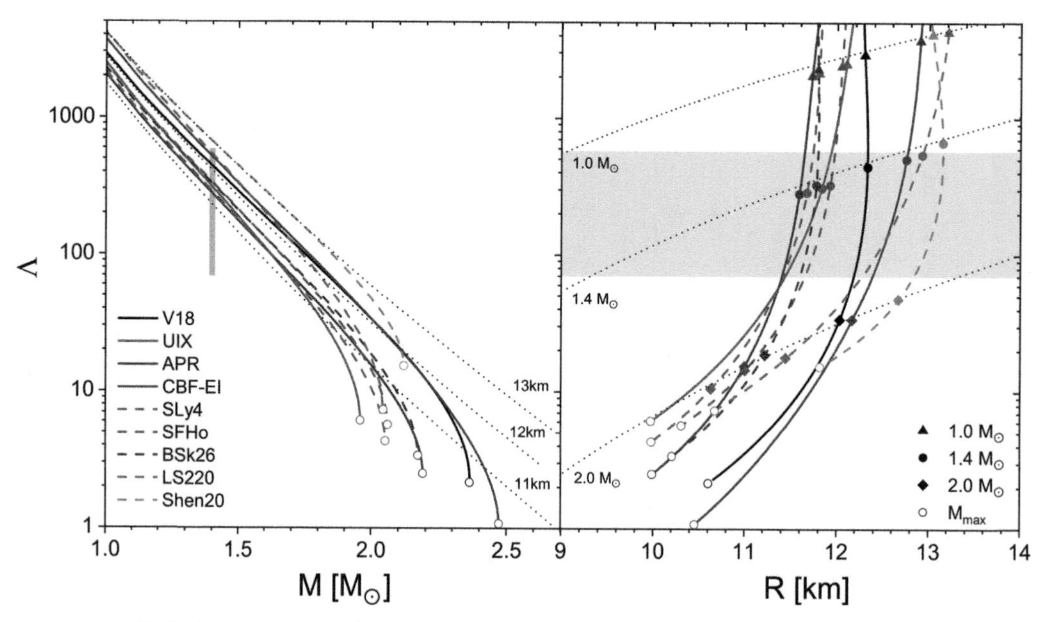

Figure 4.5 Relations between Λ, M and R for a single NS with different EoSs. The shaded grey areas are constrained by the interpretation of the GW170817 event as a symmetric NS merger [12]. The dotted curves show the predictions according to the universal fit Eq. (4.8), in comparison with the exact results (markers). See text for details.

softer. As a consequence, the maximum mass of the star is substantially reduced to values not compatible with recent observations.

This problem, known in the literature as the hyperon puzzle [199], is subject of current intensive research. Its solution requires a mechanism (or mechanisms) that could eventually provide the additional repulsion needed to make the EoS stiffer and therefore the maximum mass compatible with the current observational limits. Three different mechanisms that could provide such additional repulsion have been proposed: (i) more repulsive hyperon–hyperon interactions in relativistic density functional methods driven by either repulsive vector meson exchanges or less attractive scalar σ meson exchange, (ii) repulsive hyperonic three-body forces and (iii) a phase transition to deconfined quark matter at densities below the hyperon threshold.

The first solution, more repulsive hyperon–hyperon interactions, has been mainly explored in the context of RMF models [200–203] and is based on the well-known fact that, in a meson-exchange model of nuclear forces, vector mesons generate repulsion at short distances. If the interaction of hyperons with vector mesons is repulsive enough, then it could provide the required stiffness to explain the current pulsar mass observations. However, in order to be consistent with experimental data of hypernuclei that indicate that at least the ΛN interaction is attractive, the repulsion in the hyperonic sector is included in these models only in the hyperon-hyperon interaction through the exchange of the hidden-strangeness ϕ vector meson coupled only to the hyperons. In this way, the onset of hyperons is shifted to higher densities and NSs with maximum masses larger than $2\,M_\odot$ and a significant hyperon fraction can be successfully obtained.

The second solution proposed is based on the well-known fact that the three-nucleon forces in the nuclear Hamiltonian are fundamental ingredients that are needed to reproduce properly the properties of few-nucleon systems as well as the empirical saturation point of symmetric nuclear matter in non-relativistic many-body approaches. Therefore, it seems natural to suggest that TBFs involving one or more hyperons (i.e., NNY, NYY and YYY) could provide additional repulsion at high densities (already established in the case of three-nucleon forces) that can solve the hyperon puzzle. Indeed if TBFs involving hyperons are repulsive enough, they can make the EoS stiffer at high densities and, therefore, make the maximum mass of the star compatible with the recent observations. This idea has been explored by a number of authors [204–212]. However, due to the large uncertainties involved in the physics of hyperonic TBFs, no general consensus has been reached regarding their role in solving the hyperon puzzle.

Several authors have suggested that an early phase transition from hadronic matter to deconfined quark matter at densities below the hyperon threshold could provide a solution to the hyperon puzzle [213–216]. Therefore, massive NSs could actually be hybrid stars with a stiff quark-matter core. The question that arises then is whether quarks can provide sufficient repulsion to produce a $2\ M_{\odot}$ NS. To achieve this, quark matter should have two important and necessary features: (i) a significant overall quark-quark repulsion to maintain a stiff EoS, for example, in vector channels and (ii) a strong enough attraction in certain channels which leads to color superconductivity needed to make the deconfined quark matter phase energetically favourable over the hadronic one. Several models of hybrid stars with the necessary properties to generate $2\ M_{\odot}$ NSs have been proposed. Conversely, the observation of $2\ M_{\odot}$ NSs may also help to impose important constraints on the models of hybrid and strange stars with a quark matter core, and improve our present understanding of the hadron-quark phase transition.

An alternative way to circumvent the hyperon puzzle is to invoke the appearance of other hadronic degrees of freedom such as for instance the Δ isobar or meson condensates that shift the onset of hyperons to higher densities. The Δ isobar is often neglected in the studies of NSs because its threshold density was found to be higher than the typical densities prevalent in cores of NSs. Nevertheless, it was shown [217] that the onset of the Δ depends crucially on the density dependence of the slope of the nuclear symmetry energy, L. By using a state-of-the-art EoS and recent experimental constraints on L, the authors of Ref. [217] showed that the Δ isobar could actually appear before the hyperons in NS interiors. However, they found also that, as soon as the Δ is present, the EoS, as in the case of hyperons, becomes considerably softer and, consequently, the maximum mass is reduced to values below the current observational limit. Thus, the hyperon puzzle is effectively replaced with the so-called Δ puzzle.

The possible existence of a Bose-Einstein condensate of negative kaons in the inner core of NSs has also been extensively considered in the literature (see, e.g. [198] and references therein). As the density of stellar matter increases, the K^- chemical potential μ_{K^-} is lowered by the attractive vector meson field originating from dense nucleonic matter. When μ_{K^-} becomes smaller than the electron chemical potential μ_e, the process $e^- \to K^- + \nu_e$ becomes energetically possible. The critical density for this process was calculated to be in the range of 2.5–5 times saturation density. However, as in the case of the Δ, the appearance of the kaon condensation induces also a strong softening of the EoS and consequently leads to a reduction of the maximum mass to values below the current observational limits.

We finish this section by mentioning that hyperons can strongly influence not only the masses and radii of NSs but also their thermal evolution and gravitational instabilities. These aspects of NS physics are very sensitive to their composition, and therefore to the hyperonic content of NS interiors. The cooling of NSs is affected by the presence of hyperons because they modify the neutrino emissivity of dense matter and allow for fast cooling mechanisms [218, 219]. Furthermore, the emission of gravitational waves in hot and rapidly rotating NSs due to the so-called $r-$mode instability is affected by hyperons, because they dominate the bulk viscosity of matter as soon as they appear in the NS interior.

4.7 SUMMARY

In this chapter, we have discussed the various theoretical approaches for the nuclear equation of state and the constraints imposed on it by nuclear structure experiments and in particular recent astrophysical observations related to neutron star structure and gravitational-wave emission from binary neutron star mergers.

While the former provides information only on the low-density domain of the EoS, the latter constrain also the high-density part, which determines features like the NS maximum mass and radius. In particular, the recent first combined analyses providing simultaneous mass and radius constraints of two pulsars prove extremely valuable for this purpose and have the potential to single out the realistic theoretical models in the near future.

Our discussion was limited to purely baryonic models (nucleons and hyperons) of the EoS. For the inclusion of quark degrees of freedom, see the contribution of T. Kojo to this volume. Open problems in the current theoretical treatment of the EoS include

- The role and computation of TBF in the EoS;
- The solution of the "hyperon puzzle" in microscopic approaches and the more general question about the treatment of non-nucleonic degrees of freedom;
- The conditions and realisation of the phase transition to quark matter;
- The importance of thermal effects for the EoS;
- The NS cooling mechanisms, in particular involving pairing, consistent with the EoS.

Answers to those questions will surely improve our understanding of the various fundamental phenomena related to the astrophysics of NSs, their internal composition, the explosion mechanisms of core-collapse supernovae, the mass threshold for BH formation, the dynamics of NS binary mergers and the nucleosynthesis of heavy elements. Therefore, the current theoretical uncertainties will require significant efforts to be undertaken in these directions.

We expect that in the near future a new generation of telescopes and projects such as eXTP [220] and advanced GW detectors such as LIGO and Virgo can provide more and more precise data that can significantly contribute to probe the internal structure of compact objects, thus improving the different theoretical models.

Bibliography

[1] M. Oertel, M. Hempel, T. Klähn, and S. Typel. *Rev. Mod. Phys.*, 89(1):015007, 2017.

[2] G. F. Burgio and A. F. Fantina. *Astrophys. Space Sci. Libr.*, 457:255, 2018.

[3] G. F. Burgio, H.-J. Schulze, I. Vidaña, and J.-B. Wei. *Prog. Part. Nucl. Phys.*, 120:103879, 2021.

[4] M. Baldo. *Nuclear Methods and The Nuclear Equation of State*. World Scientific, Singapore, 1999.

[5] T. E. Riley and others. *Astrophys. J. Lett.*, 887:L21, 2019.

[6] M. C. Miller and others. *Astrophys. J. Lett.*, 887:L24, 2019.

[7] T. E. Riley et al. *Astrophys. J. Lett.*, 918(2):L27, 2021.

[8] M. C. Miller et al. *Astrophys. J. Lett.*, 918(2):L28, 2021.

[9] P. T. H. Pang, I. Tews, M. W. Coughlin, M. Bulla, C. Van Den Broeck, and T. Dietrich. *Astrophys. J.*, 922(1):14, 2021.

[10] G. Raaijmakers et al. *Astrophys. J. Lett.*, 918(2):L29, 2021.

[11] B. P. Abbott et al. *Phys. Rev. Lett.*, 119:161101, 2017.

[12] B. P. Abbott et al. *Phys. Rev. Lett.*, 121:161101, 2018.

[13] B. P. Abbott et al. *Astrophys. J. Lett.*, 892(1):L3, 2020.

[14] L. Baiotti. *Prog. Part. Nucl. Phys.*, 109:103714, 2019.

[15] J.R. Stone and P.-G. Reinhard. *Prog. Part. Nucl. Phys.*, 58(2):587–657, 2007.

[16] T.H.R. Skyrme. *Nucl. Phys.*, 9(4):615–634, 1958.

[17] D. Vautherin and D. M. Brink. *Phys. Rev. C*, 5:626–647, 1972.

[18] P Quentin and H Flocard. *Annu. Rev. Nucl. Part. Sc.*, 28(1):523–594, 1978.

[19] M. Bender, P.-H. Heenen, and P.-G. Reinhard. *Rev. Mod. Phys.*, 75:121–180, 2003.

[20] J Erler, P Klüpfel, and P-G Reinhard. *J. Phys. G*, 38(3):033101, 2011.

[21] J. Boguta and A.R. Bodmer. *Nucl. Phys. A*, 292(3):413–428, 1977.

[22] B. D. Serot and J. D. Walecka. *Adv. Nucl. Phys.*, 16:1–327, 1986.

[23] M. M. Nagels, Th. A. Rijken, and J. J. de Swart. *Phys. Rev. D*, 17:768–776, 1978.

[24] R. Machleidt. *Advances in Nuclear Physics*, pages 189–376. Springer US, 1989.

[25] B. Holzenkamp, K. Holinde and J. Speth. *Nucl. Phys. A*, 500:485, 1989.

[26] P. M. M. Maessen, T. A. Rijken and J. J. de Swart. *Phys. Rev. C*, 40:226, 1989.

[27] Th. A. Rijken, V. G. J. Stoks and Y. Yamamoto. *Phys. Rev. C*, 59:21, 1999.

[28] V. G. J. Stoks and Th. A. Rijken. *Phys. Rev. C*, 59:3009, 1999.

[29] J. Haidenbauer and U.-G. Meissner. *Phys. Rev. C*, 72:044005, 2005.

[30] S. Weinberg. *Phys. Lett. B*, 251:288–292, 1990.

[31] S. Weinberg. *Nucl. Phys. B*, 363:3–18, 1991.

[32] D. R. Entem and R. Machleidt. *Phys. Rev. C*, 68(4):041001, 2003.

[33] E. Epelbaum, H.-W. Hammer, and U.-G. Meißner. *Rev. Mod. Phys.*, 81:1773–1825, 2009.

[34] J. P. Jeukenne, A. Lejeune, and C. Mahaux. *Phys. Rep.*, 25(2):83–174, 1976.

[35] J. Goldstone. *Proceedings of the Royal Society of London Series A*, 239(1217): 267–279, 1957.

[36] B. Ter Haar and R. Malfliet. *Phys. Rev. Lett.*, 56:1237–1240, 1986.

[37] R. Brockmann and R. Machleidt. *Phys. Rev. C*, 42:1965–1980, 1990.

[38] A. Akmal, V. R. Pandharipande, and D. G. Ravenhall. *Phys. Rev. C*, 58:1804–1828, 1998.

[39] S. Fantoni and A. Fabrocini. volume 510 of *Lecture Notes in Physics, Berlin Springer Verlag*, page 119, 1998.

[40] L.P. Kadanoff and G. Baym. *Quantum Statistical Mechanics*. W.A. Benjamin Inc., 1962.

[41] W. H. Dickhoff and D. Van Neck. *Many-Body Theory Exposed! Propagator Description of Quantum Mechanics in Many-Body Systems (2ND Edition)*. World Scientific Press, 2008.

[42] S. K. Bogner, R. J. Furnstahl, and A. Schwenk. *Prog. Part. Nucl. Phys.*, 65:94–147, 2010.

[43] R. B. Wiringa, S. C. Pieper, J. Carlson, and V. R. Pandharipande. *Phys. Rev. C*, 62:014001, 2000.

[44] J. Carlson, J. Morales, V. R. Pandharipande, and D. G. Ravenhall. *Phys. Rev. C*, 68(2):025802, 2003.

[45] S. Gandolfi, A. Y. Illarionov, K. E. Schmidt, F. Pederiva, and S. Fantoni. *Phys. Rev. C*, 79(5):054005, 2009.

[46] M. Baldo and C. Maieron. *J. Phys. G*, 34:R243, 2007.

[47] H. Q. Song, M. Baldo, G. Giansiracusa, and U. Lombardo. *Phys. Rev. Lett.*, 81:1584–1587, 1998.

[48] J.-J. Lu, Z.-H. Li, C.-Y. Chen, M. Baldo, and H.-J. Schulze. *Phys. Rev. C*, 96(4):044309, 2017.

[49] J.-J. Lu, Z.-H. Li, C.-Y. Chen, M. Baldo, and H.-J. Schulze. *Phys. Rev. C*, 98(6):064322, 2018.

[50] R. Schiavilla, V. R. Pandharipande, and R. B. Wiringa. *Nucl. Phys. A*, 449:219–242, 1986.

[51] B. S. Pudliner, V. R. Pandharipande, J. Carlson, and R. B. Wiringa. *Phys. Rev. Lett.*, 74:4396–4399, 1995.

[52] W. Zuo, A. Lejeune, U. Lombardo, and J. F. Mathiot. *Nucl. Phys. A*, 706:418–430, 2002.

[53] W. Zuo, A. Lejeune, U. Lombardo, and J. F. Mathiot. *EPJA*, 14:469–475, 2002.

[54] Z. H. Li, U. Lombardo, H.-J. Schulze, and W. Zuo. *Phys. Rev. C*, 77:034316, 2008.

[55] Z. H. Li and H.-J. Schulze. *Phys. Rev. C*, 78:028801, 2008.

[56] R. B. Wiringa, V. G. J. Stoks, and R. Schiavilla. *Phys. Rev. C*, 51:38–51, 1995.

[57] V. G. J. Stoks, R. A. M. Klomp, C. P. F. Terheggen, and J. J. de Swart. *Phys. Rev. C*, 49:2950–2962, 1994.

[58] J.-B. Wei, J.-J. Lu, G. F. Burgio, Z.-H. Li, and H.-J. Schulze. *EPJA*, 56(2):63, 2020.

[59] G. E. Brown, W. Weise, G. Baym, and J. Speth. *Comm. Nucl. Part. Phys.*, 17:39–62, 1987.

[60] Z. H. Li, U. Lombardo, H.-J. Schulze, W. Zuo, L. W. Chen, and H. R. Ma. *Phys. Rev. C*, 74(4):047304, 2006.

[61] H. Müther, F. Sammarruca, and Z. Ma. *Int. J. Mod. Phys. E*, 26(03):1730001, 2017.

[62] V. R. Pandharipande and R. B. Wiringa. *Rev. Mod. Phys.*, 51:821–860, 1979.

[63] S. Fantoni and S. Rosati. *Nuovo Cimento A Serie*, 25:593–615, 1975.

[64] J. Carlson, S. Gandolfi, F. Pederiva, S. C. Pieper, R. Schiavilla, K. E. Schmidt, and R. B. Wiringa. *Rev. Mod. Phys.*, 87:1067–1118, 2015.

[65] O. Benhar and A. Lovato. *Phys. Rev. C*, 96:054301, 2017.

[66] A. Lovato, O. Benhar, S. Fantoni, A. Yu. Illarionov, and K. E. Schmidt. *Phys. Rev. C*, 83:054003, 2011.

[67] R. B. Wiringa and S. C. Pieper. *Phys. Rev. Lett.*, 89(18):182501, 2002.

[68] G. Camelio, A. Lovato, L. Gualtieri, O. Benhar, J. A. Pons, and V. Ferrari. *Phys. Rev. D*, 96:043015, 2017.

[69] T. Duguet. In *Lecture Notes in Physics, Berlin Springer Verlag*, volume 879, page 293, 2014.

[70] T. H. R. Skyrme. *Philosophical Magazine*, 1:1043–1054, 1956.

[71] M. Bender, K. Bennaceur, T. Duguet, P.-H. Heenen, T. Lesinski, and J. Meyer. *Phys. Rev. C*, 80(6):064302, 2009.

[72] N. Chamel, S. Goriely, and J. M. Pearson. *Phys. Rev. C*, 80(6):065804, 2009.

[73] J. Margueron and H. Sagawa. *J. Phys. G*, 36(12):125102, 2009.

[74] S. Goriely, N. Chamel, and J. M. Pearson. *Phys. Rev. C*, 82(3):035804, 2010.

[75] N. Chamel, J. M. Pearson, A. F. Fantina, C. Ducoin, S. Goriely, and A. Pastore. *Acta Physica Polonica B*, 46:349, 2015.

[76] F. Douchin and P. Haensel. *Astron. Astrophys.*, 380:151–167, 2001.

[77] R. B. Wiringa, V. Fiks, and A. Fabrocini. *Phys. Rev. C*, 38:1010–1037, 1988.

[78] N. Chamel, A. F. Fantina, J. M. Pearson, and S. Goriely. *Phys. Rev. C*, 84:062802, 2011.

[79] J. M. Pearson, N. Chamel, S. Goriely, and C. Ducoin. *Phys. Rev. C*, 85:065803, 2012.

[80] A. F. Fantina, N. Chamel, J. M. Pearson, and S. Goriely. *Astron. Astrophys.*, 559:A128, 2013.

[81] A. Y. Potekhin, A. F. Fantina, N. Chamel, J. M. Pearson, and S. Goriely. *Astron. Astrophys.*, 560:A48, 2013.

[82] M. Baldo, L. Robledo, P. Schuck, and X. Viñas. *J. Phys. G*, 37(6):064015, 2010.

[83] M. Baldo, L. M. Robledo, P. Schuck, and X. Viñas. *Phys. Rev. C*, 87(6):064305, 2013.

[84] B. K. Sharma, M. Centelles, X. Viñas, M. Baldo, and G. F. Burgio. *Astron. Astrophys.*, 584:A103, 2015.

[85] B. D. Serot. *Rep. Prog. Phys.*, 55:1855–1946, 1992.

[86] T. Nikšić, D. Vretenar, and P. Ring. *Prog. Part. Nucl. Phys.*, 66:519–548, 2011.

[87] G. A. Lalazissis, J. König, and P. Ring. *Phys. Rev. C*, 55:540–543, 1997.

[88] B. G. Todd-Rutel and J. Piekarewicz. *Phys. Rev. Lett.*, 95(12):122501, 2005.

[89] S. Typel. *Phys. Rev. C*, 71(6):064301, 2005.

[90] X. Roca-Maza, X. Viñas, M. Centelles, P. Ring, and P. Schuck. *Phys. Rev. C*, 84(5):054309, 2011.

[91] J. M. Lattimer and D. F. Swesty. *Nucl. Phys. A*, 535:331–376, 1991.

[92] H. Shen, H. Toki, K. Oyamatsu, and K. Sumiyoshi. *Nucl. Phys. A*, 637:435–450, 1998.

[93] Y. Sugahara and H. Toki. *Nucl. Phys. A*, 579:557–572, 1994.

[94] H. Shen, F. Ji, J. Hu, and K. Sumiyoshi. *Astrophys. J.*, 891(2):148, 2020.

[95] A. W. Steiner, J. M. Lattimer and E. F. Brown. *Astrophys. J. Lett.*, 765:L5, 2013.

[96] M. Hempel and J. Schaffner-Bielich. *Nucl. Phys. A*, 837:210–254, 2010.

[97] A.S. Botvina and I.N. Mishustin. *Nucl. Phys. A*, 843(1):98 – 132, 2010.

[98] A. R. Raduta and F. Gulminelli. *Phys. Rev. C*, 82(6):065801, 2010.

[99] S. I. Blinnikov, I. V. Panov, M. A. Rudzsky, and K. Sumiyoshi. *Astron. Astrophys.*, 535:A37, 2011.

[100] N. Buyukcizmeci et al. *Nucl. Phys. A*, 907:13–54, 2013.

[101] P. Grangé, A. Lejeune, M. Martzolff, and J.-F. Mathiot. *Phys. Rev. C*, 40:1040–1060, 1989.

[102] Z.H. Li, U. Lombardo, H.-J. Schulze, and W. Zuo. *Phys. Rev. C*, 77:034316, 2008.

[103] F. Li, J.-J. Lu, Z.-H. Li, C.-Y. Chen, G. F. Burgio, and H.-J. Schulze. *Phys. Rev. C*, 103(2):024307, 2021.

[104] H.-M. Liu, J. Zhang, Z.-H. Li, J.-B. Wei, G. F. Burgio, and H.-J. Schulze. *Phys. Rev. C*, 106(2):025801, 2022.

[105] M. Baldo, I. Bombaci, and G. F. Burgio. *Astron. Astrophys.*, 328:274–282, 1997.

[106] E. Chabanat, P. Bonche, P. Haensel, J. Meyer, and R. Schaeffer. *Nucl. Phys. A*, 635:231–256, 1998.

[107] S. Goriely, N. Chamel, and J. M. Pearson. *Phys. Rev. C*, 88:024308, 2013.

[108] D. Miśkowiec et al. *Phys. Rev. Lett.*, 72:3650–3653, 1994.

[109] P. Danielewicz, R. Lacey, and W. G. Lynch. *Science*, 298:1592, 2002.

[110] T. Krüger, I. Tews, K. Hebeler, and A. Schwenk. *Phys. Rev. C*, 88:025802, 2013.

[111] C. Drischler, R. J. Furnstahl, J. A. Melendez, and D. R. Phillips. *Phys. Rev. Lett.*, 125:202702, 2020.

[112] C. Drischler, K. Hebeler, and A. Schwenk. *Phys. Rev. C*, 93(5):054314, 2016.

[113] P. Danielewicz and J. Lee. *Nucl. Phys. A*, 922:1, 2014.

[114] J. L. Ritman et al. *Z. Phys. A*, 352:355–357, 1995.

[115] P. Russotto et al. *Phys. Rev. C*, 94(3):034608, 2016.

[116] G. Audi, A. H. Wapstra, and C. Thibault. *Nucl. Phys. A*, 729:337–676, 2003.

[117] H. De Vries, C. W. De Jager and C. De Vries. *Atom. Data Nucl. Data Tabl.*, 36:495, 1987.

[118] G. Coló, N. Van Giai, J. Meyer, K. Bennaceur and P. Bonche. *Phys. Rev. C*, 70:024307, 2004.

[119] J. Piekarewicz. *Phys. Rev. C*, 69:041301, 2004.

[120] E. Khan, J. Margueron and I. Vidaña. *Phys. Rev Lett.*, 109:092501, 2012.

[121] C. Fuchs, A. Faessler, E. Zabrodin and Yu-Ming Zheng. *Phys. Rev Lett.*, 86:1974, 2001.

[122] J.-J. Lu, Z.-H. Li, G. F. Burgio, A. Figura, and H.-J. Schulze. *Phys. Rev. C*, 100:054335, 2019.

[123] https://compose.obspm.fr.

[124] http://www.astro.sunysb.edu/dswesty/lseos.html.

[125] J. Margueron, R. Hoffmann Casali, and F. Gulminelli. *Phys. Rev. C*, 97(2):025805, 2018.

[126] S. Shlomo, V. M. Kolomietz, and G. Colò. *EPJA*, 30:23–30, 2006.

[127] J. Piekarewicz. *J. Phys. G*, 37(6):064038, 2010.

[128] B.-A. Li and X. Han. *Phys. Lett. B*, 727:276–281, 2013.

[129] I. Tews, J. M. Lattimer, A. Ohnishi, and E. E. Kolomeitsev. *Astrophys J.*, 848:105, 2017.

[130] B. T. Reed, F. J. Fattoyev, C. J. Horowitz and J. Piekarewicz. *Phys. Rev. Lett.*, 126:172502, 2021.

[131] R. Essick, I. Tews, Ph. Landry, and A. Schwenk. *Phys. Rev. Lett.*, 127(19):192701, 2021.

[132] U. Garg and others. *Nucl. Phys, A*, 788:36, 2007.

[133] A. Klimkiewicz and others. *Phys. Rev. C*, 76:051603(R), 2007.

[134] A. Carbone, G. Col'o, A. Bracco, L.-G. Cao, P.F. Bortignon, F. Camera and O. Wieland. *Phys. Rev. C*, 81:041301(R), 2010.

[135] L.-W. Chen, C. M. Ko and B.-A. Li. *Phys. Rev. Lett.*, 94:032701, 2005.

[136] D. V. Shetty, S. J. Yennello and G. A. Souliotis. *Phys. Rev. C*, 76:024606, 2007.

[137] B. A. Brown. *Phys. Rev. Lett.*, 85:5296, 2000.

[138] S. Typel and A. Brown. *Phys. Rev. C*, 64:027302, 2001.

[139] C. J. Horowitz, S. J. Pollock, P. A. Souder and R. Michaels. *Phys. Rev. C*, 63:025501, 2001.

[140] B. A. Brown, G. Shen, G. C. Hillhouse, J. Meng and A. Trzcińska. *Phys. Rev. C*, 76:034305, 2007.

[141] M. Centelles, X. Roca–Maza, X. Viñas and M. Warda. *Phys. Rev. Lett.*, 102:122502, 2009.

[142] M. Warda , X. Viñas, X. Roca–Maza and M. Centelles. *Phys. Rev. C*, 80:024316, 2010.

[143] B.-A. Li, G.-C. Yong and W. Zuo. *Phys. Rev. C*, 71:014608, 2005.

[144] C. Fuchs. *Prog. Part. Nucl. Phys.*, 56:1–103, 2006.

[145] J. Lattimer and Y. Lim. *Astrophys. J.*, 771:51, 2013.

[146] L. W. Chen, C. M. Ko, B. A. Li, and J. Xu. *Phys. Rev. C*, 82:024321, 2010.

[147] M. B. Tsang, Y. Zhang, P. Danielewicz, M. Famiano, Z. Li, W. G. Lynch, and A. W. Steiner. *Phys. Rev. Lett.*, 102:122701, 2009.

[148] X. Roca–Maza, X. Viñas, M. Centelles, B. K. Agrawal, G. Colò, N. Paar, J. Piekarewicz, and D. Vretenar. *Phys. Rev. C*, 92:064304, 2015.

[149] J. Hu, S. Bao, Y. Zhang, K. Nakazato, K. Sumiyoshi and H. Shen. *Prog. Theor. Exp. Phys.*, 043D01, 2020.

[150] T. Zhao and J. M. Lattimer. *Phys. Rev. D*, 98:063020, 2018.

[151] C. Y. Tsang, M. B. Tsang, P. Danielewiczm F. J. Fattoyev and W. G. Lynch. *Phys. Lett. B*, 796:1, 2019.

[152] B.-A. Li and M. Magno. *Phys. Rev. C*, 102:045807, 2020.

[153] A. Burrows and J. M. Lattimer. *Astrophys. J.*, 307:178, 1986.

[154] M. Prakash, I. Bombaci, M. Prakash, P. J. Ellis, R. Knorren and J. M. Lattimer. *Phys. Rep.*, 280:1, 1997.

[155] J. A. Pons, S. Reddy, M. Prakash, J. M. Lattimer, and J. A. Miralles. *Astrophys. J.*, 513(2):780–804, 1999.

[156] L. Baiotti and L. Rezzolla. *Rep. Prog. Phys.*, 80(9):096901, 2017.

[157] V. Paschalidis and N. Stergioulas. *Living Rev. Relativ.*, 20(1), 2017.

[158] A. Bauswein, H.-T. Janka, and R. Oechslin. *Phys. Rev. D*, 82:084043, 2010.

[159] K. Hotokezaka, K. Kiuchi, K. Kyutoku, T. Muranushi, Y. Sekiguchi, M. Shibata, and K. Taniguchi. *Phys. Rev. D*, 88:044026, 2013.

[160] A. Endrizzi, D. Logoteta, B. Giacomazzo, I. Bombaci, W. Kastaun, and R. Ciolfi. *Phys. Rev. D*, 98:043015, Aug 2018.

[161] G. Camelio, T. Dietrich, M. Marques, and S. Rosswog. *Phys. Rev. D*, 100(12):123001, 2019.

[162] S. Typel and H. H. Wolter. *Nucl. Phys. A*, 656(3):331–364, 1999.

[163] G. F. Burgio and H.-J. Schulze. *Astron. Astrophys.*, 518:A17, 2010.

[164] H. Togashi, K. Nakazato, Y. Takehara, S. Yamamuro, H. Suzuki, and M. Takano. *Nucl. Phys. A*, 961:78–105, 2017.

[165] A. R. Raduta, F. Nacu, and M. Oertel. *EPJA*, 57(12), dec 2021.

[166] A. Lejeune, P. Grange, M. Martzolff, and J. Cugnon. *Nucl. Phys. A*, 453(2):189–219, 1986.

[167] C. Bloch and C. De Dominicis. *Nucl. Phys. A*, 7:459–479, 1958.

[168] A. Rios, A. Polls, A. Ramos, and H. Müther. *Phys. Rev. C*, 78(4):044314, 2008.

[169] R. A. Hulse and J. H. Taylor. *Astrophys, J. Lett.*, 195:L51, 1975.

[170] P. Demorest and others. *Nature*, 467:1081, 2010.

[171] J. Antoniadis and others. *Science*, 340:1233232, 2013.

[172] H. T. Cromartie and others. *Nature Astronomy*, 4:72, 2020.

[173] J. M. Lattimer and A. W. Steiner. *Astrophys. J.*, 784:123, 2014.

[174] S. Guillot, M. Servillat, N. A. Webb and R. E. Rutledge. *Astrophys. J.*, 772:7, 2013.

[175] S. Guillot and R. E. Rutledge. *Astrophys. J. Lett.*, 796:7, 2014.

[176] M. Shibata, S. Fujibayashi, K. Hotokezakam, K. Kiuchi, K. Kyutoku, Y. Sekiguchi, and M. Tanaka. *Phys. Rev. D*, 96:123012, 2017.

[177] B. Margalit and B. D. Metzger. *Astrophys. J.*, 850:L19, 2017.

[178] L. Rezzolla, E. R. Most, and L. R. Weih. *Astrophys. J.*, 852:L25, 2018.

[179] T. Hinderer. *Astrophys. J.*, 677:1216, 2008.

[180] T. Hinderer. *Astrophys. J.*, 697:964, 2009.

[181] T. Hinderer, B. D. Lackey, R. N. Lang and J. S. Read. *Phys. Rev. D*, 81:123016, 2010.

[182] K. Yagi and N. Yunes. *Phys. Rev. D*, 88(2):023009, 2013.

[183] K. Yagi and N. Yunes. *Phys. Rep.*, 681:1–72, 2017.

[184] J. M. Lattimer and M. Prakash. *Phys. Rep.*, 442(1):109–165, 2007.

[185] S. Postnikov, M. Prakash, and J. M. Lattimer. *Phys. Rev. D*, 82:024016, 2010.

[186] J. M. Lattimer and M. Prakash. *Phys. Rep.*, 621:127–164, 2016.

[187] B. P. Abbott et al. *Phys. Rev. X*, 9:011001, 2019.

[188] A. Kurkela E. Annala, T. Gorda and A. Vuorinen. *Phys. Rev. Lett.*, 120:172703, 2018.

[189] D. Radice, A. Perego, F. Zappa, and S. Bernuzzi. *Astrophys. J.*, 852(2):L29, 2018.

[190] E. R. Most, L. R. Weih, L. Rezzolla, and J. Schaffner–Bielich. *Phys. Rev. Lett.*, 120:261103, 2018.

[191] Y. Lim and J. W. Holt. *Phys. Rev. Lett.*, 121:062701, 2018.

[192] T. Malik, N. Alam, M. Fortin, C. Providência, B. K. Agrawal, T. K. Jhan, B. Kumar and S. K. Patra. *Phys. Rev. C*, 98:035804, 2018.

[193] G. F. Burgio, A. Drago, G. Pagliara, H.-J. Schulze and J. B. Wei. *Astrophys. J.*, 860:139, 2018.

[194] D. Radice and L. Dai. *EPJA*, 55:50, 2019.

[195] M. W. Coughlin, T. Dietrich, B. Margalit, and B. D. Metzger. *Mon. Not. R. Astron. Soc.*, page L131, 2019.

[196] K. Kiuchi, K. Kyutoku, M. Shibata, and K. Taniguchi. *Astrophys. J. Lett.*, 876:L31, 2019.

[197] J. B. Wei, G. F. Burgio, H. Chen and H.-J. Schulze. *J. Phys. G*, 46:034001, 2019.

[198] I. Vidaña. *Proc. Roy. Soc. Lond. A*, 474:33, 2018.

[199] D. Chatterjee and I. Vidaña. *Eur. Phys. J. A*, 52:29, 2016.

[200] I. Bednarek, P. Haensel, J. L. Zdunik, M. Bejger and R. Mańka. *Astron. Astrophys.*, 543:A157, 2012.

[201] S. Weissenborn, D. Chatterjee and J. Schaffner–Bielich. *Phys. Rev. C*, 85:065802, 2012.

[202] M. Oertel, C. Providência, F. Gulminelli and Ad. R. Raduta. *J. Phys. G*, 42:075202, 2015.

[203] K. A. Maslov, E. E. Kolomeitsev and D. N, Voskresensky. *Phys. Lett. B*, 748:369, 2015.

[204] T. Takatsuka, S. Nishizaki and Y. Yamamoto. *Eur. Phys. J. A*, 13:213, 2002.

[205] T. Takatsuka, S. Nishizaki and R. Tamagaki. *Prog. Theor. Phys. Suppl.*, 174:80, 2008.

[206] I. Vidaña, D. Logoteta, C. Providência, A. Polls, and I. Bombaci. *Europhys. Lett.*, 94:11002, 2011.

[207] Y. Yamamoto, T. Furumoto, N. Yasutake and Th. A. Rijken. *Phys. Rev. C*, 88:022801, 2013.

[208] Y. Yamamoto, T. Furumoto, N. Yasutake and Th. A. Rijken. *Phys. Rev. C*, 90:045805, 2014.

[209] D. Lonardoni, A. Lovato, S. Gandolfi and F. Pederiva. *Phys. Rev. Lett.*, 114:092301, 2015.

[210] D. Logoteta, I. Vidaña and I. Bombaci. *Eur. Phys. J. A*, 55:207, 2019.

[211] D. Gerstung, N. Kaiser and W. Weise. *Eur. Phys. J. A*, 56:175, 2020.

[212] E. Friedman and A. Gal. *EPJ Web of Conferences*, 271:06002, 2022.

[213] S. Weissenborn, I. Sagert, G. Pagliara, M. Hempel and J. Schaffner–Bielich. *Astrophys. J. Lett.*, 740:L14, 2011.

[214] T. Klähn, D. Blaschke and R. Łastowiecki. *Phys. Rev. D*, 88:085001, 2013.

[215] L. Bonanno and A. Sedrakian. *Astron. Astrophys.*, 539:A16, 2012.

[216] R. Lastowiecki, D. Blaschke, H. Grigorian and S. Typel. *Acta Phys. Polon. Suppl.*, 5:535, 2012.

[217] A. Drago, G. Pagliara and D. Pigato. *Phys. Rev. C*, 90:065809, 2014.

[218] D. G. Yakovlev, A. D. Kaminker, O. Y. Gnedin, and P. Haensel. *Phys. Rep.*, 354: 1–155, 2001.

[219] D. G. Yakovlev and C. J. Pethick. *Annu. Rev. Astron. Astrophys.*, 42:169–210, 2004.

[220] A. L. Watts et al. *Sci. China: Phys. Mech. Astron.*, 62(2):29503, 2019.

CHAPTER 5

Relativistic Description of the Neutron Star Equation of State

Constança Providência, Tuhin Malik, Milena Bastos Albino, Márcio Ferreira

THE general behavior of the nuclear equation of state (EOS), relevant for the description of neutron stars (NS), is studied within a relativistic mean-field description of nuclear matter. Different formulations, both with density-dependent couplings and with non-linear mesonic terms, are considered and their predictions compared and discussed. A special attention is drawn to the effect on the neutron star properties of the inclusion of exotic degrees of freedom as hyperons. Properties such as the speed of sound, the trace anomaly, the proton fraction, and the onset of direct Urca processes inside neutron stars are discussed. The knowledge of the general behavior of the hadronic equation of state and the implication it has on the neutron star properties will allow to identify signatures of a deconfinement phase transition discussed in other studies.

5.1 INTRODUCTION

Neutron stars (NS) are objects with several extreme properties which make them a true laboratory for dense baryonic matter. Under the extreme conditions existing in their interior it is expected, for instance, that quark deconfinement may occur in the center of NS. These are also the ideal objects to study very asymmetric nuclear matter which cannot be tested in the laboratory. In the present multimessenger era, astrophysical observations are starting to impose some stringent constraints on the equation of state (EOS) of the high-density baryonic matter. These constraints come from the gravitational wave detection by the LIGO Virgo collaboration as the detection of the binary neutron star merger GW170817 [2] or the GW190425 [5] , from radio data [11, 19, 22, 32, 87] or the recent X-ray observations of NICER allowing a prediction of both the NS mass and radius [71, 72, 82, 85, 86].

The nuclear matter EOS at low densities is constrained not only by well-known nuclear matter properties as the binding energy, saturation density and incompressibility [70], but

DOI: 10.1201/9781003306580-5

also from *ab initio* calculations of pure neutron matter based on a chiral effective field theoretical description [23, 24, 48]. At very high densities, $\sim 40\rho_0$ where ρ_0 is the nuclear saturation density, perturbative QCD calculations have been performed and they also impose strong constraints [57, 58]. Although the pQCD EOS is determined at densities that are not attained inside neutron stars, it was shown recently that these constraints may affect in a non-trivial way the EOS inside neutron stars [47].

From the measurement of mass and radius of several NS, one expects to be able to recover the EOS. The integration of the Tolman-Oppenheimer-Volkoff (TOV) equations establishes a one-to-one relation between the mass-radius curve and the pressures-energy density function [61]. Several methods have been used to obtain the EoS from the known mass-radius curve such as Baseyan inference [79, 83, 91, 92] or neural network techniques [17, 29, 41–43, 75], see [109] for a recent review on the application of machine learning techniques to learn about QCD matter under extreme conditions. Another problem is the determination of the NS composition from the knowledge of the EOS. Several studies have been carried out with the objective of extracting the proton fraction. Starting from a Taylor expansion representation of the EOS in the parabolic approximation for the asymmetry, it was shown that the proton fraction could not be recovered from the β-equilibrium EOS [21, 52, 74]. In [74], the authors attribute the failure to the existence of multiple solutions. In [52], the reason was assigned to the occurrence of correlations between the nuclear matter parameters.

Many studies have been performed with the objective of determining the EOS of strongly interacting matter constrained by observations and well accepted *ab initio* calculations as the ones reported above. In order, to span the whole phase space that joins the low-density to the high-density constraints different interpolation schemes have been undertaken based on agnostic descriptions of the EOS. Among these, we can point out the use of a piecewise polytropic interpolation [9, 57], a spectral interpolation [62], a speed-of-sound interpolation [7, 10, 90], meta-models based on Taylor expansions [30, 31, 68, 69, 96, 105, 106] or a nonparametric inference of the EOS [27, 47, 60, 110]. These studies have been used to infer signatures of the presence of deconfined matter inside neutron stars, for instance, by analyzing the behavior of the speed of sound with density [7, 90] or the trace anomaly which may signal the restoration of conformal symmetry [8, 40]. However, the above approaches are not able determine the composition of neutron stars.

The present chapter reviews recent work developed within the framework of a relativistic mean-field (RMF) description of hadronic matter at zero temperature having as main objective the determination of the region in the neutron star mass radius diagram, and corresponding EOS, in conformity with present observations and *ab initio* constraints. A Bayesian inference will be applied in the search for the parameters of the models. In comparison with the agnostic approaches described above, our perspective has an underlying microscopic model, which allows us to discuss composition, including proton fraction or hyperon content. We consider this information completes the one obtained from the agnostic descriptions of the EOS and may bring extra clues into the interpretation of the results obtained. In the following chapters, we will review the methodology and results obtrained in the works [64–67]. In particular, we will compare outputs obtained considering the different microscopic models in order to assess the generality and the specificity of the conclusions. The microscopic models based on a Lagrangian formulation used in these works may be divided in two classes: (i) the Lagrangian density is formulated in terms of constant parameters and include non-linear mesonic terms as proposed in [14, 76]. These models are designated by NL; (ii) the Lagrangian density contains only quadratic mesonic terms and is expressed in terms of couplings that have an explicit density dependence as explored in [100, 101]. In this class, we consider two different parametrizations of the couplings, the one proposed in [101] which we designate by DDH and the one used in [64] designated as DDB. We will also discuss the limitations of this second class of models concerning the high-density behavior of the coupling to the ϱ-meson, which defines the density dependence of the symmetry energy, and we will propose a generalization that overcomes the limitation [65]. Lastly, and considering recent interest in identifying signatures of deconfinement and of imposing high - pQCD constraints, we will discuss some of the physical quantities examined, including the speed of sound, polytropic index and trace anomaly and discuss the limitations enforced by pQCD.

Some other works have been developed in the last years using a Bayesian inference approach to constraint the parameters of RMF models including, [99] where a simpler version of the NL description was considered, [95] which has restricted the $\Lambda - \omega$ couplings in hyperonic stellar matter imposing as constraints the GW and NICER observations, [13] where the authors have studied how the pure neutron matter pressure and energy per particle constrains the isovector behavior of nuclear matter, and studied several correlations between nuclear matter properties (NMP) and NS properties, [51] where the authors have constrained the NL model from the present available NS observations and tested how constraining might be the future observations programmed for eXTP [102] and STROBE-X [84].

In the present chapter, we will first present the microscopic models used to perform the study, the Bayesian inference methodology, together with the priors, the data chosen to fit the models. We next compare the behaviour of the different data sets generated, including the nuclear matter properties (NMP) and the neutron star properties, including the speed of sound and the proton fraction. The inclusion of hyperons will be discussed as well as the onset of the nucleon direct Urca processes. We will also refer to some properties that are directly connected to QCD, in particular the trace anomaly and the constraints imposed by pQCD on the generated data sets of EOS.

5.2 FORMALISM

In this section, we briefly summarize the frameworks that will be applied to describe the nuclear or hadronic matter EOS. We will start by introducing the models through the definition of the Lagrangian density. As referred in the Introduction, two different classes are considered. They define the density dependence of the EOS through completely different approaches: (i) density-dependent couplings are introduced (DDH and DDB models); (ii) non-linear mesonic terms are included (NL).

5.2.1 The model

The equation of state of nuclear matter is determined from the Lagrangian density that describes the nuclear system. The degrees of freedom include the nucleons of mass m

described by Dirac spinors Ψ, and the meson fields, the scalar isoscalar σ field, the vector isoscalar ω field, and the vector isovector ϱ field, with masses m_i, $i = \sigma,\ \omega,\ \varrho$, which describe the nuclear interaction. The parameters Γ_i or g_i, $i = \sigma,\ \omega,\ \varrho$ designate the couplings of the mesons to the nucleons. The Lagrangian density is given by

$$\begin{aligned}\mathcal{L} =& \bar{\Psi}\Big[\gamma^\mu\left(i\partial_\mu - \Gamma_\omega A_\mu^{(\omega)} - \Gamma_\varrho \boldsymbol{t}\cdot\boldsymbol{A}_\mu^{(\varrho)}\right) - (m - \Gamma_\sigma\phi)\Big]\Psi + \frac{1}{2}\left(\partial_\mu\phi\partial^\mu\phi - m_\sigma^2\phi^2\right)\\ &-\frac{1}{4}F_{\mu\nu}^{(\omega)}F^{(\omega)\mu\nu} + \frac{1}{2}m_\omega^2 A_\mu^{(\omega)}A^{(\omega)\mu} - \frac{1}{4}\boldsymbol{F}_{\mu\nu}^{(\varrho)}\cdot\boldsymbol{F}^{(\varrho)\mu\nu} + \frac{1}{2}m_\varrho^2\boldsymbol{A}_\mu^{(\varrho)}\cdot\boldsymbol{A}^{(\varrho)\mu} + \mathcal{L}_{NL},\end{aligned} \tag{5.1}$$

where the last term $\mathcal{L}_{NL}$ is null if density-dependent couplings Γ_i are chosen, or includes self-interacting and mixed meson terms if the meson–nucleon couplings are taken as constant parameters. In order to distinguish, we will designate the constant couplings by the lower case letter g_i in the NL formulation. In the above expression γ^μ and $\boldsymbol{t}$ designate, respectively, the Dirac matrices and the isospin operator. The vector meson field strength tensors are defined as $F^{(\omega,\varrho)\mu\nu} = \partial^\mu A^{(\omega,\varrho)\nu} - \partial^\nu A^{(\omega,\varrho)\mu}$.

5.2.1.1 Density-dependent description

The density-dependent models include meson–nucleon couplings Γ_i, that depend on the total nucleonic density ρ, and are defined as

$$\Gamma_i(\rho) = \Gamma_{i,0}\, h_i(x)\,, \quad x = \rho/\rho_0\,,\, i = \sigma, \omega, \varrho, \tag{5.2}$$

with $\Gamma_{i,0}$ the couplings at saturation density ρ_0. For the isoscalar mesons, σ and ω, two parametrizations h_i are considered:

$$h_i(x) = \exp[-(x^{a_i} - 1)] \tag{5.3}$$

as in [65], giving origin to the DDB sets, and

$$h_i(x) = a_M\frac{1 + b_i(x + d_i)^2}{1 + c_i(x + d_i)^2}\,, \tag{5.4}$$

as in [100, 101], and originating the DDH data sets. The ϱ-meson nucleon coupling is defined as in [101]

$$h_\varrho(x) = \exp[-a_\varrho(x - 1)]\,. \tag{5.5}$$

5.2.1.2 Non-linear meson terms

The model introduced in [77] is defined with constant couplings, which we designate by g_i, $i = \sigma,\ \omega,\ \varrho$, and, instead, includes non-linear meson terms in the Lagrangian density, which are defined by

$$\mathcal{L}_{NL} = \ -\tfrac{1}{3}bg_\sigma^3(\sigma)^3 - \tfrac{1}{4}cg_\sigma^4(\sigma)^4 + \tfrac{\xi}{4!}(g_\omega^2\omega_\mu\omega^\mu)^2 + \Lambda_\omega g_\varrho^2\boldsymbol{\varrho}_\mu\cdot\boldsymbol{\varrho}^\mu g_\omega^2\omega_\mu\omega^\mu. \tag{5.6}$$

The parameters multiplying each one of these terms b, c, ξ, Λ_ω will be fixed together with the meson–nucleon couplings g_i by imposing nuclear matter and NS observational constraints.

The parameters b, c, in front of the σ self-interacting terms control the nuclear matter incompressibility at saturation [14]. The ξ term was introduced in [94] to modulate the high density dependence of the EoS, the larger ξ the softer the EOS. The non-linear $\omega - \varrho$ term influences the density dependence of the symmetry energy [18].

The equations of motion for the meson fields are given by

$$\sigma = \frac{g_\sigma}{m^2_{\sigma,\text{eff}}} \sum_i \rho^s_i \tag{5.7}$$

$$\omega = \frac{g_\omega}{m^2_{\omega,\text{eff}}} \sum_i \rho_i \tag{5.8}$$

$$\varrho = \frac{g_\varrho}{m^2_{\varrho,\text{eff}}} \sum_i t_{3i}\rho_i, \tag{5.9}$$

where ρ^s_i and ρ_i are, respectively, the scalar density and the number density of nucleon i, and the effective meson masses are defined as

$$m^2_{\sigma,\text{eff}} = m^2_\sigma + b g^3_\sigma \sigma + c g^4_\sigma \sigma^2 \tag{5.10}$$

$$m^2_{\omega,\text{eff}} = m^2_\omega + \frac{\xi}{3!} g^4_\omega \omega^2 + 2\Lambda_\omega g^2_\varrho g^2_\omega \varrho^2 \tag{5.11}$$

$$m^2_{\varrho,\text{eff}} = m^2_\varrho + 2\Lambda_\omega g^2_\omega g^2_\varrho \omega^2. \tag{5.12}$$

The non-linear meson terms define effective meson masses that depend on the density: (i) $m_{\omega,\text{eff}}$ increases with the ω-field and, as a consequence, the ω field is not proportional to the density for a non-zero ξ, but increases with a power of ρ smaller than one; (ii) $m_{\varrho,\text{eff}}$ increases with the density ρ, and, therefore, as the density increases the ϱ field becomes weaker, resulting in a softer symmetry energy. The magnitude of the softening depends on ξ: the larger ξ the smaller the softening.

Notice that the meson equations, i.e. Eqs. (5.7), (5.8) and (5.9), are also valid for the DDB and DDH models with the replacement $m_{i,eff} \to m_i$, since in the last two descriptions non-linear terms are not present.

5.2.2 Bayesian inference procedure

The model parameters are determined within a Bayesian inference procedure, i.e. applying Bayes' theorem [45], based on observed or experimental data, designated by fit data. The Bayesian parameter optimization system is determined from four different inputs that must be given: the prior, the likelihood function, the fit data, and the sampler.

The Prior:- The prior domain in our Bayesian setup is determined from a Latin hypercube sampling, allowing the parameters of the underlying RMF model to vary so that a broad range of nuclear matter saturation properties is spanned. For each of the different RMP models considered a uniform prior is defined.

The Fit Data:- As fit data, we have considered the three RMF models (see Table 5.1): the nuclear saturation density ρ_0, the binding energy per nucleon ϵ_0, the incompressibility coefficient K_0, and the symmetry energy $J_{\text{sym},0}$, all evaluated at ρ_0. We also include the pressure of pure neutron matter (PNM) at densities of 0.08, 0.12, and 0.16 fm^{-3} from N^{3}LO calculations in chiral effective field theory (chEFT) [48], considering 2$\times$ N^{3}LO

TABLE 5.1 The constraints used as fit data in the Baseyian inference are binding energy per nucleon ϵ_0, incompressibility K_0, symmetry energy $J_{\rm sym,0}$ at the nuclear saturation density ρ_0, each with a 1σ uncertainty, the pressure of pure neutron matter (PNM) at densities of 0.08, 0.12, and 0.16 fm^{-3}, obtained from a chEFT calculation [48], considering a $2\times$ N^{3}LO uncertainty for the PNM pressure and the maximum mass of neutron stars must exceed $2M_\odot$.

Constraints			
Quantity		Value/Band	Ref
NMP [MeV]	ρ_0	0.153 ± 0.005	[101]
	ϵ_0	-16.1 ± 0.2	[26]
	K_0	230 ± 40	[88, 97]
	$J_{\rm sym,0}$	32.5 ± 1.8	[28]
PNM [MeV fm^{-3}]	$P(\rho)$	$2\times$ N^{3}LO	[48]
	$dP/d\rho$	> 0	
NS mass [$M_\odot$]	$M_{\rm max}$	> 2.0	[32]

data uncertainty. Finally, it is also required that the maximum NS mass is at least $2M_\odot$. This requirement is introduced in the likelihood with uniform probability.

The Log-Likelihood:- A log-likelihood function is optimized as a cost function for the fit data defined in Table 5.1. It is defined by the equation below, Eq. 5.13, taking into account the uncertainties σ_j associated with each data point j,

$$Log(\mathcal{L}) = -0.5 \times \sum_j \left\{ \left(\frac{d_j - m_j(\boldsymbol{\theta})}{\sigma_j} \right)^2 + Log(2\pi\sigma_j^2) \right\}. \tag{5.13}$$

The maximum NS mass is treated differently, using a step function probability.

To populate the multi-dimensional posterior, we employ the nested sampling algorithm [89], specifically the PyMultinest sampler [15, 16], which is well-suited for low-dimensional problems. The EoS data set for subsequent analyses will be generated using the full posterior, which contains 25287 EoS. The posterior obtained for the three data sets is given at the bottom of this Chapter.

5.3 RESULTS

In the following subsections, we compare the performance of the three different microscopic models used to generate the data sets within Bayesian inference calculations that consider as fit data the ones presented in Table 5.1. Both NS and NMP will be compared. We will also discuss the effect of including hyperons, as well as the proton fraction and the onset of the nucleonic direct Urca processes. Finally, the behavior of the speed of sound and trace anomaly with the baryonic density will be discussed and the compatibility with pQCD constraints will be commented.

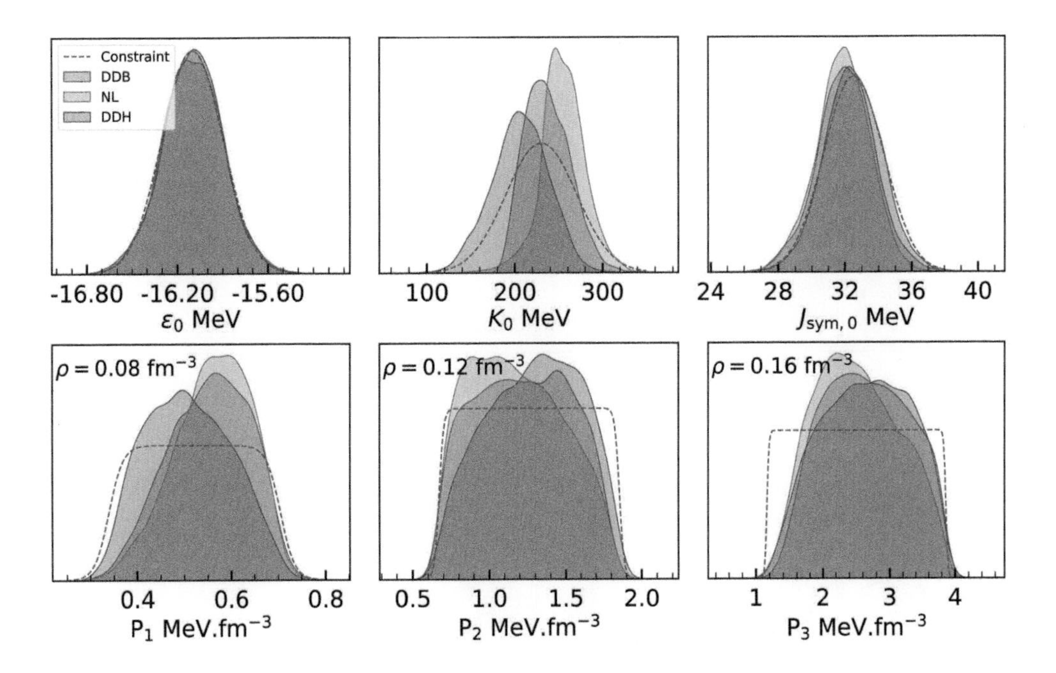

Figure 5.1 Fit data considered to constrain all EOS data set built for the present study, both for nucleonic and for hyperonic matter.

5.3.1 NL, DDB, and DDH: a comparison

A comparison of the performance of the three frameworks concerning the reproduction of the fit data is summarized in Fig. 5.1. The chosen fit data were the same for the three frameworks and are given in Table 5.1. All models reproduce the binding energy ϵ_0 and symmetry energy $J_{sym,0}$ at saturation in a similar way. The largest differences concern the incompressibility with DDH preferring smaller values and NL preferring larger ones. DDB peaks at the maximum of the fit data but with a much smaller width. Concerning the pure neutron matter (PNM), constraints, the three frameworks satisfy the constraint imposed at the larger density in a similar way, but there are differences at the lowest and intermediate densities with DDH concentrating at lower pressure values for the lowest density. These behaviors will be reflected in the NMP and NS properties.

Having verified that the three frameworks reproduce the fit data, we analyze next the NMP at saturation. This is summarized in the corner plot shown in Fig. 5.2 and in Table 5.2, where, considering the parabolic approximation for the energy of nuclear matter per particle with the isospin asymmetry $\delta = (\rho_p - \rho_n)/\rho$ at nuclear density ρ,

$$\epsilon(\rho, \delta) \simeq \epsilon(\rho, 0) + S(\rho)\delta^2, \tag{5.14}$$

the parameters corresponding to the symmetric nuclear matter energy per particle $\epsilon(\rho, 0)$ and the symmetry energy $S(\rho)$ expansion around saturation density till fourth order n are given by: (i) for the symmetric nuclear matter, the energy per nucleon $\epsilon_0 = \epsilon(\rho_0, 0)$ $(n = 0)$, the incompressibility coefficient K_0 $(n = 2)$, the skewness Q_0 $(n = 3)$, and the

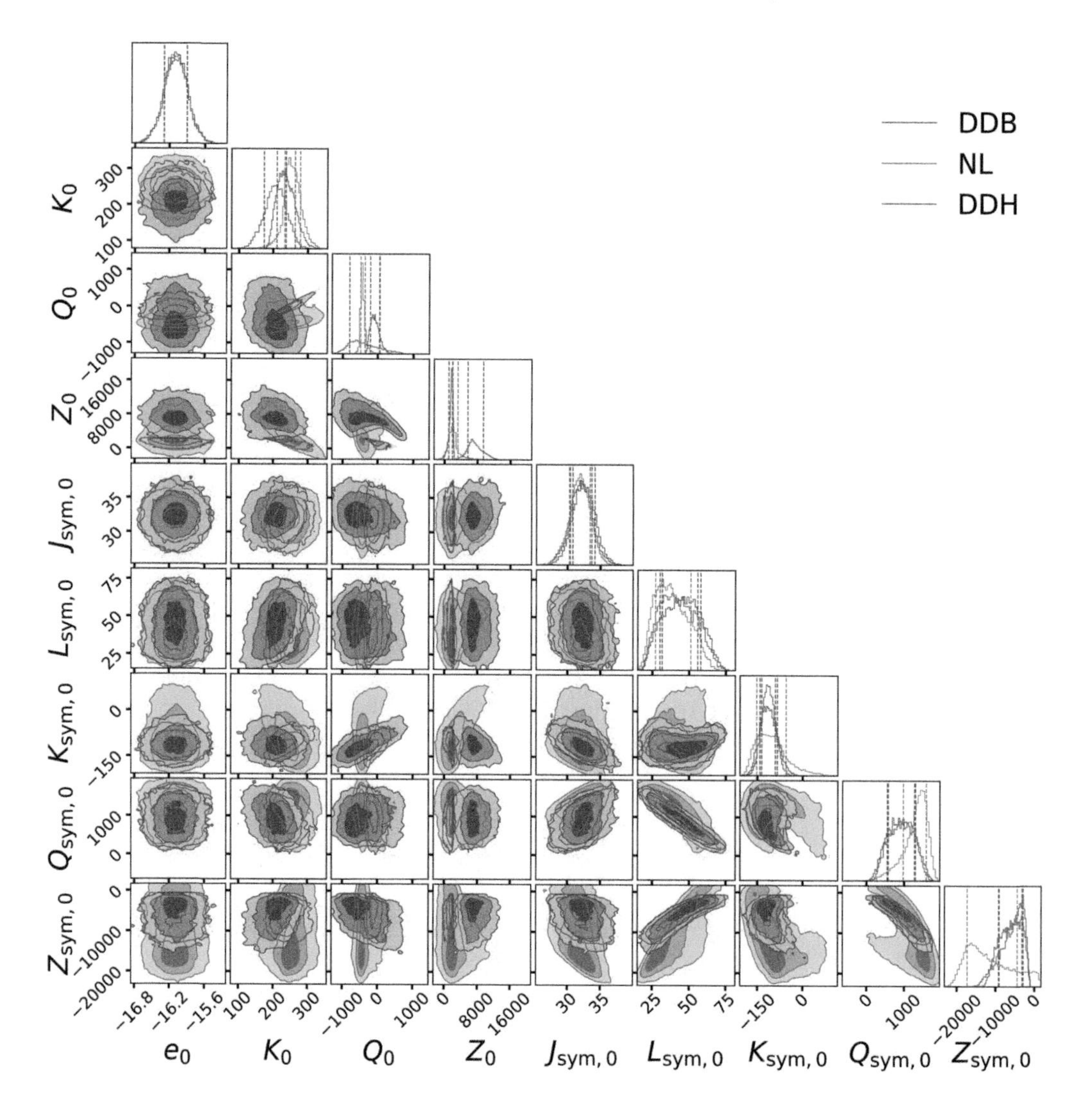

Figure 5.2 Corner plot comparing the nuclear matter properties of the three nucleonic data sets DDB. NL and DDH, in particular, the symmetric nuclear matter properties at saturation density ρ_0 defined by Eq. (5.15): binding energy ϵ_0, incompressibility K_0, skewness Q_0, and kurtosis Z_0; the symmetry energy properties at saturation defined by Eq. (5.17): symmetry energy $J_{\rm sym,0}$, slope $L_{\rm sym,0}$, incompressibility $K_{\rm sym,0}$, skewness $Q_{\rm sym,0}$, and kurtosis $Z_{\rm sym,0}$.

kurtosis Z_0 ($n = 4$), respectively, defined by

$$X_0^{(n)} = 3^n \rho_0^n \left(\frac{\partial^n \epsilon(\rho, 0)}{\partial \rho^n} \right)_{\rho_0}, \; n = 2, 3, 4; \tag{5.15}$$

(ii) for the symmetry energy, the symmetry energy at saturation $J_{\rm sym,0}$ ($n = 0$),

$$J_{\rm sym,0} = S(\rho_0) = \frac{1}{2} \left(\frac{\partial^2 \epsilon(\rho, \delta)}{\partial \delta^2} \right)_{\delta=0}, \tag{5.16}$$

TABLE 5.2 Nuclear matter properties at saturation density, median values and 90% CI, of the three data sets, DDB, NL, and DDH. Symmetric nuclear matter properties at saturation density ρ_0 defined by Eq. (5.15): binding energy ϵ_0, incompressibility K_0, skewness Q_0, and kurtosis Z_0. Symmetry energy properties at saturation defined by Eq. (5.17): symmetry energy $J_{\text{sym},0}$, slope $L_{\text{sym},0}$, incompressibility $K_{\text{sym},0}$, skewness $Q_{\text{sym},0}$, and kurtosis $Z_{\text{sym},0}$.

Model			ρ_0	ϵ_0	K_0	Q_0	Z_0	$J_{\text{sym},0}$	$L_{\text{sym},0}$	$K_{\text{sym},0}$	$Q_{\text{sym},0}$	$Z_{\text{sym},0}$
			fm^{-3}	MeV								
DDB	median		0.152	-16.10	235	-90	1585	32.05	42	-114	935	-5941
	90 % CI	min	0.142	-16.43	199	-262	486	29.15	25	-149	364	-10751
		max	0.164	-15.76	282	162	2043	34.81	63	-76	1434	-2128
NL	median		0.152	-16.10	254	-440	1952	31.89	37	-109	1367	-12613
	90 % CI	min	0.145	-16.43	213	-516	243	29.08	23	-171	629	-19118
		max	0.160	-15.77	297	-247	5295	34.41	58	-3	1710	-394
DDH	median		0.156	-16.10	206	-460	7189	32.44	45	-114	930	-5215
	90 % CI	min	0.144	-16.43	150	-978	4459	29.68	25	-157	412	-11529
		max	0.167	-15.78	257	395	10908	35.24	65	-64	1491	-2078

the slope $L_{\text{sym},0}$ ($n = 1$), the curvature $K_{\text{sym},0}$ ($n = 2$), the skewness $Q_{\text{sym},0}$ ($n = 3$), and the kurtosis $Z_{\text{sym},0}$ ($n = 4$), respectively, defined as

$$X^{(n)}_{\text{sym},0} = 3^n \rho_0^n \left(\frac{\partial^n S(\rho)}{\partial \rho^n}\right)_{\rho_0}, \quad n = 1, 2, 3, 4. \tag{5.17}$$

Some comments are in order: (i) as discussed before the incompressibility of DDH models peaks at a lower values than the other two, which present a similar behavior, and spreads over a larger range of values; (ii) concerning the skewness and kurtosis, which define the high-density behavior of the EOS, DDH presents a very widespread for the skewness from low negative to high positive values, and the kurtosis takes the largest values, to compensate the low incompressibility values it may take. This is necessary for the model to satisfy the $2M_\odot$ constraint imposed. Concerning the other two models, DDB presents the most restricted distribution which we can identify as a subset of the one NL defines, that is disjoint from the set defined by DDH for the kurtosis; it is interesting to verify that NL may take small and even negative values of the kurtosis; (iii) concerning the symmetry energy the distribution presented by the three models for the symmetry energy and slope at saturation is similar. However, there are differences in the higher order parameters, in particular, $K_{sym,0}$ and $Z_{sym,0}$: the two models with density-dependent coupling, DDH and DDB, behave in a similar way but NL spreads along a wider range of values and $K_{sym,0}$ may take positive values and $Z_{sym,0}$ takes very large negative values. The differences encountered are in part due to the fact that for DDH and DDB models the coupling of the ϱ-meson to

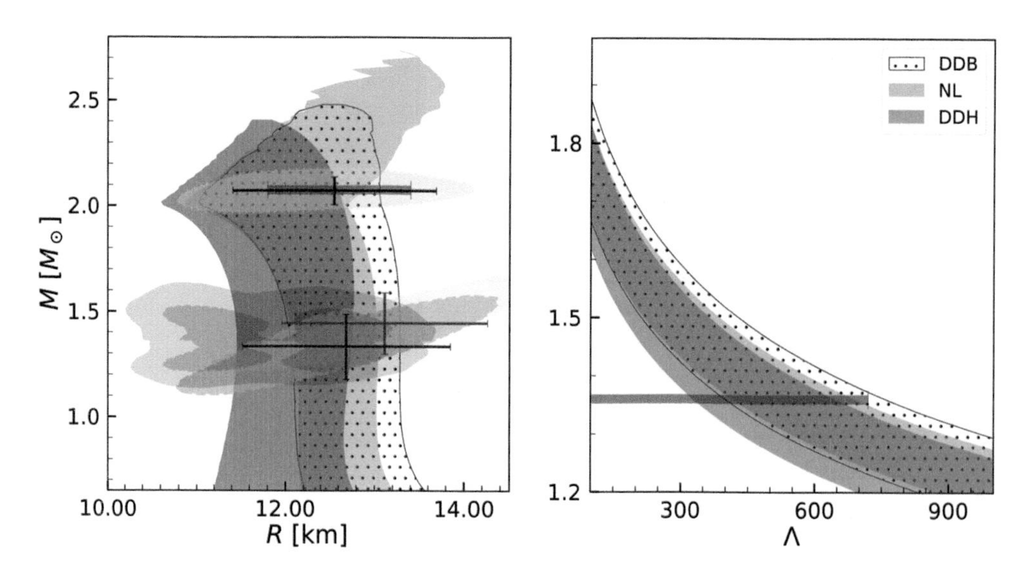

Figure 5.3 NS mass-radius (left) and tidal deformability-mass regions obtained from the 90% CI for the conditional probabilities $P(R|M)$ and $P(\Lambda|M)$ for DDB (dotted red), DDH (blue) and NL (green) frameworks. The blue horizontal bar on the left panel indicates the 90% CI radius for the pulsar PSR J0740+6620 with $M = 2.08 M_\odot$ obtained combining observational data from GW170817 and NICER as well as nuclear saturation properties [72]. The gray shaded regions indicate the 90% (solid) and 50% (dashed) CI of the LIGO/Virgo analysis for high mass (top) and low mass (bottom) components of the NS binary that originated the GW170817 event [4]. The NICER 1σ (68%) credible zone of the 2-D mass-radii posterior distribution for the PSR J0030+0451 (lilas and light green) [71, 85], and the PSR J0740 + 6620 (light orange) [72, 86] are also included. The horizontal (radius) and vertical (mass) error bars reflect the 1σ credible interval derived for NICER data's 1-D marginalized posterior distribution.

the nucleons tends to zero at sufficiently large densities. Generalizing the parametrization of the ϱ-meson coupling will allow to go beyond this limitation. We will come back to this problem in one of the following sections.

NS properties, as the mass and radius, are determined from the integration of the Tolmann-Oppenheimer-Volkoff equations for spherical stars in statiscal equilibrium [78, 98], see [46] for a review. The tidal deformabilities Λ, quantities that are obtained from the detection of gravitational waves [1], are obtained integrating the equations obtained in [49].

The radius and tidal deformability for NS with a given mass have been calculated within the three frameworks and the results are plotted in Fig. 5.3, on the left side the radius-mass and on the right side the tidal deformability-mass. Results of several observations, in particular, from the LIGO Virgo Collaboration for the GW170817 [3] and from NICER [71, 72, 85, 86] for the pulsars PSR J0030+0451 and PSR J0740 + 6620, have been included. The three data sets show different properties which reflect the different NMP the different sets have as discussed before. The main conclusions that can be drawn are: (i)

NL data set is the most restricted for low mass stars presenting intermediate radii mostly between 12 and 13 km. DDH, the data set with smallest K_0 values, presents the smallest radii for low mass stars, ~11.5–12.5 km, while the DDB set predicts the largest radii, ~12–13.5 km. The density dependence of the EOS at high densities is strongly influenced by the non-linear terms in the NL data set and the function that defines the density dependence of the meson couplings in the other two sets, DDB and DDH. DDH data set is soft at low densities so that low mass stars have a quite small radius, but at large densities becomes stiff to allow maximum mass stars with almost $2.5M_\odot$. The DDB data set allows for larger maximum masses than DDH; however, NL data set attains the largest masses, close to $2.75M_\odot$. All data sets agree with the presently available NS observations. In the right panel, the tidal deformabilities are plotted as a function of the mass for the three data sets. Their behavior follows the one obtained for the radii, with DDH having the smallest values and DDB the highest. Only some models of DDB are outside the 90% CI obtained from GW170817 value for a $1.36M_\odot$ star (see the blue horizontal bar).

The corner plot shown in Fig. 5.4 involving some NS properties allows some interesting conclusions: (i) there is some correlation between the radius and the tidal deformability of a $1.4M_\odot$ stars and the star maximum mass; (ii) the central speed of sound squared of the maximum mass star is clearly model dependent: for DDH model, $c_s^2 \sim 0.7 - 0.8$ is pratically constant and quite high, while for the other two models c_s^2 can be as low as 0.45 or even lower and as high as 0.75; (iii) it is precisely DDH with the largest speed of sound that predicts the smallest radii for $1.4M_\odot$ stars.

5.3.2 Including hyperons

In the inner core of a NS, non-nucleonic degrees of freedom may set in. In the present section, we will discuss the onset of hyperons. As in [65], we will introduce only two hyperons, the neutral Λ-hyperon and the negatively charged Ξ^--hyperon. These two hyperons are the ones that appear in the largest fractions, either because of having the smallest hyperon mass as the Λ, or because of being negatively charged (the Ξ^-), and, therefore, favorably replace the electrons and reduce the total pressure of the system. Both hyperons have an attractive potential in symmetric nuclear matter, and form hypernuclei. The binding energy of hyperons in hypernuclei has been used to fit the couplings of the hyperons to mesons in the RMF description of hadronic matter [33, 81]. Although the mass of the Σ^--hyperon is smaller than the one of Ξ^-, it interacts repulsively with nuclear matter, as the non-existence of Σ-hypernuclei seems to show [44]. As a consequence, in NS matter its onset occurs at larger densities than the Ξ^- onset [35, 37, 93, 104].

The introduction of hyperons requires a generalization of the Dirac term of Eq. (5.6) to include Λ and Ξ^- besides protons and neutrons,

$$\mathcal{L}_D = \sum_{j=p,n,\Lambda,\Xi^-} \bar{\Psi}_j \Big[\gamma^\mu \left(i\partial_\mu - \Gamma_{\omega,j} A_\mu^{(\omega)} - \Gamma_{\varrho,j} \boldsymbol{t}_j \cdot \boldsymbol{A}_\mu^{(\varrho)}\right) - (m - \Gamma_{\sigma,j}\phi)\Big]\Psi \qquad (5.18)$$

For the couplings of the hyperons to the vector-mesons, we consider the SU(6) values for the vector isoscalar mesons, $g_{\omega\Xi} = \frac{1}{3} g_{\omega N} = \frac{1}{2} g_{\omega\Lambda}$ and $g_{\phi\Xi} = 2 g_{\phi\Lambda} = -\frac{2\sqrt{2}}{3} g_{\omega N}$ and for the isovector ϱ-meson, $g_{\varrho\Xi} = g_{\varrho N}$. In this last case, the hyperon isospin also defines the strength of the coupling. Having assumed these values for the couplings of

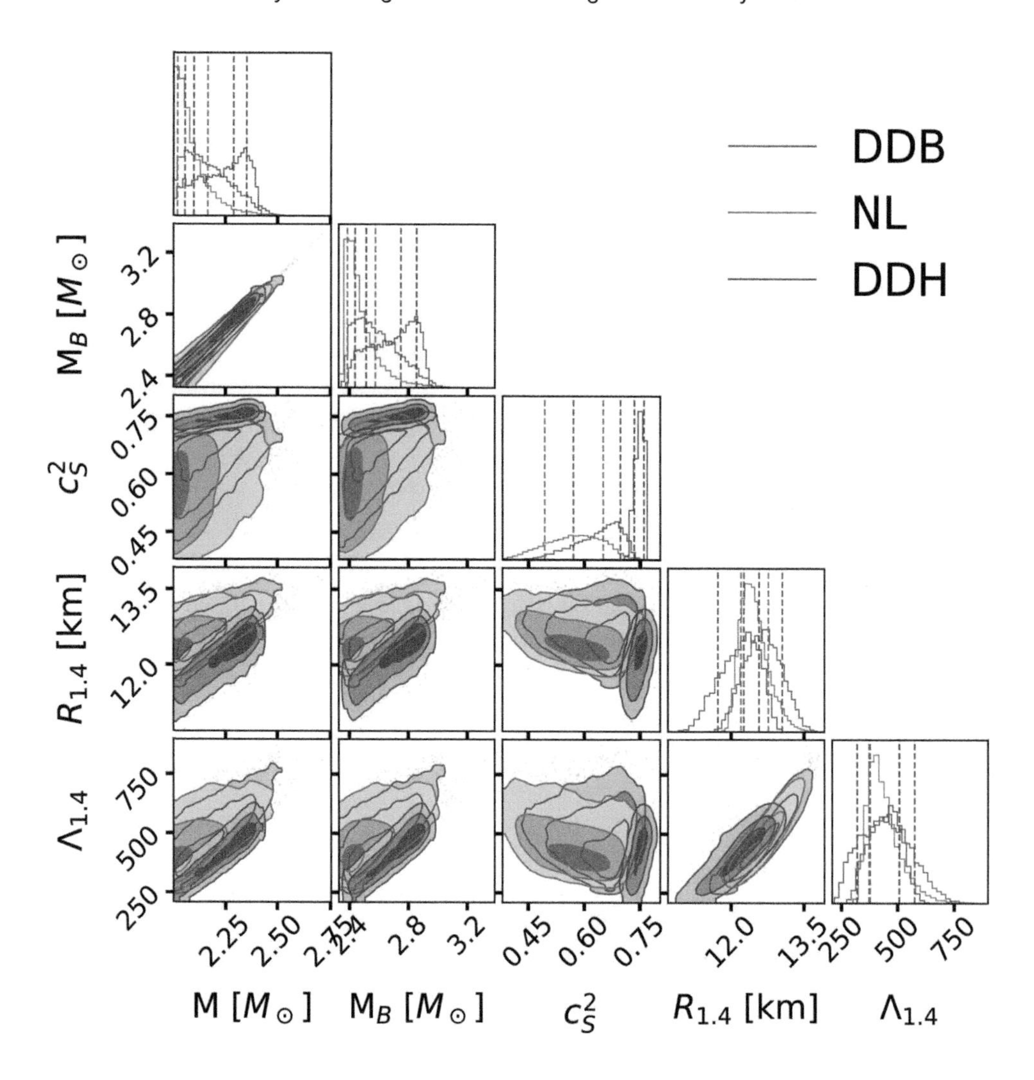

Figure 5.4 Corner plot comparing the following NS properties obtained with the three data sets, DDB, NL and DDH: maximum gravitational mass M, maximum baryonic mass M_B and respective central speed of sound squared c_s^2, radius R and tidal deformability Λ of a 1.4 $M_\odot$ star.

the hyperons to the vector mesons, the coupling to the σ-meson is fitted to hypernuclei properties [33, 34, 81]. In general, we express the couplings to the mesons as a fraction of the nucleon couplings, $g_{m\,i} = x_{m\,i}\, g_\sigma$ with $m = \sigma,\ \omega,\ \varrho$ and $i = \Lambda$ and Ξ^-. Considering the results of the fits done in [33, 34, 81], values between 0.609 and 0.622 were determined for the fraction $x_{\sigma\Lambda}$, and will be adopted in the present study. For the fraction $x_{\sigma\Xi^-}$, the range 0.309 to 0.321 will be used, as determined from fits to the binding energy of Ξ^- in the hypernuclei ${}^{15}_{\Xi}\mathrm{C}$ and ${}^{12}_{\Xi}\mathrm{Be}$ [37].

We have performed calculations for hyperonic stars within two models, DDB and NL, imposing the same fit data that was considered to constrain the EOS data sets of nucleon

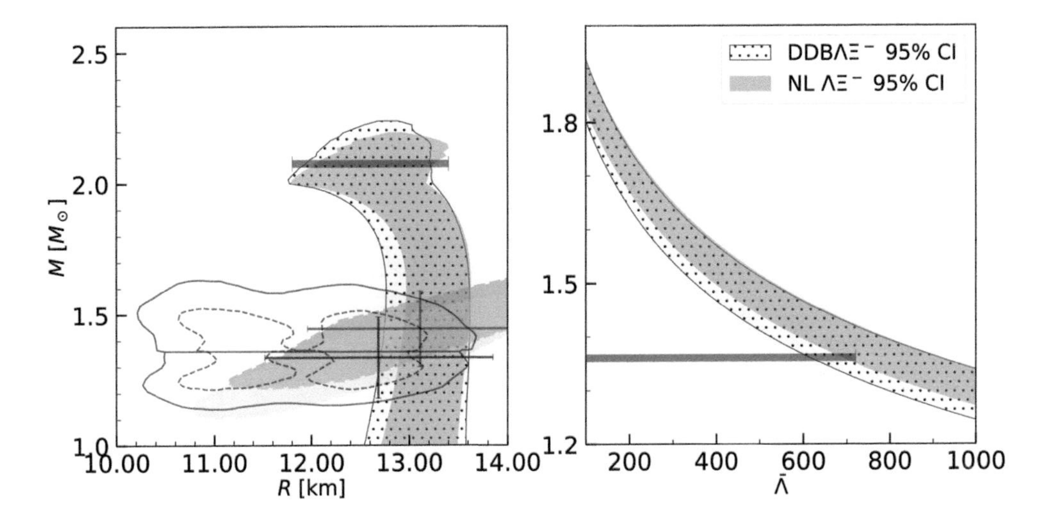

Figure 5.5 The 90% CI region for the hyperon data sets DDB-hyp (dotted red) and NL-hyp (green) derived using the conditional probabilities $P(R|M)$ (left) and $P(\Lambda|M)$ (right). The lines in the left panel indicate the 90% (solid) and 50% (dashed) CI for the binary components of the GW170817 event [4]. Also shown is the 1σ (68%) credible 2-D posterior distribution in the mass-radii domain from the millisecond pulsar PSR J0030+0451 (cyan and yellow) [71, 85] obtained from the NICER x-ray data. The horizontal (radius) and vertical (mass) error red bars reflect the 1σ credible interval derived for the same NICER data's 1-D marginalized posterior distribution. The blue bars represent the radius of the PSR J0740+6620 at $2.08 M_{\odot}$ (left panel) and the tidal deformability at 1.36 $M_{\odot}$ (right panel) [3].

matter, and which is summarized in Table 5.1. Chemical equilibrium dictates that:

$$\mu_\Lambda = \sqrt{(m_\Lambda^*)^2 + k_{F\Lambda}^2} + x_{\omega\Lambda}\, g_\omega \omega = \mu_n \tag{5.19}$$

$$\mu_\Xi = \sqrt{(m_\Xi^*)^2 + k_{F\Xi}^2} + x_{\omega\Xi}\, g_\omega \omega - \frac{1}{2} g_\varrho \varrho = \mu_n + \mu_e, \tag{5.20}$$

where m_i^* is the effective mass of hyperon i and k_{Fi} its Fermi momentum. Charge neutrality imposes that $\rho_p = \rho_\Xi + \rho_e + \rho_\mu$.

In Fig. 5.5, predictions obtained for the NS radius (left) and tidal deformability (right) for different NS masses are plotted. In Table 5.4, the median and the 90% CI nuclear matter properties of both data sets are summarized and in Table 5.3.2 some NS properties are given, in particular, the median and the 90%CI of the maximum mass, respective, baryonic mass, radius, central speed of sound squared and central baryonic density, together with the radius and tidal deformability of a $1.4 M\odot$ star.

We first discuss the effect on the nuclear matter properties of including hyperons, comparing results of Tables 5.2 and 5.4. Isovector properties are essentially not affected for the DDB data set, and only slightly for the NL data set reflected in a small increase of the different properties. Isoscalar properties are the mostly affected: the incompressibility K_0 suffers an increase of 15%-20%, and the median skewness becomes positive. The reason

TABLE 5.3 The Neutron star properties, median values and 90% CI, of the three nucleon data sets, DDB, NL and DDH. The following properties are given: the maximum mass $M_{\rm max}$ and respective baryon mass $M_{\rm B,max}$, radius $R_{\rm max}$, speed of the sound squared at the center c_s^2 and central baryonic density ρ_c, and the radius and tidal deformability of the $1.4M_\odot$ star, $R_{1.4}$ and $\Lambda_{1.4}$.

Model			$M_{\rm max}$	$M_{\rm B,max}$	$R_{\rm max}$	$R_{1.4}$	$\Lambda_{1.4}$	C_s^2	ρ_c
			$M_\odot$		km		...	c^2	fm^{-3}
DDB	median		2.148	2.567	11.13	12.66	466	0.649	1.002
	90 % CI	min	2.022	2.396	10.54	12.04	334	0.520	0.865
		max	2.366	2.857	11.85	13.28	648	0.718	1.121
NL	median		2.062	2.446	10.92	12.44	423	0.576	1.051
	90 % CI	min	2.006	2.370	10.52	12.08	347	0.446	0.904
		max	2.260	2.715	11.70	13.03	582	0.685	1.127
DDH	median		2.242	2.712	10.97	12.21	423	0.750	0.986
	90 % CI	min	2.037	2.439	10.15	11.46	273	0.727	0.887
		max	2.380	2.898	11.52	12.75	546	0.763	1.170

TABLE 5.4 Nuclear matter properties at saturation density, median values and 90% CI, of the two data sets including hyperons, DDB-hyp and NL-hyp. Symmetric nuclear matter properties at saturation density ρ_0 defined by Eq. (5.15): binding energy ϵ_0, incompressibility K_0, skewness Q_0 and kurtosis Z_0. Symmetry energy properties at saturation defined by Eq. (5.17): symmetry energy $J_{\rm sym,0}$, slope $L_{\rm sym,0}$, incompressibility $K_{\rm sym,0}$, skewness $Q_{\rm sym,0}$ and kurtosis $Z_{\rm sym,0}$.

Model			ρ_0	ϵ_0	K_0	Q_0	Z_0	$J_{\rm sym,0}$	$L_{\rm sym,0}$	$K_{\rm sym,0}$	$Q_{\rm sym,0}$	$Z_{\rm sym,0}$
			fm^{-3}	MeV								
DDB-hyp	median		0.152	-16.09	272	130	1425	32.15	43	-98	966	-6713
	90% CI	min	0.147	-16.39	247	-16	803	29.48	26	-127	354	-12178
		max	0.157	-15.79	309	349	1680	34.83	65	-59	1453	-2723
NL-hyp	median		0.150	-16.09	296	-117	2105	31.85	42	-70	1312	-13592
	90% CI	min	0.144	-16.41	270	-246	-405	29.16	31	-127	895	-18989
		max	0.157	-15.76	341	104	3078	34.44	57	-12	1607	-3893

TABLE 5.5 NS properties, the median and the 90% CI, of the data sets with hyperons, DDB-hyp and NL-hyp. The following properties are given: the maximum mass $M_{\max}$ and respective baryonic mass $M_{\mathrm{B,max}}$, radius $R_{\max}$, speed of the sound squared at the center c_s^2 and central baryonic density ρ_c, and the radius and tidal deformability of the $1.4M_{\odot}$ star, $R_{1.4}$ and $\Lambda_{1.4}$.

Model			$M_{\max}$	$M_{\mathrm{B,max}}$	$R_{\max}$	$R_{1.4}$	$\Lambda_{1.4}$	C_s^2	ρ_c
			$M_{\odot}$		km		...	c^2	fm^{-3}
NL-hyp	median		2.024	2.357	11.82	13.22	659	0.47	0.920
	90 % CI	min	2.003	2.329	11.55	12.97	595	0.41	0.860
		max	2.083	2.433	12.20	13.52	758	0.51	0.968
NL-hyp	median		2.040	2.385	11.73	13.11	610	0.48	0.932
	90 % CI	min	1.992	2.322	11.48	12.76	526	0.44	0.871
		max	2.130	2.501	12.08	13.51	743	0.50	0.964

for this change is the fact that the presence of the onset of hyperons relieves the pressure inside the NS and the condition that $2M_{\odot}$ stars must be described obliges the EOS to be harder, mainly affecting the isoscalar channel of the EOS.

The implication of the hardening of the EOS is that larger NS radii are predicted (compare Fig. 5.5 left with 5.3 left). The median values of the radius of $1.4\,M_{\odot}$ stars reflect clearly this effect: they increase from 12.66 (12.44) km for DDB (NL) to 14.22 (13.11) km, i.e. more than ~ 0.5 km or $\sim 5\%$. Measurements of the NS radius with an uncertainty smaller than 5%, such as the ones programmed with eXTP [102] and STROBE-X [84], could distinguish between these two scenarios. Also the tidal deformability is strongly affected increasing its median value from 466 (423) to 650 (610), respectively, for DDB (NL), and the constraint imposed by GW170817 is essentially not satisfied (see Fig. 5.5 middle panel). Another important property that distinguishes both scenarios is the NS maximum mass that decreases from a maximum value at 90% CI of 2.37 (2.26) $M_{\odot}$ for DDB (NL) to 2.08 (2.13) $M_{\odot}$. Concerning the NS properties in the center of the star it is pointed out the decrease of the speed of sound, its square decreasing essentially to values of the order of 0.5 or below in the presence of hyperons, when it takes values of the order of 0.6 going up to ~ 0.7 if only nucleon matter is considered. In Sec. 5.3.4, the speed of sound in matter with hyperons will be compared with the one obtained with nucleonic models.

5.3.3 Onset of direct Urca

In this subsection, the dependence on the baryon density of the proton fraction of β-equilibrium matter will be discussed. We consider two of the microscopic models discussed in Sec. 5.3.1, DDB and NL. Within the same models, the onset of hyperons and its influence on the proton fraction will also be commented using the data sets described in the previous section, Sec. 5.3.2.

The ϱ-meson coupling to the nucleon in DDB and DDH data sets decreases exponentially with the density. At high densities it approaches zero, and, as a consequence, these

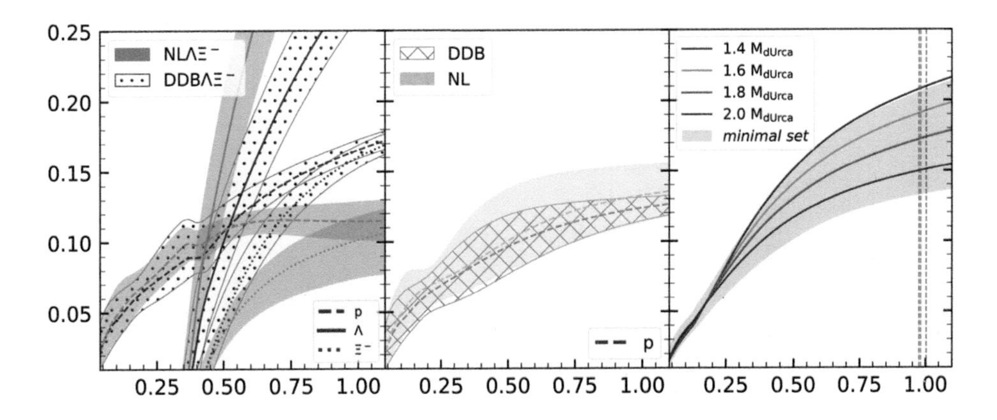

Figure 5.6 Proton and hyperons fractions for data sets DDB-hyp (red dotted) and NL-hyp (blue) (left panel), proton fractions for the data sets DDB (slashed region) and NL (pink region) (middle panel), proton fraction defining the minimal set compatible with chEFT PNM calculations at 2σ obtained varying the parameter y introduced in Eq. (5.21) (pink region), or fixing y to a given M_{dUrca}, i.e. 1.4, 1.6, 1.8, 2.0 $M_{\odot}$ (full lines).

models allow for very asymmetric matter at high densities since the symmetry energy is low. Therefore, the opening of nucleon direct Urca processes [107, 108] does not occur inside NS [36, 37, 39]. This is clearly seen from the middle panel in Fig. 5.6, where the median and 90% CI bands of the proton fraction are plotted as a function of the baryonic density for the data sets DDB (slashed band) and for NL (pink band): at large densities the proton fraction for DDB is smaller and narrower than the NL proton fraction. The smaller width is also an indication that the g_ϱ coupling of all models tends to the same value, zero, at high densities. NL models, however, span a wider range of proton fractions, and in particular, the opening of direct Urca processes may occur in some models.

In the left panel of the same figure, the fraction of protons is plotted together with the Λ and Ξ^- hyperon fractions for matter including hyperons. The Λ-hyperon is the first hyperon to set in just above twice saturation density, while the Ξ^- sets in just below 3 ρ_0. The onset of the Λ-hyperon implies a decrease of the neutron fraction, decreasing the pressure caused by this species and, therefore, the system energy. As a consequence, the proton fraction also decreases (see discussion in [67]). As soon as the Ξ^--hyperon sets in the proton fraction suffers an increase to compensate for the negatively charged hyperon. This behavior is well illustrated in the left panel of 5.6. The effect of much stronger for the DDB-hyp data set because the ϱ-meson coupling is weaker and, therefore, the repulsive term that enters the Ξ chemical potential is weaker, see Eq. (5.20). The coupling g_ϱ (Γ_ϱ) varies at 90% CI within the range $[9.55, 14.60]$ ($[6.97, 8.78]$) for NL (DDB) at saturation density. Including hyperons in the model these values change only slightly to $[9.79, 14.31]$ for NL-hyp and $[7.13, 8.58]$ for DDB-hyp at saturation density.

It was shown in Sec. 5.3.1, in particular, with the corner plot 5.2, that while DDB and DDH data sets differ a lot when comparing the symmetric nuclear matter properties, the symmetry energy properties are very similar considering all orders of the Taylor expansion studied. In order to overcome the special feature of these models of not allowing for

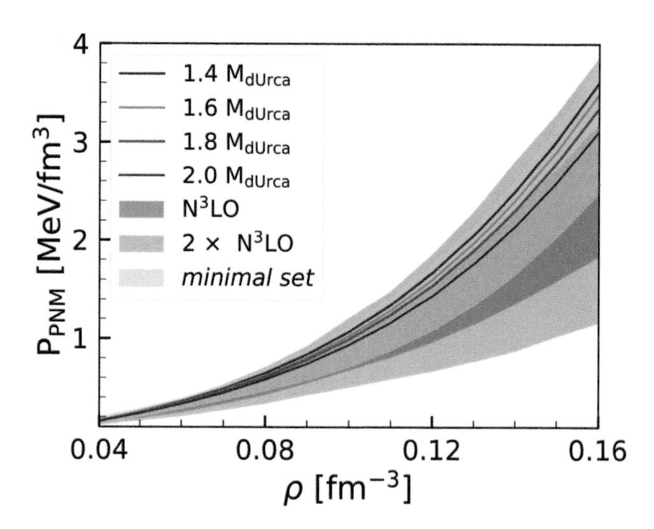

Figure 5.7 Pure neutron matter pressure as a function of the baryon density for the DDB model using Eq. (5.21) to define the ϱ-meson coupling: the minimal set compatible with chEFT PNM calculations at 2σ obtained varying the parameter y (pink region), or fixing y to a given M_{dUrca}, i.e. 1.4, 1.6, 1.8, 2.0 $M_\odot$ (full lines). The dark (light) gray bands define the chEFT pressure at $1\sigma(2\sigma)$ from [48].

nucleon direct Urca processes, in [64] a generalization of the ϱ-meson coupling was proposed including a new parameter y. For the function $h_\varrho(x)$ that defines the density dependence of the coupling Γ_ϱ, see Eq. (5.21), we consider

$$h_\varrho(x) = y \, \exp[-a_\varrho(x-1)] + (y-1) \, , \quad 0 < y \leq 1 \, . \tag{5.21}$$

Allowing y, imposing the constraints defined in Table 5.1, we have generated the PNM pink band in Fig. 5.7. In the same figure, the chEFT PNM EOS from [48] is also included considering 1 σ (dark gray) and 2 σ (light gray), as well as the PNM EOS imposing that M_{dUrca} is 1.4, 1.6, 1.8 and 2.0 $M_\odot$, where we designate by M_{dUrca}, the mass of the star where nucleon direct Urca processes set in its center. In fact, the new parameter y must be determined from NS properties that are sensitive to the high density behavior of the symmetry energy, such as the proton fraction. In particular, the onset of nucleon direct Urca (dUrca) processes is an appropriate observation and was used in [64] to constraint y. In Fig. 5.6 right panel, we show the proton fraction (full lines) corresponding to different M_{dUrca}. This was possible by choosing the adequate y. The pink region spans the proton fraction compatible with PNM chEFT calculation at 2σ, already defined in Fig. 5.7. These constraints derived from pure neutron matter exclude dUrca processes from stars with a mass $\lesssim 1.4 \, M_\odot$ at 2σ. If we restrict ourselves to 1σ, M_{dUrca} rises to a value above 1.6 $M_\odot$. These results are in agreement with the analysis performed in [12], where it is concluded that NS cooling curves seem to indicate that $\mathrm{M_{dUrca}} \sim 1.6 - 1.8 \, \mathrm{M}_\odot$.

5.3.4 Speed of sound, trace anomaly, and pQCD constraints

Lately, some discussion has been concentrated on the behavior of the speed of sound with density. This quantity, which is directly related to the dependence of the pressure on the energy density, is sensitive to the onset of new degrees of freedom and first-order phase transitions. In particular, at high densities, it is expected that matter is deconfined and exhibits conformal symmetry with the square of the speed of sound being equal to 1/3. One of the present great interests is to identify possible signatures of the presence of deconfined quark matter inside NS.

The general behavior of the speed of sound squared obtained from agnostic descriptions of the EOS of baryonic matter, that has been constrained by low-density pure neutron matter *ab-initio* calculations [23, 24, 48] and the pQCD EOS at densities of the order $\gtrsim 40\rho_0$, and by NS observations, includes a steep increase until an energy density of the order of $\sim$ 500 MeV/fm is attained, followed by a decrease or flattening, approaching 1/3 at high densities [7, 8, 10, 47, 56, 90], see also the discussion in [54].

In Fig. 5.8, the top panels of the three columns show the behavior of the speed of sound squared for the three data sets DDB, NL, and DDH, in particular, the 68% and 95% CI are shown. The different sets present a different behavior: for set DDH c_s^2 increases monotonically with a small dispersion, and attains values close to 0.8 for densities of the order of 1 fm^{-3}; set NL is on the other extreme, and above $\rho \sim 0.3$ fm^{-3} shows a quite large dispersion including a flattening or slight decrease, never attaining values above 0.7 and presenting values that can go below 0.4; DDB shows an intermediate behavior, not so extreme as DDH, but also showing a monotonic increase.

The NL data set contains EOS with quite different behaviors at high densities, controlled by the ω^4 term. In the left panel of Fig. 5.9, the speed of sound squared is plotted for different ranges of the parameter ξ, for set 1 $\xi \in [0.0, 0.004]$, for set 2 $\xi \in [0.004, 0.015]$ and for set 3 $\xi \in [0.014, 0.04]$. This parameter controls the contribution of the ω^4 term in the Lagrangian density, and as discussed in [66, 77], in the high density limit it makes the speed of sound squared go to 1/3. This indicates that a quite large range of values of the speed of sound squared is possible considering just nuclear degrees of freedom. In the middle panel, c_s^2 for the NL-hyp set has also been included. This set presents a peak above $2\rho_0$, when the hyperons set in, followed by a monotonous increase of the speed of sound, attaining values $c_s^2 \lesssim 0.6$ at 1 fm^{-3}.

Several quantities have been proposed as indicators of the presence of deconfined matter, including the polytropic index $\gamma = d\ln P/d\ln\epsilon$ [10], which takes the value 1 in conformal matter, the trace anomaly scaled by the energy density introduced in [40] $\Delta = 1/3 - P/\epsilon$ which should approach zero in the conformal limit, and the derived quantity proposed in [8] $d_c = \sqrt{\Delta^2 + \Delta'^2}$, where $\Delta' = c_s^2\,(1/\gamma - 1)$ is the logarithmic derivative of Δ with respect to the energy density, which approaches zero in the conformal limit. In the following, we will discuss how these quantities behave when we consider the different EOS data sets introduced in the present study. This will allow to identify properties that are still present in hadronic matter from properties that totally characterize deconfined matter.

In middle and bottom lines of Fig. 5.8, the polytropic index γ and the trace anomaly Δ are plotted as a function of the baryonic density in units of the saturation density. The horizontal line in the γ panels identifies the value 1.75 that has been proposed as indicating the transition to deconfined quark matter [10]. For all models the polytropic index γ increases until $\sim \rho_0$, followed by a monotonous decrease that goes below 1.75 at a density above ~ 0.4 fm^{-3}. The behavior of the three sets is quite consistent, and it seems to indicate that

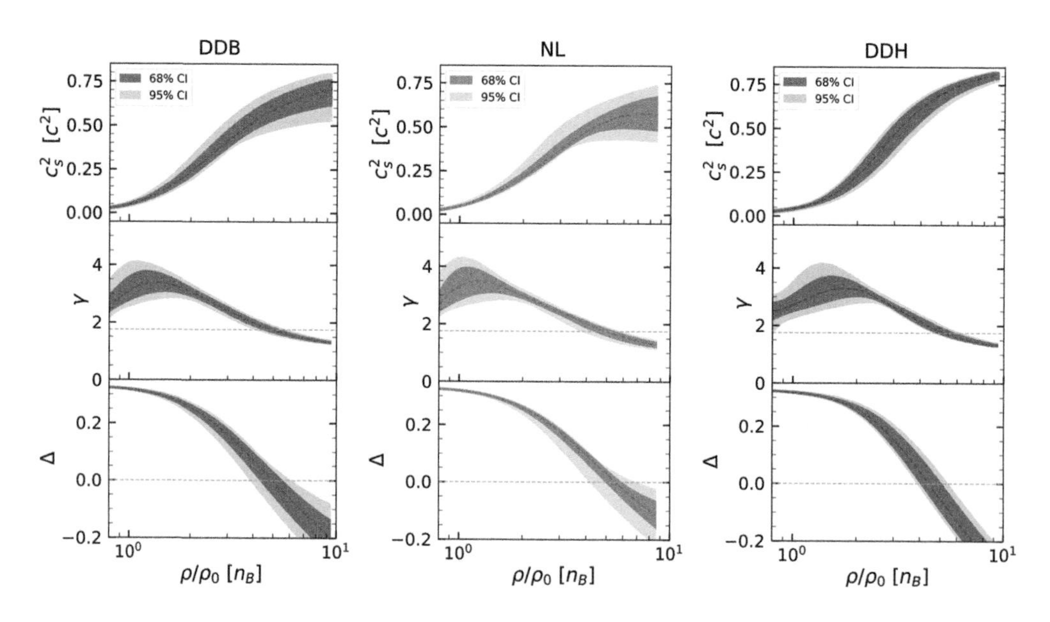

Figure 5.8 The speed of sound squared c_s^2, the polytropic index $\gamma = d\ln P/d\ln\epsilon$ and the trace anomaly $\Delta = 1/3 - P/\epsilon$ for the three data sets, DDB, NL, and DDH. The horizontal lines in the γ plots identifies the value 1.75.

a value of $\gamma < 1.75$ is not enough to identify a phase transition to deconfined matter. The normalized trace anomaly shows a behavior similar to the one discussed in [40], where results from several studies [6, 25, 47, 82] have been compared, and it crosses the zero axis at densities of the order of 0.4-0.8 fm^{-3}, becoming afterward negative. At sufficiently high densities this quantity should tend to the pQCD values that are slightly positive. Considering the models studied, for the NL data set (and even DDB) Δ shows a change of slope around 1 fm^{-3}, which could match a positive trace anomaly in finite density QCD.

We have added to Fig. 5.9 the right panel where the effective nucleon mass is plotted as a function of the baryon density. It is seen that the mass decreases quite fast with density and at $\sim$ 1 fm^{-3} the effective masses are below 300 MeV, and may even reach ~ 100 MeV for some NL samples. This corresponds to an approximate chiral symmetry restoration and could be the explanation for a behavior similar to the one expected for deconfined matter.

We complete this discussion with Fig. 5.10, where the ratio of the pressure to the free particle pressure p/p_{free} is plotted as a function of the baryon chemical potential and the quantity d_c defined in [8] as a function of the baryon density. For all sets, the ratio p/p_{free} saturates and even decreases for chemical potentials greater than 1300 MeV, after attaining a value of the order of 0.3. Although the dispersion is quite large, this ratio takes values approximately 0.1 smaller than the ones obtained in [8]. In [8], it is proposed that $d_c < 0.2$ could identify the presence of deconfined matter. In fact, d_c never goes below 0.2 for DDH, and for DDB the median stays above 0.2 although values below 0.2 are compatible with the 95% CI. For the NL data set, the median of d_c may take values below 0.2 for densities above 1 fm^{-3}, and it will be interesting to understand the reason for this behavior.

As already referred before, several studies based on an agnostic description of the EOS constrain the generated EOS imposing at high densities the pQCD EOS. The baryon

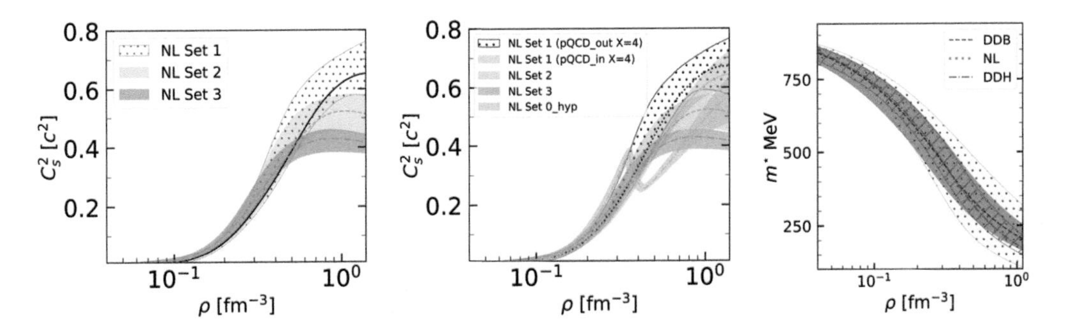

Figure 5.9 The figure displays the median and 95% credible interval of the square of sound velocity (c_s^2) for the NL model (left and middle panels) and the effective mass (right) for DDB, NL, and DDH models as a function of baryon density. The left panel highlights three distinct intervals of the parameter ξ : NL Set 1, NL Set 2, and NL Set 3. In the plot, NL Set 1 is represented by a black dotted region and corresponds to ξ values within the interval [0, 0.004]. NL Set 2 is represented by an orange region and encompasses ξ values within the range [0.004, 0.015]. NL Set 3 is depicted in blue and represents ξ values within the interval [0.015, 0.04]. Each set of EOS contains a comparable number of samples, approximately 18,000 samples, providing a robust statistical basis for the displayed results. In the middle panel, Set 1 was divided into two parts: green (black dotted) EoS that satisfy (do not satisfy) pQCD constraints with X=4. In this panel the c_s^2 for the NL-hyp set is also shown (pink band). The right panel was obtained with the data sets presented in Sec. 5.3.1, which contain $\sim 15,000$ to 17,000 samples.

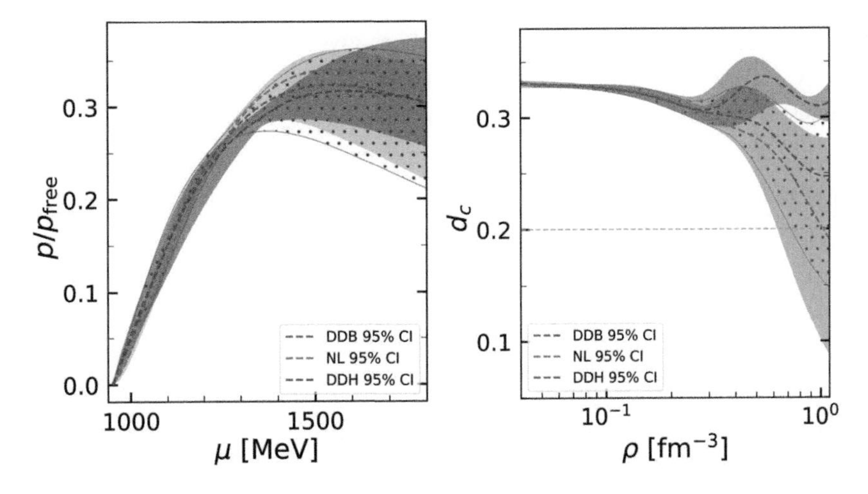

Figure 5.10 The figure illustrates the pressure normalized by the free pressure with respect to the baryon chemical potential μ, and the relationship between d_c and ρ for three data sets: DDB, NL, and DDH, arranged from left to right. The median values are represented by lines, while the 95% confidence interval regions are depicted as shaded bands.

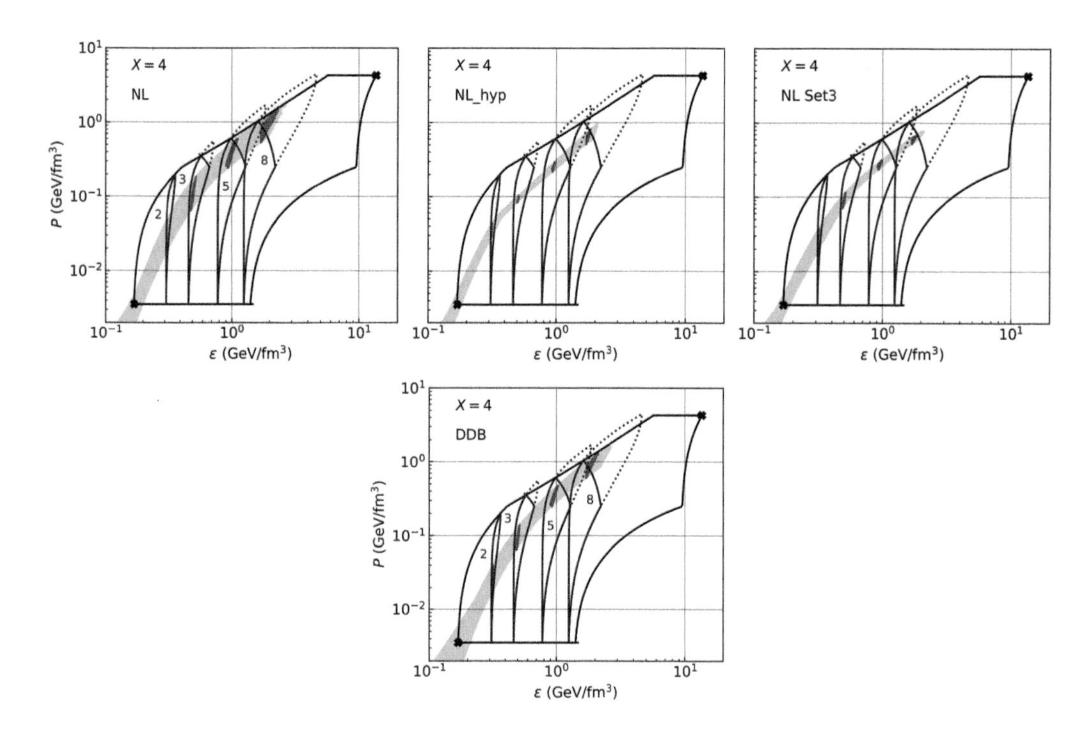

Figure 5.11 The pressure versus the energy density is shown for NL, NL-hyp, and NL restricted to $\xi \in [0.015 : 0.04]$ (set 3) (top line, from left to right) and DDB (bottom line). The constraints from Ref. [55] that ensure stability, causality, and thermodynamic consistency delimit the region inside the black solid line. The application of the constraints specifically to some baryon number densities, $n = 2$, 3, 5, and 8 n_s, defines the regions enclosed by the solid (dotted) blue lines that satisfy (do not satisfy) the pQCD constraints, respectively, from left to right (where n_s=0.16 fm^{-3}). Results for the most constraining renormalization scale parameter, $X = 4$, [58] are given. The green and red dots represent, respectively, the models in our sets that satisfy and do not satisfy pQCD constraints. Notice that the central density of the maximum mass stars is $\rho_c \lesssim 7\, n_s$.

density for which pQCD EOS is defined, $\gtrsim 40\rho_0$, is outside the range of densities where the RMF models defined in Sec. 5.2.1 are valid. Using thermodynamic relations and causality [55] showed that pQCD EOS imposes constraints at densities existing in the interior of NS, in particular, for densities $\gtrsim 2.2\rho_0$. In the following, we study how these constraints affect our different data sets. In order to build the geometric construction proposed in [55], we consider the same constraints these authors chose although other choice could have been done, in particular, at low densities, since our EOS may be considered well constrained until $2\rho_0$, the density above which non-nucleon degrees of freedom may set in. Note also that the pQCD constraints depend on the renormalization scale [57]: we show results in Fig. 5.11 for the scale imposing the strongest constraints, $X = 4$, [66]. The analysis was performed for sets NL, NL-hyp, NL restricted to $\xi \in [0.015 : 0.04]$ (labelled as set 3) and DDB and the following conclusions may be drawn: a) the constraints are satisfied for $\xi > 0.015$, i.e. set 3; b) only a few EOS from the set NL-hyp do not satisfy the constraints at $8n_s$. Notice, however, that at 90% CI the largest density in the center of NS within this set is $\sim 6n_s$, and, therefore all hyperonic stars are compatible with pQCD. In [66], it was shown that for smaller QCD renormalization scales the total set satisfies the pQCD constraints up to $8n_s$; c) several EOS from the set NL do not satisfy the pQCD constraints at densities $\sim 8n_s$ or even $\sim 5n_s$. These models have $\xi < 0.004$, and the highest maximum masses. If these EOS are removed from the NL set, the absolute maximum mass drops from $\sim 2.75M_\odot$ to $\sim 2.5M_\odot$ for models that satisfy pQCD with $X = 1$, and to $\sim 2.15M_\odot$ if $X = 4$. In the middle panel of Fig. 5.9, the speed of sound squared for the NL EOS with $\xi < 0.004$, has been divided into two subsets, according to their capacity to satisfying (green band) or not (black dotted band) the pQCD constraints. The EOS that satisfy these constraints present smaller values of c_s^2 at high densities. The bottom line of Fig. 5.11, we also show results for the DDB set, again taking the most constraining QCD scale: some EOS do not satisfy the 8 n_s constraints, but at 90% CI no star with a central density above $7n_s$ was obtained.

5.4 CONCLUSIONS

It was an objective of the present work to analyze in a critical way the capacity that RMF models have to describe hadronic matter, and the overall implications when they are used to extract nuclear matter properties from NS observations. As RMF models, we have considered two of the main frameworks frequently used, RMF models including non-linear mesonic terms with constant coupling parameters, designated as NL [14, 50, 66, 77, 94], and models with coupling parameters with an explicit dependence on the density, which do not include mesonic terms beyond quadratic terms [59, 65, 100, 101]. Other relativistic mean-field approaches have been left out, such as the chiral invariant nuclear model discussed in [53] and developed later in [20, 73, 80, 103], or the inclusion of the isovector scalar meson as studied in [63]. They will be considered in a future work.

A set of fit-data has been imposed, constituted by some nuclear matter properties, the pure neutron matter pressure obtained within a chEFT, and a maximum star mass above $2M_\odot$. A Bayesian inference formalism was applied to determine the coupling parameters probability distribution, and from these the NMP and NS properties were calculated. We have shown the mass-radius domain spanned by the posterior of the three data sets are not totally coincident, with the DDH framework predicting smaller radii, DDB larger radii, and the NL larger maximum masses, all at 90% confidence intervals.

The inclusion of hyperons in these models was also discussed. It was shown that hyperons do not exclude $2M_\odot$ stars, although maximum masses are much smaller than the ones attained with nucleonic models. However, the radius of canonical stars are larger if hyperonic degrees of freedom are introduced. This is due to the fact that in order to attain a $2M_\odot$ mass, and since the onset of hyperons softens de EOS, the nuclear matter parameters describing the symmetric nuclear matter EOS have to be larger. This confirms similar conclusions drawn in [38, 67].

The behavior of the proton fraction with density inside NS was also discussed. It was shown that the frameworks DDH and DDB have too small high-density couplings to the ϱ-meson and as a consequence no nucleonic direct Urca processes are predicted inside NS, as already discussed in [36, 37, 64]. This limitation of the models with density-dependent couplings was overcome with a generalization of the ϱ-meson coupling. The new parameter introduced may be constrained by observations on the cooling of NS. In fact, in [64] it was shown that above three times saturation density the symmetry energy is strongly correlated with the mass of NS characterized by the onset of nucleonic direct Urca processes at their center. Constraints from chEFT seem to rule out the direct Urca onset inside NS with a mass below 1.4 $M_{\odot}$.

We have analyzed several EOS properties as the speed of sound, the trace anomaly, and the consistence of the RMF EOS with the predictions of pQCD. It was shown that within DDH and DDB models the speed of sound are monotonically increasing functions of the density, while within the NL model, the speed of sound behavior is sensitive to the coupling of ω^4 term, and may present a maximum followed by a decreasing tendency with density. Its behavior may be confused with the onset of a non-nucleonic degree of freedom, as discussed in [9, 10, 90]. The different behaviors of the three frameworks reflect the different functionals that define the EOS of each one and the lack of constraining high-density observations or experimental data. It was shown that the three models predict values of the polytropic index below 1.75 for densities above 0.4 to 0.7 fm^{-3} and that the trace anomaly becomes negative for these densities. It was also discussed that the quantity related to the trace anomaly and its derivative introduced in [8], d_c, takes values generally above 0.2, a limit proposed in [8] as defining a transition to deconfined quark matter, although within the models NL and DDB values below 0.2 are not excluded at densities above $\gtrsim 0.6$ fm^{-3}. Notice, however, while in [8] this quantity may take values above 0.5, for the present three models it never takes values above 0.35.

Using thermodynamic and causality arguments together with low density nuclear matter and high-density pQCD constraints, a discussion similar to the one proposed in [55] was developed, and some of the models within RMF that do not satisfy the high density constraints have been identify. These are models with a very stiff high density EOS, although still causal.

ACKNOWLEDGMENTS

This work was partially supported by national funds from FCT (Fundação para a Ciência e a Tecnologia, I.P, Portugal) under Projects No. UIDP/04564/2020, No. UIDB/-04564/2020 and 2022.06460.PTDC and No. POCI-01-0145-FEDER-029912. The authors acknowledge the Laboratory for Advanced Computing at the University of Coimbra (https://www.uc.pt/lca) for providing HPC resources that have contributed to the research results reported within this paper.

TABLE OF MODEL PARAMETERS

The posterior parameters for the three models DDB, NL and DDH, obtained in Sec. 5.2.2 are given in Table 5.6.

TABLE 5.6 Based on the posterior distribution for DDB, NL, and DDH restricted to nucleonic degrees of freedom, the median values and the 90% confidence intervals (CI) for the parameters were calculated. Please refer to Section 5.2.2 for the specific terminology used for parameter names. The masses of the nucleon, ω meson, and ρ meson in all models are 939 MeV, 783 MeV, and 763 MeV, respectively. However, the σ meson mass is fixed to 500 MeV for NL and 550 MeV for DDB and DDH.

DDB				NL				DDH			
Parameters	median	90% CI		Parameters	median	90% CI		Parameters	median	90% CI	
		min	max			min	max			min	max
g_σ	9.024	8.170	10.059	g_σ	8.454	8.010	9.691	g_σ	8.827	8.146	9.322
g_ω	10.761	9.413	12.313	g_ω	9.915	9.084	12.167	g_ω	10.475	9.378	11.224
g_ρ	3.954	3.485	4.389	g_ρ	12.193	9.546	14.599	g_ρ	3.976	3.475	4.434
a_σ	0.080	0.054	0.113	B	4.586	2.205	6.903	a_σ	1.247	1.158	1.499
a_ω	0.039	0.004	0.105	C	-1.985	-4.627	3.530	b_σ	1.392	0.585	4.082
a_ρ	0.542	0.318	0.703	ξ	0.004	0.000	0.016	c_σ	1.867	0.775	6.011
				Λ	0.064	0.036	0.103	d_σ	0.423	0.235	0.656
								a_ω	1.215	1.022	1.663
								b_ω	7.544	1.876	14.074
								c_ω	9.546	2.307	19.140
								d_ω	0.187	0.132	0.380
								a_ρ	0.500	0.304	0.720

Bibliography

[1] B. P. Abbott et al. Multi-messenger Observations of a Binary Neutron Star Merger. *Astrophys. J. Lett.* 848(2):L12.

[2] B. P. Abbott et al. GW170817: Observation of Gravitational Waves from a Binary Neutron Star Inspiral. *Phys. Rev. Lett.*, 119(16):161101, 2017.

[3] B. P. Abbott et al. GW170817: Measurements of neutron star radii and equation of state. *Phys. Rev. Lett.*, 121(16):161101, 2018.

[4] B. P. Abbott et al. Properties of the binary neutron star merger GW170817. *Phys. Rev. X*, 9(1):011001, 2019.

[5] B. P. Abbott et al. GW190425: Observation of a Compact Binary Coalescence with Total Mass $\sim 3.4 M_{\odot}$. *Astrophys. J. Lett.*, 892(1):L3, 2020.

[6] Mohammad Al-Mamun, Andrew W. Steiner, Joonas Nättilä, Jacob Lange, Richard O'Shaughnessy, Ingo Tews, Stefano Gandolfi, Craig Heinke, and Sophia Han. Combining Electromagnetic and Gravitational-Wave Constraints on Neutron-Star Masses and Radii. *Phys. Rev. Lett.*, 126(6):061101, 2021.

[7] Sinan Altiparmak, Christian Ecker, and Luciano Rezzolla. On the Sound Speed in Neutron Stars. *Astrophys. J. Lett.*, 939(2):L34, 2022.

[8] Eemeli Annala, Tyler Gorda, Joonas Hirvonen, Oleg Komoltsev, Aleksi Kurkela, Joonas Nättilä, and Aleksi Vuorinen. Strongly interacting matter exhibits deconfined behavior in massive neutron stars. *Nat. Comm.*,14:8451, 2023.

[9] Eemeli Annala, Tyler Gorda, Evangelia Katerini, Aleksi Kurkela, Joonas Nättilä, Vasileios Paschalidis, and Aleksi Vuorinen. Multimessenger constraints for ultra-dense matter. *Phys. Rev. X*, 12: 011058, 2022.

[10] Eemeli Annala, Tyler Gorda, Aleksi Kurkela, Joonas Nättilä, and Aleksi Vuorinen. Evidence for quark-matter cores in massive neutron stars. *Nature Phys.*, 16(9):907–910, 2020.

[11] J. Antoniadis, P. C. C. Freire, N. Wex, T. M. Tauris, R. S. Lynch, M. H. van Kerkwijk, M. Kramer, C. Bassa, V. S. Dhillon, T. Driebe, J. W. T. Hessels, V. M. Kaspi, V. I. Kondratiev, N. Langer, T. R. Marsh, M. A. McLaughlin, T. T. Pennucci, S. M. Ransom, I. H. Stairs, J. van Leeuwen, J. P. W. Verbiest, and D. G. Whelan. A Massive Pulsar in a Compact Relativistic Binary. *Science*, 340:448, April 2013.

[12] M. V. Beznogov and D. G. Yakovlev. Statistical theory of thermal evolution of neutron stars – II. Limitations on direct Urca threshold. *Mon. Not. Roy. Astron. Soc.*, 452(1):540–548, 2015.

[13] Mikhail V. Beznogov and Adriana R. Raduta. Bayesian inference of the dense matter equation of state built upon covariant density functionals. *Phys. Rev. C*, 107(4):045803, 2023.

[14] J. Boguta and A. R. Bodmer. Relativistic Calculation of Nuclear Matter and the Nuclear Surface. *Nucl. Phys. A*, 292:413–428, 1977.

[15] J. Buchner, A. Georgakakis, K. Nandra, L. Hsu, C. Rangel, M. Brightman, A. Merloni, M. Salvato, J. Donley, and D. Kocevski. X-ray spectral modelling of the AGN obscuring region in the CDFS: Bayesian model selection and catalogue. *Astron. Astrophys.*, 564:A125, 2014.

[16] Johannes Buchner. Nested sampling methods *Statist. Surv.*, 17:169, 2023.

[17] Valéria Carvalho, Márcio Ferreira, Tuhin Malik, and Constança Providência. Decoding Neutron Star Observations: Revealing Composition through Bayesian Neural Networks. *hys. Rev. D*, 108:043031, 2023.

[18] Rafael Cavagnoli, Debora P. Menezes, and Constanca Providencia. Neutron star properties and the symmetry energy. *Phys. Rev. C*, 84:065810, 2011.

[19] H. T. Cromartie et al. Relativistic Shapiro delay measurements of an extremely massive millisecond pulsar. *Nature Astron.*, 4(1):72–76, 2019.

[20] Constanca da Providencia, Joao da Providencia, and Steven A. Moszkowski. EOS of nuclear matter within a generalised NJL model. *Ser. Adv. Quant. Many Body Theor.*, 6:242–245, 2002.

[21] Pedro Barata de Tovar, Márcio Ferreira, and Constança Providência. Determination of the symmetry energy from the neutron star equation of state. *Phys. Rev. D*, 104(12):123036, 2021.

[22] Paul Demorest, Tim Pennucci, Scott Ransom, Mallory Roberts, and Jason Hessels. Shapiro Delay Measurement of A Two Solar Mass Neutron Star. *Nature*, 467:1081–1083, 2010.

[23] C. Drischler, K. Hebeler, and A. Schwenk. Chiral interactions up to next-to-next-to-next-to-leading order and nuclear saturation. *Phys. Rev. Lett.*, 122(4):042501, 2019.

[24] C. Drischler, J. A. Melendez, R. J. Furnstahl, and D. R. Phillips. Quantifying uncertainties and correlations in the nuclear-matter equation of state. *Phys. Rev. C*, 102(5):054315, 2020.

[25] Christian Drischler, Sophia Han, and Sanjay Reddy. Large and massive neutron stars: Implications for the sound speed within QCD of dense matter. *Phys. Rev. C*, 105(3):035808, 2022.

[26] M. Dutra, O. Lourenço, S. S. Avancini, B. V. Carlson, A. Delfino, D. P. Menezes, C. Providência, S. Typel, and J. R. Stone. Relativistic Mean-Field Hadronic Models under Nuclear Matter Constraints. *Phys. Rev. C*, 90(5):055203, 2014.

[27] Reed Essick, Philippe Landry, and Daniel E. Holz. Nonparametric Inference of Neutron Star Composition, Equation of State, and Maximum Mass with GW170817. *Phys. Rev. D*, 101(6):063007, 2020.

[28] Reed Essick, Philippe Landry, Achim Schwenk, and Ingo Tews. Detailed examination of astrophysical constraints on the symmetry energy and the neutron skin of Pb208 with minimal modeling assumptions. *Phys. Rev. C*, 104(6):065804, 2021.

[29] M. árcio Ferreira and Constança Providência. Unveiling the nuclear matter EoS from neutron star properties: a supervised machine learning approach. *JCAP*, 07:011, 2021.

[30] Márcio Ferreira, M. Fortin, Tuhin Malik, B. K. Agrawal, and Constança Providência. Empirical constraints on the high-density equation of state from multimessenger observables. *Phys. Rev. D*, 101(4):043021, 2020.

[31] Márcio Ferreira and Constança Providência. Constraints on high density equation of state from maximum neutron star mass. *Phys. Rev. D*, 104(6):063006, 2021.

[32] E. Fonseca et al. Refined Mass and Geometric Measurements of the High-mass PSR J0740+6620. *Astrophys. J. Lett.*, 915(1):L12, 2021.

[33] M. Fortin, S. S. Avancini, C. Providência, and I. Vidaña. Hypernuclei and massive neutron stars. *Phys. Rev. C*, 95(6):065803, 2017.

[34] M. Fortin, M. Oertel, and C. Providência. Hyperons in hot dense matter: what do the constraints tell us for equation of state? *Publ. Astron. Soc. Austral.*, 35:44, 2018.

[35] M. Fortin, C. Providencia, A. R. Raduta, F. Gulminelli, J. L Zdunik, P. Haensel, and M. Bejger. Neutron star radii and crusts: uncertainties and unified equations of state. *Phys. Rev. C*, 94(3):035804, 2016.

[36] M. Fortin, C. Providencia, A. R. Raduta, F. Gulminelli, J. L Zdunik, P. Haensel, and M. Bejger. Neutron star radii and crusts: uncertainties and unified equations of state. *Phys. Rev. C*, 94(3):035804, 2016.

[37] M. Fortin, Adriana R. Raduta, Sidney Avancini, and Constança Providência. Relativistic hypernuclear compact stars with calibrated equations of state. *Phys. Rev. D*, 101(3):034017, 2020.

[38] M. Fortin, J. L. Zdunik, P. Haensel, and M. Bejger. Neutron stars with hyperon cores: stellar radii and equation of state near nuclear density. *Astron. Astrophys.*, 576:A68, 2015.

[39] Morgane Fortin, Adriana R. Raduta, Sidney Avancini, and Constança Providência. Thermal evolution of relativistic hyperonic compact stars with calibrated equations of state. *Phys. Rev. D*, 103(8):083004, 2021.

[40] Yuki Fujimoto, Kenji Fukushima, Larry D. McLerran, and Michal Praszalowicz. Trace Anomaly as Signature of Conformality in Neutron Stars. *Phys. Rev. Lett.*, 129(25):252702, 2022.

[41] Yuki Fujimoto, Kenji Fukushima, and Koichi Murase. Methodology study of machine learning for the neutron star equation of state. *Phys. Rev. D*, 98(2):023019, 2018.

[42] Yuki Fujimoto, Kenji Fukushima, and Koichi Murase. Mapping neutron star data to the equation of state using the deep neural network. *Phys. Rev. D*, 101:054016, 2020.

[43] Yuki Fujimoto, Kenji Fukushima, and Koichi Murase. Extensive studies of the neutron star equation of state from the deep learning inference with the observational data augmentation. *J. High Energy Phys.*, 2021(3):1–46, 2021.

[44] A. Gal, E. V. Hungerford, and D. J. Millener. Strangeness in nuclear physics. *Rev. Mod. Phys.*, 88(3):035004, 2016.

[45] Andrew Gelman, John B Carlin, Hal S Stern, David B Dunson, Aki Vehtari, Donald B Rubin, John Carlin, Hal Stern, Donald Rubin, and David Dunson. *Bayesian Data Analysis Third edition*. CRC Press, Boca Raton, Florida, 2013.

[46] Norman K. Glendenning. *Compact Stars*. Springer, Berlin, 1997.

[47] Tyler Gorda, Oleg Komoltsev, and Aleksi Kurkela. Ab-initio QCD calculations impact the inference of the neutron-star-matter equation of state. 4 2022.

[48] K. Hebeler, J. M. Lattimer, C. J. Pethick, and A. Schwenk. Equation of state and neutron star properties constrained by nuclear physics and observation. *Astrophys. J.*, 773:11, 2013.

[49] Tanja Hinderer. Tidal Love numbers of neutron stars. *Astrophys. J.*, 677:1216–1220, 2008.

[50] C. J. Horowitz and J. Piekarewicz. Neutron star structure and the neutron radius of Pb-208. *Phys. Rev. Lett.*, 86:5647, 2001.

[51] Chun Huang, Geert Raaijmakers, Anna L. Watts, Laura Tolos, and Constança Providência. Constraining fundamental nuclear physics parameters using neutron star mass-radius measurements I: Nucleonic models. 3 2023.

[52] Sk Md Adil Imam, N. K. Patra, C. Mondal, Tuhin Malik, and B. K. Agrawal. Bayesian reconstruction of nuclear matter parameters from the equation of state of neutron star matter. *Phys. Rev. C*, 105(1):015806, 2022.

[53] V. Koch, T. S. Biro, J. Kunz, and U. Mosel. A Chirally Invariant Fermionic Field Theory for Nuclear Matter. *Phys. Lett. B*, 185:1–5, 1987.

[54] Toru Kojo. QCD equations of state and speed of sound in neutron stars. *AAPPS Bull.*, 31(1):11, 2021.

[55] Oleg Komoltsev and Aleksi Kurkela. How Perturbative QCD Constrains the Equation of State at Neutron-Star Densities. *Phys. Rev. Lett.*, 128(20):202701, 2022.

[56] Aleksi Kurkela. Thoughts about the utility of perturbative QCD in the cores of neutron stars. In *15th Conference on Quark Confinement and the Hadron Spectrum*, 11 2022.

[57] Aleksi Kurkela, Eduardo S. Fraga, Jürgen Schaffner-Bielich, and Aleksi Vuorinen. Constraining neutron star matter with Quantum Chromodynamics. *Astrophys. J.*, 789:127, 2014.

[58] Aleksi Kurkela, Paul Romatschke, and Aleksi Vuorinen. Cold Quark Matter. *Phys. Rev. D*, 81:105021, 2010.

[59] G. A. Lalazissis, T. Niksic, D. Vretenar, and P. Ring. New relativistic mean-field interaction with density-dependent meson-nucleon couplings. *Phys. Rev. C*, 71:024312, 2005.

[60] Philippe Landry and Reed Essick. Nonparametric inference of the neutron star equation of state from gravitational wave observations. *Phys. Rev. D*, 99(8):084049, 2019.

[61] Lee Lindblom. Determining the Nuclear Equation of State from Neutron-Star Masses and Radii. *Astrophys. J.*, 398:569, October 1992.

[62] Lee Lindblom. Spectral Representations of Neutron-Star Equations of State. *Phys. Rev. D*, 82:103011, 2010.

[63] B. Liu, V. Greco, V. Baran, M. Colonna, and M. Di Toro. Asymmetric nuclear matter: The Role of the isovector scalar channel. *Phys. Rev. C*, 65:045201, 2002.

[64] Tuhin Malik, B. K. Agrawal, and Constança Providência. Inferring the nuclear symmetry energy at suprasaturation density from neutrino cooling. *Phys. Rev. C*, 106(4):L042801, 2022.

[65] Tuhin Malik, Márcio Ferreira, B. K. Agrawal, and Constança Providência. Relativistic Description of Dense Matter Equation of State and Compatibility with Neutron Star Observables: A Bayesian Approach. *Astrophys. J.*, 930(1):17, 2022.

[66] Tuhin Malik, Márcio Ferreira, Milena Bastos Albino, and Constança Providência. Spanning the full range of neutron star properties within a microscopic description. 1 2023.

[67] Tuhin Malik and Constança Providência. Bayesian inference of signatures of hyperons inside neutron stars. 5 2022.

[68] Jérôme Margueron, Rudiney Hoffmann Casali, and Francesca Gulminelli. Equation of state for dense nucleonic matter from metamodeling. I. Foundational aspects. *Phys. Rev. C*, 97(2):025805, 2018.

[69] Jérôme Margueron, Rudiney Hoffmann Casali, and Francesca Gulminelli. Equation of state for dense nucleonic matter from metamodeling. II. Predictions for neutron star properties. *Phys. Rev. C*, 97(2):025806, 2018.

[70] Jérôme Margueron, Rudiney Hoffmann Casali, and Francesca Gulminelli. Equation of state for dense nucleonic matter from metamodeling. I. Foundational aspects. *Phys. Rev.*, C97(2):025805, 2018.

[71] M. C. Miller et al. PSR J0030+0451 Mass and Radius from $NICER$ Data and Implications for the Properties of Neutron Star Matter. *Astrophys. J. Lett.*, 887(1):L24, 2019.

[72] M. C. Miller et al. The Radius of PSR J0740+6620 from NICER and XMM-Newton Data. *Astrophys. J. Lett.*, 918(2):L28, 2021.

[73] I. N. Mishustin, L. M. Satarov, and W. Greiner. How far is normal nuclear matter from the chiral symmetry restoration? *Phys. Rep.*, 391:363–380, 2004.

[74] Chiranjib Mondal and Francesca Gulminelli. Can we decipher the composition of the core of a neutron star? *Phys. Rev. D*, 105:083016, 2021.

[75] Filip Morawski and Michał Bejger. Neural network reconstruction of the dense matter equation of state derived from the parameters of neutron stars. *Astron. Astrophys.*, 642:A78, 2020.

[76] Horst Mueller and Brian D. Serot. Relativistic mean field theory and the high density nuclear equation of state. *Nucl. Phys. A*, 606:508–537, 1996.

[77] Horst Mueller and Brian D. Serot. Relativistic mean field theory and the high density nuclear equation of state. *Nucl. Phys. A*, 606:508–537, 1996.

[78] J. R. Oppenheimer and G. M. Volkoff. On Massive neutron cores. *Phys. Rev.*, 55:374–381, 1939.

[79] Feryal Ozel, Gordon Baym, and Tolga Guver. Astrophysical Measurement of the Equation of State of Neutron Star Matter. *Phys. Rev. D*, 82:101301, 2010.

[80] Helena Pais, Débora P. Menezes, and Constança Providência. Neutron stars: From the inner crust to the core with the (extended) Nambu–Jona-Lasinio model. *Phys. Rev. C*, 93(6):065805, 2016.

[81] Constança Providência, Morgan Fortin, Helena Pais, and Aziz Rabhi. Hyperonic stars and the symmetry energy. 11 2018.

[82] G. Raaijmakers, S. K. Greif, K. Hebeler, T. Hinderer, S. Nissanke, A. Schwenk, T. E. Riley, A. L. Watts, J. M. Lattimer, and W. C. G. Ho. Constraints on the Dense Matter Equation of State and Neutron Star Properties from NICER's Mass–Radius Estimate of PSR J0740+6620 and Multimessenger Observations. *Astrophys. J. Lett.*, 918(2):L29, 2021.

[83] Carolyn A. Raithel, Feryal Ozel, and Dimitrios Psaltis. From Neutron Star Observables to the Equation of State: An Optimal Parametrization. *Astrophys. J.*, 831(1):44, 2016.

[84] Paul S. *et al.* Ray. STROBE-X: X-ray Timing and Spectroscopy on Dynamical Timescales from Microseconds to Years. *arXiv e-prints*, page arXiv:1903.03035, March 2019.

[85] Thomas E. Riley et al. A $NICER$ View of PSR J0030+0451: Millisecond Pulsar Parameter Estimation. *Astrophys. J. Lett.*, 887(1):L21, 2019.

[86] Thomas E. Riley et al. A NICER View of the Massive Pulsar PSR J0740+6620 Informed by Radio Timing and XMM-Newton Spectroscopy. *Astrophys. J. Lett.*, 918(2):L27, 2021.

[87] Roger W. Romani, D. Kandel, Alexei V. Filippenko, Thomas G. Brink, and WeiKang Zheng. PSR J1810+1744: Companion Darkening and a Precise High Neutron Star Mass. *Astrophys. J. Lett.*, 908(2):L46, 2021.

[88] Shlomo, S., Kolomietz, V. M., and Colò, G. Deducing the nuclear-matter incompressibility coefficient from data on isoscalar compression modes. *Eur. Phys. J. A*, 30(1):23–30, 2006.

[89] John Skilling. Nested Sampling. In Rainer Fischer, Roland Preuss, and Udo Von Toussaint, editors, *Bayesian Inference and Maximum Entropy Methods in Science and Engineering: 24th International Workshop on Bayesian Inference and Maximum Entropy Methods in Science and Engineering*, volume 735 of *American Institute of Physics Conference Series*, pages 395–405, November 2004.

[90] Rahul Somasundaram, Ingo Tews, and Jérôme Margueron. Perturbative QCD and the neutron star equation of state. *Phys. Rev. C*, 107(5):L052801, 2023.

[91] Andrew W. Steiner, James M. Lattimer, and Edward F. Brown. The Equation of State from Observed Masses and Radii of Neutron Stars. *Astrophys. J.*, 722:33–54, 2010.

[92] Andrew W. Steiner, James M. Lattimer, and Edward F. Brown. The Neutron Star Mass-Radius Relation and the Equation of State of Dense Matter. *Astrophys. J. Lett.*, 765:L5, 2013.

[93] J. R. Stone, V. Dexheimer, P. A M. Guichon, A. W. Thomas, and S. Typel. Equation of state of hot dense hyperonic matter in the Quark–Meson-Coupling (QMC-A) model. *Mon. Not. Roy. Astron. Soc.*, 502(3):3476–3490, 2021.

[94] Y. Sugahara and H. Toki. Relativistic mean field theory for unstable nuclei with nonlinear sigma and omega terms. *Nucl. Phys. A*, 579:557–572, 1994.

[95] Xiangdong Sun, Zhiqiang Miao, Baoyuan Sun, and Ang Li. Astrophysical Implications on Hyperon Couplings and Hyperon Star Properties with Relativistic Equations of States. *Astrophys. J.*, 942(1):55, 2023.

[96] Hoa Dinh Thi, Chiranjib Mondal, and Francesca Gulminelli. The Nuclear Matter Density Functional under the Nucleonic Hypothesis. *Universe*, 7(10):373, 2021.

[97] et al Todd-Rutel. Neutron-Rich Nuclei and Neutron Stars: A New Accurately Calibrated Interaction for the Study of Neutron-Rich Matter. *Phys. Rev. Lett.*, 95:122501, 2005.

[98] Richard C. Tolman. Static solutions of Einstein's field equations for spheres of fluid. *Phys. Rev.*, 55:364–373, 1939.

[99] Silvia Traversi, Prasanta Char, and Giuseppe Pagliara. Bayesian Inference of Dense Matter Equation of State within Relativistic Mean Field Models using Astrophysical Measurements. *Astrophys. J.*, 897:165, 2020.

[100] S. Typel, G. Ropke, T. Klahn, D. Blaschke, and H. H. Wolter. Composition and thermodynamics of nuclear matter with light clusters. *Phys. Rev. C*, 81:015803, 2010.

[101] S. Typel and H. H. Wolter. Relativistic mean field calculations with density dependent meson nucleon coupling. *Nucl. Phys. A*, 656:331–364, 1999.

[102] Anna L. *et al.* Watts. Dense matter with eXTP. *Science China Physics, Mechanics, and Astronomy*, 62(2):29503, February 2019.

[103] Si-Na Wei, Wei-Zhou Jiang, Rong-Yao Yang, and Dong-Rui Zhang. Symmetry energy and neutron star properties in the saturated Nambu–Jona-Lasinio model. *Phys. Lett. B*, 763:145–150, 2016.

[104] S. Weissenborn, D. Chatterjee, and J. Schaffner-Bielich. Hyperons and massive neutron stars: the role of hyperon potentials. *Nucl. Phys. A*, 881:62–77, 2012.

[105] Wen-Jie Xie and Bao-An Li. Bayesian Inference of High-density Nuclear Symmetry Energy from Radii of Canonical Neutron Stars. *Astrophys. J.*, 883:174, 2019.

[106] Wen-Jie Xie and Bao-An Li. Bayesian Inference of the Symmetry Energy of Superdense Neutron-rich Matter from Future Radius Measurements of Massive Neutron Stars. *Astrophys. J.*, 899(1):4, 2020.

[107] D. G. Yakovlev, A. D. Kaminker, Oleg Y. Gnedin, and P. Haensel. Neutrino emission from neutron stars. *Phys. Rept.*, 354:1, 2001.

[108] Dima G. Yakovlev and C. J. Pethick. Neutron star cooling. *Ann. Rev. Astron. Astrophys.*, 42:169–210, 2004.

[109] Kai Zhou, Lingxiao Wang, Long-Gang Pang, and Shuzhe Shi. Exploring QCD matter in extreme conditions with Machine Learning. 3 2023.

[110] Wenjie Zhou, Jinniu Hu, Ying Zhang, and Hong Shen. Nonparametric Model for the Equations of State of a Neutron Star from Deep Neural Network. *Astrophys. J.*, 950(2):186, 2023.

CHAPTER 6

Inference of Microscopic Nuclear Interactions and the Equation of State from Multimessenger Astrophysics

Rahul Somasundaram, Ingo Tews

6.1 INTRODUCTION

Neutron stars provide a unique opportunity to study cold nuclear matter, and the interactions within, at the highest densities in the universe. During the last decade, groundbreaking astrophysical observations of these fascinating objects have revolutionized the field of dense-matter physics.

Radio observations of rotating neutron stars have enabled astronomers to find the heaviest systems observed to date [17, 39, 42, 62], reaching masses beyond twice the mass of our sun. These discoveries alone ruled out many models of the nuclear equation of state (EOS) that are too "soft," i.e., those that only provide low to moderate pressure at the densities explored in neutron stars. Such soft EOS include models with strong softening phase transitions, such as first-order phase transitions to weakly interacting quark matter. While such transitions have not been ruled out, the observation of heavy neutron stars severely limits their possible parameter space.

Another exciting new opportunity to electromagnetically observe neutron stars was provided by the Neutron Star Interior Composition Explorer (NICER) [24, 25], a NASA mission to measure the time-dependent X-ray flux, referred to as a pulse profile, from fast-rotating neutron stars. The time dependence of these signals depends, among other things, on the star's compactness, providing constraints on the masses and radii of the objects [127, 128, 154, 155]. This inference depends on the modeling of the electromagnetic emission,

DOI: 10.1201/9781003306580-6

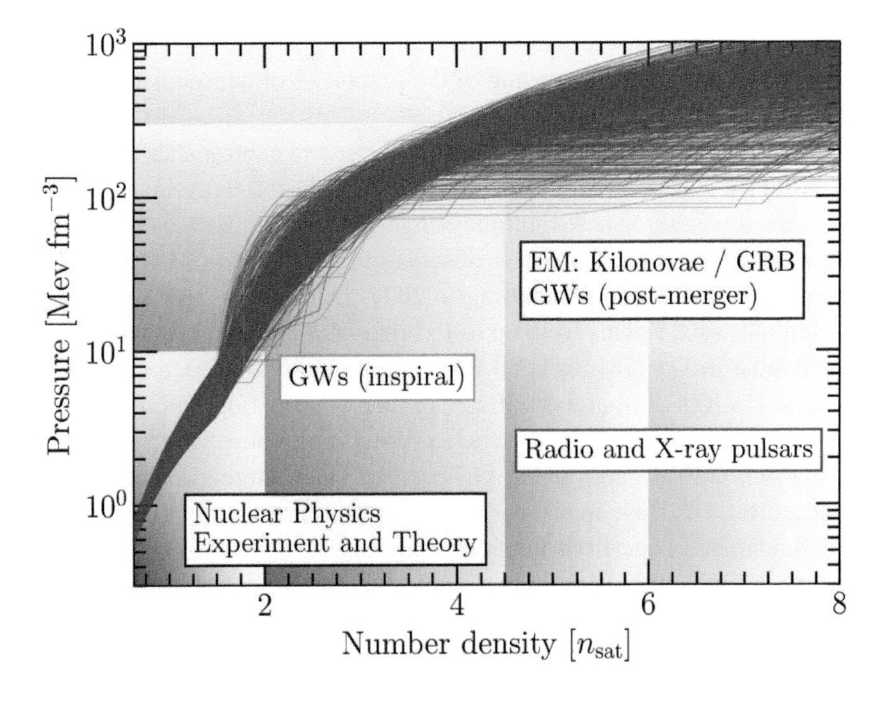

Figure 6.1 Several possible EOS models (blue lines) can be constrained at low densities by using nuclear-theory calculations and at high densities by different astrophysical messengers: gravitational waves from neutron-star inspirals, data from radio and x-ray pulsars, and electromagnetic signals associated with NS mergers. The boundaries indicated here are not strict but depend on the EOS and properties of the studied system. Reprinted from Ref. [136] with permission. © Springer Nature (2023).

in particular the location and properties of hot spots on the star's surface, but the NICER collaboration typically employs several models to explore this source of uncertainty. So far, the NICER collaboration provided mass-radius measurements of two neutron stars, both of which are consistent with each other and favor "stiff" EOS, i.e., those that provide higher relatively pressures at neutron-star densities.

Finally, a new chapter in neutron-star observations was started when a new window on the universe was opened: the observation of gravitational waves by the LIGO-Virgo collaboration. Shortly after the revolutionary observation of gravitational waves from a binary black-hole merger [2], the LIGO-Virgo collaboration observed gravitational waves from a binary neutron-star collision [3]. For the latter, its electromagnetic counterpart was also observed [4]. Gravitational waves are sensitive to the so-called "tidal deformability" of the stars, which measures how easy a star can be deformed by the gravitational field of its companion: A small, compact neutron star is much more difficult to deform than a large, fluffy one. The measurement of tidal deformability, therefore, serves as a probe for the radii of neutron stars. The observations of the first neutron-star merger, GW170817, have provided robust constraints on the nuclear EOS, preferring smaller neutron stars and softer EOS [5, 16, 28, 41, 43, 103, 104, 147, 148, 172]. The combination of all of these data has limited the space for the EOS, by removing very soft EOS that disagree with electromagnetic observations and very stiff EOS that disagree with gravitational waves. This has provided a better understanding of the properties of dense nuclear matter beyond nuclear saturation density.

While these observations invigorated dense-matter and neutron-star theory, the upcoming decade will be a decade of unprecedented opportunity for dense-matter physics. In the next 5–10 years, a wealth of new multimessenger data on neutron stars and their macroscopic properties is expected from new observation campaigns of LIGO, O4 from May 2023 till the end of 2024, and O5 starting in 2027. During O4, LIGO is expected to observe gravitational-wave signals from up to 10 binary NS mergers [37] while its sensitivity will almost double in O5 [18]. Beyond that, third-generation detectors, such as Cosmic Explorer in the US [60, 153] and the Einstein Telescope in Europe [146], will enable us to observe almost all neutron-star mergers occurring in the observable universe. This will initiate a data-rich era for neutron-star physics and offer unprecedented opportunities to probe the properties of dense nuclear matter with high precision, allowing us to address some of the fundamental questions in nuclear physics.

Among these, we will be able to better understand the features of microscopic interactions among nucleons in high-density matter, with a particular focus on high neutron-to-proton ratios. In the past, microscopic models for interactions among nucleons were constructed on a phenomenological basis, by proposing operator structures, linking the momenta, spins and isospins of interacting nucleons, in an ad hoc manner in order to describe available experimental data. In the past two decades, this phenomenological approach was replaced by the modern theory describing nuclear interactions, chiral effective field theory (EFT) [54, 117]. Chiral EFT is a systematic theory based on a low-momentum expansion of nucleonic interactions consistent with all symmetries of quantum Chromodynamics (QCD). Because of this, it provides a well-motivated order-by-order expansion for the relevant operators describing nuclear interactions, providing a clear scheme for how to improve interactions. Chiral EFT's systematic expansion also enables us to estimate theoretical uncertainties. However, the EFT momentum expansion breaks down with increasing nucleon momenta, i.e., at higher densities, as new degrees of freedom become important. This is very relevant for neutron stars because new degrees of freedom, such as deconfined quark matter, might play an important role in describing neutron-star structure. However, we do not know at which densities this breakdown occurs. Future astrophysical data will enable us to better understand the limits of applicability of EFT interactions describing dense matter in terms of nucleons [57]. They will also enable us to estimate the strength of the various interaction operators in chiral EFT [157]. In this contribution, we will discuss these two prospects.

Gaining a better understanding of nuclear interactions from multimessenger observations of neutron stars will help constrain microscopic physics at all densities and temperatures explored in nuclear systems, enabling us to build a consistent theory. For example, this will have direct implications for the physics of atomic nuclei at neutron-rich extremes that are probed in experiments at the Facility for Rare-Isotope Beams (FRIB). FRIB will measure the structure of neutron-rich nuclei close to the neutron dripline, where neutrons in atomic nuclei become unbound. The structure of these nuclei is determined by the same microscopic nuclear interactions, importantly among neutrons, and is extremely relevant for our understanding of the creation of heavy elements in the rapid neutron-capture process (r-process) [130]. Furthermore, a better understanding of nuclear interactions will alleviate a major source of uncertainty in astrophysical modeling [38, 149], allowing us to analyze the impact of thermal effects, neutrino interactions, and nuclear reactions in neutron-star mergers and the r-process, further elucidating the role that neutron-star mergers play in nucleosynthesis. To realize this enormous discovery potential in this upcoming data-rich era, high-fidelity nuclear-theory calculations for dense matter *with reliable uncertainty estimates* are crucial.

In this contribution, we will review the state of the art for nuclear theory describing the EOS at low densities (Sec. 6.2), explain how we can model uncertainties at high densities (Sec. 6.3), and finally discuss what we can learn about the dense-matter EOS and nuclear interactions from current and future astrophysical observations (Sec. 6.4).

6.2 NUCLEAR THEORY FOR THE EQUATION OF STATE AT LOW DENSITIES

6.2.1 Chiral Effective Field Theory

We presently cannot study many-body nuclear systems, such as atomic nuclei or nucleonic matter in the core of neutron stars, directly from first principles dictated by QCD. Hence, effective approaches to describe interactions are usually employed when calculating properties of nuclear systems. In low-energy nuclear physics, nucleonic degrees of freedom (d.o.f.), protons and neutrons, are the relevant ones, and effective approaches model the nuclear interactions between them, given by a Hamiltonian that can be written as

$$H = T + V_{NN} + V_{3N} + V_{AN}\,, \tag{6.1}$$

with T being the kinetic energy of the nucleons, V_{NN} the two-nucleon (NN) interaction, V_{3N} being the three-nucleon ($3N$) interaction, and V_{AN} containing all higher-nucleon (AN) interactions.

It is a long-standing problem in nuclear physics to model these interaction contributions, and many highly accurate phenomenological interaction models have been developed in the past decades. To study interactions between two nucleons, for example the Argonne, Nijmegen, and CD-Bonn potentials [118, 167, 184] were developed and guided by the quantitative reproduction of experimental NN scattering data. However, it also also been known for a long time that 3N interactions play an important role in nuclear physics [63], and the investigation of 3N interactions is at the forefront of nuclear-theory research. Examples of phenomenological 3N forces are the Tucson-Melbourne, Urbana, and Illinois potentials [32, 145, 156]. While these forces have led to remarkable results in nuclear physics, they are not built on a systematically improvable framework and were constructed phenomenologically. Hence, it is not clear how to connect similar physical processes in the NN and 3N interactions, how to improve both NN and 3N forces in a systematic way, and how to provide theoretical uncertainty estimates for calculations with these interactions. However, solutions to all of these issues are crucial to make use of a data-rich era in nuclear physics.

In the past 30 years, chiral EFT [134, 179, 181, 182] has been introduced and consolidated as a more fundamental approach to derive nuclear interactions. Importantly, it

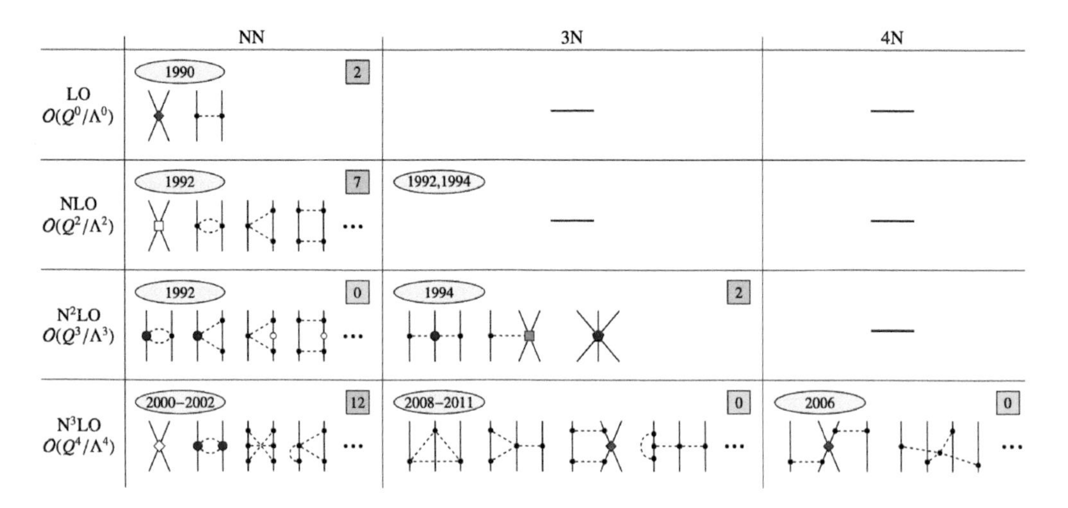

Figure 6.2 Contributions to the NN, 3N, and four-nucleon (4N) forces up to N^3LO in delta-less chiral EFT in Weinberg power counting. Solid (dashed) lines denote nucleons (pions). The figure indicates when the corresponding terms were derived (yellow ellipses) and how many new couplings appear at each order and sector (red squares). Reprinted from Ref. [80] with permission. © Elsevier (2021).

addresses the previously mentioned shortcomings, see also reviews in Refs. [55, 76, 144]. Chiral EFT provides a systematic framework to construct NN, 3N, and AN interactions consistent with the symmetries of QCD. It uses a separation of scales between nucleon momenta p, typically of the order of the pion mass, and the breakdown scale Λ_b, which is expected to be of the order of the ρ-meson mass, $\Lambda_b \approx 600$ MeV [47, 53], to expand nuclear interactions in powers of $Q \equiv p/\Lambda_b$, see Fig. 6.2. The physics beyond the breakdown scale is integrated out and described by a set of short-distance contact interactions, the strength of which is determined by low-energy couplings (LECs) that are typically fit to available NN scattering data and light nuclei (but see, e.g., Ref. [49] for an approach fitting to many-body systems). At each order in the EFT, only a finite number of diagrams enters the description of the interaction according to a chosen power-counting scheme, of which the Weinberg power counting has been predominant [181, 182], see Fig. 6.2. However, other power countings have been introduced in the past, see, e.g., Refs. [92–94, 111, 112, 132, 186]. In addition, there are chiral EFT interactions that include explicit intermediate delta-isobar states, see, e.g., Refs. [90, 141–143]. These interactions are called "delta-full," while interactions where the delta isobar is integrated out are called "delta-less." Chiral EFT, together with a power counting scheme, provides systematically improvable nuclear Hamiltonians that can then be employed in computational many-body methods.

Importantly, 3N interactions naturally emerge in the EFT expansion and are consistent with the NN sector, i.e., the same interaction vertices appearing in both sectors are governed by the same LECs. Four-nucleon interactions also emerge in the EFT scheme but were found to be negligible at the current level of precision [99]. Similarly, AN forces can be expected to be negligible based on the EFT expansion. Importantly, electroweak interactions can also be derived within the EFT and provide consistent electroweak currents [96–98, 139, 140].

Chiral EFT has been extremely successful in describing the structure of atomic nuclei and properties of dense nuclear matter and NSs, see e.g., Refs. [45, 72, 74, 78, 81, 86, 108, 135, 168, 171, 172, 183] for a small selection of works. In particular, calculations in the framework of chiral EFT have shown that 3N forces are very important for the correct description of nuclear systems, see for example Refs. [77, 79, 131, 135]. The key feature making chiral EFT crucially important for the multimessenger era is that it provides theoretical uncertainty estimates due to the truncation of its order-by-order expansion [44, 47, 53]. Quantifying and propagating these theoretical uncertainties, e.g., in dense nucleonic matter [109, 115, 173], is key for meaningful comparisons between competing nuclear-theory predictions and constraints from nuclear experiments and neutron-star observations, that are nowadays facilitated by Bayesian Inference in a statistically rigorous way [43, 57, 58, 88]. Only if uncertainties in nuclear interactions as well as astrophysical modeling are considered can one take full advantage of the great variety of data anticipated in the next decade, and arrive at robust conclusions about nuclear forces and the nuclear EOS.

To illustrate some of the concepts discussed above, we will briefly discuss the structure of V_{NN} within the chiral EFT framework. Details regarding V_{3N} can be found, for example, in the previously mentioned reviews. In general, chiral EFT interactions can be decomposed into short-range contact pieces and long-range pieces mediated by one- and multiple-pion exchanges,

$$V^{(\nu)} = V^{(\nu)}_{\text{cont}} + V^{(\nu)}_{\pi} , \tag{6.2}$$

where ν is the chiral EFT order, indicating the power of the expansion parameter Q^{ν}. The leading-order (LO) contact interactions are momentum independent, and are given by

$$V^{(0)}_{\text{cont}} = C_S + C_T \boldsymbol{\sigma}_1 \cdot \boldsymbol{\sigma}_2 , \tag{6.3}$$

where $\boldsymbol{\sigma}_i$ are the spin operators for the two nucleons. The parameters C_S and C_T are the two LECs that govern the strength of the LO contact interaction and are typically fit to NN scattering data. We note that the form of the LO contact interaction as given in Eq. (6.3) constitutes one possible choice of operators. It can be shown that different operator sets are linearly dependent and contain the same physics [87]. This freedom in choosing the operator basis essentially arises from the fact that nucleons, being fermions, have antisymmetric wavefunctions, and is referred to as the Fierz Rearrangement Freedom (FRF). We will not discuss further details here, but refer the interested reader to Refs. [67, 87] and references therein for more details.

At next-to-leading order (NLO), the contact interaction is now momentum dependent. For initial and final relative momenta $\mathbf{p}$ and $\mathbf{p}'$, momentum transfer $\mathbf{q} = \mathbf{p}' - \mathbf{p}$, and momentum transfer in the exchange channel $\mathbf{k} = (\mathbf{p}' + \mathbf{p})/2$, the NLO interaction is given by

$$\begin{aligned} V^{(2)}_{\text{cont}} &= C_1 \boldsymbol{q}^2 + C_2 \boldsymbol{q}^2 \boldsymbol{\tau}_1 \cdot \boldsymbol{\tau}_2 + C_3 \boldsymbol{q}^2 \boldsymbol{\sigma}_1 \cdot \boldsymbol{\sigma}_2 \\ &+ C_4 \boldsymbol{q}^2 \boldsymbol{\sigma}_1 \cdot \boldsymbol{\sigma}_2 \boldsymbol{\tau}_1 \cdot \boldsymbol{\tau}_2 + \frac{i}{2} C_5 \left(\boldsymbol{\sigma}_1 + \boldsymbol{\sigma}_2\right) \cdot \left(\boldsymbol{q} \times \boldsymbol{k}\right) \\ &+ C_6 \left(\boldsymbol{\sigma}_1 \cdot \boldsymbol{q}\right) \left(\boldsymbol{\sigma}_2 \cdot \boldsymbol{q}\right) + C_7 \left(\boldsymbol{\sigma}_1 \cdot \boldsymbol{q}\right) \left(\boldsymbol{\sigma}_2 \cdot \boldsymbol{q}\right) \boldsymbol{\tau}_1 \cdot \boldsymbol{\tau}_2 , \end{aligned} \tag{6.4}$$

where, again using FRF, a subset of 7 out of 14 operators has been chosen. This choice, employed for example in Refs. [66, 67], is made in order to construct local chiral EFT interactions, i.e. interactions that are independent of the momentum **k**. However, many other chiral EFT interactions in the literature trade isospin-dependent operators for **k** dependencies, see, e.g., Ref. [52]. The couplings C_1 to C_7 are the LECs that govern the contact interaction at NLO and are also typically fit to NN scattering data. There are no new contacts that appear in next-to-next-to-leading order, and we refer the reader to Refs. [52, 152, 162] for more information at the contact interactions at fourth order.

The long-range contributions to the chiral EFT Hamiltonian, mediated by pion exchanges, can be decomposed as [52, 160]

$$\begin{aligned} V_\pi = & V_C(r) + W_C(r)\boldsymbol{\tau}_1 \cdot \boldsymbol{\tau}_2 \\ & + (V_S(r) + W_S(r)\boldsymbol{\tau}_1 \cdot \boldsymbol{\tau}_2)\,\boldsymbol{\sigma}_1 \cdot \boldsymbol{\sigma}_2 \\ & + (V_T(r) + W_T(r)\boldsymbol{\tau}_1 \cdot \boldsymbol{\tau}_2)\,S_{12} \\ & + (V_{LS}(r) + W_{LS}(r)\boldsymbol{\tau}_1 \cdot \boldsymbol{\tau}_2)\,\boldsymbol{L} \cdot \boldsymbol{S}\,, \end{aligned} \tag{6.5}$$

without loss of generality. Here r is the distance between the two nucleons, $S_{12} = (3\boldsymbol{\sigma}_1 \cdot \hat{\boldsymbol{r}}\,\boldsymbol{\sigma}_2 \cdot \hat{\boldsymbol{r}} - \boldsymbol{\sigma}_1 \cdot \boldsymbol{\sigma}_2)$ is the tensor operator, and $\boldsymbol{S} = (\boldsymbol{\sigma}_1 + \boldsymbol{\sigma}_2)/2$ is the total spin operator. Depending on the chiral order, one- and multiple-pion exchanges contribute to the different terms. At LO, only the one-pion exchange (OPE) appears and is given by [66, 67, 160],

$$W_S(r) = \frac{g_A^2 m_\pi^2}{48\pi F_\pi^2} \frac{e^{-x}}{r} \tag{6.6}$$

$$W_T(r) = \frac{g_A^2}{48\pi F_\pi^2} \frac{e^{-x}}{r^3} (3 + 3x + x^2) \tag{6.7}$$

where $x = m_\pi r$, m_π is the pion mass, g_A is the axial-vector coupling constant, and F_π is the pion decay constant. At NLO and beyond, V_π receives contribution from two-pion exchange (TPE) diagrams. The leading three-pion exchange contribution appears at $\mathrm{N^3LO}$ but can be shown to be negligibly small [52]. The strength of these multi-pion exchange potentials is determined by the πN couplings which are constrained by chiral symmetry and can be determined by analyses of low-energy πN-scattering.

In Fig. 6.3, we show the result of an order-by-order fit of the LECs to neutron-proton phase shifts in the two S-wave channels. At each order, the bands represent a Bayesian estimate of the theoretical truncation uncertainties at the 95% confidence level, see Sec. 6.2.3 for a discussion of theoretical uncertainties in microscopic models. Going to larger orders in the chiral EFT, from LO up to next-to-next-to-next-to-leading order ($\mathrm{N^3LO}$), we see a considerable improvement in the reproduction of phase shift data, as well as a significant reduction in the theoretical uncertainty bands, indicating a good convergence of the EFT in these channels. Note that, at any given order, the theoretical uncertainties widen with increasing energy in the laboratory frame, in line with our expectation that chiral EFT generally performs better at low energy scales and worsens gradually as the energy approaches the EFT breakdown scale. This figure highlights the value of chiral EFT over traditional approaches: one can estimate theoretical uncertainties, and by going to higher orders, the description of nuclear systems can be systematically improved.

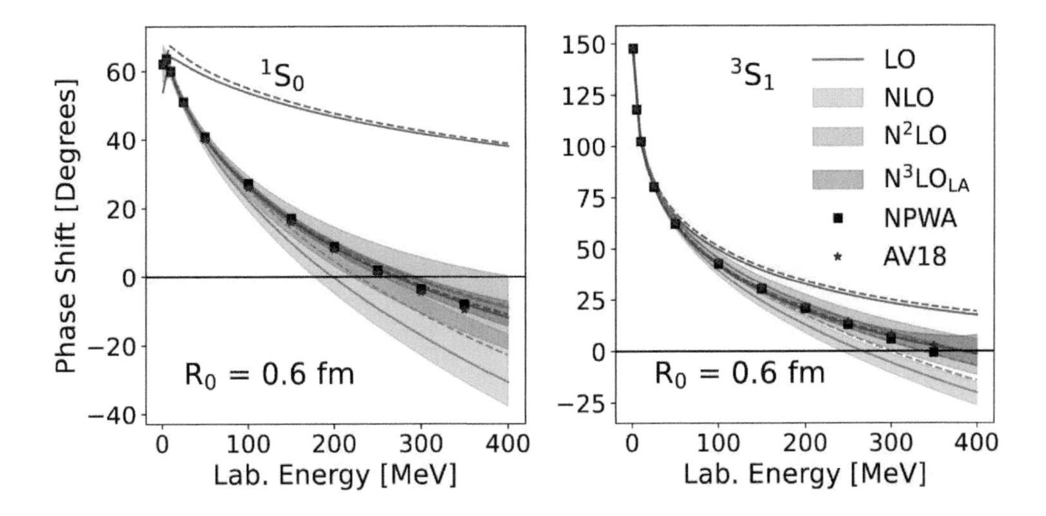

Figure 6.3 Neutron-Proton phase shifts in the spin singlet (left) and triplet (right) S-wave channels. The black squares show the phase shift data from Ref. [166]. Results are shown for different chiral orders from LO up to N^{3}LO. The bands represent a Bayesian estimate of the truncation uncertainties in each order. The solid lines represent the maximum of the Bayesian posterior distribution and the dashed lines result from a least-squares fit. Reprinted from Ref. [162] with permission. © APS (2024).

To employ chiral EFT interactions in many-body calculations, we need to apply regulators that remove high-momentum divergences in the many-body system, usually by cutting interactions for nucleon momenta above a chosen cutoff Λ_c. Chiral EFT interactions can be sorted into three categories according to the regularization scheme employed: local interactions up to N^{3}LO [67, 141, 142, 162, 169], semi-local interactions up to N^{4}LO [152], or nonlocal interactions up to N^{4}LO [50–52]. The regulators introduce artifacts to the study of many-body systems [48, 115, 170] that increase the theoretical uncertainty and need to be quantified. Both the breakdown scale Λ_b and the cutoff Λ_c limit the range of applicability of chiral EFT interactions. While the latter is a calculation tool and can be varied, the former is more fundamental and indicates that the description of dense matter in terms of nucleon contact interactions and pion exchanges is not sufficient to describe high-density physical processes. This limitation is valid not only for chiral EFT but also for most phenomenological models based on the same degrees of freedom. However, the systematic nature of chiral EFT allows us to estimate this breakdown scale from data and to make statements about the convergence of the chiral expansion [57].

6.2.2 Constraints on the Equation of State

Nuclear interactions determine the properties of all nuclear systems, from the structure of atomic nuclei to dense nuclear matter. The latter is characterized by the EOS, which relates density, pressure, energy, temperature, and the composition of strongly-interacting matter.

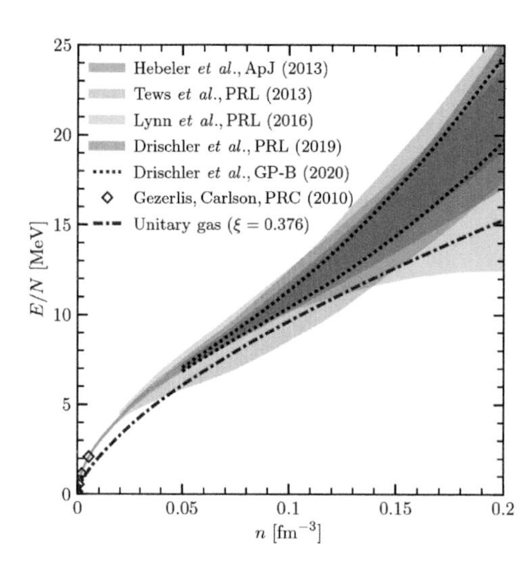

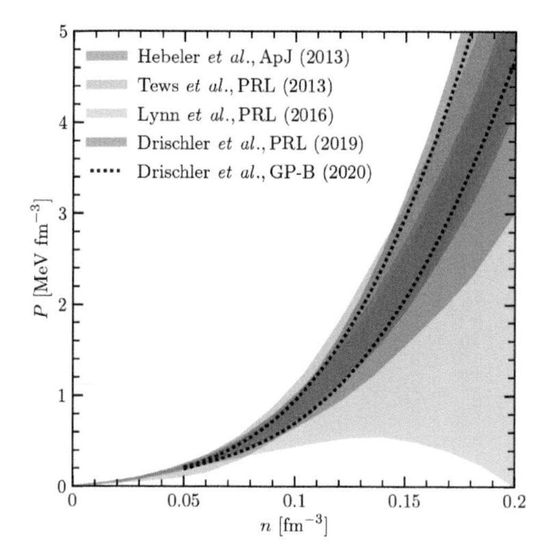

Figure 6.4 Energy per particle E/A (left) and pressure P (right) for different calculations of pure neutron matter using various many-body methods and chiral EFT interactions: MBPT [45, 78, 171] and QMC [115]. In addition, the uncertainty band for a Gaussian-process anaylsis [44] based on the calculations of Ref. [45] is shown. In the left panel, QMC results with phenomenological interactions of Ref. [68] as well as the conjectured unitary-gas bound [175] are shown, too. Reprinted from Ref. [89] with permission. © APS (2021).

For most neutron stars, it is sufficient to know the relation of energy density ε and pressure p as thermal energies are much lower than Fermi energies in the system.

At densities up to a few times the nuclear saturation density, the relevant d.o.f. are nucleons, and their interactions can be described using microscopic many-body methods with nuclear-potential models as input as discussed above. For the former, the commonly employed methods are quantum Monte Carlo (QMC) [31, 116], many-body perturbation theory (MBPT) [46], the coupled-cluster (CC) method [75], or the self-consistent Green's Function (SCGF) method [29, 30]. For the latter, nowadays one commonly employs interactions from chiral EFT; see Sec. 6.2.1. In this contribution, we will not explain the different many-body methods in detail and instead refer the reader to the various references.

Chiral EFT has been used to calculate the EOS of nucleonic matter up to about twice the nuclear saturation density with theoretical uncertainties [47, 173]. Fig. 6.4 shows a comparison of the results for the energy per particle E/A and pressure P as functions of the baryon number density n from several chiral EFT calculations in pure neutron matter. Pure neutron matter is the simplest dense-matter system but relevant to neutron stars because below $\sim 2n_{\rm sat}$ neutron stars contain only about $\sim 5\%$ protons. Furthermore, pure neutron matter is extremely difficult to probe in terrestrial laboratories, see e.g. Refs. [9, 158], because nuclear experiments use atomic nuclei with similar numbers of protons and neutrons, and hence, they mainly probe nuclear matter closer to symmetry. Therefore, microscopic many-body calculations are an ideal way to study pure neutron matter and provide valuable constraints for neutron-star studies. In Fig. 6.4, E/A is shown for the MBPT calculations of Refs. [45, 78, 171] and QMC calculations of Ref. [115]. For all calculations, uncertainty bands are provided following different prescriptions and we refer the reader to the different publications for details. However, the ability to provide these uncertainty bands is one of the main benefits of chiral EFT calculations (see Sec.6.2.3), and extremely important when comparisons to data are performed. While there are slight differences in the various results and the associated uncertainties, all calculations agree well for E/A.

From $E/A(n)$, ε follows as $\varepsilon(n) = n \cdot (E/A(n) + m_N)$, where m_N is the nucleon mass. From $E/A(n)$, one can also calculate the pressure of matter as

$$P(n) = n^2 \frac{\partial E/A(n)}{\partial n} . \tag{6.8}$$

The pressure is a key quantity in neutron-star physics as it directly enters the Tolman-Oppenheimer-Volkoff (TOV) [133, 177] equations, i.e., the structure equations for neutron stars. Again, different many-body calculations based on chiral EFT give consistent results for the pressure of pure neutron matter, see the right panel of Fig. 6.4. However, small differences in the calculations are more pronounced in the pressure, which is a derivative of the energy per particle. In particular, QMC calculations require local chiral EFT calculations as input whose regulator artifacts can reduce the pressure, leading to a broadening of the uncertainty band.

As explained before, chiral EFT breaks down when nucleon momenta are too high, close to or beyond the breakdown scale. Several studies have shown that this happens around twice nuclear saturation density [47, 173] for currently used chiral EFT Hamiltonians. This breakdown might be indicated by a broadening of the uncertainty bands but it remains unclear at which densities exactly this happens. We will discuss ways on how to use observational data to answer this question later in this contribution.

6.2.3 Theoretical Uncertainties in Microscopic Descriptions of the Equation of State

All theoretical descriptions of nuclear systems suffer from uncertainties mainly due to (a) an incomplete knowledge of nuclear interactions, i.e., the truncation of the interaction's operator basis, (b) uncertainties due to the regulator scheme and scale, (c) uncertainties in LECs, and (d) the computational method to solve the nuclear many-body system. Among these, the truncation uncertainty (a) is most dominant but can be estimated within chiral EFT. We will explain this source of uncertainty in more detail below but also refer to reader to Refs. [53, 64] for more details.

One can express any observable X, such as $E/A(n)$, as a sum of contributions at all orders of the EFT,

$$X = X_0 \sum_{k=0}^{\infty} c_k Q^k , \tag{6.9}$$

where X_0 sets the scale for X and is typically the leading-order result, and the c_k are expansion coefficients in the EFT expansion. Due to the truncation of this series at a certain order $k_{\max}$, one introduces an uncertainty,

$$\Delta X = X - X_0 \sum_{k=0}^{k_{\max}} c_k Q^k = X_0 \sum_{k=k_{\max}+1}^{\infty} c_k Q^k \,. \tag{6.10}$$

Form any order-by-order calculation in chiral EFT, one can calculate the series of c_k with $k \leq k_{\max}$, and use these results to estimate the distribution of unknown c_k [47, 53]. Any calculation with microscopic nuclear interaction models contains such uncertainties but only when using EFT approaches can they be estimated as only these approaches provide a series of c_k.

Another important source of uncertainty is due to the regulator scheme and scale. Regulator artifacts in general depend on $(p/\Lambda_c)^n$. If Λ_c is large, then uncertainties due to the regular are small. However, the cutoff values in present chiral EFT interactions explore a small range of values between 400 and 500 MeV, which can lead to non-negligible regulator artifacts [48, 113–115]. The uncertainty due to the regulators can be estimated by varying the cutoff.

Finally, as the LECs are adjusted to data with experimental errors, all LECs carry an uncertainty (c). This uncertainty is largest for the 3N LECs and studies have shown that the 3N LEC values can depend significantly on the choice of fit systems [174]. The 3N pion-exchange LECs, however, have been well constrained using the Roy-Steiner formalism [83, 84]. We will not discuss many-body methods and their method uncertainties here, but after great computational advances in the past years the latter are usually much smaller than the uncertainties due to nuclear interactions.

The uncertainty bands in Fig. 6.4 were obtained using different approaches. For studies before 2015, they were typically extracted from combining regulator, LEC, and many-body uncertainties, whereas truncation uncertainties were not estimated. Since Ref. [53], however, the truncation uncertainty is regularly estimated and is accounted for in the results after 2015. Nevertheless, uncertainty bands between different EFT calculations agree very well.

6.3 MODELS FOR THE HIGH-DENSITY EQUATION OF STATE

At higher densities, where chiral EFT breaks down, new d.o.f., such as hyperons or quarks, might appear [12, 15, 110, 164]. These d.o.f. necessarily need to be taken into account when inferring properties of the EOS or nuclear interactions from data. In the past, one commonly assumed a certain composition for neutron-star matter, mainly a nucleonic one, and employed relatively simple mean-field models for the EOS. However, more recently, composition agnostic schemes have been developed that can be used to describe matter above a chosen EFT breakdown density without any strong model assumptions that might bias the data analysis [16, 57, 70, 78, 103, 147, 150, 172]. We will briefly review these approaches.

6.3.1 Extensions under the nucleonic hypothesis

Unlike the computational methods mentioned in Sec. 6.2.2, that have been developed to solve the many-body Schroedinger equation for nucleonic matter, commonly employed models for neutron-star matter typically rely on the mean-field approximations. As far as nucleonic mean-field models are concerned, both relativistic [85, 180] and non-relativistic [33, 34, 121] versions have been developed. For the latter, the Hamiltonian

$$H = T(\rho_n, \rho_p) + V(\rho_n, \rho_p) \tag{6.11}$$

is decomposed into the kinetic term $T(\rho_n, \rho_p)$, with ρ_τ being the neutron and proton densities, and an interaction term $V(\rho_n, \rho_p)$ that depends on the non-relativistic model in question. For Skyrme type of models, which are among the most common non-relativistic models that are used in astrophysical applications, V consists primarily of a zero range contact interaction V_0 that reads [33, 34]

$$V_0 = \frac{1}{4}[(2 + x_0)\rho^2 - (2x_0 + 1)(\rho_p^2 + \rho_n^2)] \, , \tag{6.12}$$

where $\rho = \rho_p + \rho_n$ and x_0 is a fit parameter. Other terms involving more parameters, such as effective mass terms, finite-range terms, spin-orbit interactions, etc., also contribute to the potential V, see Ref. [33, 34] for more details. The parameters of these Skyrme functionals are fit to the saturation properties of symmetric nuclear matter and/or the properties of heavy atomic nuclei. Based on the details of the calibration procedure and the exact form of the Skyrme Hamiltonian, over 200 Skyrme functionals exist, see Ref. [121] for a list of a few of them.

Recently, a more model-agnostic approach for non-relativistic nuclear energy density functionals has been developed under the metamodeling framework. This is a density functional approach, similar to the Skyrme approach [33, 34], that allows one to incorporate nuclear physics knowledge directly encoded in terms of the Nuclear Empirical Parameters (NEPs). These parameters are defined via a Taylor expansion of the energy per particle in symmetric matter, $e_{\rm sat}$ and the symmetry energy, $e_{\rm sym}$ about saturation density, $n_{\rm sat}$,

$$e_{\rm sat}(n) = E_{\rm sat} \qquad\qquad + \frac{1}{2}K_{\rm sat}x^2 + \frac{1}{3!}Q_{\rm sat}x^3 + \frac{1}{4!}Z_{\rm sat}x^4 + \dots \, , \tag{6.13}$$

$$e_{\rm sym}(n) = E_{\rm sym} + L_{\rm sym}x + \frac{1}{2}K_{\rm sym}x^2 + \frac{1}{3!}Q_{\rm sym}x^3 + \frac{1}{4!}Z_{\rm sym}x^4 + \dots \, , \tag{6.14}$$

where $x \equiv (n - n_{\rm sat})/(3n_{\rm sat})$ is the expansion parameter and $e_{\rm sym}(n) = e_{\rm PNM}(n) - e_{\rm sat}(n)$ is the difference between the energies of pure neutron matter energy and symmetric matter. In the metamodel, the potential energy $V(\rho_n, \rho_p)$ is expanded in a Taylor series around saturation density, in both symmetric and pure neutron matter. The NEPs can, thus, be mapped directly to the coefficients of the expansion of the potential energy in the metamodel. Therefore, by varying the empirical parameters within their uncertainties, the metamodel is able to reproduce the EOS predicted by a large number of existing nucleonic models [121, 122], such as the Skyrme SLy4 interaction [33, 34] that is shown in Fig. 6.5 for example.

We will end this subsection with a brief note on relativistic mean-field models. The first attempt to go beyond a nonrelativistic treatment of nuclear matter was the relativistic mean-field (RMF) approach initiated by Walecka and collaborators [180], which is based on meson exchanges between nucleons whose wave functions are solutions of the in-medium Dirac equation. In this framework, nucleons move in attractive (scalar fields)

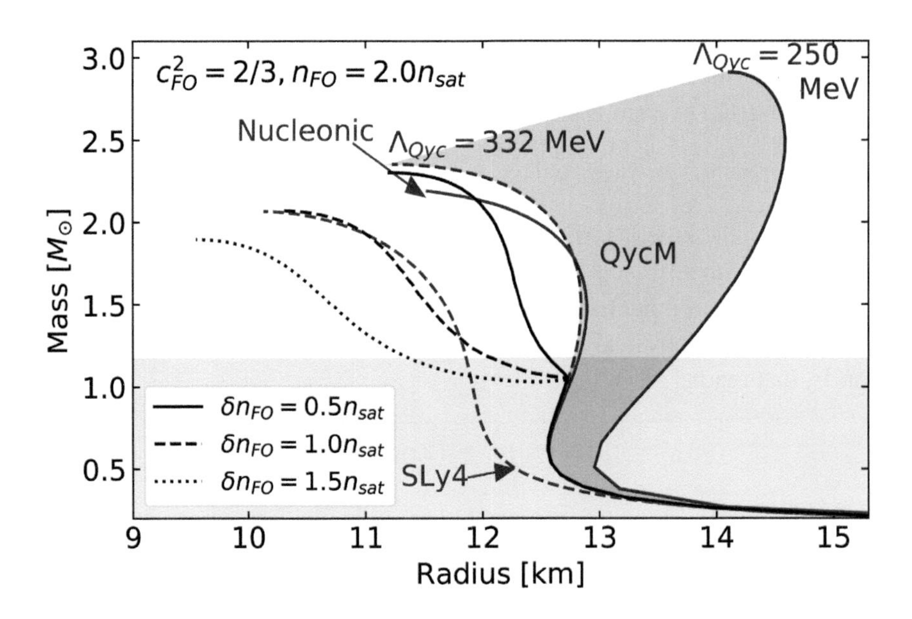

Figure 6.5 Mass-Radius curves for different models, including those that describe phase transitions. The dashed red line is the SLy4 EOS [33, 34] whereas the solid red line is a nucleonic EOS constructed using the metamodel [121] in Ref. [163]. The blue curves correspond to the Quarkyonic matter model [120, 125] and the black lines represent EOS with a first-order phase transition modelled using the constant sound speed model of Ref. [14].

and in repulsive (vector fields) backgrounds. This provides both the "Walecka" saturation mechanism, that results from the interplay between the scalar and vector fields, and the correct magnitude of the spin-orbit potential. The parameters describing the meson-nucleon couplings are adjusted to the saturation properties of nuclear matter and/or ground state properties of atomic nuclei, such as binding energies, charge radii, etc [36, 61, 95]. They can also be adjusted to calculations based on chiral EFT Hamitlonians [13]. RMF models, with suitable refinements, have also widely been used in studies of neutron stars and finite nuclei [36, 61, 85, 151].

6.3.2 Phase transitions in NSs

Due to the high densities explored in the cores of neutron stars, matter might exhibit exotic phases that are not realized anywhere else in the universe. Proposals for the existence of quark matter (QM) in the cores of neutron stars have a long history, starting from the early works by Seidov [161], Bodmer [23], and Witten [185]. In particular, Ref. [185] explored the strange quark matter hypothesis, stating that the absolute ground-state of matter may be composed of up (u), down (d), and strange (s) quarks instead of only the observed u, d matter that builds nucleons. Since then, questions regarding the nature of the transition to QM, its location in the space of thermodynamic variables, and implications for the resulting EOS have attracted a lot of attention, see, for example, Refs. [11, 14, 19–22, 35, 129, 187] for a small selection. The theoretical description of QM has made use of different models such as the simple MIT bag model but also more advanced field-theory based NJL models, see Ref. [27] for a review of the latter. Besides these "microscopic approaches," more agnostic types of modeling have also been proposed [14, 187]. In such approaches, the phase transition is described in terms of thermodynamical quantities instead of microphysical coupling constants.

The transition from nucleonic matter to QM occurs via a *phase transition.* Different kinds of phase transitions are possible in dense matter [73, 163]. Most models assume an abrupt first-order phase transition (FOPT) which creates a discontinuity in the sound speed as a function of the density, but also other transitions, such as a hyperonization, have been proposed. Such phase transitions typically soften the EOS and, therefore, produce smaller NS radii. Fig. 6.5 shows an example of three EOSs with FOPTs. Furthermore, the transition could be fully analytic, as observed at finite temperature and zero chemical potential [26]. An example of such a crossover is the transition to Quarkyonic matter [124], which has been shown to naturally explain certain features of neutron-star phenomenology [125]. In Fig. 6.5, we show examples of the Quarkyonic-matter EOS.

6.3.3 Composition agnostic extensions

The possible presence of exotic phases of matter in the cores of neutron stars presents a source of uncertainty, as their effects on the EOS can be quite drastic. Accurately modeling this uncertainty based on microphysical models alone is unfeasible due to the existence of many different proposals for exotic phases that have never been observed before. Therefore, the era of multimessenger astronomy has seen the rise of models for the high-density EOS that are agnostic regarding the composition of NS matter. These agnostic approaches are flexible enough to encapsulate the various nucleonic models discussed in Sec. 6.3.1 while also including all possible behavior from non-nucleonic exotic phases of matter. Examples of such composition agnostic approaches include the polytropic representation of the EOS [16, 78, 150], spectral EOS representations [105–107], models based on Gaussian Processes [57, 103], and the parametric sound-speed extension scheme [70, 147, 164, 172].

In this contribution, we will present results from both the sound-speed extension scheme as well as the Gaussian process approach. Both of these use the speed of sound c_S to generate a large number of EOS models that are constrained at low density by nuclear theory, e.g., chiral EFT calculations. They are constrained at high densities only by the requirements that $c_S \geq 0$, to account for the fact that the pressure in neutron stars is increasing as one increases the density, and $c_S \leq c$ with c being the speed of light, accounting for causality. Hence, these approaches take into account uncertainties at low densities but also account for EOS uncertainties at high densities. While we refer the reader to the individual publications on the details of these approaches, the EOS can be reconstructed from each sampled speed-of-sound curve. By inverting the expression $c_S^2 = dp/d\epsilon = (nd\mu)/(\mu dn)$, one obtains the chemical potential μ, the pressure and the energy density as

$$\log\left(\frac{\mu(n)}{\mu_i}\right) = \int_{n_i}^{n} \frac{c_S^2(n')}{n'} dn' , \tag{6.15}$$

$$p(n) = p(n_i) + \int_{n_i}^{n} c_S^2(n')\mu(n')dn' , \tag{6.16}$$

$$\epsilon(n) = \epsilon(n_i) + \int_{n_i}^{n} \mu(n')dn' , \tag{6.17}$$

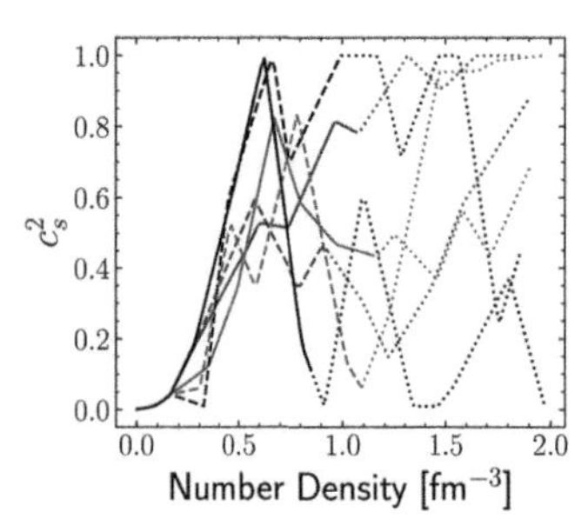

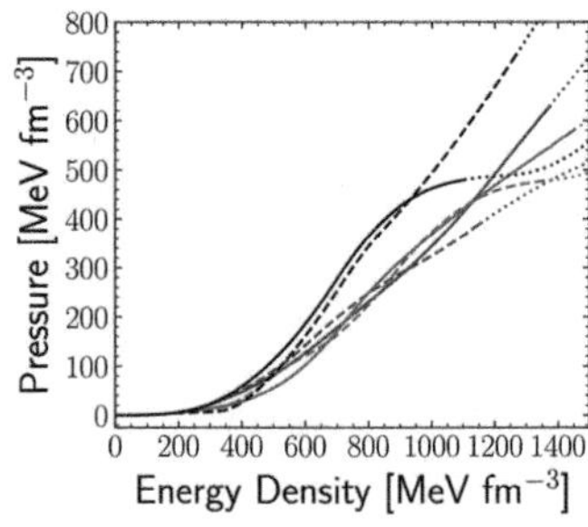

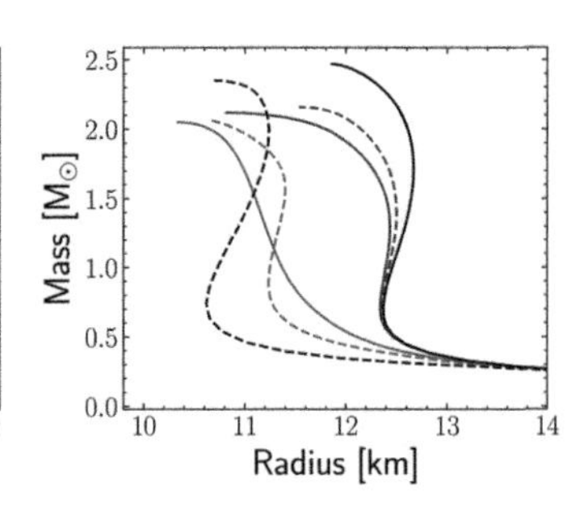

Figure 6.6 Examples for the speed of sound as function of baryon number density, the resulting EOS, and mass radius curves from the parametric sound-speed extension scheme. Reprinted from from Ref. [164] with permission. © APS (2023).

provided that these quantities are known at a given density n_i. Then, the TOV equations can be solved for each EOS sample to determine neutron-star radii and dimensionless tidal deformabilities as functions of masses; see for instance Ref. [172] for more details. Fig. 6.6 illustrates the speed of sound model.

6.4 INFERENCE OF MICROSCOPIC NUCLEAR PHYSICS INFORMATION FROM ASTROPHYSICAL OBSERVATIONS

So far, we have discussed how to use chiral EFT to constrain the nuclear EOS at low densities and how to model the high-density EOS in a model-agnostic way. This forward modeling of the EOS is useful to constrain the limits of neutron-star structure. However, we have entered an era of neutron-star multimessenger astrophysics that will allow us to use current and forthcoming data to infer properties of the EOS with high accuracy. These data might also enable us to infer properties of nuclear Hamiltonians by observing their impact in a high-density, neutron-rich environment that is realized nowhere else in the universe, providing exciting data points to constrain these interactions. For example, future data will allow us to constrain properties of 3N forces [157, 159] and test the range of applicability of microscopic models based on nucleonic d.o.f. from chiral EFT [57]. These goals can be achieved by comparing EOS calculations with multimessenger observations of neutron stars [43, 58, 128, 148]: data from observations of isolated neutron stars, e.g., x-ray pulse profiles [127, 128, 154, 155] or radio signals [17, 42, 62], as well as gravitational-wave data [5]. In this section, we will discuss some recent results in this direction and the possible impact of future data from next-generation observatories.

6.4.1 Current astrophysical and experimental data

Astrophysical data from neutron-star observations have been used to place constraints on the nuclear EOS. Here, we will describe the results obtained within the Nuclear Physics Multimessenger Astrophysics (NMMA) framework [43, 136]. NMMA is a modular open-source PYTHON toolkit that uses Bayesian inference to infer the EOS when analyzing astrophysical and experimental data. NMMA samples over a pre-computed set of EOS when analyzing data from NASA's NICER mission, radio observations of heavy pulsars, and

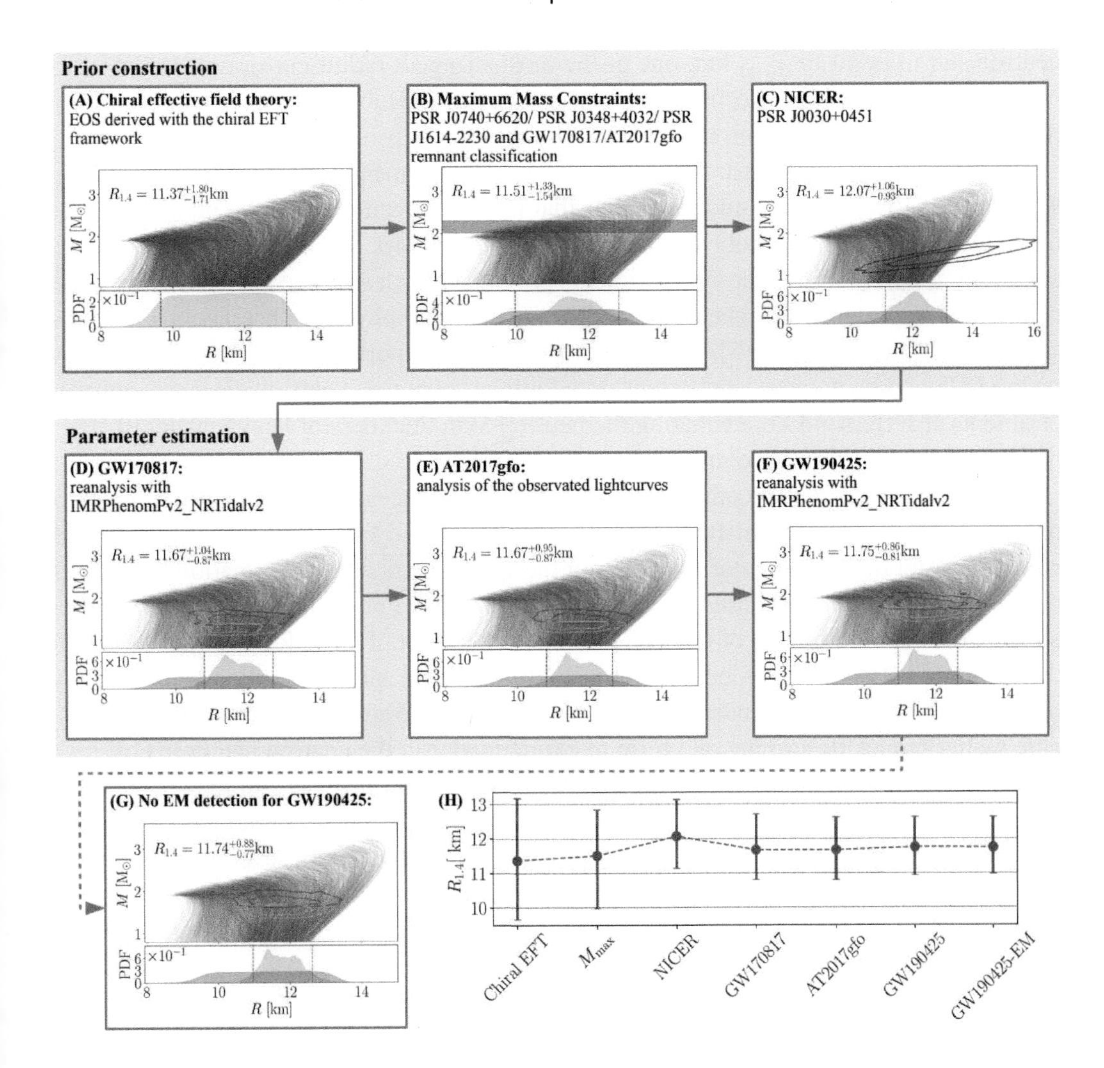

Figure 6.7 Constraints on the neutron-star EOS where each panel includes the effect of different multimessenger observations of neutron stars, as indicated in each panel. The lower panels show the probability distribution functions for the radius of a 1.4 $M_{\rm sol}$ neutron star, with the 90% confidence range indicated by dashed lines. Allowed (disallowed) EOSs are shown as blue (gray) lines. The final panel (H) shows the radius constraint at each step of the analysis at the 90% confidence level. Reprinted from Ref. [43] with permission. © AAAS (2020).

performs a full simultaneous Bayesian parameter estimation of multimessenger GW signals and the associated EM counterparts. In addition, NMMA has been extended to include data from nuclear experiments like heavy-ion collisions [88]. The NMMA framework, explained in detail in Ref. [136], has been used to address several interesting nuclear-physics questions. Importantly, it has been used to constrain the nuclear EOS [43, 88, 136], see Fig. 6.7. In its most recent form, performing a simultaneous inference of the EOS from gravitational-wave and electromagnetic data, NMMA finds the radius of a typical 1.4 $M_{\rm sol}$

neutron star to be $11.98^{+0.35}_{-0.40}$ km, one of the most stringent radius constraints to date. The NMMA framework has also been used to constrain the Hubble constant [43], finding a result in good agreement with the Planck Cosmic Microwave Background analysis [8].

Besides these exciting astrophysical data, new experimental data points provide constraints on the EOS. Dense nuclear matter that is nearly symmetric in the number of neutrons and protons can be probed in heavy-ion collisions [40, 88], e.g., in ultra-relativistic collisions at RHIC [165]. These experiments provide us with a bridge between the regimes of low and high densities, where nuclear theory and astrophysical observations are most sensitive, respectively [40, 88]. Hence, they provide an important complementary window on the EOS and on nuclear interactions. In addition, in the past years, nuclear structure experiments at Jefferson Lab, extracting the neutron-skin thickness of heavy nuclei [9, 151], have provided stringent benchmarks for nuclear theory.

These results, as well as others, highlight the importance of combining different astrophysical messengers, information from nuclear theory, and from experiment [56, 58, 88] when inferring the neutron-star EOS. Using these complementary information, it is possible to forward-model properties of astrophysical systems. For example, NMMA was used to study the maximum mass of neutron stars and the impact of NICER observations [138]), the nature of the compact-object merger GW190814 [176] and was applied to explain EM observations of gamma-ray bursts [100]. It was also used to study properties of the quarkyonic matter model, describing one form of exotic matter in the neutron-star core [137].

However, in this contribution, we are interested in the inverse problem: using multimessenger astrophysics to infer properties of microscopic nuclear interactions. As discussed before, the nuclear Hamiltonian is a complicated object with several LECs that are adjusted to experimental data from NN scattering or light nuclei. For a reasonable description of nuclear systems, Hamiltonians in chiral EFT need to be at least at next-to-next-to-leading order in the expansion, including nine NN LECs, two 3N LECs, and several pion-nucleon LECs that are commonly adjusted to their Roy-Steiner values. That leaves at least 11 parameters that need to be adjusted to data or inferred from observations.

In Ref. [157], the possibility of inferring different 3N contributions to the nuclear Hamiltonian from astrophysical data was tested while taking into account all EOS uncertainties discussed previously, see also Refs. [123, 159, 178] for studies that inferred 3N contributions without modeling these uncertainties. Two EOS distributions were computed from two different nuclear Hamiltonians that describe atomic nuclei and symmetric nuclear matter similarly well but differ in pure neutron matter [115]. For each Hamiltonian, the EOS uncertainty was computed as described in Sec. 6.2.3. Then, a general composition-agnostic sound-speed extension was attached beyond two times the nuclear saturation density, where it is assumed that chiral EFT interactions break down, to account for the uncertainties in the high-density EOS. This leads to two different predictions for neutron-star masses and radii, see Fig. 6.8. While accounting for current nuclear-physics uncertainties in this way, it was tested if current astrophysical data would enable us to distinguish between these different interactions. Unfortunately, current constraints, mainly from GW1701817, proved uninformative for answering this question which can be understood from the fact that current gravitational-wave information and EFT calculations up to twice nuclear saturation density lead to similar radius constraints [28]. Hence, the data are not precise enough to distinguish between two different nuclear Hamiltonians from chiral

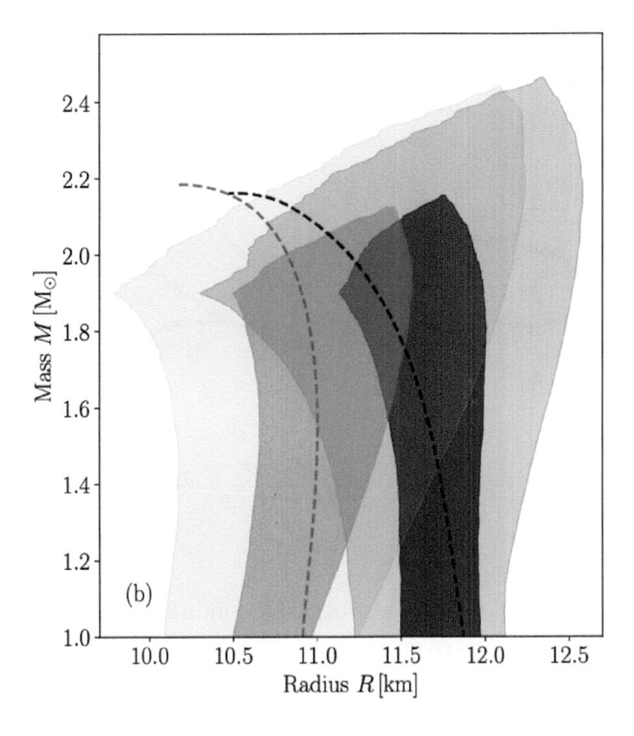

Figure 6.8 The mass-radius for EOS distributions based on two different chiral EFT Hamiltonians that differ in the 3N forces. The bands are given at 50% and 90% confidence. Reprinted from Ref. [157] with permission. ©APS (2023).

EFT, but only provides an upper bound on the radius/tidal deformability that both nuclear Hamiltonians are in agreement with. Other available neutron-star merger observations, such as GW190425 [6], are even less informative. Hence, current observations cannot help us to constrain individual pieces of nuclear Hamiltonians. As we will see later, however, next-generation gravitational-wave observatories, such as Cosmic Explorer [59, 60, 153] and the Einstein Telescope [119, 146], will enable such constraints.

While current astrophysical constraints still carry sizeable uncertainties, making direct constraints of nuclear Hamiltonians impossible, they can still provide valuable information on the density up to which EFT calculations can be trusted. One way of addressing such a problem is to perform Bayesian model comparison which requires the computation of the so-called Bayes factor [57, 102]. For this, one calculates the conditional probability $p(d_{\text{astro}}|h)$ which is a measure of how well the astrophysical data d_{astro} is described by a set of model assumptions h. Then, the ratio of evidences for two different sets of model assumptions h_1 and h_2 results in the Bayes factor $\mathcal{B}$, i.e. $\mathcal{B} = p(d_{\text{astro}}|h_1)/p(d_{\text{astro}}|h_2)$. Therefore, a Bayes factor $\mathcal{B} \gg 1$ would indicate a significant preference by the data for model h_1 and vice versa for $\mathcal{B} \ll 1$. Using Gaussian processes to model the neutron-star EOS, Ref. [57] compared the Bayesian model evidence for Gaussian processes trained only on astrophysical data (theory agnostic) to Gaussian processes trained on astrophysical data and different chiral EFT calculations: the MBPT calculation of Ref. [171] and the QMC calculation of Ref. [173], as well as an artificially stiffer and a softer result based

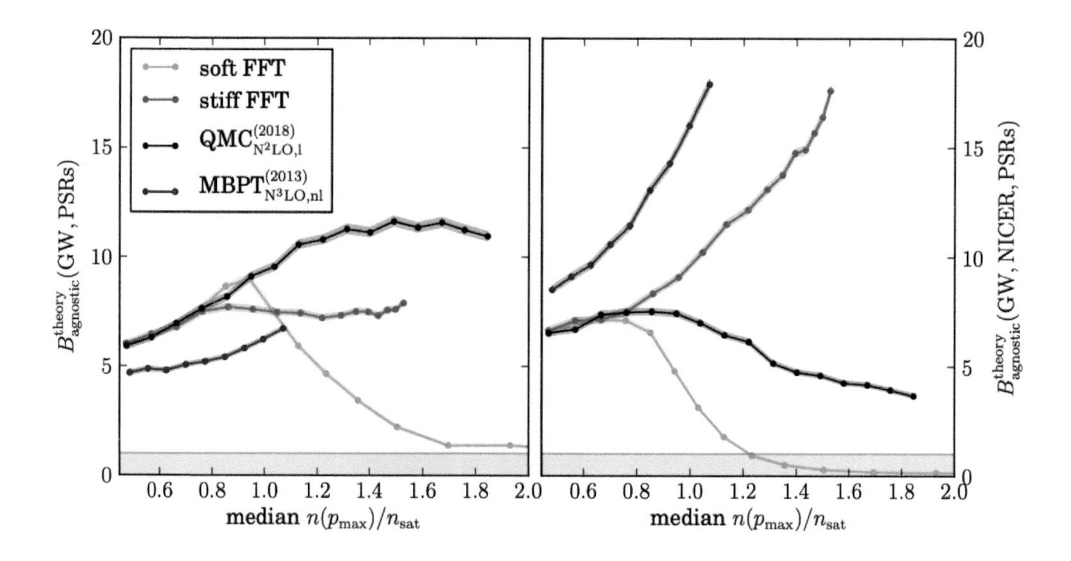

Figure 6.9 Bayes factors for Gaussian processes conditioned on astrophysical data and different chiral EFT calculations over a theory agnostic analysis plotted as function of the maximal density up to which theory calculations are used. The astrophysical data include massive pulsars and gravitational-wave data (left) and massive pulsars, gravitational-wave and NICER data (right). Both panels show two different chiral EFT calculations [171, 173] as well as artificial theories that are soft (yellow) or stiff (green). Reprinted from Ref. [57] with permission. ©APS (2020).

on the latter calculation. These Bayes factors were then calculated as a function of the density up to which an EFT calculations would be enforced, see Fig. 6.9. Ref. [57] found that if only heavy pulsar and gravitational-wave observations were considered, EFT input from the QMC calculation would be most favored, up to about twice the nuclear saturation density. The other EFT calculation, as well as the stiffer theory were slightly disfavored, while the softer theory was heavily disfavored compared to the QMC calculation. This aligns with the observation mentioned before that gravitational-wave and chiral EFT information up to twice the saturation density agree very well [28], but that different calculations/Hamiltonians cannot be distinguished currently. However, the Bayes factors change when the NICER observations was also accounted for: in this case, the stiffer models were preferred.

Although the breakdown density of chiral EFT Hamiltonians can be estimated by studying the order-by-order convergence of chiral EFT calculations [44, 47], additional uncertainties in these estimations (e.g., due to the broad range of momenta in a nuclear many-body system) warrant the comparison between several approaches. An approach like the one discussed here can be used to estimate the breakdown density, by studying up to which density a theoretical prediction reproduces the observed astrophysical data. This, in turn, manifests as a local maximum in Fig. 6.9, with stronger evidence for a breakdown for a more pronounced maximum. The benefit of this approach is that it can also be used to test the breakdown of nuclear Hamiltonians in general. Furthermore, this data-driven

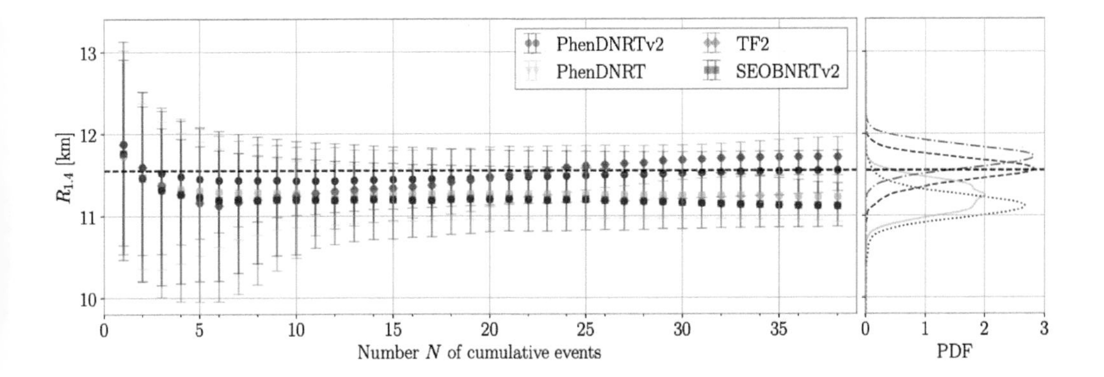

Figure 6.10 Radius constraints for a $1.4M_{\rm sol}$ neutron star versus number of combined gravitational-wave signals for different waveform models, compared to the value injected with PhenDNRTv2 (left), as well as the posterior distribution for all waveform models when all signals are combined (right). The signals were simulated using a network of Advanced LIGO and Advanced Virgo at design sensitivity. Reprinted from Ref. [101] with permission. ©APS (2022).

approach based on Gaussian processes is general enough so that modeling systematics at high densities are not important.

Presently, in all cases, Gaussian processes trained on EFT calculations are preferred over theory-agnostic ones but individual data points (in this case NICER) can change results considerably. Hence, constraints will need to become stronger, with smaller uncertainties and less dependence on individual data points, to draw robust conclusions. In the next Section, we will discuss the potential of future observations.

6.4.2 Potential of future observations

We are entering an era that will offer an unprecedented wealth of astrophysical data on the EOS. Gravitational-wave detectors such as LIGO and Virgo will likely be observing several NS mergers per year [37], reducing the uncertainty in the cold EOS with every observation [101]. In Fig. 6.10, the prediction for the radius of a $1.4M_{\rm sol}$ neutron star as a function of number of gravitational-wave observations for a network of interferometers consisting of Advanced LIGO and Advanced Virgo at design sensitivity [1, 7] is shown. The radius uncertainty improves with the number of detections and halves for about 10 detections compared to the current level. If more neutron-star mergers were observed, the statistical uncertainties of the neutron-star radius at $1.4M_{\rm sol}$ would decrease to ± 250 m at 90% confidence, as reported in Ref. [101]. At this level, waveform systematics will become important and need to be addressed carefully, see Fig. 6.10.

As mentioned before, during O4, LIGO might optimistically observe gravitational-wave signals from up to 10 binary neutron-star mergers [37] and its sensitivity will almost double in O5 [18]. After LIGO completes O5, it could potentially undergo upgrades that enhance its sensitivity such as A$\#$ [71] or Voyager [10]. But even with these detectors, we

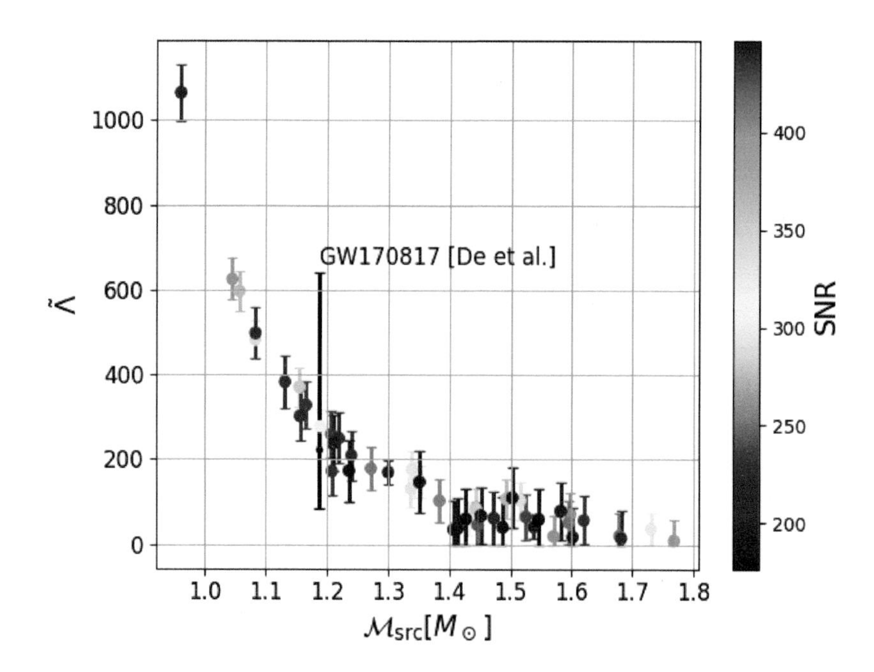

Figure 6.11 Chirp masses and binary tidal deformabilities for the 50 loudest events from a simulation of one year worth of third-generation binary neutron-star merger observations. Each event is colored according to its SNR. We show GW170817 [41] with SNR ~ 32 for comparison. Simulation data provided by Philippe Landry.

might need to accrue many years worth of observations to be able to constrain the radius of typical neutron stars to within better than five percent.

The future of gravitational-wave astronomy lies in the next, the third, generation of detectors, that is expected to have a sensitivity that is improved by an order of magnitude with respect to current instruments. In the United States, the third generation gravitational-wave observatory concept is called Cosmic Explorer [59, 60, 153], which is expected to be a network of two detectors similar to LIGO but with 20 km or 40 km arms. In Europe, the triangular Einstein Telescope [82, 146] is scheduled to be built underground in the 2030s. When these detectors come online, they are expected to deliver many high-precision measurements of gravitational waves with signal-to-noise ratios 1–2 orders of magnitude larger than those for GW170817. A network of two Cosmic Explorers and one Einstein Telescope might observe a couple of 1000s of binary neutron-star merger events per year, with over 100 of them having a signal-to-noise ratio (SNR) of over a 100. In comparison, GW170817, the loudest signal observed to data, had a SNR ~ 32. Fig. 6.11 shows the 50 loudest events from a simulation of a one-year run of the expected next-generation GW detector network, with GW170817 shown for comparison. In this simulation, ~ 5000 events were detected, and many more than the 50 loudest events will allow us to place tight constraints on the cold dense-matter EOS and its nuclear physics.

Using such data, Ref. [157] discussed above also studied the impact that next-generation detectors can have on distinguishing between different models for 3N forces. While current astrophysical observations were not sufficiently precise to distinguish

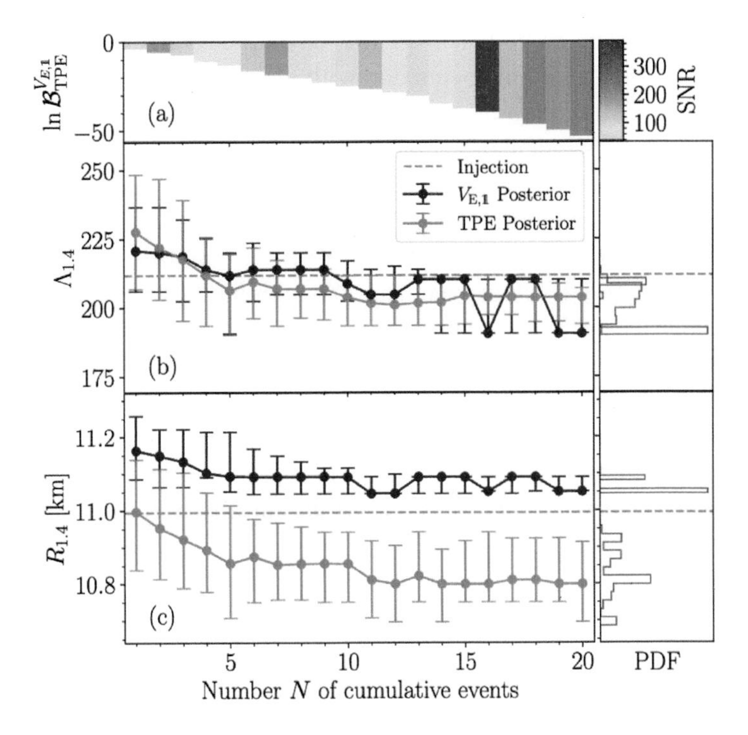

Figure 6.12 Posteriors on the radius and the tidal deformability of a $1.4M_{sol}$ neutron star and Bayes Factors for an injection of an EOS using one 3N Hamiltonian (orange), and recovered with the injection and an additional 3N Hamiltonian (blue), as a function of number of events. Events are color-coded using the event's SNR. Reprinted from Ref. [157] with permission. ©APS (2023).

between different 3N forces (see previous Section), next-generation observations will enable us to do so. For this study, Ref. [157] chose one representative EOS from each nuclear Hamiltonian set to employ in two injection studies, modeling detections with the future Einstein Telescope. For both injections, a recovery with both nuclear Hamiltonians was then performed. Fig. 6.12 shows the results of one recovery as a function of the number of gravitational-wave detections with the Einstein Telescope. For each recovery, Ref. [157] calculated Bayes factors in favor of the two different Hamiltonians, and found log Bayes factors ranging from 4 to 50, depending on the EOS that was used for the injection. This showed that given current nuclear-physics uncertainties it is clearly possible to distinguish nuclear Hamiltonians from observations with third-generation gravitational-wave detectors, underlining the potential of multimessenger astronomy for nuclear physics and to resolve degeneracies in predictions of different nuclear-physics models. The large differences in Bayes factors are due to the high cost of these studies, limiting their scope, e.g., only one injection per Hamiltonian was done. Significantly better constraints could be achieved in actual science runs in future.

Another tantalizing possibility is the direct inference of LECs in the chiral Hamiltonian, that determine the nature of nuclear forces, from such observations. However, a full Bayesian analysis that results in posterior distributions on the LECs requires fast computation of the nuclear EOS which is currently not feasible due to the computational expense of solving the Schroedinger equation with many-body methods. Nevertheless, recently, steps have been taken to develop emulator tools whose purpose is to reproduce expensive nuclear-physics calculations on a much shorter time scale, albeit at the cost of adding numerical uncertainties in model predictions due to the emulator itself. These emulators have been applied to modeling few-body nuclear systems with microscopic interactions [65, 126], but also to covariant density functional theories [69]. Furthermore, for the first time, Ref. [91] developed emulators for computing the nuclear matter EOS using the coupled-cluster many-body approach. Emulators for the EOS, applicable to other many-body methods, would be highly beneficial in the field of multimessenger astronomy. Indeed, they would enable a complete propagation of nuclear uncertainties to neutron star observables while simultaneously providing a means to constrain nuclear interactions with statistically meaningful uncertainties. This field of research promises to be a very exciting one in the years to come, enabling us to realize the full discovery potential of multimessenger astrophysics.

6.5 SUMMARY AND OUTLOOK

The next decade will be a "Golden Age" of multimessenger neutron-star observations. It will provide the opportunity to pin down the nuclear EOS, infer the existence of exotic forms of matter, and constrain the strong interactions that govern the most extreme environments in the universe. To fully realize this enormous discovery potential and answer fundamental question in nuclear physics, it is crucial to enhance the amount of information we can extract from multimessenger observations. This will be possible with third-generation gravitational-wave detectors, such as Cosmic Explorer and the Einstein Telescope.

Furthermore, for analyses of these data, high-fidelity nuclear-physics inputs with quantified uncertainties are key. This requires a broad improvement of dense-matter theory over a wide range of densities, temperatures, and isospin asymmetries. Finally, it will be important to improve the modeling of electromagnetic emissions from astrophysical events, such as kilonovae that might accompany neutron-star merger signals, as these signals truly probe the heaviest systems in the universe: the merger remnants short times before their collapse to black holes. To achieve these goals, it is important for different communities (nuclear theory, nuclear experiment, data science, astrophysical modeling, and simulations, computational science) to come together and work in interdiscplinary teams to tackle the many different aspects of multimessenger astrophysics. This will enable us to answer fundamental question in nuclear science, and infer microscopic nuclear interactions and the nuclear EOS. We live in truly exciting times.

Bibliography

[1] J. Aasi et al. Advanced LIGO. *Class. Quant. Grav.*, 32:074001, 2015.

[2] B. P. Abbott et al. Observation of Gravitational Waves from a Binary Black Hole Merger. *Phys. Rev. Lett.*, 116(6):061102, 2016.

[3] B. P. Abbott et al. GW170817: Observation of Gravitational Waves from a Binary Neutron Star Inspiral. *Phys. Rev. Lett.*, 119(16):161101, 2017.

[4] B. P. Abbott et al. Multi-messenger Observations of a Binary Neutron Star Merger. *Astrophys. J. Lett.*, 848(2):L12, 2017.

[5] B. P. Abbott et al. GW170817: Measurements of neutron star radii and equation of state. *Phys. Rev. Lett.*, 121(16):161101, 2018.

[6] B.P. Abbott et al. GW190425: Observation of a Compact Binary Coalescence with Total Mass $\sim 3.4 M_{\odot}$. *Astrophys. J. Lett.*, 892:L3, 2020.

[7] F. Acernese et al. Advanced Virgo: a second-generation interferometric gravitational wave detector. *Class. Quant. Grav.*, 32(2):024001, 2015.

[8] R. Adam et al. Planck 2015 results. X. Diffuse component separation: Foreground maps. *Astron. Astrophys.*, 594:A10, 2016.

[9] D. Adhikari and others (PREX collaboration). Accurate Determination of the Neutron Skin Thickness of ^{208}Pb through Parity-Violation in Electron Scattering. *Phys. Rev. Lett.*, 126(17):172502, 2021.

[10] R. X. Adhikari et al. A cryogenic silicon interferometer for gravitational-wave detection. *Class. Quant. Grav.*, 37(16):165003, 2020.

[11] Mark Alford, Matt Braby, M. W. Paris, and Sanjay Reddy. Hybrid stars that masquerade as neutron stars. *Astrophys. J.*, 629:969–978, 2005.

[12] Mark G. Alford. Color superconducting quark matter. *Ann. Rev. Nucl. Part. Sci.*, 51:131–160, 2001.

[13] Mark G. Alford, Liam Brodie, Alexander Haber, and Ingo Tews. Relativistic mean-field theories for neutron-star physics based on chiral effective field theory. *Phys. Rev. C*, 106(5):055804, 2022.

[14] Mark G. Alford, Sophia Han, and Madappa Prakash. Generic conditions for stable hybrid stars. *Phys. Rev. D*, 88(8):083013, 2013.

[15] Mark G. Alford and Armen Sedrakian. Compact stars with sequential QCD phase transitions. *Phys. Rev. Lett.*, 119(16):161104, 2017.

[16] Eemeli Annala, Tyler Gorda, Aleksi Kurkela, and Aleksi Vuorinen. Gravitational-wave constraints on the neutron-star-matter Equation of State. *Phys. Rev. Lett.*, 120(17):172703, 2018.

[17] John Antoniadis, Paulo C.C. Freire, Norbert Wex, Thomas M. Tauris, Ryan S. Lynch, et al. A Massive Pulsar in a Compact Relativistic Binary. *Science*, 340:6131, 2013.

[18] Vishal Baibhav, Emanuele Berti, Davide Gerosa, Michela Mapelli, Nicola Giacobbo, Yann Bouffanais, and Ugo N. Di Carlo. Gravitational-wave detection rates for compact binaries formed in isolation: LIGO/Virgo O3 and beyond. *Phys. Rev. D*, 100(6):064060, 2019.

[19] Andreas Bauswein, Niels-Uwe F. Bastian, David B. Blaschke, Katerina Chatziioannou, James A. Clark, Tobias Fischer, and Micaela Oertel. Identifying a first-order phase transition in neutron star mergers through gravitational waves. *Phys. Rev. Lett.*, 122(6):061102, 2019.

[20] Andreas Bauswein, Sebastian Blacker, Vimal Vijayan, Nikolaos Stergioulas, Katerina Chatziioannou, James A. Clark, Niels-Uwe F. Bastian, David B. Blaschke, Mateusz Cierniak, and Tobias Fischer. Equation of state constraints from the threshold binary mass for prompt collapse of neutron star mergers. *Phys. Rev. Lett.*, 125(14):141103, 2020.

[21] Sanjin Benic, David Blaschke, David E. Alvarez-Castillo, Tobias Fischer, and Stefan Typel. A new quark-hadron hybrid equation of state for astrophysics - I. High-mass twin compact stars. *Astron. Astrophys.*, 577:A40, 2015.

[22] D. Blaschke, S. Fredriksson, H. Grigorian, A. M. Oztas, and F. Sandin. The Phase diagram of three-flavor quark matter under compact star constraints. *Phys. Rev. D*, 72:065020, 2005.

[23] A. R. Bodmer. Collapsed nuclei. *Phys. Rev. D*, 4:1601–1606, 1971.

[24] Slavko Bogdanov et al. Constraining the Neutron Star Mass–Radius Relation and Dense Matter Equation of State with NICER. I. The Millisecond Pulsar X-Ray Data Set. *Astrophys. J. Lett.*, 887(1):L25, 2019.

[25] Slavko Bogdanov et al. Constraining the Neutron Star Mass–Radius Relation and Dense Matter Equation of State with $NICER$. II. Emission from Hot Spots on a Rapidly Rotating Neutron Star. *Astrophys. J. Lett.*, 887(1):L26, 2019.

[26] Szabocls Borsanyi, Zoltan Fodor, Christian Hoelbling, Sandor D. Katz, Stefan Krieg, and Kalman K. Szabo. Full result for the QCD equation of state with 2+1 flavors. *Phys. Lett. B*, 730:99–104, 2014.

[27] Michael Buballa. NJL model analysis of quark matter at large density. *Phys. Rept.*, 407:205–376, 2005.

[28] Collin D. Capano, Ingo Tews, Stephanie M. Brown, Ben Margalit, Soumi De, Sumit Kumar, Duncan A. Brown, Badri Krishnan, and Sanjay Reddy. Stringent constraints on neutron-star radii from multimessenger observations and nuclear theory. *Nature Astron.*, 4(6):625–632, 2020.

[29] Arianna Carbone, Andrea Cipollone, Carlo Barbieri, Arnau Rios, and Artur Polls. Self-consistent Green's functions formalism with three-body interactions. *Phys. Rev. C*, 88(5):054326, 2013.

[30] Arianna Carbone, Arnau Rios, and Artur Polls. Symmetric nuclear matter with chiral three-nucleon forces in the self-consistent Green's functions approach. *Phys. Rev. C*, 88:044302, 2013.

[31] J. Carlson, S. Gandolfi, F. Pederiva, Steven C. Pieper, R. Schiavilla, K. E. Schmidt, and R. B. Wiringa. Quantum Monte Carlo methods for nuclear physics. *Rev. Mod. Phys.*, 87:1067, 2015.

[32] J. Carlson, V. R. Pandharipande, and Robert B. Wiringa. Three-nucleon interaction in 3-body, 4-body, and infinite-body systems. *Nucl. Phys. A*, 401:59–85, 1983.

[33] E. Chabanat, P. Bonche, P. Haensel, J. Meyer, and R. Schaeffer. A Skyrme parametrization from subnuclear to neutron star densities. 2. Nuclei far from stablities. *Nucl. Phys. A*, 635:231–256, 1998. [Erratum: Nucl.Phys.A 643, 441–441 (1998)].

[34] E. Chabanat, J. Meyer, P. Bonche, R. Schaeffer, and P. Haensel. A Skyrme parametrization from subnuclear to neutron star densities. *Nucl. Phys. A*, 627:710–746, 1997.

[35] N. Chamel, A. F. Fantina, J. M. Pearson, and S. Goriely. Maximum mass of neutron stars with exotic cores. *Astron. Astrophys.*, 553:A22, 2013.

[36] Wei-Chia Chen and J. Piekarewicz. Building relativistic mean field models for finite nuclei and neutron stars. *Phys. Rev. C*, 90(4):044305, 2014.

[37] Alberto Colombo, Om Sharan Salafia, Francesco Gabrielli, Giancarlo Ghirlanda, Bruno Giacomazzo, Albino Perego, and Monica Colpi. Multi-messenger Observations of Binary Neutron Star Mergers in the O4 Run. *Astrophys. J.*, 937(2):79, 2022.

[38] Michael W Coughlin, Tim Dietrich, Zoheyr Doctor, Daniel Kasen, Scott Coughlin, Anders Jerkstrand, Giorgos Leloudas, Owen McBrien, Brian D Metzger, Richard O'Shaughnessy, and Stephen J Smartt. Constraints on the neutron star equation of state from at2017gfo using radiative transfer simulations. *Mon. Not. R. Astron. Soc.*, 480(3):3871–3878, 2018.

[39] H. T. Cromartie et al. Relativistic Shapiro delay measurements of an extremely massive millisecond pulsar. *Nature Astron.*, 4(1):72–76, 2019.

[40] Pawel Danielewicz, Roy Lacey, and William G. Lynch. Determination of the equation of state of dense matter. *Science*, 298:1592–1596, 2002.

[41] Soumi De, Daniel Finstad, James M. Lattimer, Duncan A. Brown, Edo Berger, and Christopher M. Biwer. Tidal Deformabilities and Radii of Neutron Stars from the Observation of GW170817. *Phys. Rev. Lett.*, 121(9):091102, 2018.

[42] Paul Demorest, Tim Pennucci, Scott Ransom, Mallory Roberts, and Jason Hessels. Shapiro Delay Measurement of A Two Solar Mass Neutron Star. *Nature*, 467: 1081–1083, 2010.

[43] Tim Dietrich, Michael W. Coughlin, Peter T. H. Pang, Mattia Bulla, Jack Heinzel, Lina Issa, Ingo Tews, and Sarah Antier. Multimessenger constraints on the neutron-star equation of state and the Hubble constant. *Science*, 370(6523):1450–1453, 2020.

[44] C. Drischler, R. J. Furnstahl, J. A. Melendez, and D. R. Phillips. How Well Do We Know the Neutron-Matter Equation of State at the Densities Inside Neutron Stars? A Bayesian Approach with Correlated Uncertainties. *Phys. Rev. Lett.*, 125(20):202702, 2020.

[45] C. Drischler, K. Hebeler, and A. Schwenk. Chiral interactions up to next-to-next-to-next-to-leading order and nuclear saturation. *Phys. Rev. Lett.*, 122(4):042501, 2019.

[46] C. Drischler, J. W. Holt, and C. Wellenhofer. Chiral Effective Field Theory and the High-Density Nuclear Equation of State. *Ann. Rev. Nucl. Part. Sci.*, 71:403–432, 2021.

[47] C. Drischler, J. A. Melendez, R. J. Furnstahl, and D. R. Phillips. Quantifying uncertainties and correlations in the nuclear-matter equation of state. *Phys. Rev. C*, 102(5):054315, 2020.

[48] A. Dyhdalo, R. J. Furnstahl, K. Hebeler, and I. Tews. Regulator Artifacts in Uniform Matter for Chiral Interactions. *Phys. Rev. C*, 94(3):034001, 2016.

[49] A. Ekström, G. R. Jansen, K. A. Wendt, G. Hagen, T. Papenbrock, B. D. Carlsson, C. Forssén, M. Hjorth-Jensen, P. Navrátil, and W. Nazarewicz. Accurate nuclear radii and binding energies from a chiral interaction. *Phys. Rev. C*, 91(5):051301, 2015.

[50] D. R. Entem and R. Machleidt. Accurate charge dependent nucleon nucleon potential at fourth order of chiral perturbation theory. *Phys. Rev. C*, 68:041001, 2003.

[51] D. R. Entem, R. Machleidt, and Y. Nosyk. High-quality two-nucleon potentials up to fifth order of the chiral expansion. *Phys. Rev. C*, 96(2):024004, 2017.

[52] E. Epelbaum, W. Glockle, and Ulf-G. Meissner. The Two-nucleon system at next-to-next-to-next-to-leading order. *Nucl. Phys. A*, 747:362–424, 2005.

[53] E. Epelbaum, H. Krebs, and U. G. Meißner. Improved chiral nucleon-nucleon potential up to next-to-next-to-next-to-leading order. *Eur. Phys. J. A*, 51(5):53, 2015.

[54] Evgeny Epelbaum, Hans-Werner Hammer, and Ulf-G. Meissner. Modern Theory of Nuclear Forces. *Rev. Mod. Phys.*, 81:1773–1825, 2009.

[55] Evgeny Epelbaum, Hermann Krebs, and Patrick Reinert. High-precision nuclear forces from chiral EFT: State-of-the-art, challenges and outlook. *Front. in Phys.*, 8:98, 2020.

[56] Reed Essick, Philippe Landry, Achim Schwenk, and Ingo Tews. Detailed examination of astrophysical constraints on the symmetry energy and the neutron skin of Pb208 with minimal modeling assumptions. *Phys. Rev. C*, 104(6):065804, 2021.

[57] Reed Essick, Ingo Tews, Philippe Landry, Sanjay Reddy, and Daniel E. Holz. Direct Astrophysical Tests of Chiral Effective Field Theory at Supranuclear Densities. *Phys. Rev. C*, 102(5):055803, 2020.

[58] Reed Essick, Ingo Tews, Philippe Landry, and Achim Schwenk. Astrophysical Constraints on the Symmetry Energy and the Neutron Skin of Pb-208 with Minimal Modeling Assumptions. *Phys. Rev. Lett.*, 127(19):192701, 2021.

[59] Matthew Evans et al. Cosmic Explorer: A Submission to the NSF MPSAC ngGW Subcommittee. arXiv:2306.13745 [astro-ph.IM], 2023.

[60] Matthew Evans and others (Cosmic Explorer Consortium). A Horizon Study for Cosmic Explorer: Science, Observatories, and Community. arXiv:2109.09882 [astro-ph.IM], 2021.

[61] F. J. Fattoyev, C. J. Horowitz, J. Piekarewicz, and G. Shen. Relativistic effective interaction for nuclei, giant resonances, and neutron stars. *Phys. Rev. C*, 82:055803, 2010.

[62] E. Fonseca et al. Refined Mass and Geometric Measurements of the High-mass PSR J0740+6620. *Astrophys. J. Lett.*, 915(1):L12, 2021.

[63] J. Fujita and H. Miyazawa. Pion Theory of Three-Body Forces. *Prog. Theor. Phys.*, 17:360–365, 1957.

[64] R. J. Furnstahl, N. Klco, D. R. Phillips, and S. Wesolowski. Quantifying truncation errors in effective field theory. *Phys. Rev. C*, 92(2):024005, 2015.

[65] A. J. Garcia, C. Drischler, R. J. Furnstahl, J. A. Melendez, and Xilin Zhang. Wave-function-based emulation for nucleon-nucleon scattering in momentum space. *Phys. Rev. C*, 107(5):054001, 2023.

[66] A. Gezerlis, I. Tews, E. Epelbaum, M. Freunek, S. Gandolfi, K. Hebeler, A. Nogga, and A. Schwenk. Local chiral effective field theory interactions and quantum Monte Carlo applications. *Phys. Rev. C*, 90(5):054323, 2014.

[67] A. Gezerlis, I. Tews, E. Epelbaum, S. Gandolfi, K. Hebeler, A. Nogga, and A. Schwenk. Quantum Monte Carlo Calculations with Chiral Effective Field Theory Interactions. *Phys. Rev. Lett.*, 111(3):032501, 2013.

[68] Alexandros Gezerlis and J. Carlson. Low-density neutron matter. *Phys. Rev. C*, 81:025803, 2010.

[69] Pablo Giuliani, Kyle Godbey, Edgard Bonilla, Frederi Viens, and Jorge Piekarewicz. Bayes goes fast: Uncertainty Quantification for a Covariant Energy Density Functional emulated by the Reduced Basis Method. arXiv:2209.13039 [nucl-th], 2022..

[70] S. K. Greif, G. Raaijmakers, K. Hebeler, A. Schwenk, and A. L. Watts. Equation of state sensitivities when inferring neutron star and dense matter properties. *Mon. Not. Roy. Astron. Soc.*, 485(4):5363–5376, 2019.

[71] Ish Gupta et al. Characterizing Gravitational Wave Detector Networks: From A$^\sharp$ to Cosmic Explorer. arXiv:2307.10421 [gr.qc], 2023.

[72] P. Gysbers, G. Hagen, J. D. Holt, G. R. Jansen, T. D. Morris, P. Navrátil, T. Papenbrock, S. Quaglioni, A. Schwenk, S. R. Stroberg, and K. A. Wendt. Discrepancy between experimental and theoretical β-decay rates resolved from first principles. *Nature Phys.*, 15(5):428–431, 2019.

[73] P. Haensel, A. Y. Potekhin, and D. G. Yakovlev. *Neutron stars 1: Equation of state and structure*. Springer, New York, 2007.

[74] G. Hagen, A. Ekström, C. Forssén, G. R. Jansen, W. Nazarewicz, T. Papenbrock, K. A. Wendt, S. Bacca, N. Barnea, B. Carlsson, C. Drischler, K. Hebeler, M. Hjorth-Jensen, M. Miorelli, G. Orlandini, A. Schwenk, and J. Simonis. Neutron and weak-charge distributions of the ^{48}Ca nucleus. *Nature Phys.*, 12(2):186–190, 2015.

[75] G. Hagen, T. Papenbrock, A. Ekström, K. A. Wendt, G. Baardsen, S. Gandolfi, M. Hjorth-Jensen, and C. J. Horowitz. Coupled-cluster calculations of nucleonic matter. *Phys. Rev. C*, 89(1):014319, 2014.

[76] H. W. Hammer, S. König, and U. van Kolck. Nuclear effective field theory: status and perspectives. *Rev. Mod. Phys.*, 92(2):025004, 2020.

[77] Hans-Werner Hammer, Andreas Nogga, and Achim Schwenk. Three-body forces: From cold atoms to nuclei. *Rev. Mod. Phys.*, 85:197, 2013.

[78] K. Hebeler, J. M. Lattimer, C. J. Pethick, and A. Schwenk. Equation of state and neutron star properties constrained by nuclear physics and observation. *Astrophys. J.*, 773:11, 2013.

[79] K. Hebeler and A. Schwenk. Chiral three-nucleon forces and neutron matter. *Phys. Rev. C*, 82:014314, 2010.

[80] Kai Hebeler. Three-nucleon forces: Implementation and applications to atomic nuclei and dense matter. *Phys. Rept.*, 890:1–116, 2021.

[81] H. Hergert, S. K. Bogner, T. D. Morris, A. Schwenk, and K. Tsukiyama. The In-Medium Similarity Renormalization Group: A Novel Ab Initio Method for Nuclei. *Phys. Rept.*, 621:165–222, 2016.

[82] S. Hild et al. Sensitivity Studies for Third-Generation Gravitational Wave Observatories. *Class. Quant. Grav.*, 28:094013, 2011.

[83] Martin Hoferichter, Jacobo Ruiz de Elvira, Bastian Kubis, and Ulf-G. Meißner. Matching pion-nucleon Roy-Steiner equations to chiral perturbation theory. *Phys. Rev. Lett.*, 115(19):192301, 2015.

[84] Martin Hoferichter, Jacobo Ruiz de Elvira, Bastian Kubis, and Ulf-G. Meißner. Roy–Steiner-equation analysis of pion–nucleon scattering. *Phys. Rept.*, 625:1–88, 2016.

[85] C. J. Horowitz and Brian D. Serot. Selfconsistent Hartree Description of Finite Nuclei in a Relativistic Quantum Field Theory. *Nucl. Phys. A*, 368:503–528, 1981.

[86] Baishan Hu et al. Ab initio predictions link the neutron skin of ^{208}Pb to nuclear forces. *Nature Phys.*, 18(10):1196–1200, 2022.

[87] L. Huth, I. Tews, J. E. Lynn, and A. Schwenk. Analyzing the Fierz Rearrangement Freedom for Local Chiral Two-Nucleon Potentials. *Phys. Rev. C*, 96(5):054003, 2017.

[88] S. Huth, P. T. H. Pang, I. Tews, T. Dietrich, A. Le Fevre, A. Schwenk, W. Trautmann, K. Agarwal, M. Bulla, M. W. Coughlin, and C. Van Den Broeck. Constraining Neutron-Star Matter with Microscopic and Macroscopic Collisions. *Nature*, 606:276–280, 2022.

[89] S. Huth, C. Wellenhofer, and A. Schwenk. New equations of state constrained by nuclear physics, observations, and QCD calculations of high-density nuclear matter. *Phys. Rev. C*, 103(2):025803, 2021.

[90] W. G. Jiang, A. Ekström, C. Forssén, G. Hagen, G. R. Jansen, and T. Papenbrock. Accurate bulk properties of nuclei from $A = 2$ to ∞ from potentials with Δ isobars. *Phys. Rev. C*, 102(5):054301, 2020.

[91] W. G. Jiang, C. Forssén, T. Djärv, and G. Hagen. Emulating *ab initio* computations of infinite nucleonic matter. arXiv:2212.13216 [nucl-th], 2022.

[92] David B. Kaplan, Martin J. Savage, and Mark B. Wise. Nucleon - nucleon scattering from effective field theory. *Nucl. Phys. B*, 478:629–659, 1996.

[93] David B. Kaplan, Martin J. Savage, and Mark B. Wise. A New expansion for nucleon-nucleon interactions. *Phys. Lett. B*, 424:390–396, 1998.

[94] David B. Kaplan, Martin J. Savage, and Mark B. Wise. Two nucleon systems from effective field theory. *Nucl. Phys. B*, 534:329–355, 1998.

[95] T. Klahn et al. Constraints on the high-density nuclear equation of state from the phenomenology of compact stars and heavy-ion collisions. *Phys. Rev. C*, 74:035802, 2006.

[96] S. Kolling, E. Epelbaum, H. Krebs, and U. G. Meissner. Two-pion exchange electromagnetic current in chiral effective field theory using the method of unitary transformation. *Phys. Rev. C*, 80:045502, 2009.

[97] S. Kolling, E. Epelbaum, H. Krebs, and U. G. Meissner. Two-nucleon electromagnetic current in chiral effective field theory: One-pion exchange and short-range contributions. *Phys. Rev. C*, 84:054008, 2011.

[98] Hermann Krebs. Nuclear Currents in Chiral Effective Field Theory. *Eur. Phys. J. A*, 56(9):234, 2020.

[99] T. Krüger, I. Tews, K. Hebeler, and A. Schwenk. Neutron matter from chiral effective field theory interactions. *Phys. Rev. C*, 88:025802, 2013.

[100] N. Kunert, S. Antier, V. Nedora, M. Bulla, P.T.H. Pang, S. Anand, M.W. Coughlin, I. Tews, J. Barnes, M. Pilloix, W. Kiendrebeogo, and T. Dietrich. Model selection for GRB 211211A through multi-wavelength analyses. arXiv:2301.02049 [astro-ph.HE], 2023.

[101] Nina Kunert, Peter T. H. Pang, Ingo Tews, Michael W. Coughlin, and Tim Dietrich. Quantifying modeling uncertainties when combining multiple gravitational-wave detections from binary neutron star sources. *Phys. Rev. D*, 105(6):L061301, 2022.

[102] Philippe Landry and Kabir Chakravarti. Prospects for constraining twin stars with next-generation gravitational-wave detectors. arXiv:2212.09733 [astro-ph.HE], 2022.

[103] Philippe Landry and Reed Essick. Nonparametric inference of the neutron star equation of state from gravitational wave observations. *Phys. Rev. D*, 99(8):084049, 2019.

[104] Philippe Landry, Reed Essick, and Katerina Chatziioannou. Nonparametric constraints on neutron star matter with existing and upcoming gravitational wave and pulsar observations. *Phys. Rev. D*, 101(12):123007, 2020.

[105] Lee Lindblom. Spectral Representations of Neutron-Star Equations of State. *Phys. Rev.*, D82:103011, 2010.

[106] Lee Lindblom and Nathaniel M. Indik. A Spectral Approach to the Relativistic Inverse Stellar Structure Problem. *Phys. Rev. D*, 86:084003, 2012.

[107] Lee Lindblom and Nathaniel M. Indik. Spectral Approach to the Relativistic Inverse Stellar Structure Problem II. *Phys. Rev. D*, 89(6):064003, 2014. [Erratum: Phys.Rev.D 93, 129903 (2016)].

[108] D. Lonardoni, S. Gandolfi, J. E. Lynn, C. Petrie, J. Carlson, K. E. Schmidt, and A. Schwenk. Auxiliary field diffusion Monte Carlo calculations of light and medium-mass nuclei with local chiral interactions. *Phys. Rev. C*, 97(4):044318, 2018.

[109] D. Lonardoni, I. Tews, S. Gandolfi, and J. Carlson. Nuclear and neutron-star matter from local chiral interactions. *Phys. Rev. Res.*, 2:022033, May 2020.

[110] Diego Lonardoni, Alessandro Lovato, Stefano Gandolfi, and Francesco Pederiva. Hyperon Puzzle: Hints from Quantum Monte Carlo Calculations. *Phys. Rev. Lett.*, 114(9):092301, 2015.

[111] B. Long and U. van Kolck. Renormalization of Singular Potentials and Power Counting. *Annals Phys.*, 323:1304–1323, 2008.

[112] Bingwei Long and C. J. Yang. Renormalizing Chiral Nuclear Forces: Triplet Channels. *Phys. Rev. C*, 85:034002, 2012.

[113] A. Lovato, I. Bombaci, D. Logoteta, M. Piarulli, and R. B. Wiringa. Benchmark calculations of infinite neutron matter with realistic two- and three-nucleon potentials. *Phys. Rev. C*, 105(5):055808, 2022.

[114] Alessandro Lovato, Omar Benhar, Stefano Fantoni, and Kevin E. Schmidt. Comparative study of three-nucleon potentials in nuclear matter. *Phys. Rev. C*, 85:024003, 2012.

[115] J. E. Lynn, I. Tews, J. Carlson, S. Gandolfi, A. Gezerlis, K. E. Schmidt, and A. Schwenk. Chiral Three-Nucleon Interactions in Light Nuclei, Neutron-α Scattering, and Neutron Matter. *Phys. Rev. Lett.*, 116(6):062501, 2016.

[116] J. E. Lynn, I. Tews, S. Gandolfi, and A. Lovato. Quantum Monte Carlo Methods in Nuclear Physics: Recent Advances. *Ann. Rev. Nucl. Part. Sci.*, 69:279–305, 2019.

[117] R. Machleidt and D. R. Entem. Chiral effective field theory and nuclear forces. *Phys. Rept.*, 503:1–75, 2011.

[118] R. Machleidt, F. Sammarruca, and Y. Song. Nonlocal nature of the nuclear force and its impact on nuclear structure. *Phys. Rev. C*, 53(4):R1483–R1487, 1996.

[119] Michele Maggiore et al. Science Case for the Einstein Telescope. *JCAP*, 03:050, 2020.

[120] J. Margueron, H. Hansen, P. Proust, and G. Chanfray. Quarkyonic stars with isospin-flavor asymmetry. *Phys. Rev. C*, 104(5):055803, 2021.

[121] Jerome Margueron, Rudiney Hoffmann Casali, and Francesca Gulminelli. Equation of state for dense nucleonic matter from metamodeling. I. Foundational aspects. *Phys. Rev. C*, 97(2):025805, 2018.

[122] Jerome Margueron, Rudiney Hoffmann Casali, and Francesca Gulminelli. Equation of state for dense nucleonic matter from metamodeling. II. Predictions for neutron star properties. *Phys. Rev. C*, 97(2):025806, 2018.

[123] Andrea Maselli, Andrea Sabatucci, and Omar Benhar. Constraining three-nucleon forces with multimessenger data. *Phys. Rev. C*, 103(6):065804, 2021.

[124] Larry McLerran and Robert D. Pisarski. Phases of dense quarks at large nc. *Nuclear Physics A*, 796(1):83–100, 2007.

[125] Larry McLerran and Sanjay Reddy. Quarkyonic Matter and Neutron Stars. *Phys. Rev. Lett.*, 122(12):122701, 2019.

[126] J. A. Melendez, C. Drischler, A. J. Garcia, R. J. Furnstahl, and Xilin Zhang. Fast & accurate emulation of two-body scattering observables without wave functions. *Phys. Lett. B*, 821:136608, 2021.

[127] M. C. Miller and others (NICER collaboration). PSR J0030+0451 Mass and Radius from $NICER$ Data and Implications for the Properties of Neutron Star Matter. *Astrophys. J. Lett.*, 887(1):L24, 2019.

[128] M. C. Miller and others (NICER collaboration). The Radius of PSR J0740+6620 from NICER and XMM-Newton Data. *Astrophys. J. Lett.*, 918(2):L28, 2021.

[129] Elias R. Most, L. Jens Papenfort, Veronica Dexheimer, Matthias Hanauske, Stefan Schramm, Horst Stöcker, and Luciano Rezzolla. Signatures of quark-hadron phase transitions in general-relativistic neutron-star mergers. *arXiv: 1807.03684*, 2018.

[130] M. R. Mumpower, R. Surman, G. C. McLaughlin, and A. Aprahamian. The impact of individual nuclear properties on r-process nucleosynthesis. *Prog. Part. Nucl. Phys.*, 86:86–126, 2016. [Erratum: Prog.Part.Nucl.Phys. 87, 116–116 (2016)].

[131] P. Navratil, V. G. Gueorguiev, J. P. Vary, W. E. Ormand, and A. Nogga. Structure of A=10-13 nuclei with two plus three-nucleon interactions from chiral effective field theory. *Phys. Rev. Lett.*, 99:042501, 2007.

[132] A. Nogga, R. G. E. Timmermans, and U. van Kolck. Renormalization of one-pion exchange and power counting. *Phys. Rev. C*, 72:054006, 2005.

[133] J. R. Oppenheimer and G. M. Volkoff. On massive neutron cores. *Phys. Rev.*, 55:374–381, 1939.

[134] C. Ordonez, L. Ray, and U. van Kolck. The Two nucleon potential from chiral Lagrangians. *Phys. Rev. C*, 53:2086–2105, 1996.

[135] Takaharu Otsuka, Toshio Suzuki, Jason D. Holt, Achim Schwenk, and Yoshinori Akaishi. Three-body forces and the limit of oxygen isotopes. *Phys. Rev. Lett.*, 105:032501, 2010.

[136] Peter T. H. Pang, Tim Dietrich, Michael W. Coughlin, Mattia Bulla, Ingo Tews, Mouza Almualla, Tyler Barna, Weizmann Kiendrebeogo, Nina Kunert, Gargi Mansingh, Brandon Reed, Niharika Sravan, Andrew Toivonen, Sarah Antier, Robert O. VandenBerg, Jack Heinzel, Vsevolod Nedora, Pouyan Salehi, Ritwik Sharma, Rahul Somasundaram, and Chris Van Den Broeck. NMMA: A nuclear-physics and multi-messenger astrophysics framework to analyze binary neutron star mergers. arXiv:2205.08513 [astro-ph.HE], 2022.

[137] Peter T. H. Pang, Lars Sivertsen, Rahul Somasundaram, Tim Dietrich, Srimoyee Sen, Ingo Tews, Michael Coughlin, and Chris Van Den Broeck. Probing Quarkyonic Matter in Neutron Stars with the Bayesian Nuclear-Physics Multi-Messenger Astrophysics Framework. arXiv:2308.15067 [nucl-th], 2023.

[138] Peter T. H. Pang, Ingo Tews, Michael W. Coughlin, Mattia Bulla, Chris Van Den Broeck, and Tim Dietrich. Nuclear Physics Multimessenger Astrophysics Constraints on the Neutron Star Equation of State: Adding NICER's PSR J0740+6620 Measurement. *Astrophys. J.*, 922(1):14, 2021.

[139] S. Pastore, L. Girlanda, R. Schiavilla, M. Viviani, and R. B. Wiringa. Electromagnetic Currents and Magnetic Moments in (chi)EFT. *Phys. Rev. C*, 80:034004, 2009.

[140] S. Pastore, R. Schiavilla, and J. L. Goity. Electromagnetic two-body currents of one- and two-pion range. *Phys. Rev. C*, 78:064002, 2008.

[141] M. Piarulli, A. Baroni, L. Girlanda, A. Kievsky, A. Lovato, E. Lusk, L.E. Marcucci, S. C. Pieper, R. Schiavilla, M. Viviani, and R.B. Wiringa. Light-nuclei spectra from chiral dynamics. *Phys. Rev. Lett.*, 120(5):052503, 2018.

[142] M. Piarulli, L. Girlanda, R. Schiavilla, R. Navarro Pérez, J. E. Amaro, and E. Ruiz Arriola. Minimally nonlocal nucleon-nucleon potentials with chiral two-pion exchange including Δ resonances. *Phys. Rev. C*, 91(2):024003, 2015.

[143] Maria Piarulli, Luca Girlanda, Rocco Schiavilla, Alejandro Kievsky, Alessandro Lovato, Laura E. Marcucci, Steven C. Pieper, Michele Viviani, and Robert B. Wiringa. Local chiral potentials with Δ-intermediate states and the structure of light nuclei. *Phys. Rev. C*, 94(5):054007, 2016.

[144] Maria Piarulli and Ingo Tews. Local Nucleon-Nucleon and Three-Nucleon Interactions Within Chiral Effective Field Theory. *Front. in Phys.*, 7:245.

[145] Steven C. Pieper, V. R. Pandharipande, Robert B. Wiringa, and J. Carlson. Realistic models of pion exchange three nucleon interactions. *Phys. Rev. C*, 64:014001, 2001.

[146] M. Punturo, M. Abernathy, F. Acernese, B. Allen, N. Andersson, et al. The Einstein Telescope: A third-generation gravitational wave observatory. *Class. Quant. Grav.*, 27:194002, 2010.

[147] G. Raaijmakers et al. Constraining the dense matter equation of state with joint analysis of NICER and LIGO/Virgo measurements. *Astrophys. J. Lett.*, 893(1):L21, 2020.

[148] G. Raaijmakers, S. K. Greif, K. Hebeler, T. Hinderer, S. Nissanke, A. Schwenk, T. E. Riley, A. L. Watts, J. M. Lattimer, and W. C. G. Ho. Constraints on the Dense Matter Equation of State and Neutron Star Properties from NICER's Mass–Radius Estimate of PSR J0740+6620 and Multimessenger Observations. *Astrophys. J. Lett.*, 918(2):L29, 2021.

[149] David Radice, Albino Perego, Kenta Hotokezaka, Steven A. Fromm, Sebastiano Bernuzzi, and Luke F. Roberts. Binary Neutron Star Mergers: Mass Ejection, Electromagnetic Counterparts and Nucleosynthesis. *Astrophys. J.*, 869(2):130, 2018.

[150] Jocelyn S. Read, Benjamin D. Lackey, Benjamin J. Owen, and John L. Friedman. Constraints on a phenomenologically parameterized neutron- star equation of state. *Phys. Rev.*, D79:124032, 2009.

[151] Brendan T. Reed, F. J. Fattoyev, C. J. Horowitz, and J. Piekarewicz. Implications of PREX-2 on the Equation of State of Neutron-Rich Matter. *Phys. Rev. Lett.*, 126(17):172503, 2021.

[152] P. Reinert, H. Krebs, and E. Epelbaum. Semilocal momentum-space regularized chiral two-nucleon potentials up to fifth order. *Eur. Phys. J. A*, 54(5):86, 2018.

[153] David Reitze et al. Cosmic Explorer: The U.S. Contribution to Gravitational-Wave Astronomy beyond LIGO. *Bull. Am. Astron. Soc.*, 51(7):035, 2019.

[154] Thomas E. Riley and others (NICER collaboration). A NICER View of PSR J0030+0451: Millisecond Pulsar Parameter Estimation. *Astrophys. J. Lett.*, 887(1):L21, 2019.

[155] Thomas E. Riley and others (NICER collaboration). A NICER View of the Massive Pulsar PSR J0740+6620 Informed by Radio Timing and XMM-Newton Spectroscopy. *Astrophys. J. Lett.*, 918(2):L27, 2021.

[156] M. R. Robilotta and H. T. Coelho. Taming the Two Pion Exchange Three Nucleon Potential. *Nucl. Phys. A*, 460:645–674, 1986.

[157] Henrik Rose, Nina Kunert, Tim Dietrich, Peter T. H. Pang, Rory Smith, Chris Van Den Broeck, Stefano Gandolfi, and Ingo Tews. Revealing the strength of three-nucleon interactions with the proposed Einstein Telescope. *Phys. Rev. C*, 108(2):025811, 2023.

[158] P. Russotto et al. Results of the ASY-EOS experiment at GSI: The symmetry energy at suprasaturation density. *Phys. Rev. C*, 94(3):034608, 2016.

[159] Andrea Sabatucci, Omar Benhar, Andrea Maselli, and Costantino Pacilio. Sensitivity of neutron star observations to three-nucleon forces. *Phys. Rev. D*, 106(8):083010, 2022.

[160] S. K. Saha, D. R. Entem, R. Machleidt, and Y. Nosyk. Local position-space two-nucleon potentials from leading to fourth order of chiral effective field theory. *Phys. Rev. C*, 107(3):034002, 2023.

[161] Z. F. Seidov. The Stability of a Star with a Phase Change in General Relativity Theory. *Sov. Astron.*, 15:347, October 1971.

[162] Rahul Somasundaram, Joel E. Lynn, Lukas Huth, Achim Schwenk, and Ingo Tews. Maximally local two-nucleon interactions at fourth order in delta-less chiral effective field theory. arXiv:2306.13579 [nucl-th], 2023.

[163] Rahul Somasundaram and Jerome Margueron. Impact of massive neutron star radii on the nature of phase transitions in dense matter. *EPL*, 138(1):14002, 2022.

[164] Rahul Somasundaram, Ingo Tews, and Jerome Margueron. Investigating signatures of phase transitions in neutron-star cores. *Phys. Rev. C*, 107(2):025801, 2023.

[165] Agnieszka Sorensen and others (Community White Paper). Dense Nuclear Matter Equation of State from Heavy-Ion Collisions. arXiv:2301.13253 [nucl-th], 2023.

[166] V. G. J. Stoks, R. A. M. Klomp, M. C. M. Rentmeester, and J. J. de Swart. Partial wave analaysis of all nucleon-nucleon scattering data below 350-MeV. *Phys. Rev. C*, 48:792–815, 1993.

[167] V. G. J. Stoks, R. A. M. Klomp, C. P. F. Terheggen, and J. J. de Swart. Construction of high quality N N potential models. *Phys. Rev. C*, 49:2950–2962, 1994.

[168] S. R. Stroberg, J. D. Holt, A. Schwenk, and J. Simonis. $AbInitio$ Limits of Atomic Nuclei. *Phys. Rev. Lett.*, 126(2):022501, 2021.

[169] I. Tews, S. Gandolfi, A. Gezerlis, and A. Schwenk. Quantum Monte Carlo calculations of neutron matter with chiral three-body forces. *Phys. Rev. C*, 93(2):024305, 2016.

[170] I. Tews, L. Huth, and A. Schwenk. Large-cutoff behavior of local chiral effective field theory interactions. *Phys. Rev. C*, 98(2):024001, 2018.

[171] I. Tews, T. Krüger, K. Hebeler, and A. Schwenk. Neutron matter at next-to-next-to-next-to-leading order in chiral effective field theory. *Phys. Rev. Lett.*, 110(3):032504, 2013.

[172] I. Tews, J. Margueron, and S. Reddy. Critical examination of constraints on the equation of state of dense matter obtained from GW170817. *Phys. Rev. C*, 98(4):045804, 2018.

[173] Ingo Tews, Joseph Carlson, Stefano Gandolfi, and Sanjay Reddy. Constraining the speed of sound inside neutron stars with chiral effective field theory interactions and observations. *Astrophys. J.*, 860(2):149, 2018.

[174] Ingo Tews, Zohreh Davoudi, Andreas Ekström, Jason D. Holt, and Joel E. Lynn. New Ideas in Constraining Nuclear Forces. *J. Phys. G*, 47(10):103001, 2020.

[175] Ingo Tews, James M. Lattimer, Akira Ohnishi, and Evgeni E. Kolomeitsev. Symmetry Parameter Constraints from a Lower Bound on Neutron-matter Energy. *Astrophys. J.*, 848(2):105, 2017.

[176] Ingo Tews, Peter T. H. Pang, Tim Dietrich, Michael W. Coughlin, Sarah Antier, Mattia Bulla, Jack Heinzel, and Lina Issa. On the Nature of GW190814 and Its Impact on the Understanding of Supranuclear Matter. *Astrophys. J. Lett.*, 908(1):L1, 2021.

[177] Richard C. Tolman. Static solutions of einstein's field equations for spheres of fluid. *Phys. Rev.*, 55:364–373, 1939.

[178] Lucas Tonetto, Andrea Sabatucci, and Omar Benhar. Impact of three-nucleon forces on gravitational wave emission from neutron stars. *Phys. Rev. D*, 104(8):083034, 2021.

[179] U. van Kolck. Few nucleon forces from chiral Lagrangians. *Phys. Rev. C*, 49:2932–2941, 1994.

[180] J. D. Walecka. A theory of highly condensed matter. *Annals of Physics*, 83:491–529, 1974.

[181] Steven Weinberg. Nuclear forces from chiral Lagrangians. *Phys. Lett. B*, 251:288–292, 1990.

[182] Steven Weinberg. Effective chiral Lagrangians for nucleon - pion interactions and nuclear forces. *Nucl. Phys. B*, 363:3–18, 1991.

[183] F. Wienholtz, D. Beck, K. Blaum, C. Borgmann, M. Breitenfeldt, R. B. Cakirli, S. George, F. Herfurth, J. D. Holt, M. Kowalska, S. Kreim, D. Lunney, V. Manea, J. Menéndez, D. Neidherr, M. Rosenbusch, L. Schweikhard, A. Schwenk, J. Simonis, J. Stanja, R. N. Wolf, and K. Zuber. Masses of exotic calcium isotopes pin down nuclear forces. *Nature*, 498(7454):346–349, 2013.

[184] Robert B. Wiringa, V. G. J. Stoks, and R. Schiavilla. An Accurate nucleon-nucleon potential with charge independence breaking. *Phys. Rev. C*, 51:38–51, 1995.

[185] Edward Witten. Cosmic Separation of Phases. *Phys. Rev. D*, 30:272–285, 1984.

[186] C. J. Yang, A. Ekström, C. Forssén, and G. Hagen. Power counting in chiral effective field theory and nuclear binding. *Phys. Rev. C*, 103(5):054304, 2021.

[187] J. L. Zdunik and P. Haensel. Maximum mass of neutron stars and strange neutron-star cores. *Astron. Astrophys.*, 551:A61, 2013.

CHAPTER 7

Quark-Hadron Crossover in Neutron Stars

Toru Kojo

IT has been conjectured that highly compressed nuclear matter transforms into quark matter after baryons spatially overlap. Conventionally nuclear and quark matter have been distinguished by the first-order phase transition. Alternative descriptions to such a sharp nuclear-quark matter boundary have been developed since ~ 2010, driven by dramatic progress in neutron star observations and improved computations for low- and high-density constraints based on the quantum chromodynamics. The new development has started by specifying the domain of applicability for purely nucleonic and purely quark descriptions, and it turns out that there is a density interval where both descriptions are invalid so that the effective degrees of freedom in the system are not clear-cut. This density interval has the direct relevance to the structure of neutron stars with the mass $1.4\ M_\odot$-$2.2\ M_\odot$ ($M_\odot$: solar mass). In this chapter, we discuss the microphysics in the transient regime, taking the quark-hadron continuity as our baseline.

7.1 QUANTUM CHROMODYNAMICS IN DENSE MATTER: OVERVIEW

7.1.1 Effective degrees of freedom

The fundamental theory of strong interaction, quantum chromodynamics (QCD) [72, 127], describes the dynamics of quarks as matters and gluons as gauge bosons, both of which have the color charges associated with the $SU(3)$ color symmetry [58]. The structure of QCD looks similar to quantum electrodynamics (QED) for electrons and photons, but a distinct feature of QCD is that gluons, unlike photons, have charges to which gauge bosons couple. The gluons directly couple one another leading to highly nonlinear equations of motion. The solutions contain topological configurations [27].

The nonlinearity in QCD is already nontrivial at the classical level, but the physics at the quantum level is even richer. In general, the effective (running) coupling constants among particles change with the distance scale due to the quantum fluctuations involved in a given process. In QED, the photon exchange interactions among charged particles are screened by virtual electron-positron creations and annihilations; the interaction becomes

DOI: 10.1201/9781003306580-7

weaker at longer distance because more screening is possible. In QCD, the corresponding interactions are gluon exchanges which include screening due to quark-antiquark fluctuations. But the self-coupling among gluons in addition induces "anti-screening" virtual processes, with which the interactions become stronger at longer distance. In QCD, this anti-screening dominates over the screening effects. The scale evolution of couplings is described by the renormalization group equation.

The dynamics in QCD become strongly coupled at long distance or low energy, while at short distance or high energy QCD can be described by the weak coupling picture (asymptotic freedom) [72, 127]. The former strong coupling regime turns the effective degrees of freedom from colored quarks and gluons into "color white (or singlet)" hadrons, mesons, and baryons. This trapping of quarks and gluons is called color confinement, whose description has been still under active researches since the proposal of QCD. Another very important phenomenon is the chiral symmetry breaking (ChSB) which generates the dynamical quark mass together with the "almost massless" Nambu-Goldstone (NG) bosons [70, 75, 122]. In the standard view, normal hadrons (except NG bosons) are made of constituent quarks confined by long-range confining forces, with the residual short-range correlations which depend on hadron species under consideration [41]. Hadrons are the main actors in nuclear physics at low energy.

The most powerful method to analyze the strong coupling regime of QCD, in which the effective degrees of freedom are not very obvious, is the lattice Monte Carlo simulations (for a review, [68]) which evaluate the QCD partition function in very efficient algorithms. The current lattice QCD can reproduce hadron properties in good precision; the hadron masses, the form factors, and forces between baryons including strangeness have been calculated. In particular, the lattice played crucial roles in establishing the crossover transition in hot QCD [17], in which a heated hadron resonance gas (HRG) continuously transforms into a quark-gluon plasma (QGP). Unfortunately, this powerful method cannot be applied to cold dense matter of QCD, due to the notorious sign problem in which the integrand of the partition functions is not positive. There have been no reliable methods of calculations for the transient regime from nuclear to quark matter.

7.1.2 Scales: from hadron structures to dense matter

It is useful to summarize the characteristic scales relevant for various effective degrees of freedom (Fig.7.1). First we consider scales for hadrons [112]. The quark and gluon degrees of freedom in weak coupling picture is often used from high energy down to the energy scale of $\sim$ 1-2 GeV, and such scale resolves the structure of constituent quarks. At lower energy scale $\lesssim \Lambda_\chi \sim 1$ GeV or distance scale $\gtrsim 0.2$ fm, the constituent quark picture may be used. The scale Λ_χ is often used as a scale characteristic to the ChSB [66]. At even lower energy $\lesssim \Lambda_{\rm QCD} \sim 0.2$-$0.3$ GeV or distance scale $\gtrsim 1$ fm, the confining forces become very important to hold quarks within typical hadron sizes of ~ 0.5-0.8 fm (e.g., proton charge radius $\langle (r_p^{\rm charge})^2 \rangle^{1/2} \simeq 0.84$ fm or pion size $\langle r_\pi^2 \rangle^{1/2} \simeq 0.66$ fm [146]). In addition to these scales, we note current quark masses for up- and down-quarks are $m_{u,d} \sim 5$ MeV, strange quark $m_s \sim 100$ MeV, which are enhanced to $M_{u,d} \sim 300$ MeV and $M_s \sim 500$ MeV by the dynamical ChSB. The nucleon mass is roughly $M_N \sim N_c M_{u,d}$ where $N_c = 3$ is the number of colors.

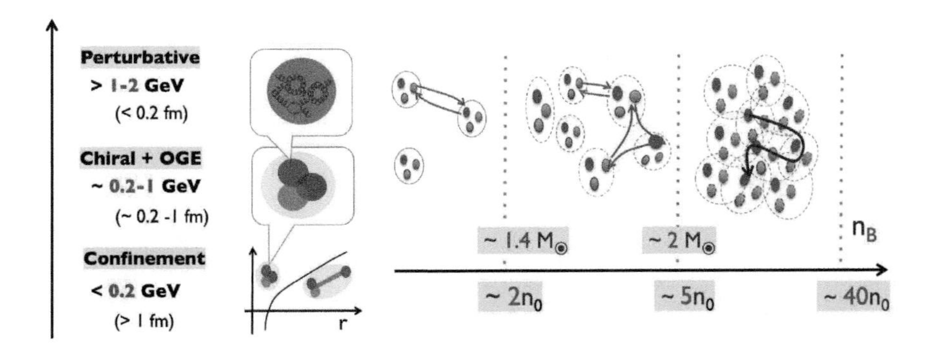

Figure 7.1 Scales for the hadron structure and dense matter. The OGE is the abbreviation of the one-gluon-exchange.

Next we examine these scales in terms of baryon density n_B in dense matter. The convenient unit is the nuclear saturation density, $n_0 \simeq 0.16\,\mathrm{fm}^{-3}$, which is about the nucleon density in typical nuclei. At high density, the cores of nucleons are supposed to touch one another around $n_B^{\rm overlap} \simeq$ 4-7n_0, as estimated from $n_B^{\rm overlap} = 1/(4\pi r_{\rm core}^3/3)$ with the baryon core radius of 0.5-0.8 fm (see, e.g., [25]). At this scale, the density is close to the density inside of a single nucleon, and it is natural to expect the importance of the quark degrees of freedom. Assuming the quark Fermi sea of u, d, s quarks, the quark Fermi momentum is $p_F \simeq 400$ MeV at $n_B = 5n_0$. Yet the density is not high enough for weak coupling pictures; indeed, the perturbative QCD (pQCD) calculations indicate that weak coupling picture is valid only for $n_B \gtrsim 40n_0$ or $p_F \sim 1$ GeV [100] where we may resolve the structure of constituent quarks formed by partonic quarks and gluons. Although not addressed in this review, the possible crossover from the constituent quark regime to the partonic regime is an interesting issue to be addressed. In the context of neutron stars, the core density of neutron stars is supposed to be $\simeq$ 2-7n_0 for which we must discuss QCD in the strongly correlated regime. Such regime is the main subject of this review.

In a dilute regime of dense matter, we take nucleons as natural effective degrees of freedom. The nuclear scale as consequences of nuclear interactions is not easy to understand from the QCD point of view; there are intricate cancellations among various physical effects. At long distance nucleons interact through exchanges of mesons, such as π, σ, ω, and so on [110]. QCD tells that the natural magnitude of the potential is $O(100)$ MeV. At $n_B \sim n_0$, the Fermi momentum of nucleons are ~ 260 MeV and the kinetic energy of nucleons is ~ 40 MeV. Yet the typical binding energy is much smaller, O(1-10) MeV, meaning that there must be large cancellation among the potentials and kinetic energy. Further complications arise from many-body forces. Nuclear many-body calculations have established the importance of three-body forces which cannot be described as the sum of two-body forces. It is difficult to determine the strength of three-body forces precisely as it requires three baryons to meet. While three-nucleon forces affect the binding energy by few MeV at $n_B \simeq n_0$, the magnitude of three-body contributions radically increases with density, and becomes substantial for $\gtrsim 2n_0$ (see, e.g., Ref. [6]). The $2n_0$ is often regarded as the upper bound for the validity range of nuclear calculations.

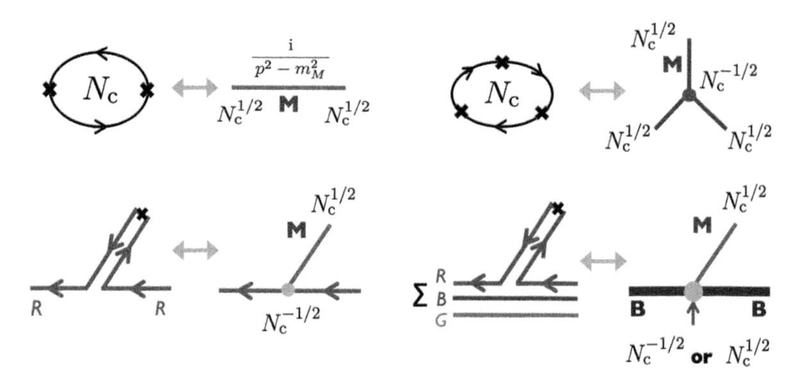

Figure 7.2 N_c counting and duality between quark and hadronic graphs. Meson-meson interactions become weaker for processes with more mesons. Meson-baryon couplings depend on how quark-meson couplings are added. For additive couplings, the sum of quark-meson couplings yield a factor N_c, making meson-baryon interactions strong.

The density interval intermediate between the nuclear and quark matter, $n_B \simeq$ 2-5n_0, is probably in the regime where neither of pure nuclear nor quark matter descriptions is valid [25]. Nuclear descriptions are problematic because of the importance of many-body forces, while the density seems not enough for quark matter descriptions. The interplay between nuclear and quark matter physics is the main topic in this review and will be elaborated in later sections.

7.1.3 Duality between hadron and quark dynamics: $1/N_c$ expansion

In order to discuss the interplay between hadron and quark dynamics, ideally we would like to express hadrons in terms of quarks and gluons and then discuss interactions among those hadrons in terms of quark exchanges. Although explicitly performing such calculations remains very difficult, one can still classify processes by looking into the quark dynamics. For this purpose, it is useful to consider the $1/N_c$ expansion with $N_c(=3)$ being the number of colors [144].

Utilizing the N_c counting, one can characterize quark-meson, meson-meson, meson-baryon couplings in terms of N_c [156]. The estimates rely on the quark-hadron duality in the sense that a process can be described both in terms of quarks or hadrons. For example, consider a correlation function (Fig.7.2). A quark composite operator $J_M^\dagger \sim \bar{q}\Gamma_M q$ creates a meson, and the meson is annihilated by another composite operator J_M after some space-time interval,

$$\langle 0|J_M(p)J_M^\dagger(p)|0\rangle \sim \langle 0|J_M|M\rangle \frac{\mathrm{i}}{p^2 - m_M^2} \langle M|J_M^\dagger|0\rangle \,. \tag{7.1}$$

The LHS in terms of quark graphs is $O(N_c)$, because red-, green-, and blue quarks all contribute. Then the operator-state coupling $\lambda_M = \langle M|J_M^\dagger|0\rangle$ is estimated to be $\sim N_c^{1/2}$. Next, we consider a three meson vertex g_{3M} which appears in a three point function,

$$\langle 0|J_1(p_1)J_2(p_2)J_3(p_3)|0\rangle \sim \lambda_1\lambda_2\lambda_3 \frac{\mathrm{i}}{p_1^2 - m_1^2} \frac{\mathrm{i}}{p_1^2 - m_1^2} \frac{\mathrm{i}}{p_1^2 - m_1^2} g_{3M} \,. \tag{7.2}$$

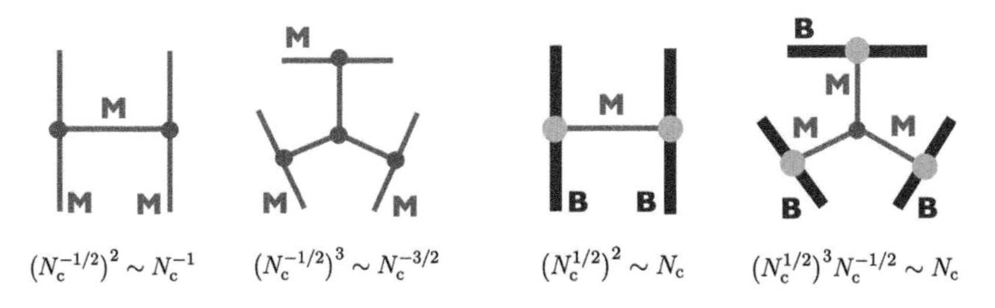

Figure 7.3 $\mathcal{N}_M$-meson forces vs $\mathcal{N}_B$-baryon forces. For mesons, increasing $\mathcal{N}_M$ reduces the strength of interactions. In contrast, for baryons, many-body forces can be as strong as two-body forces.

The LHS is again $\sim N_c$, while $\lambda_1\lambda_2\lambda_3 \sim N_c^{-3/2}$, so that g_{3M} must be $\sim N_c^{-1/2}$ which is regarded as small. Repeating similar discussions, one can find a four meson vertex to be $\sim N_c^{-1}$, a five meson vertex $\sim N_c^{-3/2}$, and so on. These arguments tell that the couplings among mesons are small. For more-body meson forces, the couplings are weaker; many-body forces among mesons are negligible.

For nuclear forces described by meson exchanges, it is convenient to first consider the quark-meson coupling. We have estimated $\langle M|J_M^\dagger|0\rangle = \langle M|\bar{q}\Gamma_M q|0\rangle = \sum_{c=R,G,...}\langle M|\bar{q}_c\Gamma_M q_c|0\rangle \sim N_c^{1/2}$. Thus for a given color c, the quark-meson coupling is $\langle M|\bar{q}_c\Gamma_M q_c|0\rangle \sim N_c^{-1/2}$. We can use this estimate for nuclear forces [83]. Nucleons contain N_c-quarks, so if contribution from each quark is *constructively* added with the same sign, then the nucleon-meson vertex is enhanced to $\sim N_c^{1/2}$. Meanwhile, if quark contributions are added *destructively*, the nucleon-meson vertex remains the same order as the quark-meson coupling, $\sim N_c^{-1/2}$. Thus, the nucleon-meson vertex strongly depends on the channel. For instance, the ω meson couples to the baryon charge of a nucleon for which quark contributions are additive; $g_{\omega NN} \sim N_c^{1/2}$ is large. In contrast, the ρ meson couples to isospin charge for which quark contributions are not additive but rather cancel; $g_{\rho NN} \sim N_c^{-1/2}$ is small. In terms of nuclear forces, the meson vertices appear as $\sim g_{XNN}^2$, so the magnitude varies from N_c to N_c^{-1}; the hierarchy of nuclear forces is characterized by $1/N_c^2$ expansion, not by $1/N_c$.

One of significant differences between meson-meson and baryon-baryon interactions is the many-body forces (Fig.7.4). For mesons many-body forces are suppressed. At temperature where hadron resonances are dominated by mesons, the HRG model, as the sum of non-interacting particles, is known to work well, and this is consistent with the $1/N_c$ counting. On the other hand, the situation can be completely different for many-baryon systems. For baryons, $\mathcal{N}$-nucleon forces are mediated by $\mathcal{N}$-mesons which meet at a vertex with the strength $\sim N_c^{1-\mathcal{N}/2}$. The suppression factor $N_c^{-\mathcal{N}/2}$ can be compensated by the baryon-meson vertices of $(N_c^{1/2})^{\mathcal{N}}$. Hence, $\mathcal{N}$-nucleon forces are $\sim N_c$, and can be as strong as two-nucleon forces. The importance of many-body forces implies that there are many-quark exchanges among baryons, so it would indicate that quark degrees of freedom are important before baryons overlap. In the following, this point will be elaborated further.

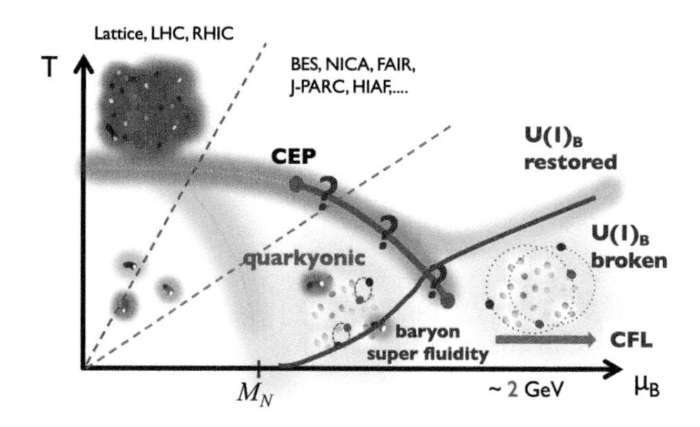

Figure 7.4 A conjectured phase diagram of QCD.

7.2 QCD PHASE STRUCTURES

The study of dense matter in QCD is also an important element to complete the phase diagram in QCD (for a review, Ref. [64], and for future directions, Ref. [106]). Soon after the asymptotic free nature of QCD is recognized in 1970s, several applications to finite temperature and/or density have appeared.

The phases of matter can be solidly characterized by the symmetries of states. The symmetry is broken by condensations of macroscopic degrees of freedom. For instance, the chiral symmetry breaking can be characterized by chiral condensates, $\langle 0|\bar{q}q|0\rangle \sim \Lambda_{\rm QCD}^3$ ($|0\rangle$: the ground state), made of quark-antiquark pairs [75, 122]. In chiral symmetric phases, the condensate is vanishing. In QCD, however, the chiral symmetry is slightly broken by the current quark mass, so $\langle 0|\bar{q}q|0\rangle$ is always finite. Thus, the distinction between the chiral symmetric and broken phases is not exact. But even without such exact criterion, m one can still classify phases by focusing on the characteristic properties.

7.2.1 Deconfinement and chiral transitions at finite temperature

Many theoretical and experimental works have been dedicated to the finite temperature deconfinement transition from a hadronic resonance gas (HRG) to a quark-gluon plasma (QGP). Theoretical studies have been performed in both weak coupling [28, 73, 151] and strong coupling regimes [128, 143], including extensive studies of functional approaches [47, 52] and lattice QCD simulations [1]. In the experimental side, relativistic heavy ion experiments, in which high-energy nuclei collide and convert the collision energy into heat, have established the picture that the QGP is nearly perfect liquid with the small viscosity (or strong collectivity of fluids, see a review Ref. [34]).

In spite of different looks in degrees of freedom, i.e., hadrons in HRG and quark-gluons in QGP, the transition turns out to be crossover. The (pseudo-)critical temperature T_c (for chiral crossover) is $\simeq$ 156.5 MeV, where the energy density is $\simeq$ 0.42 GeV/fm^3 and the entropy density $\simeq 3.7\,\mathrm{fm}^{-3}$ [26]. Remarkably, the lattice results for the EOS can be very well reproduced by the HRG models up to temperature slightly above T_c [13]. The key for this deconfinement transition is the entropic effects [128, 143]. Hadrons with the masses (m_H) much larger than the temperature, up to $m_H \sim 2.5$ GeV, are important to reproduce the EOS. Although each single hadron does not contribute much because of the Boltzmann suppression factor $\sim e^{-m_H/T}$, there are many hadron species to compensate such a suppression factor. The system is then saturated by hadrons and then color-electric flux condenses, allowing quarks and gluons to float around without staying within a specific hadron.

The finite temperature transition also accompanies the chiral symmetry restoration. The vacuum chiral condensate $\langle 0|\bar{q}q|0\rangle$ and the scalar charge in a hadron H, $\langle H|\bar{q}q|H\rangle$, have the opposite sign. Hence, as more hadrons become available, the magnitude of chiral condensates inevitably decreases [67]. Eventually the magnitude of each contribution also takes place. In this sense, the deconfinement and chiral symmetry restoration are related. Such interplay has been efficiently described by quark models coupled to the Polyakov loop [63, 130], a measure of background color fields introduced by thermally excited hadrons.

7.2.2 Toward high density: the critical endpoint of the chiral phase transition

The above discussions are applied for zero baryon density or baryon chemical potential, $\mu_B = 0$. The lattice simulations for the Taylor coefficients of EOS [26, 29, 30], measured at $\mu_B = 0$ to avoid the sign problem, indicate that the crossover continues to $\mu_B/T \sim 2$. At larger μ_B/T, whether transitions remain crossover or becomes the discontinuous phase transitions is not known. If at some high μ_B a discontinuous phase transition begins to occur, there must be a critical endpoint (CEP) [18, 35]. Establishing the existence of the CEP is an important subject being studied in the bean energy scan (BES) program at RHIC [107], where the collision energies are systematically reduced to achieve "baryon stopping" for highly compressed matter.

If the CEP is established, we expect the 1st order chiral phase transition persists to higher density and lower temperature, perhaps entering the territory of neutron star (NS) physics. Whether the first order phase transition exists in NS matter or not has the crucial impacts on the structure of NS. Theoretical predictions are model dependent. In quark models, whether models contain the density-density repulsion type interactions (similar to the ω exchange) [99] or not have very large impacts on the strength of discontinuity in the chiral first order transition [86]; at large density the chiral restoration is driven by increase in baryon density, but the repulsion tempers the growth of density, smoothing out the chiral transition. For sufficiently large repulsion, the discontinuous phase transitions entirely disappear [32]. The phase transition generally softens EOS and is not very favored by the current NS and nuclear constraints [31]. There is also a proposal that the first order line is terminated also at low-temperature side. If true, there would be double CEPs [76].

7.2.3 Color superconductivity

Because of the asymptotic freedom, QCD matter is expected to be weakly interacting at very high temperature and/or high density where typical distance scales between particles become small. Even such situation, however, the phase structure can be nontrivial. In case of dense matter, there can be many soft excitations of quarks around the Fermi surface, so that even small interactions cause dramatic effects.

One of nontrivial proposals is the color superconductivity (CSC) in which quark-pairs (diquarks) condense and form diquark condensates $\langle qq\rangle$ (for a review, Ref. [11]). Then quarks acquire the energy gaps Δ in excitations that suppress quark contributions to dynamical transport and thermal effects. The mechanism is similar to the condensation of electron-electron pairs (Cooper pairs) in superconductors. The idea of CSC was proposed by Bailin-Love [19] in 1980s within weak coupling treatments, but the impact of the diquark condensation is regarded as small. The CSC has become very popular after strong coupling models suggested that the magnitude of the gaps can be the order of $\Delta \sim 100$ MeV and may have impacts on the physics of NSs [9, 129, 140].

The concept of diquarks has been popular in hadron physics [81, 82]. There are so-called good and bad diquarks which may be formed inside of baryons. The most energetically favored channel is that two quarks form colour-antitriplet state $\bar{\mathbf{3}}_c$, flavor-antitriplet $\bar{\mathbf{3}}_f$, spin-singlet, and spatially S-wave. The small kinetic energy favors the S-wave. The color $\bar{\mathbf{3}}_c$ and spin-singlet is favored due to color-electric and magnetic interactions. In a good diquark, the color charge of $\bar{\mathbf{3}}_c$ is made smaller than a simple sum of twice of quark color charges $\mathbf{3}_c$ and hence produces less color-electric fields. In the color charge of $\bar{\mathbf{3}}_c$ channel, the color-magnetic interactions become attractive for the spin-singlet channel. With these color, spin, and spatial states, the flavor must be anti-triplet. The difference between nucleons and delta-baryons is often discussed in terms of the mass difference between good and bad diquarks.

For high density where the mass difference between flavors can be neglected, the most stable state is the color-flavor-locked (CFL) state in which ud-, ds-, and su-diquarks all condense [10]. Gauge bosons acquire the Meissner masses, and u, d, s-quarks all acquire the gaps. At lower density, the mass differences between u, d- and s-quarks become important and the pairing can be more asymmetric. There have been several attempts to find the signature of the CSC from NS observations. Later we discuss the possible impacts of the diquark pairing to the NS EOS.

7.2.4 Continuity from baryon superfluid to the CFL phase

The diquark condensates $\langle qq \rangle$ in the previous section are actually subtle objects as they are not gauge invariant and hence not observables. Schematically, $\langle qq \rangle \sim \rho \langle e^{i\Theta} \rangle$ where $\rho \neq 0$ is the magnitude and Θ is the colored phase whose average leads to $\langle e^{i\Theta} \rangle = 0$ [48]. Hence $\langle qq \rangle (= 0)$ cannot be used as an order parameter to distinguish the symmetry between different phases. Nevertheless, the $\rho \neq 0$ leaves some dynamical consequences differing from the phase with $\rho \neq 0$.

Instead of qq operators, one can imagine gauge invariant operator $BB \sim (qqq)(qqq)$ in which each (qqq) is made color singlet. This dibaryon operator BB is not invariant under $U(1)_B$ symmetry transformation, so that its condensation $\langle BB \rangle \neq 0$ unambiguously means the SSB of the $U(1)_B$ symmetry. Meanwhile, one can regard it as the product of three diquark operators, $BB \sim (qq)(qq)(qq)$, so condensed (qq)'s in the CSC phase should yield $\langle BB \rangle \sim \rho^3 \neq 0$. In this case, the colored phases of diquarks cancel one another so that the phase average does not vanish.

The $U(1)_B$ symmetry is the exact symmetry in QCD, unlike the chiral symmetry. Hence, its SSB unambiguously distinguishes the $U(1)_B$ symmetric and broken phases. For instance, the ($U(1)_B$ symmetric) QGP at high density and the CSC phase cannot be continued by crossover; there must be phase transitions.

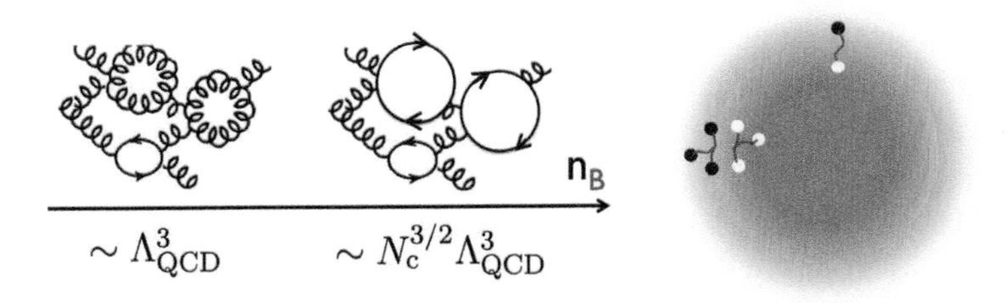

Figure 7.5 Quantum fluctuations (Left) and the quark Fermi sea with hadronic Fermi surface (Right) in Quarkyonic matter.

What is uncertain is the distinction between $U(1)_B$ broken baryonic matter (baryon superfluid phase) and the CSC [135]; they cannot be distinguished by the SSB of $U(1)_B$. The CSC (or Higgs phase) and confined phases look different in light of the color $SU(3)$ symmetry. But it has been known from the work of Fradkin and Shenkar [55] that, in QCD (as Yang-Mills plus matter fields in fundamental representations), at strong coupling the Higgs phase can be smoothly connected to the confined phase. Thus, if baryonic matter transforms the CFL quark matter at a sufficiently low density where the interactions remain strong, we may not find any phase transition; we would instead find crossover.

7.2.5 Quarkyonic matter

The previous discussions on the quark-hadron continuity in the context of superfluidity are largely based on the symmetry, and the dynamical aspects are not explicitly touched. For example, the gap in the baryon superfluidity is usually regarded as $O(1\text{-}10)$ MeV, while the gap in the CSC can be $O(10\text{-}100)$ MeV. This in turn leads to very different transition temperature of $U(1)_B$ broken phases. To smoothly connect the baryon superfluid and CSC phases, one needs to explain the rapid growth of gaps.

There is another line of thought concerning the quark-hadron continuity scenario. The argument is originally initiated in studies of dense QCD with a large number of colors, $N_c \to \infty$ [117]. In this large N_c limit, quantum fluctuations of gluons dominate over those of quarks (Fig.7.5). This in turn means that the quark screening to the confining forces is suppressed by a factor $\sim 1/N_c$ [144, 156]. At finite density or finite quark chemical potential $\mu_Q = \mu_B/N_c$, the confining forces are Debye screened only at $\mu_Q \sim q_F \sim N_c^{1/2}\Lambda_{\rm QCD}$, while it is supposed that baryons overlap and form quark matter around $\mu_Q \sim \Lambda_{\rm QCD}$ [117]. Thus, these estimates suggest the existence of quark matter with gluons confining for $\Lambda_{\rm QCD} \lesssim \mu_Q \lesssim N_c^{1/2}\Lambda_{\rm QCD}$ (for general d-spatial dimensions, the upper bound is $\sim N_c^{1/(d-1)}$ [89]). This state of matter is called *Quarkyonic* matter, with the name suggesting a matter intermediate between quark and baryonic matter. Here, $N_c = 3$ does not quite look a good expansion parameter, but what is relevant here is that the conception of Quarkyonic matter introduces a new way to look at dense QCD matter between baryonic and quark matter.

In Quarkyonic matter, the concepts of quark matter (or quark Fermi sea) and confining gluons are reconciled by having the quark Fermi sea filled by quark states and baryonic states around the edge of the quark Fermi sea. The Fermi sea with colored states saturated is automatically color-singlet, while for incompletely occupied levels quarks states must be suitably superposed to make the resulting states color-singlet. In this picture, in dilute regime we have only baryons, and, as density increases, the quark Fermi sea surrounded by baryonic momentum shell gradually develops. When this quark Fermi sea in the bulk becomes very large, we reach matter with quarks in the bulk but baryons near the edge. This continuous evolution of the quark Fermi sea bridges the gap between nuclear and quark matter descriptions. In Sec. 7.6, the Quarkyonic matter concepts will be used to explain the rapid stiffening of EOS.

7.3 NEUTRON STAR EQUATIONS OF STATE: GENERAL REMARKS

The bulk aspect of states of matter is reflected in equations of state (EOS). At high temperature and low density, the QCD EOS has been studied by lattice simulations and heavy ion experiments.

Meanwhile, dense matter beyond $n_B \sim n_0$ and at low temperature is not realized in terrestrial experiments; heavy ion experiments can create compressed nuclei but the matter is not quite cold. Neutron stars (NSs), which are extremely dense objects with the masses of 1-$2M_\odot$ ($M_\odot$: the solar mass) and the radii 10-13 km, are the only known places to accommodate cold, dense matter beyond the nuclear regime (for reviews, Refs. [25, 92, 103]). The large gravity associated with the energy density can hold matter very dense. When the gravity is greater than the pressure increased with the matter compression, such a dense object collapses into a black hole (BH). Accordingly there are the maximum mass and the maximum baryon density that NSs can achieve. Stiff EOS, defined as large pressure P at given energy density ε, leads to large maximum masses of NSs.

7.3.1 Mass-Radius relations and EOS

The structure of a NS can be calculated by solving the Einstein equation coupled to EOS [138]. For a spherical symmetric and non-rotating NS, one can reduce the Einstein equation with the EOS into the Tolman-Oppenheimer-Volkov (TOV) equation. The TOV equation determines the density and energy distributions in stars. For a given density at the core ($n_{\rm core} \equiv n_B(r = 0)$), matter is integrated until the matter pressure becomes zero. The condition $P(r = R) = 0$ defines the radius and the mass of a star. Changing $n_{\rm core}$, we obtain a family of the mass-radius (M-R) relations.

For a given EOS, a typical M-R curve has the following shape (Fig.7.6). NSs with low $n_{\rm core} \lesssim n_0$ have low masses $\lesssim M_\odot$ and large radii $R \gtrsim 13$ km. In this domain, small increase of M reduces R radically; this occurs for the crust part made of dilute matter of nuclei, since a small change in gravity is enough to radically compress dilute matter. This regime continues until the crust part becomes very thin with the thickness of $\lesssim 0.5$ km, and, for $n_{\rm core} \gtrsim n_0$, the radius is large determined by the size of the core. Because of the stiff core which is hard to compress, R no longer changes much for increase in M. The turning points of M-R curves are largely determined by nuclear EOS; for stiff EOS the radii tend to be larger. This regime of growing M with small changes in R continues to the maximum mass, or get terminated by first-order phase transitions which introduce kinks in M-R curves. The kinks reflect the discontinuous softening of EOS occurring

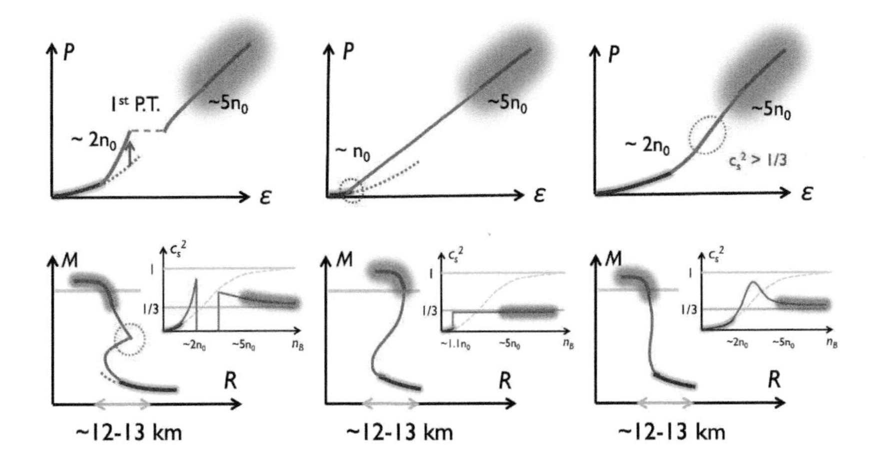

Figure 7.6 Relations among EOS, sound velocity, and M-R curves. Provided that a nuclear EOS is given, three types of EOS have been typically considered [92].

in first-order transitions (P is constant but ε increases). For the relationship between c_s^2 and M-R, see, e.g., Refs. [8, 145].

7.3.2 Observational constraints

Here we quickly summarize important NS constraints on EOS. The first precise measurement of $2M_\odot$ NS, announced in 2010 for PSR J1614-2230 [42], can be regarded as the beginning of new era for NS studies. Its initial estimate was $1.97 \pm 0.04 M_\odot$, and later the accumulation of data improves the accuracy and the resulting estimate is $1.928 \pm 0.017 M_\odot$ [54]. In 2013, the existence of $2M_\odot$ NS was further supported by finding of PSR J0348+0432 with $2.01 \pm 0.04 M_\odot$ [16]. Since then, even heavier NS, PSR J0740+6620 with $2.08 \pm 0.07 M_\odot$ [40, 53] and PSR J0952-0607 with $2.35 \pm 0.17 M_\odot$ [134] have been found. These discoveries of $2M_\odot$ NS demand high-density EOS at $\gtrsim 4n_0$ to be very stiff and have changed our guideline to construct EOS.

Yet, the $2M_\odot$ constraints alone are not powerful enough to get insights on the properties of matter. In particular, unless we constrain the radii of NS, we may start with very stiff nuclear EOS with which the EOS after 1st PT occurs can remain stiff. What limits the stiffness at low-density $\sim$ 1-$2n_0$ is the nuclear laboratory constraints and the radii of $\simeq 1.4 M_\odot$ NS, denoted as $R_{1.4}$. The first accurate constraints came from gravitational waves (GWs) in the historic NS-NS merger event GW170817 [2]. The upperbound on the tidal deformability was obtained and it can be translated into the constraint on $R_{1.4} \lesssim 13.4$ km. Combining the GW data and nuclear constraints around n_0 prefer lower radii, $R_{1.4} \simeq 12 \pm 1$ km [3, 15]. With these constraints on low-density EOS, the 2 $M_\odot$ constraints are highly nontrivial. We also note that the electromagnetic signals (kilonovae signals) from NS-NS merger event suggest that the maximum mass should not be too large, $M_{\max} \lesssim 2.35\ M_\odot$ [22, 114, 131, 139].

There are also new development of simultaneous measurements of M and R from pulsars. The NICER keeps track of the time evolution of hot spots on the surface of pulsars.

The measurements of the rotation period, surface velocity, and gravitational redshift allow us to resolve M, R, and the rotation frequency $\omega_{\rm rot}$. The two independent analyses in NICER announced 12.35 ± 0.75 km [119] and $12.39^{+1.30}_{-0.98}$ km [133] for $2.08 M_\odot$ NS. This means that the radii hardly change from $1.4 M_\odot$ to $2.1 M_\odot$ NS, or density from $\sim 2n_0$ to $\sim 5n_0$. Radical 1st PT is unlikely for this density interval as it leads to rapid the shrinkage of radii.

Current samples for the simultaneous measurements of M and R are still limited, but in the future the detection of GWs (planned for NEMO [4] in 2030- and the third generation detectors [51, 111] in 2035-) will be daily events and allow us statistical and systematic analyses. In particular, GWs from the dynamic stage of NS-NS merger event, which has not been observed yet due to large noises in current detectors, will constrain R for $\gtrsim 2M_\odot$ [21, 60, 79, 84, 121] and discriminate typical scenarios shown in Fig.7.6.

7.3.3 General constraints on EOS

There are general conditions that any EOS must satisfy. To describe EOS we first have to choose thermodynamic variables. In studies of NSs, we often write energy density ε as a function of baryon density n_B, charge density $n_{\rm em}$, lepton density n_l, entropy density s, and so on. From the energy density $\varepsilon(n_B, n_{\rm em}, n_l, s, \cdots)$, one can derive all thermodynamic relations by taking derivatives with respect to $n_B, n_{\rm em}, \cdots$. For instance, one can compute chemical potentials and temperature as

$$\mu_B = \frac{\partial \varepsilon}{\partial n_B}, \quad \mu_{\rm em} = \frac{\partial \varepsilon}{\partial n_{\rm em}}, \quad \mu_l = \frac{\partial \varepsilon}{\partial n_l}, \quad T = \frac{\partial \varepsilon}{\partial s}, \cdots \tag{7.3}$$

Or one can Legendre transform the energy density into pressure as

$$P(\mu_B, \mu_{\rm em}, \mu_l, T, \cdots) = \varepsilon - \mu_B n_B - \mu_{\rm em} n_{\rm em} - \mu_l n_l - sT - \cdots. \tag{7.4}$$

Starting with the pressure, densities are calculated as

$$n_B = \frac{\partial P}{\partial \mu_B}, \quad n_{\rm em} = \frac{\partial P}{\partial \mu_{\rm em}}, \quad n_l = \frac{\partial P}{\partial \mu_l}, \quad s = \frac{\partial P}{\partial T}, \cdots \tag{7.5}$$

In the QCD community, the expression of P is more often used.

The physically acceptable pressure must satisfy several conditions. The pressure must be concave in any variations with respect to chemical potentials and temperature. For example, the susceptibility must be positive,

$$\chi_B = \frac{\partial^2 P}{\partial \mu_B^2} = \left\langle \left(n_B - \langle n_B \rangle \right)^2 \right\rangle > 0, \tag{7.6}$$

where the bracket indicates the statistical average. We can do the same for variables made of the linear combination of $(\mu_B, \mu_{\rm em}, \cdots)$, so the concave condition must be applied to any directions of variables. We call this condition the thermodynamic stability.

Another important condition is the causality condition that the sound velocity of matter must be less than the light velocity. In the hydrodynamics of compressible fluids, a small perturbation to an equilibrated matter leads to a wave equation of pressure propagation [78]

$$(\partial_t^2 - c_s^2 \vec{\nabla}^2)\delta p = 0, \qquad c_s^2 = \left. \frac{\partial P}{\partial \varepsilon} \right|_s, \tag{7.7}$$

where c_s is the adiabatic speed of sound. We call the condition $c_s \leq c_{\rm light}$ $(= 1)$ the causality condition. Which densities to be fixed depends on matter under considerations. For neutron star matter we impose the charge neutrality condition, $n_{\rm em} = 0$, and calculate c_s^2 along such a line in the $(\mu_B, \mu_{\rm em}, n_l)$ space to satisfy the condition. We note that, for heavy ion collision experiments, the proton fraction is $\simeq 0.4$-0.5 and hence the domain explored is different from NS matter.

The thermodynamic stability and causality conditions are usually satisfied if we compute EOS in a single model [69]. But in practice, each model has its own range of applicability so that one must patch different models. Such patchworks are powerfully constrained by the above-mentioned conditions and are hence technically very nontrivial; EOS at different density regions constrain one another. In the context of QCD, nuclear and quark matter constrain each other, providing us with hints on the properties of transitions between these two regimes.

7.4 NUCLEAR EQUATIONS OF STATE: OVERALL TRENDS

Neutron star EOS is a long-standing subject and there have been huge amount of calculations for nuclear matter. It is impossible to cover all of them (for a recent review, Ref. [101]), but there is a certain qualitative tendency which is intrinsic to the effective degrees of freedom. In this section, we briefly go over those trends.

7.4.1 Nuclear matter around saturation density

Properties of nuclear matte are largely inferred by extrapolating the properties of laboratory nuclei to infinite matter. The energy per particle around $n_B \simeq n_0$ is expressed as

$$\hat{E}(n_B, Y_p) = \hat{E}(n_B, 1/2) + S_2(n)(1 - 2Y_p)^2 + S_3(n)(1 - 2Y_p)^3 + \cdots . \tag{7.8}$$

EOS around $Y_p \simeq 1/2$ is constrained by laboratory nuclei which are almost isospin symmetric. A commonly used expression is power series around n_0,

$$\hat{E}(n_B, 1/2) = -B + \frac{K_{1/2}}{18}(u-1)^2 + \frac{Q_{1/2}}{162}(u-1)^2 + \cdots , \tag{7.9}$$

where we use $u \equiv n_B/n_0$. The linear term $(u - 1)$ is absent because the matter is self-bound, $\partial \hat{E}/\partial n_B|_{n_0} = 0$ at $Y_p = 1/2$. The nuclear saturation density, $n_0 \simeq 0.16\,{\rm fm}^{-3}$, is extracted by observing that the core density of nuclei becomes universal for larger nuclei. Another important quantity is the binding energy per nucleon, $B \simeq$ 15-16 MeV, whose experimental status is less certain than the determination of n_0 because of the systematic error associated with the incomplete separation between the bulk and surface contributions. The nuclear incompressibility K is about $K \simeq 230 \pm 20$ MeV [5, 148].

EOS for asymmetric matter with $Y_p \sim 0$ is a necessary input for NS matter. The difference between EOS at $Y_p = 0$ and $Y_p = 0.5$, $\tilde{S}(n_B) = \varepsilon(n_B, Y_p = 0) - \varepsilon(n_B, Y_p = 0.5)$, is called the "symmetry energy." But $\tilde{S}$ cannot be directly accessed from experiments as neutron-rich nuclei are unstable against the β-decay. In practice, we usually treat $S_2(n)$ as the symmetry energy, expanding it around n_0,

$$S_2(n) = S_V + \frac{L}{3}(u-1) + \frac{K_{\rm sym}}{18}(u-1)^2 + \frac{Q_{\rm sym}}{162}(u-1)^2 + \cdots \tag{7.10}$$

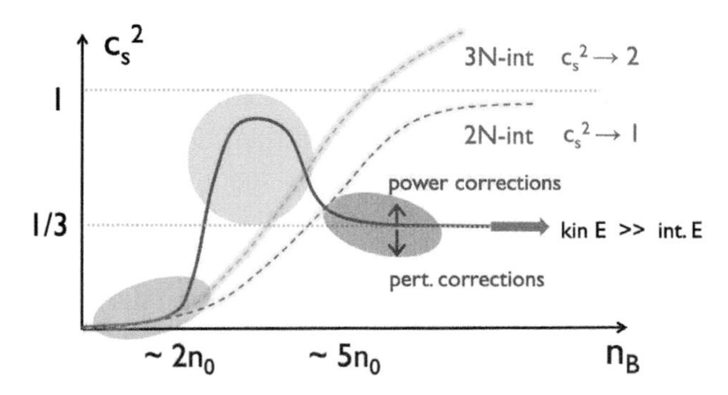

Figure 7.7 Squared sound velocity c_s^2 inferred for QCD together with c_s^2 for nuclear matter with two- or three-body (contact) interactions.

As for S_V and L, neutron matter calculations, CREX and PREX neutron skin measurements, and NS constraints, give consistent trends; the preferred ranges are $S_V \simeq 31 \pm 2$ MeV and its slope is $L \simeq 50 \pm 10$ MeV (for a recent summary, see Ref. [102]).

Reproducing nuclear matter properties remains difficult and important problems in nuclear many-body theories. Approaches firmly based on experiments are many-body calculations with two- and three-nucleon interactions as inputs. Two nucleon forces have been well constrained by nucleon-nucleon scattering at energy below the pion production threshold. Meanwhile, it has been challenging to accurately determine three-nucleon forces. They have been estimated by the spectroscopy of light nuclei [126]. A cleaner way to determine three-body force is to utilize the proton-deutron (p-d) scattering [136].

The systematic approach to build nuclear forces is based on the Chiral Effective Field Theory (ChEFT) in which one can introduce a hierarchy of processes based on the chiral symmetry and the power counting of momenta [152]. Its application to nuclear forces was initiated by Weinberg [153–155] and has been elaborated by including higher order contributions [49, 77]. The ChEFT descriptions for scattering experiments and light nuclei have been quite successful [50, 109]. Many-body calculations based on the ChEFT nuclear forces are largely consistent with the empirical constraints around $n_B \simeq n_0$ [36, 44]. If one extrapolates the EOS to $n_B \simeq 1.5 n_0$, the sound velocity is $c_s^2 \sim 0.1$ and the radii of $1.4 M_\odot$ NS is about $R_{1.4} \simeq$ 11.5-12.5 km [44, 147].

7.4.2 High-density extrapolation of nuclear matter

Nuclear matter EOS is often extrapolated to even higher density. In doing so, three-body forces are arranged to satisfy the $2M_\odot$ constraints. To see the importance of many-body forces, we consider a simple parametrization [142],

$$\varepsilon_N(n_B) = m_N n_B + c_0 n_B^{5/3}/m_N + d n_B^{\mathcal{N}} \,, \tag{7.11}$$

where the first term is the mass energy density, the second kinetic term, and the last term responsible for $\mathcal{N}$-body contributions. The kinetic energy is suppressed by a factor $1/N_c^2$ compared to the mass energy. The pressure is calculated as

$$P = n_B^2 \frac{\partial(\varepsilon_N/n_B)}{\partial n_B} = c_0' n_B^{5/3}/m_N + d(\mathcal{N} - 1) n_B^{\mathcal{N}} \,, \tag{7.12}$$

where the contribution from the large mass energy density drops out. For this reason, the pressure in nuclear matter is basically small, unless the interaction plays the essential role.

The coefficients d for two- and three-body forces are constrained by nuclear constraints at $n_B \sim n_0$, and are not very large. But for $n_B \gg n_0$, the many-body forces can grow very large. Now suppose the interaction terms become dominant in both ε and P. Then $P \sim (\mathcal{N}-1)\varepsilon$, and the sound velocity is

$$c_s^2 \sim \mathcal{N} - 1 \,. \tag{7.13}$$

This tells us the asymptotic behavior of EOS. If two-body forces become dominant, then $\mathcal{N} = 2$ so that $c_s^2 \to 1$ at large density. If three-body forces dominate, $c_s^2 \to 2$. In the latter case, EOS becomes very stiff just before reaching $c_s^2 = 1$, but eventually violates the causality constraint.

Nuclear EOS such as ChEFT that are consistent with nuclear constraints are soft at low density, and $c_s^2 \sim 0.1$-0.2 for $n_B \lesssim 2n_0$. The c_s^2 develops with power growth and becomes stiff EOS around $n_B \sim 4n_0$ and soon violates the causality around ~ 5-$6n_0$. To avoid such problems, there must be intricate cancellations among various many-body forces which may be achieved in functional methods [43]. But perhaps the necessity of such complicated calculations is the sign that one should use another effective degree of freedom; quark descriptions seem more appropriate for such density.

7.5 QUARK MATTER AND UNIFIED EQUATIONS OF STATE

Quark matter often appears in the context of softening of EOS [124]. For hybrid models which include hadronic matter at low density and quark matter at high density, two states of matter are distinguished by the first-order phase transition (1st PT). At the transition, the baryon density jumps to the density where quark descriptions become more natural. In order to construct such a hybrid model, quark matter EOS must be soft; otherwise, the model cannot allow the appearance of the 1st PT [97]. This sort of arguments also give impressions that the appearance of exotic phases would lead to softening and therefore would be disfavored by the NS constraints.

On the other hand, leaving aside hybrid models, it is known that one can arrange quark EOS to be very stiff. Indeed, EOS used for purely quark stars are EOS of such kinds. The simplest form is the ideal gas of massless quarks ($N_{\rm c} = N_{\rm f} = 3$),

$$P(\mu_q) = \frac{N_{\rm c} N_{\rm f}}{12\pi^2}\mu_q^4 - B\,, \qquad \mu_q = \mu_B / N_{\rm c}\,, \tag{7.14}$$

with B being a normalization constant of EOS, called the bag constant. Taking derivatives, one can readily derive $\varepsilon = 3P + 4B$. The EOS leads to $c_s^2 = 1/3$ much larger than in nuclear EOS for $n_B \lesssim 2n_0$. Then, whether such quark EOS is soft or stiff entirely depends on the normalization of EOS, i.e., the bag constant B. For a smaller B, the P is larger and ε is smaller, realizing stiff EOS.

7.5.1 Evolution of quark matter models and hybrid EOS

The possible availability of quark matter in stars was discussed before QCD was established [80]. Soon after the discovery of the asymptotic freedom, hybrid descriptions of nuclear-quark matter appeared [23, 39]. By construction with the 1st PT, the quark EOS was regarded as soft.

The first serious application of perturbative QCD was carried out for massless quarks in Ref. [57] to three-loop order, together with the resummation of plasmon contributions that is necessary to remove the IR divergences from the thermodynamic potential. These authors pointed out that quark EOS with large α_s and large B is similar to EOS with small α_s and small B, and the resulting EOS can be very stiff [56].

The early ideas of self-bound quark stars, which are entirely made of deconfined quarks, were improved and became very popular after the work [157]. It was pointed out that deconfined strange quark matter may be more stable than nuclear matter without strange quarks. The decay path from nuclear to strange quark matter is suppressed by the energy barrier which should be large if many-body states in each phase are radically different. The scaling of M-R of quark stars in terms of B was also found,

$$M_{\rm max} \simeq 2.0 M_\odot \left(\frac{56\,{\rm MeV/fm^3}}{B} \right)^{1/2}, \qquad R|_{M_{\rm max}} \simeq 10.7\,{\rm km} \times \frac{M_{\rm max}}{2M_\odot} . \tag{7.15}$$

For a reasonable value of B, the maximum mass can be as large as $2M_\odot$. Thus, quark matter EOS can be regarded as stiff if B is small enough. For such value of B, the onset of deconfined quark matter is $\mu_q \simeq 274$ MeV or $\mu_B \simeq 822$ MeV, considerably lower than the nucleon mass. At the onset, the baryon density jumps from zero to $\simeq 1.7n_0$, perhaps a bit too low to apply deconfined quark matter models.

Ref. [7] analyzed hybrid models using schematic parametrization of quark matter models with various non-perturbative effects. The bag constant, perturbative α_s corrections, mass corrections, and possible contributions from the CSC condensates are considered, and they found that realistic nuclear EOS and parametrized quark matter models yield very similar EOS. Hence hybrid stars "masquerade" NSs. Researches along this direction are further elaborated by using dynamical quark models [87, 88] with which the density evolution of various condensates and effective quark masses can be calculated. In the period of these studies, many works considered quark matter at strong coupling, although nuclear and quark matter were treated as distinct.

7.5.2 Modern unified EOS: three window models

7.5.2.1 *ChEFT-pQCD interpolation*

Since ~ 2010, studies of quark matter have become increasingly more quantitative, in line with the dramatic improvement in NS observations. At very high density, pQCD analyses have been improved by introducing uncertainty estimates for the renormalization scale dependence or the convergence test of the α_s expansion. The analyses have determined the breakdown scale of weak coupling regime to be $n_B \simeq 40n_0$ or $\mu_q \sim 1$ GeV [100]. The current state-of-art calculations are N2LO plus soft N3LO contributions in the α_s expansion [71] (see also Refs. [59] for other types of resummation of Feynman diagrams).

The above breakdown scale is much higher than the density $n_B \sim 5n_0$ where baryons overlap, implying the presence of strongly correlated quark matter for 5-40n_0. Interpolation between nuclear matter EOS based on the chiral EFT and pQCD EOS is performed to construct EOS for $\sim$ 1-40n_0 [14]. One of important observations is that, if we start using quark

EOS at sufficiently low density close to n_0, the unified EOS with $c_s^2 \simeq 1/3$ can satisfy the available NS constraints, and such EOS leads to substantial amount of quark matter core. The microscopic interplay between nuclear and quark matter was not addressed, though; the question is how c_s^2 quickly reaches the conformal value $c_s^2 \simeq 1/3$.

7.5.2.2 Masuda-Hatsuda-Takatsuka models

To describe quark matter in NSs, it is necessary to consider some model at strong coupling and its connection to hadronic matter. A phenomenological but concrete model of hadron-to-quark crossover, "three-window" model, was first proposed by Masuda-Hatsuda-Takatsuka [115]. They tried to solve the "hyperon puzzle" — hadronic EOS with hyperons is too soft to account for the existence of $2M_\odot$ NS. Near the onset, hyperons *at rest* add much energy but very little pressure. For quark EOS, the Nambu-Jona-Lasinio (NJL) model with strangeness is used as a representative of quark models with non-perturbative effects (e.g., the ChSB). The model does not represent confinement, so its use is limited to high-density where baryons are supposed to overlap. The hadronic and quark matter are interpolated (or *superposed*) as [116]

$$\varepsilon(n_B) = w_- \varepsilon_H(n_B) + w_+ \varepsilon_Q(n_B)\,, \qquad w_\pm = \frac{1}{2}\left[\left(1 \pm \tanh\left(\frac{n_B - n_c}{\Gamma}\right)\right)\right], \tag{7.16}$$

where ε_H and ε_Q are hadronic and quark EOS, respectively, and $w_\pm$ is some weight functions with $n_c \sim 3n_0$ and $\Gamma \sim n_0$. Using reasonable size of effective density–density repulsion (vector repulsion) added to the standard NJL model, the model satisfies the $2M_\odot$ constraints. The important view behind this modeling is that quark EOS can be stiff. Assuming crossover, we implicitly could discard the high-density part of hadronic EOS which is not trustable; in this standpoint quark EOS no longer has to be softer than *extrapolated* hadronic EOS [97].

Two more aspects of crossover models deserve further comments. Smooth interpolation between soft nuclear and stiff quark EOS demands that the pressure increases rapidly as a function of energy density, and hence $c_s^2 = \partial P/\partial \varepsilon$ makes a peak with $c_s^2 > 1/3$ before the quark matter regime is established to yield $c_s^2 \simeq 1/3$. This sound velocity peak is novel; it has not been seen in purely nucleonic models and may be thought of distinct features related to quark degrees of freedom. Another important finding is that, while the appearance of the strangeness drastically softens EOS in hadronic matter, it does not cause much softening in quark matter models.

7.5.2.3 QHC models

Three window models of Masuda-Hatsuda-Takatsuka are further elaborated into QHC (Quark-Hadron-Crossover) EOS series [24, 25, 94, 97] with detailed account for interaction effects such as vector repulsion and attractive diquark pairing within the NJL type model. In the QHC, the unified EOS is constructed by using nuclear EOS $P_H(\mu_B)$ for $n_B \lesssim 2n_0$, quark EOS $P_Q(\mu_B)$ for $n_B \gtrsim 5n_0$, and parametrize EOS for $2n_0 \lesssim n_B \lesssim 5n_0$ as

$$P_{\text{inter}}(\mu_B) = \sum_{n=0}^{5} c_n \mu_B^n\,, \tag{7.17}$$

where at each boundary $P_{\rm inter}$ is matched to P_H and P_Q up to second derivatives. Hence there are six boundary conditions that uniquely fix constants c_n's. The main difference from the hyperbolic interpolation in Eq.(7.16) is that the QHC model does not allow quark (nuclear) EOS to affect EOS at $n_B \lesssim 5n_0$ ($n_B \gtrsim 2n_0$), and hence is free from any artefacts that the *extrapolated* EOS would cause.

For nuclear EOS, the Akmal-Pandharipande-Ravenhall (APR) EOS [6], Togashi EOS [149], and N3LO ChEFT EOS [45] have been used. Using these well-established models, the properties of NS with $\sim 1.4M_\odot$ as well as nuclear constraints at $n_B \sim n_0$ are automatically satisfied. For each nuclear EOS, the thermodynamic stability and causality impose powerful constraints on the range of quark matter EOS, or in practice on the range of effective coupling constants. Demanding the maximum NS mass to be $\gtrsim 2M_\odot$, it turns out that the effective couplings must be those expected from the non-perturbative scale $\Lambda_{\rm QCD}$ [141].

Important observations found in the QHC modeling is that attractive correlations can stiffen EOS. In fact, whether repulsive or attractive forces stiffen EOS or not depends on the density density dependence. To see this, we consider a simple parametrization of the energy density as [97].

$$\varepsilon(n_B) = a n_B^{4/3} + b n_B^{\mathcal{N}} \,, \tag{7.18}$$

where the first term is the relativistic kinetic energy of quarks and the second is interactions. From this, one can derive the pressure as

$$P = \frac{\varepsilon}{3} + b\Big(\mathcal{N} - \frac{4}{3}\Big) n_B^{\mathcal{N}} \,, \tag{7.19}$$

where we have absorbed the parameter a in favor of ε. The first term gives the conformal limit, see also Ref. [38, 61, 113] for discussions in the context of the trace anomaly $\langle T^\mu_\mu \rangle \neq 0$ that is the measure of the conformality. There are two types of stiffening to increase P at a given ε. For the $\mathcal{N} > 4/3$ case, EOS is stiffened by a repulsive interaction with $b > 0$; this case should be easy to imagine from two-body repulsion in dilute regime, proportional to $\sim +n_B^2$. Meanwhile, less known is the case with $\mathcal{N} < 4/3$ where the attractive interaction with $b < 0$ stiffens EOS. An example is the BCS pairing occurring near the quark Fermi surface. The phase space for the pairs is $\sim 4\pi p_F^2 \Delta \sim n_B^{2/3}$, provided Δ weakly depends on n_B. Then the pairing reduces the energy density by $\sim \Delta^2 n_B^{2/3}$. See, e.g., Refs. [33, 38, 97] for the impact of pairing effects on EOS.

The mechanism to stiffen EOS by attractive correlations near the quark Fermi surface is potentially very important because it would draw one's attention to studies of various exotic phases based on quark degrees of freedom. Near the Fermi surface quark excitations do not cost much energy while there are many channels in quark-quark or quark-quarkhole interactions [95]. Furthermore, provided the validity of the quark-hadron continuity picture, the knowledge of hadron physics should be transferred to the physics at $n_B \simeq$ 2-$5n_0$. Studies of baryon-baryon interactions at short distance are useful in this respect, see, e.g., Refs. [123, 125, 137] and Ref. [74] for lattice calculations.

The interpolation scheme dictated above can be also used to constrain the composition and various condensates in the crossover domain [96, 120]. To do so, we generalize our target of interpolation from the thermodynamic potential to generating functional, i.e., the thermodynamic potential in the presence of external fields coupled to our target quantities, e.g., u-quark density. We compute nuclear and quark matter EOS in such external fields and then demand, for sufficiently small external fields, that they can be smoothly interpolated. With such *unified generating functionals*, we differentiate the functional with respect to the external fields and get the expectation values of our target quantities for entire density interval.

7.6 QUARK-HADRON DUALITY FOR CROSSOVER

Phenomenological models of crossover in the previous sections use hadronic and quark matter EOS as boundary conditions to infer the physics in the crossover region. While they give useful insights, more direct descriptions of the microphysics are called for.

Difficulties in describing the transient regime arise from the identification of effective degrees of freedom. If we consider models including both baryons and quarks, we have to resolve the problem of double counting [46], as baryons are made of quarks. In field theory computations in practice, the double counting results in UV divergences in thermodynamic quantities and must be handled exactly at given orders of computations [91]. But attempts to use only quark degrees of freedom require complicated computations. In addition, those baryons interact by meson exchanges which can be interpreted as quark exchanges. This means that quarks float around among baryons, without sticking to a specific baryon.

In this section, we push forward the above-mentioned picture and present models of the quark-hadron *duality* which allows us to express EOS in either degrees of freedom, without double counting. For attempts to realize the duality in terms of hadronic variables, see, e.g., Refs. [104, 108, 132].

7.6.1 Mode-by-mode percolation

An idea of percolation was introduced by Baym [20] and subsequently taken up in Ref. [37] and later in Ref. [105]. A quantum description of quark percolation, including the Anderson localization [12], was proposed in Ref. [65]. Quark wavefunctions are studied in many-baryon systems whose interactions are given by quark exchanges at moderate distance. Each baryon is made of the valence quark core surrounded by meson clouds. Meson clouds of σ, ω, ρ,... are regarded as outgoing and incoming valence quarks from the core region (Fig.7.8), while a pion cloud is treated differently as they are not simple quark-antiquark

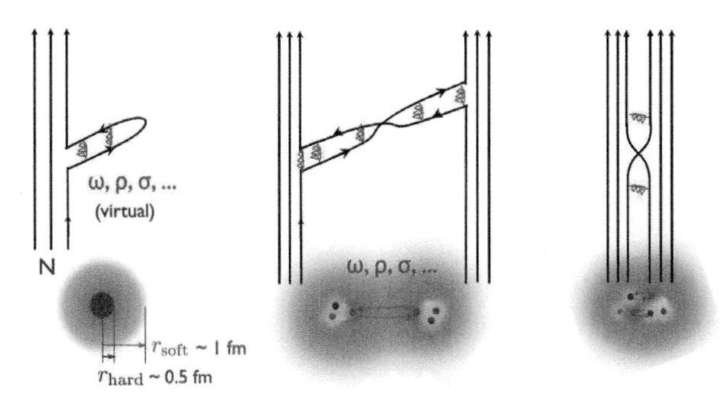

Figure 7.8 Meson clouds around the valence core, meson (quark) exchanges associated with the overlap of meson clouds, and quark exchanges at short distance.

pair of valence quarks. For mesons in the former type, their clouds around the core as well as the exchanges among baryons may be understood in terms of fluctuating valence quarks. The number of valence quarks is large, N_c, in a baryon and hence interactions induced by valence quark exchanges can be strong.

Now let us ask when a quark propagates from one side of a system to the other side through quark exchanges among baryons. It should certainly occur when baryons occupy whole volume of the system. But even before that, such propagation may occur occasionally if baryons geometrically make a chain connecting one side of the system to the other side. "Classical percolation" describes the situation that the probability to find such geometric connected baryons becomes finite in the thermodynamic limit. The classical percolation occurs as the 1D chain, and at higher density as the 2D plane and eventually as the 3D volume. If we discretize space into cubic grids with the bond size $r_{\rm soft} \sim r_{\rm cloud}$ and then place baryons on sites randomly, the classical percolation occurs when baryons occupy $\simeq 0.34$ of the total volume [150]. If we set $r_{\rm cloud} = 0.7$ fm, the classical percolation first occurs at [65]

$$n_B^{\rm cl-perc} \simeq 0.34/(4\pi r_{\rm cloud}^3/3) \simeq 1.5 n_0 \,. \tag{7.20}$$

This density is rather low and close to the nuclear territory.

This geometrical connection of baryons, however, does not immediately mean that quark states are widely spread or deconfined over such geometry, since (i) quark exchanges between neighboring baryons do not occur with the 100% probability and (ii) quantum destructive interference effects, similar to the Anderson localization, suppress the transmission of quarks at long distance. These effects can be studied in a schematic hamiltonian for quarks in a given spatial distribution of baryons [85]

$$H = \sum_n E_n |n\rangle\langle n| + \sum_{n\neq m} |n\rangle V_{nm} \langle m| \,, \tag{7.21}$$

where $V_{nm} = -V$ is a hopping term while E_n is an onsite energy (chosen to $6V$ to set the minimal quark energy to zero) which is finite in baryons and infinite outside of baryons. Since baryons are regarded as heavy, here the Born-Oppenheimer approximation is assumed to calculate quark spectra. For simplicity, distributions of baryons are given randomly, and quark spectra for a given hamiltonian are averaged over baryon distributions.

More and more baryons occupy the sites, quarks can have more extended wavefunctions. With quantum mechanical treatments of quarks, it turns out that the quark delocalization ("deconfinement") in a specific direction occurs when baryons occupy $\simeq 0.44$ of the total volume, or at $n_B \simeq 1.8 n_0$, a bit higher density than the classical percolation of baryons; after paths for quarks are opened, only specific modes can propagate in long distance. Including these quantum effects, the density for the quantum percolation of quarks is still close to the nuclear territory.

7.6.2 Quark saturation: inevitable and rapid stiffening of EOS

The percolation model in the previous section gives us insights on the quark delocalization. But relying on the Born-Oppenheimer type descriptions limits our discussions to relatively low density. Indeed, as discussed shortly, quark states at low momenta are saturated around

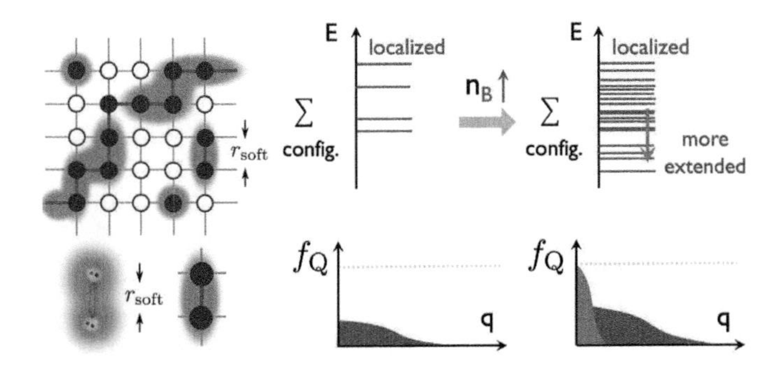

Figure 7.9 Quark energy levels with a variety of baryon configurations. At low density, quarks are localized within a finite spatial domain. States with broader spatial extension (or smaller momenta) appear at higher density.

2-3n_0 beyond which baryons (if exist) are forced to be relativistic; as a consequence no obvious time scale separation between baryon and quark dynamics.

7.6.2.1 Duality relations

Instead of directly treating many-body states of baryons and quarks, we characterize them by occupation probability of states, $f_B(k)$ for baryons and $f_Q(q)$ for quarks *in a given color* [62, 93, 98]. These distributions are related through the sum rule

$$f_Q(\mathbf{q}) = \int_{\mathbf{k}} f_B(\mathbf{k})\varphi(\mathbf{q}-\mathbf{k}/N_c)\,, \tag{7.22}$$

where φ is a quark momentum distribution (normalization: $\int_{\mathbf{q}} \varphi(\mathbf{q}) = 1$) in a baryon with the momentum $\mathbf{k}$, with the average and variance of momenta

$$\langle \mathbf{q} \rangle_{\mathbf{k}} = \mathbf{k}/N_c\,, \quad \langle (\mathbf{q} - \langle \mathbf{q} \rangle)^2 \rangle_{\mathbf{k}} = \Lambda^2 \sim \Lambda_{\rm QCD}^2\,, \quad \langle O \rangle_{\mathbf{k}} \equiv \int_{\mathbf{q}} O\varphi(\mathbf{q}-\mathbf{k}/N_c)\,. \tag{7.23}$$

It is important to note that, at low density with small $|\mathbf{k}|$, the quark average momentum is small but the variation is large; quarks inside of a baryon have high momenta of $\sim \Lambda_{\rm QCD}$ but their momenta point to different directions one another, leaving only a small total baryon momentum.

The sum rules and the normalization condition of φ lead to (each baryon contains one red quark, one blue quark, and so on)

$$n_B = \int_{\mathbf{q}} f_Q(\mathbf{q}) = \int_{\mathbf{k}} f_B(\mathbf{k})\,. \tag{7.24}$$

The baryon number can be counted in terms of either quarks or baryons. The energy of a single baryon is the sum of average quark energies

$$E_B(\mathbf{k}) = N_c \langle E_Q(\mathbf{q}) \rangle_{\mathbf{k}} = N_c \langle E_Q(\mathbf{q}) \rangle_{\mathbf{k}=0} + \frac{k_i k_j}{2N_c} \left\langle \frac{\partial^2 E_Q}{\partial q_i \partial q_j} \right\rangle_{\mathbf{k}=0} + \cdots \tag{7.25}$$

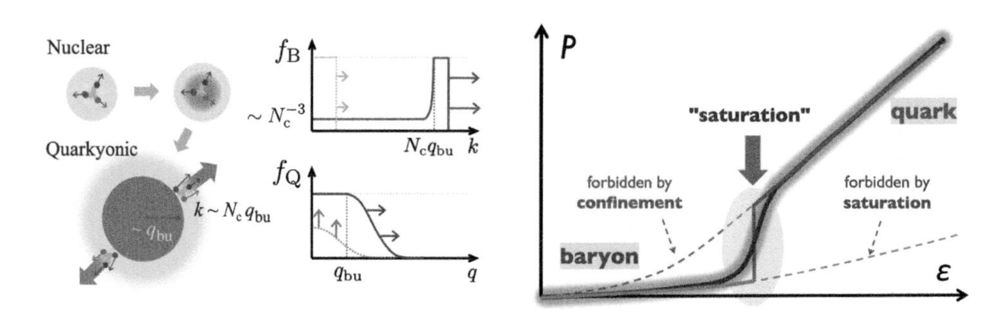

Figure 7.10 (Left) and the quark Fermi sea with hadronic Fermi surface (Right) in Quarkyonic matter.

The kinetic energy is suppressed by a factor $1/N_c^2$ compared to the mass. With this baryon energy, the single particle contribution to energy density is

$$\varepsilon_0 = \int_{\mathbf{k}} E_B(\mathbf{k}) f_B(\mathbf{k}) = N_c \int_{\mathbf{k},\mathbf{q}} E_Q(\mathbf{q}) \varphi(\mathbf{q} - \mathbf{k}/N_c) f_B(\mathbf{k}) = N_c \int_{\mathbf{q}} E_Q(\mathbf{q}) f_Q(\mathbf{q}), \tag{7.26}$$

which can be expressed in either baryonic or quark distributions.

7.6.2.2 Quark saturation

The above simple description can capture basic properties of matter from baryonic matter at low density to quark matter at high-density modulo details of interactions. In the low-density limit, one can look at the expression in terms of f_B, finding $\varepsilon \simeq E_B(0) n_B \sim O(N_c)$ and $P = n_B^2 \partial(\varepsilon/n_B)/\partial n_B \sim O(1/N_c)$, that reflects the softness of baryonic EOS. In the high-density limit, one can look at the expression in terms of f_Q, inferring $f_Q(q) \sim \Theta(q_F - q)$ with q_F being the quark Fermi momentum with which $P \simeq \varepsilon/3$. But what is most interesting is the transient regime where we are unsure which degrees of freedom to be used.

To see the most dramatic phenomenon in the present description, we consider an ideal baryon gas characterized by $f_B(k) = \Theta(k_F - k)$ where baryonic states are occupied up to $k = k_F$ with the probability 1. For $k_F \ll N_c \Lambda$, one can calculate f_Q expanding φ by $\mathbf{k}/N_c$,

$$f_Q(\mathbf{q}) \simeq \varphi(\mathbf{q}) \int_{\mathbf{k}} f_B(\mathbf{k}) = n_B \varphi(\mathbf{q}) \rightarrow f_Q(0) \sim (k_F/\Lambda)^3 . \tag{7.27}$$

We note that the magnitude of f_Q grows as $\sim n_B$, but for $k_F \gtrsim \Lambda$ the quark state at $q = 0$ is fully saturated; the ideal baryon gas regime must be definitely terminated; otherwise, the constraint from the Pauli principle, $f_Q \leq 1$, would be violated.

After the quark saturation, descriptions in terms of f_B must become exotic; baryons can no longer occupy low momentum states as in ideal gas and must occupy high momentum states. At density where quark states are saturated to q_{sat}, baryons occupy states to $\sim N_c q_{\text{sat}}$ since N_c quarks with momenta of $\mathbf{q} \sim q_{\text{sat}} \mathbf{n}$ move collectively in the same direction. But most baryon states must be under occupied; if all states up to $k \sim N_c q_{sat}$ were occupied, the baryon number would be $n_B \sim (N_c q_{sat})^3$, but this estimate is inconsistent with the estimate from the quark computation, $n_B \sim q_{sat}^3$. The consistency requires that baryons occupy states with the probability $\sim 1/N_c^3$ [90]. The only exception is the edge region where the occupation probability can be $\simeq 1$ but its momentum thickness Δ_B is $\sim 1/N_c^2$. The physical reason for the system to favor this momentum shell with the probability ~ 1 (instead of $\sim 1/N_c^3$) is to make whole momentum distribution as compact as possible for a given n_B.

Taking the above discussions into account, a schematic distribution for f_B and the resulting EOS are parametrized as

$$\varepsilon_0(n_B) = \int_k f_B(k) E_B(k)\,, \quad f_B(k) = \frac{1}{N_c^3}\Theta(k_{bu} - k) + \Theta(k_{sh} - k)\Theta(k - k_{bu}), \quad (7.28)$$

with $k_{bu} = N_c q_{sat}$ and $k_{sh} = k_{bu} + \Delta_B$. This distribution has been analytically derived by solving a dynamical equation for a conveniently chosen φ [62].

This form of EOS is essentially equivalent to a model originally postulated by McLerran and Reddy based on the quarkyonic matter picture [118],

$$\varepsilon = \int_q \Theta(q_{sat} - q) E_Q(k) + \int_k \Theta(k_{sh} - k)\Theta(k - k_{bu}) E_B(k)\,, \quad (7.29)$$

which is a hybrid model using quarks and baryons in different domains of momenta. The original hybrid model has raised questions concerning double counting [46], but the derivation based on the quark-hadron duality has removed the worries about such problems.

7.6.2.3 Rapid stiffening of EOS

Rapid stiffening of EOS after the quark saturation can be expressed either by baryonic or quark descriptions. In baryonic descriptions, baryons quickly become relativistic as the quark saturation pushes baryons into high momenta so that the EOS becomes relativistic and stiff, even though the density is relatively low, a few times of n_0. Without the quark saturation, baryons would become relativistic only when $p_F \sim M_B$ or $n_B \sim 100 n_0$. On the other hand, in quark descriptions, the EOS is soft while f_Q scales as $f_Q \sim n_B \varphi$, but after the saturation f_Q behaves as $f_Q \sim \Theta(q_f - q) + \Theta(q - q_f) f_{tail}(q)$, leading to $P \sim \varepsilon/3$. In the former regime the energy per particle ε/n_B is nearly constant, while in the latter ε/n_B grows with n_B. Since $P = n_B^2 \partial(\varepsilon/n_B)/\partial n_B$, around the quark saturation the energy per particle starts to grow rapidly and hence does the pressure.

An important observation is that, in the pre-saturation regime, relativistic quarks contribute to the energy density through the baryon mass, but do not contribute to the pressure because confinement prevents the "mechanical" pressure of quarks from contributing thermodynamic pressure. In the post-saturation, the quarks manifestly contribute to the pressure through the quark Pauli blocking effects. Hence, before and after the quark saturation the energy density changes smoothly, while the pressure grows rapidly in the transient regime. This feature leads to a peak in sound velocity $c_s^2 = \partial P/\partial \varepsilon$.

Bibliography

[1] Gert Aarts et al. Phase Transitions in Particle Physics - Results and Perspectives from Lattice Quantum Chromo-Dynamics. In *Phase Transitions in Particle Physics: Results and Perspectives from Lattice Quantum Chromo-Dynamics*, 1 2023.

[2] B. P. Abbott et al. GW170817: Observation of Gravitational Waves from a Binary Neutron Star Inspiral. *Phys. Rev. Lett.*, 119(16):161101, 2017.

[3] B. P. Abbott et al. GW170817: Measurements of neutron star radii and equation of state. *Phys. Rev. Lett.*, 121(16):161101, 2018.

[4] K. Ackley et al. Neutron Star Extreme Matter Observatory: A kilohertz-band gravitational-wave detector in the global network. *Publ. Astron. Soc. Austral.*, 37:e047, 2020.

[5] B. K. Agrawal, S. Shlomo, and V. Kim Au. Nuclear matter incompressibility coefficient in relativistic and nonrelativistic microscopic models. *Phys. Rev. C*, 68:031304, Sep 2003.

[6] A. Akmal, V. R. Pandharipande, and D. G. Ravenhall. The Equation of state of nucleon matter and neutron star structure. *Phys. Rev. C*, 58:1804–1828, 1998.

[7] Mark Alford, Matt Braby, M. W. Paris, and Sanjay Reddy. Hybrid stars that masquerade as neutron stars. *Astrophys. J.*, 629:969–978, 2005.

[8] Mark G. Alford, Sophia Han, and Madappa Prakash. Generic conditions for stable hybrid stars. *Phys. Rev. D*, 88(8):083013, 2013.

[9] Mark G. Alford, Krishna Rajagopal, and Frank Wilczek. QCD at finite baryon density: Nucleon droplets and color superconductivity. *Phys. Lett. B*, 422:247–256, 1998.

[10] Mark G. Alford, Krishna Rajagopal, and Frank Wilczek. Color flavor locking and chiral symmetry breaking in high density QCD. *Nucl. Phys. B*, 537:443–458, 1999.

[11] Mark G. Alford, Andreas Schmitt, Krishna Rajagopal, and Thomas Schäfer. Color superconductivity in dense quark matter. *Rev. Mod. Phys.*, 80:1455–1515, 2008.

[12] P. W. Anderson. Absence of diffusion in certain random lattices. *Phys. Rev.*, 109:1492–1505, Mar 1958.

[13] Anton Andronic, Peter Braun-Munzinger, Krzysztof Redlich, and Johanna Stachel. Decoding the phase structure of QCD via particle production at high energy. *Nature*, 561(7723):321–330, 2018.

[14] Eemeli Annala, Tyler Gorda, Aleksi Kurkela, Joonas Nättilä, and Aleksi Vuorinen. Evidence for quark-matter cores in massive neutron stars. *Nature Phys.*, 16(9): 907–910, 2020.

[15] Eemeli Annala, Tyler Gorda, Aleksi Kurkela, and Aleksi Vuorinen. Gravitational-wave constraints on the neutron-star-matter Equation of State. *Phys. Rev. Lett.*, 120(17):172703, 2018.

[16] John Antoniadis et al. A Massive Pulsar in a Compact Relativistic Binary. *Science*, 340:6131, 2013.

[17] Y. Aoki, G. Endrodi, Z. Fodor, S. D. Katz, and K. K. Szabo. The Order of the quantum chromodynamics transition predicted by the standard model of particle physics. *Nature*, 443:675–678, 2006.

[18] M. Asakawa and K. Yazaki. Chiral Restoration at Finite Density and Temperature. *Nucl. Phys. A*, 504:668–684, 1989.

[19] D. Bailin and A. Love. Superfluidity and Superconductivity in Relativistic Fermion Systems. *Phys. Rept.*, 107:325, 1984.

[20] R. Balian, M. Rho, and G. Ripka. *Nuclear Physics with Heavy Ions and Mesons.* Ions Lourds Et Mésons en Physique Nucléaire. North-Holland Publishing Company, 1978.

[21] Andreas Bauswein, Niels-Uwe F. Bastian, David B. Blaschke, Katerina Chatziioannou, James A. Clark, Tobias Fischer, and Micaela Oertel. Identifying a first-order phase transition in neutron star mergers through gravitational waves. *Phys. Rev. Lett.*, 122(6):061102, 2019.

[22] Andreas Bauswein, Oliver Just, Hans-Thomas Janka, and Nikolaos Stergioulas. Neutron-star radius constraints from GW170817 and future detections. *Astrophys. J. Lett.*, 850(2):L34, 2017.

[23] G. Baym and S. A. Chin. Can a Neutron Star Be a Giant MIT Bag? *Phys. Lett. B*, 62:241–244, 1976.

[24] Gordon Baym, Shun Furusawa, Tetsuo Hatsuda, Toru Kojo, and Hajime Togashi. New Neutron Star Equation of State with Quark-Hadron Crossover. *Astrophys. J.*, 885:42, 2019.

[25] Gordon Baym, Tetsuo Hatsuda, Toru Kojo, Philip D. Powell, Yifan Song, and Tatsuyuki Takatsuka. From hadrons to quarks in neutron stars: a review. *Rept. Prog. Phys.*, 81(5):056902, 2018.

[26] A. Bazavov et al. Chiral crossover in QCD at zero and non-zero chemical potentials. *Phys. Lett. B*, 795:15–21, 2019.

[27] A. A. Belavin, Alexander M. Polyakov, A. S. Schwartz, and Yu. S. Tyupkin. Pseudoparticle Solutions of the Yang-Mills Equations. *Phys. Lett. B*, 59:85–87, 1975.

[28] Jean-Paul Blaizot and Edmond Iancu. The Quark gluon plasma: Collective dynamics and hard thermal loops. *Phys. Rept.*, 359:355–528, 2002.

[29] D. Bollweg, J. Goswami, O. Kaczmarek, F. Karsch, Swagato Mukherjee, P. Petreczky, C. Schmidt, and P. Scior. Taylor expansions and Padé approximants for cumulants of conserved charge fluctuations at nonvanishing chemical potentials. *Phys. Rev. D*, 105(7):074511, 2022.

[30] Szabolcs Borsanyi, Zoltan Fodor, Jana N. Guenther, Ruben Kara, Sandor D. Katz, Paolo Parotto, Attila Pasztor, Claudia Ratti, and Kalman K. Szabo. QCD Crossover at Finite Chemical Potential from Lattice Simulations. *Phys. Rev. Lett.*, 125(5):052001, 2020.

[31] Len Brandes, Wolfram Weise, and Norbert Kaiser. Evidence against a first-order phase transition in neutron star cores: impact of new data. 6 2023.

[32] Nino M. Bratovic, Tetsuo Hatsuda, and Wolfram Weise. Role of Vector Interaction and Axial Anomaly in the PNJL Modeling of the QCD Phase Diagram. *Phys. Lett. B*, 719:131–135, 2013.

[33] Jens Braun and Benedikt Schallmo. From quarks and gluons to color superconductivity at supranuclear densities. *Phys. Rev. D*, 105(3):036003, 2022.

[34] Wit Busza, Krishna Rajagopal, and Wilke van der Schee. Heavy Ion Collisions: The Big Picture, and the Big Questions. *Ann. Rev. Nucl. Part. Sci.*, 68:339–376, 2018.

[35] Adam Bzdak, Shinichi Esumi, Volker Koch, Jinfeng Liao, Mikhail Stephanov, and Nu Xu. Mapping the Phases of Quantum Chromodynamics with Beam Energy Scan. *Phys. Rept.*, 853:1–87, 2020.

[36] J. Carlson, S. Gandolfi, F. Pederiva, Steven C. Pieper, R. Schiavilla, K. E. Schmidt, and R. B. Wiringa. Quantum Monte Carlo methods for nuclear physics. *Rev. Mod. Phys.*, 87:1067, 2015.

[37] T. Celik, F. Karsch, and H. Satz. A PERCOLATION APPROACH TO STRONGLY INTERACTING MATTER. *Phys. Lett. B*, 97:128–130, 1980.

[38] Ryuji Chiba and Toru Kojo. Sound velocity peak and conformality in isospin QCD, arxiv: 2304.13920. 4 2023.

[39] John C. Collins and M. J. Perry. Superdense Matter: Neutrons Or Asymptotically Free Quarks? *Phys. Rev. Lett.*, 34:1353, 1975.

[40] H. T. Cromartie et al. Relativistic Shapiro delay measurements of an extremely massive millisecond pulsar. *Nature Astron.*, 4(1):72–76, 2019.

[41] A. De Rujula, Howard Georgi, and S. L. Glashow. Hadron Masses in a Gauge Theory. *Phys. Rev. D*, 12:147–162, 1975.

[42] Paul Demorest, Tim Pennucci, Scott Ransom, Mallory Roberts, and Jason Hessels. Shapiro Delay Measurement of A Two Solar Mass Neutron Star. *Nature*, 467:1081–1083, 2010.

[43] Matthias Drews and Wolfram Weise. Functional renormalization group studies of nuclear and neutron matter. *Prog. Part. Nucl. Phys.*, 93:69–107, 2017.

[44] C. Drischler, R. J. Furnstahl, J. A. Melendez, and D. R. Phillips. How Well Do We Know the Neutron-Matter Equation of State at the Densities Inside Neutron Stars? A Bayesian Approach with Correlated Uncertainties. *Phys. Rev. Lett.*, 125(20):202702, 2020.

[45] Christian Drischler, Sophia Han, James M. Lattimer, Madappa Prakash, Sanjay Reddy, and Tianqi Zhao. Limiting masses and radii of neutron stars and their implications. *Phys. Rev. C*, 103(4):045808, 2021.

[46] Dyana C. Duarte, Saul Hernandez-Ortiz, Kie Sang Jeong, and Larry D. McLerran. Quarkyonic effective field theory, quark-nucleon duality, and ghosts. *Phys. Rev. D*, 104(9):L091901, 2021.

[47] N. Dupuis, L. Canet, A. Eichhorn, W. Metzner, J. M. Pawlowski, M. Tissier, and N. Wschebor. The nonperturbative functional renormalization group and its applications. *Phys. Rept.*, 910:1–114, 2021.

[48] S. Elitzur. Impossibility of Spontaneously Breaking Local Symmetries. *Phys. Rev. D*, 12:3978–3982, 1975.

[49] E. Epelbaum, H. Krebs, and U. G. Meißner. Improved chiral nucleon-nucleon potential up to next-to-next-to-next-to-leading order. *Eur. Phys. J. A*, 51(5):53, 2015.

[50] Evgeny Epelbaum, Hans-Werner Hammer, and Ulf-G. Meissner. Modern Theory of Nuclear Forces. *Rev. Mod. Phys.*, 81:1773–1825, 2009.

[51] Matthew Evans et al. A Horizon Study for Cosmic Explorer: Science, Observatories, and Community. 9 2021.

[52] Christian S. Fischer. QCD at finite temperature and chemical potential from Dyson–Schwinger equations. *Prog. Part. Nucl. Phys.*, 105:1–60, 2019.

[53] E. Fonseca et al. Refined Mass and Geometric Measurements of the High-mass PSR J0740+6620. *Astrophys. J. Lett.*, 915(1):L12, 2021.

[54] Emmanuel Fonseca et al. The NANOGrav Nine-year Data Set: Mass and Geometric Measurements of Binary Millisecond Pulsars. *Astrophys. J.*, 832(2):167, 2016.

[55] Eduardo H. Fradkin and Stephen H. Shenker. Phase Diagrams of Lattice Gauge Theories with Higgs Fields. *Phys. Rev. D*, 19:3682–3697, 1979.

[56] Barry Freedman and Larry D. McLerran. Quark Star Phenomenology. *Phys. Rev. D*, 17:1109, 1978.

[57] Barry A. Freedman and Larry D. McLerran. Fermions and Gauge Vector Mesons at Finite Temperature and Density. 3. The Ground State Energy of a Relativistic Quark Gas. *Phys. Rev. D*, 16:1169, 1977.

[58] H. Fritzsch, Murray Gell-Mann, and H. Leutwyler. Advantages of the Color Octet Gluon Picture. *Phys. Lett. B*, 47:365–368, 1973.

[59] Yuki Fujimoto and Kenji Fukushima. Equation of state of cold and dense QCD matter in resummed perturbation theory. *Phys. Rev. D*, 105(1):014025, 2022.

[60] Yuki Fujimoto, Kenji Fukushima, Kenta Hotokezaka, and Koutarou Kyutoku. Gravitational Wave Signal for Quark Matter with Realistic Phase Transition. *Phys. Rev. Lett.*, 130(9):091404, 2023.

[61] Yuki Fujimoto, Kenji Fukushima, Larry D. McLerran, and Michal Praszalowicz. Trace Anomaly as Signature of Conformality in Neutron Stars. *Phys. Rev. Lett.*, 129(25):252702, 2022.

[62] Yuki Fujimoto, Toru Kojo, and Larry D. McLerran. Momentum Shell in Quarkyonic Matter from Explicit Duality: A Solvable Model Analysis. 6 2023.

[63] Kenji Fukushima. Chiral effective model with the Polyakov loop. *Phys. Lett. B*, 591:277–284, 2004.

[64] Kenji Fukushima and Tetsuo Hatsuda. The phase diagram of dense QCD. *Rept. Prog. Phys.*, 74:014001, 2011.

[65] Kenji Fukushima, Toru Kojo, and Wolfram Weise. Hard-core deconfinement and soft-surface delocalization from nuclear to quark matter. *Phys. Rev. D*, 102(9):096017, 2020.

[66] J. Gasser and H. Leutwyler. Chiral Perturbation Theory to One Loop. *Annals Phys.*, 158:142, 1984.

[67] J. Gasser and H. Leutwyler. Light Quarks at Low Temperatures. *Phys. Lett. B*, 184:83–88, 1987.

[68] Christof Gattringer and Christian B. Lang. *Quantum chromodynamics on the lattice*, volume 788. Springer, Berlin, 2010.

[69] Lorenzo Gavassino, Marco Antonelli, and Brynmor Haskell. Thermodynamic Stability Implies Causality. *Phys. Rev. Lett.*, 128(1):010606, 2022.

[70] Jeffrey Goldstone, Abdus Salam, and Steven Weinberg. Broken Symmetries. *Phys. Rev.*, 127:965–970, 1962.

[71] Tyler Gorda, Aleksi Kurkela, Risto Paatelainen, Saga Säppi, and Aleksi Vuorinen. Soft Interactions in Cold Quark Matter. *Phys. Rev. Lett.*, 127(16):162003, 2021.

[72] David J. Gross and Frank Wilczek. Ultraviolet Behavior of Nonabelian Gauge Theories. *Phys. Rev. Lett.*, 30:1343–1346, 1973.

[73] Najmul Haque, Aritra Bandyopadhyay, Jens O. Andersen, Munshi G. Mustafa, Michael Strickland, and Nan Su. Three-loop HTLpt thermodynamics at finite temperature and chemical potential. *JHEP*, 05:027, 2014.

[74] Tetsuo Hatsuda. Lattice quantum chromodynamics and baryon-baryon interactions. *Front. Phys. (Beijing)*, 13(6):132105, 2018.

[75] Tetsuo Hatsuda and Teiji Kunihiro. QCD phenomenology based on a chiral effective Lagrangian. *Phys. Rept.*, 247:221–367, 1994.

[76] Tetsuo Hatsuda, Motoi Tachibana, Naoki Yamamoto, and Gordon Baym. New critical point induced by the axial anomaly in dense QCD. *Phys. Rev. Lett.*, 97:122001, 2006.

[77] K. Hebeler, S. K. Bogner, R. J. Furnstahl, A. Nogga, and A. Schwenk. Improved nuclear matter calculations from chiral low-momentum interactions. *Phys. Rev. C*, 83:031301, 2011.

[78] W. A. Hiscock and L. Lindblom. Stability and causality in dissipative relativistic fluids. *Annals Phys.*, 151:466–496, 1983.

[79] Yong-Jia Huang, Luca Baiotti, Toru Kojo, Kentaro Takami, Hajime Sotani, Hajime Togashi, Tetsuo Hatsuda, Shigehiro Nagataki, and Yi-Zhong Fan. Merger and Post-merger of Binary Neutron Stars with a Quark-Hadron Crossover Equation of State. *Phys. Rev. Lett.*, 129(18):181101, 2022.

[80] N. Itoh. Hydrostatic Equilibrium of Hypothetical Quark Stars. *Prog. Theor. Phys.*, 44:291, 1970.

[81] Robert L. Jaffe. Multi-Quark Hadrons. 1. The Phenomenology of (2 Quark 2 anti-Quark) Mesons. *Phys. Rev. D*, 15:267, 1977.

[82] Robert L. Jaffe. Perhaps a Stable Dihyperon. *Phys. Rev. Lett.*, 38:195–198, 1977. [Erratum: Phys.Rev.Lett. 38, 617 (1977)].

[83] David B. Kaplan and Martin J. Savage. The Spin flavor dependence of nuclear forces from large n QCD. *Phys. Lett. B*, 365:244–251, 1996.

[84] Atul Kedia, Hee Il Kim, In-Saeng Suh, and Grant J. Mathews. Binary neutron star mergers as a probe of quark-hadron crossover equations of state. *Phys. Rev. D*, 106(10):103027, 2022.

[85] Scott Kirkpatrick and Thomas P. Eggarter. Localized states of a binary alloy. *Phys. Rev. B*, 6:3598–3609, Nov 1972.

[86] Masakiyo Kitazawa, Tomoi Koide, Teiji Kunihiro, and Yukio Nemoto. Chiral and color superconducting phase transitions with vector interaction in a simple model. *Prog. Theor. Phys.*, 108(5):929–951, 2002. [Erratum: Prog.Theor.Phys. 110, 185–186 (2003)].

[87] T. Klahn, D. Blaschke, F. Sandin, C. Fuchs, A. Faessler, H. Grigorian, G. Ropke, and J. Trumper. Modern compact star observations and the quark matter equation of state. *Phys. Lett. B*, 654:170–176, 2007.

[88] T. Klähn, R. Łastowiecki, and D. B. Blaschke. Implications of the measurement of pulsars with two solar masses for quark matter in compact stars and heavy-ion collisions: A Nambu–Jona-Lasinio model case study. *Phys. Rev. D*, 88(8):085001, 2013.

[89] Toru Kojo. A (1+1) dimensional example of Quarkyonic matter. *Nucl. Phys. A*, 877:70–94, 2012.

[90] Toru Kojo. Delineating the properties of matter in cold, dense QCD. *AIP Conf. Proc.*, 2127(1):020023, 2019.

[91] Toru Kojo. Zero point energy of composite particles: The medium effects. *Phys. Rev. D*, 101(3):036001, 2020.

[92] Toru Kojo. QCD equations of state and speed of sound in neutron stars. *AAPPS Bull.*, 31(1):11, 2021.

[93] Toru Kojo. Stiffening of matter in quark-hadron continuity. *Phys. Rev. D*, 104(7): 074005, 2021.

[94] Toru Kojo, Gordon Baym, and Tetsuo Hatsuda. Implications of NICER for Neutron Star Matter: The QHC21 Equation of State. *Astrophys. J.*, 934(1):46, 2022.

[95] Toru Kojo, Yoshimasa Hidaka, Larry McLerran, and Robert D. Pisarski. Quarkyonic Chiral Spirals. *Nucl. Phys. A*, 843:37–58, 2010.

[96] Toru Kojo, Defu Hou, Jude Okafor, and Hajime Togashi. Phenomenological QCD equations of state for neutron star dynamics: Nuclear-2SC continuity and evolving effective couplings. *Phys. Rev. D*, 104(6):063036, 2021.

[97] Toru Kojo, Philip D. Powell, Yifan Song, and Gordon Baym. Phenomenological QCD equation of state for massive neutron stars. *Phys. Rev. D*, 91(4):045003, 2015.

[98] Toru Kojo and Daiki Suenaga. Peaks of sound velocity in two color dense QCD: Quark saturation effects and semishort range correlations. *Phys. Rev. D*, 105(7):076001, 2022.

[99] Teiji Kunihiro. Quark number susceptibility and fluctuations in the vector channel at high temperatures. *Phys. Lett. B*, 271:395–402, 1991.

[100] Aleksi Kurkela, Paul Romatschke, and Aleksi Vuorinen. Cold Quark Matter. *Phys. Rev. D*, 81:105021, 2010.

[101] J. M. Lattimer. Neutron Stars and the Nuclear Matter Equation of State. *Ann. Rev. Nucl. Part. Sci.*, 71:433–464, 2021.

[102] James M. Lattimer. Constraints on Nuclear Symmetry Energy Parameters. *Particles*, 6:30–56, 2023.

[103] James M. Lattimer and Maddapa Prakash. Neutron Star Observations: Prognosis for Equation of State Constraints. *Phys. Rept.*, 442:109–165, 2007.

[104] Hyun Kyu Lee, Yong-Liang Ma, Won-Gi Paeng, and Mannque Rho. Cusp in the symmetry energy, speed of sound in neutron stars and emergent pseudo-conformal symmetry. *Mod. Phys. Lett. A*, 37(03):2230003, 2022.

[105] Stefano Lottini and Giorgio Torrieri. A percolation transition in Yang-Mills matter at finite number of colors. *Phys. Rev. Lett.*, 107:152301, 2011.

[106] Alessandro Lovato et al. Long Range Plan: Dense matter theory for heavy-ion collisions and neutron stars. arXiv:2211.02224 [nucl-th], 2022.

[107] Xiaofeng Luo and Nu Xu. Search for the QCD Critical Point with Fluctuations of Conserved Quantities in Relativistic Heavy-Ion Collisions at RHIC : An Overview. *Nucl. Sci. Tech.*, 28(8):112, 2017.

[108] Yong-Liang Ma and Mannque Rho. Pseudoconformal structure in dense baryonic matter. *Phys. Rev. D*, 99(1):014034, 2019.

[109] R. Machleidt and D. R. Entem. Chiral effective field theory and nuclear forces. *Phys. Rept.*, 503:1–75, 2011.

[110] R. Machleidt, K. Holinde, and C. Elster. The Bonn Meson Exchange Model for the Nucleon Nucleon Interaction. *Phys. Rept.*, 149:1–89, 1987.

[111] Michele Maggiore et al. Science Case for the Einstein Telescope. *JCAP*, 03:050, 2020.

[112] Aneesh Manohar and Howard Georgi. Chiral Quarks and the Nonrelativistic Quark Model. *Nucl. Phys. B*, 234:189–212, 1984.

[113] Michał Marczenko, Larry McLerran, Krzysztof Redlich, and Chihiro Sasaki. Reaching percolation and conformal limits in neutron stars. *Phys. Rev. C*, 107(2):025802, 2023.

[114] Ben Margalit and Brian D. Metzger. Constraining the Maximum Mass of Neutron Stars From Multi-Messenger Observations of GW170817. *Astrophys. J. Lett.*, 850(2):L19, 2017.

[115] Kota Masuda, Tetsuo Hatsuda, and Tatsuyuki Takatsuka. Hadron-Quark Crossover and Massive Hybrid Stars with Strangeness. *Astrophys. J.*, 764:12, 2013.

[116] Kota Masuda, Tetsuo Hatsuda, and Tatsuyuki Takatsuka. Hadron–quark crossover and massive hybrid stars. *PTEP*, 2013(7):073D01, 2013.

[117] Larry McLerran and Robert D. Pisarski. Phases of cold, dense quarks at large N(c). *Nucl. Phys. A*, 796:83–100, 2007.

[118] Larry McLerran and Sanjay Reddy. Quarkyonic Matter and Neutron Stars. *Phys. Rev. Lett.*, 122(12):122701, 2019.

[119] M. C. Miller et al. The Radius of PSR J0740+6620 from NICER and XMM-Newton Data. *Astrophys. J. Lett.*, 918(2):L28, 2021.

[120] Takuya Minamikawa, Toru Kojo, and Masayasu Harada. Chiral condensates for neutron stars in hadron-quark crossover: From a parity doublet nucleon model to a Nambu–Jona-Lasinio quark model. *Phys. Rev. C*, 104(6):065201, 2021.

[121] Elias R. Most, L. Jens Papenfort, Veronica Dexheimer, Matthias Hanauske, Stefan Schramm, Horst Stöcker, and Luciano Rezzolla. Signatures of quark-hadron phase transitions in general-relativistic neutron-star mergers. *Phys. Rev. Lett.*, 122(6):061101, 2019.

[122] Yoichiro Nambu and G. Jona-Lasinio. Dynamical Model of Elementary Particles Based on an Analogy with Superconductivity. 1. *Phys. Rev.*, 122:345–358, 1961.

[123] M. Oka and K. Yazaki. Nuclear Force in a Quark Model. *Phys. Lett. B*, 90:41–44, 1980.

[124] F. Ozel. Soft equations of state for neutron-star matter ruled out by EXO 0748-676. *Nature*, 441:1115–1117, 2006.

[125] Aaron Park, Su Houng Lee, Takashi Inoue, and Tetsuo Hatsuda. Baryon–baryon interactions at short distances: constituent quark model meets lattice QCD. *Eur. Phys. J. A*, 56(3):93, 2020.

[126] Steven C. Pieper, V. R. Pandharipande, Robert B. Wiringa, and J. Carlson. Realistic models of pion exchange three nucleon interactions. *Phys. Rev. C*, 64:014001, 2001.

[127] H. David Politzer. Reliable Perturbative Results for Strong Interactions? *Phys. Rev. Lett.*, 30:1346–1349, 1973.

[128] Alexander M. Polyakov. Thermal Properties of Gauge Fields and Quark Liberation. *Phys. Lett. B*, 72:477–480, 1978.

[129] R. Rapp, Thomas Schäfer, Edward V. Shuryak, and M. Velkovsky. Diquark Bose condensates in high density matter and instantons. *Phys. Rev. Lett.*, 81:53–56, 1998.

[130] Claudia Ratti, Michael A. Thaler, and Wolfram Weise. Phases of QCD: Lattice thermodynamics and a field theoretical model. *Phys. Rev. D*, 73:014019, 2006.

[131] Luciano Rezzolla, Elias R. Most, and Lukas R. Weih. Using gravitational-wave observations and quasi-universal relations to constrain the maximum mass of neutron stars. *Astrophys. J. Lett.*, 852(2):L25, 2018.

[132] Mannque Rho. Dense Baryonic Matter Predicted in "Pseudo-Conformal Model". *Symmetry* 15(6):1271, 2023.

[133] Thomas E. Riley et al. A NICER View of the Massive Pulsar PSR J0740+6620 Informed by Radio Timing and XMM-Newton Spectroscopy. *Astrophys. J. Lett.*, 918(2):L27, 2021.

[134] Roger W. Romani, D. Kandel, Alexei V. Filippenko, Thomas G. Brink, and WeiKang Zheng. PSR J0952-0607: The Fastest and Heaviest Known Galactic Neutron Star. 7 2022.

[135] Thomas Schäfer and Frank Wilczek. Continuity of quark and hadron matter. *Phys. Rev. Lett.*, 82:3956–3959, 1999.

[136] K. Sekiguchi et al. Complete set of precise deuteron analyzing powers at intermediate energies: comparison with modern nuclear force predictions. *Phys. Rev. C*, 65:034003, 2002.

[137] Takayasu Sekihara and Taishi Hashiguchi. Short-range baryon-baryon potentials in constituent quark model revisited. 4 2023.

[138] S.L. Shapiro and S.A. Teukolsky. *Black Holes, White Dwarfs, and Neutron Stars: The Physics of Compact Objects*. Wiley, 2008.

[139] Masaru Shibata, Sho Fujibayashi, Kenta Hotokezaka, Kenta Kiuchi, Koutarou Kyutoku, Yuichiro Sekiguchi, and Masaomi Tanaka. Modeling GW170817 based on numerical relativity and its implications. *Phys. Rev. D*, 96(12):123012, 2017.

[140] D. T. Son. Superconductivity by long range color magnetic interaction in high density quark matter. *Phys. Rev. D*, 59:094019, 1999.

[141] Yifan Song, Gordon Baym, Tetsuo Hatsuda, and Toru Kojo. Effective repulsion in dense quark matter from nonperturbative gluon exchange. *Phys. Rev. D*, 100(3):034018, 2019.

[142] Kohsuke Sumiyoshi, Toru Kojo, and Shun Furusawa. *Equation of State in Neutron Stars and Supernovae*, pages 1–51. 2023.

[143] Leonard Susskind. Lattice Models of Quark Confinement at High Temperature. *Phys. Rev. D*, 20:2610–2618, 1979.

[144] Gerard ’t Hooft. A Planar Diagram Theory for Strong Interactions. *Nucl. Phys. B*, 72:461, 1974.

[145] Hung Tan, Veronica Dexheimer, Jacquelyn Noronha-Hostler, and Nicolas Yunes. Finding Structure in the Speed of Sound of Supranuclear Matter from Binary Love Relations. *Phys. Rev. Lett.*, 128(16):161101, 2022.

[146] M. Tanabashi *et al.* Review of particle physics. *Phys. Rev. D*, 98:030001, Aug 2018.

[147] I. Tews, J. Margueron, and S. Reddy. Critical examination of constraints on the equation of state of dense matter obtained from GW170817. *Phys. Rev. C*, 98(4):045804, 2018.

[148] B. G. Todd-Rutel and J. Piekarewicz. Neutron-rich nuclei and neutron stars: A new accurately calibrated interaction for the study of neutron-rich matter. *Phys. Rev. Lett.*, 95:122501, Sep 2005.

[149] H. Togashi, K. Nakazato, Y. Takehara, S. Yamamuro, H. Suzuki, and M. Takano. Nuclear equation of state for core-collapse supernova simulations with realistic nuclear forces. *Nucl. Phys. A*, 961:78–105, 2017.

[150] S. Torquato and Y. Jiao. Effect of dimensionality on the continuum percolation of overlapping hyperspheres and hypercubes. ii. simulation results and analyses. *J. Chem. Phys*, 137:074106, 2012.

[151] A. Vuorinen. The Pressure of QCD at finite temperatures and chemical potentials. *Phys. Rev. D*, 68:054017, 2003.

[152] Steven Weinberg. Phenomenological Lagrangians. *Physica A*, 96(1-2):327–340, 1979.

[153] Steven Weinberg. Nuclear forces from chiral Lagrangians. *Phys. Lett. B*, 251:288–292, 1990.

[154] Steven Weinberg. Effective chiral Lagrangians for nucleon - pion interactions and nuclear forces. *Nucl. Phys. B*, 363:3–18, 1991.

[155] Steven Weinberg. Three body interactions among nucleons and pions. *Phys. Lett. B*, 295:114–121, 1992.

[156] Edward Witten. Baryons in the 1/n Expansion. *Nucl. Phys. B*, 160:57–115, 1979.

[157] Edward Witten. Cosmic Separation of Phases. *Phys. Rev. D*, 30:272–285, 1984.

CHAPTER 8

Bulk Viscosity in Dense Nuclear Matter

Steven P. Harris

NEUTRON star mergers are an extreme environment where, within milliseconds, the cold, dense matter in neutron stars is dramatically heated to some of the hottest temperatures encountered in astrophysics. Matter in neutron star mergers is located solidly in the interior of the QCD phase diagram, and therefore, the observation and simulation of these mergers provides an opportunity to learn about the strong interaction at a set of densities and temperatures that compliment those obtained in relativistic heavy-ion collisions. Matter during the merger of two neutron stars is far from hydrostatic equilibrium, allowing for the study of not only the equation of state of dense matter, but of transport properties as well. Perhaps the most likely transport property to play a role in neutron star mergers is the bulk viscosity, which acts to resist changes in density. The bulk viscosity of matter at high densities has long been anticipated to damp oscillations in isolated neutron stars, and the study of its role in neutron star mergers is just beginning.

In this chapter, I describe bulk viscosity as a general concept, and then focus on bulk viscosity in the dense matter present in compact objects. While this review is focused on bulk viscosity in the conditions present in neutron star mergers, I present a history of bulk viscosity research in dense matter, from its role in damping radial oscillations in neutron stars through its current applications in neutron star mergers. The majority of the chapter consists of calculations of the bulk viscosity from Urca processes in generic neutron-proton-electron (npe) matter, and then in dense matter containing muons ($npe\mu$ matter) as well. I make several approximations in these calculations to keep the focus on the concepts. More precise calculations exist in the literature, to which I refer the reader. One concept I attempt to elucidate is the thermodynamic behavior of a fluid element throughout an oscillation and how that leads to bulk-viscous dissipation. I conclude with a discussion of the recent research into the role of weak interactions and bulk viscosity in neutron star mergers.

DOI: 10.1201/9781003306580-8

8.1 BULK VISCOSITY AND TRANSPORT IN DENSE MATTER

The nature of matter at the highest densities is a question of fundamental interest. This matter can be studied in the laboratory with heavy-ion collisions, but also in astrophysical environments. Neutron stars, first theoretically proposed in the 1930s [148], and observed as pulsars in the 1960s [126], contain matter of densities up to several times nuclear saturation density ($n_0 \equiv 0.16 \text{ fm}^{-3}$). While the matter in isolated neutron stars is cold, in the sense that the temperature is much less than the Fermi energies of the constituent particles (termed *degenerate*), when two neutron stars merge, shocks heat up the neutron star matter to temperatures of tens of MeV[1]. Neutron stars and neutron star mergers, together with heavy-ion collisions and nuclear experiments, probe complimentary regions of the QCD phase diagram [48].

8.1.1 Advantages of studying transport

It is most common to study the equation of state (EoS) of dense matter, a relationship between its energy density ε and the pressure P. This is important in its own right, but also because it leads to the prediction of masses and radii of neutron stars (through the solution of the Tolman-Oppenheimer-Volkoff (TOV) equations), the tidal deformability of a neutron star, and the oscillation frequencies of isolated neutron stars and differentially rotating neutron star merger remnants. However, the EoS $\varepsilon = \varepsilon(P)$ does not tell us the full nature of dense matter, because there is no unique relationship between the degrees of freedom of the dense matter and the EoS. This "masquerade problem" implies that a hybrid star (a neutron star with a quark core) can have a similar mass and radius as a neutron star containing no quark matter [4]. Therefore, it is important to study other aspects of dense matter beyond the EoS, including transport properties. In condensed matter physics, one might study the electrical conductivity, thermal conductivity, or the heat capacity of a sample, which would provide insight into the low-energy degrees of freedom around the Fermi surface [101]. In dense matter, transport properties like the thermal conductivity, specific heat, and both shear and bulk viscosity have been calculated and, if deemed to be significant, implemented in simulations of neutron star cooling [147], core-collapse supernovae [105], and neutron star mergers (Ref. [61] and Sec. 8.5 of this chapter.) A fairly recent review article by Schmitt & Shternin [128] provides a comprehensive summary of transport in cold, dense matter.

8.1.2 Bulk viscosity in all types of fluids

In this chapter, I discuss the bulk viscosity of dense matter. Bulk viscosity arises in a fluid that experiences a change in its density, causing some internal degree of freedom within the matter to be pushed out of equilibrium. In response, internal processes "turn on" in order to restore equilibrium. In the chemistry literature, this negative-feedback effect is known as Le Chatelier's principle [67]. Throughout the duration of the expansion or contraction of the system, the (irreversible) internal processes that attempt to restore equilibrium dissipate energy from the motion of the matter and turn it into heat. The amount of bulk-viscous

[1]In this chapter, I use natural units, where $c = \hbar = k_B = 1$. In these units, 1 MeV $\approx 1.2 \times 10^{10}$ K.

dissipation depends on the relaxation timescale relative to the timescale of the density change. If the relaxation timescale is very short, then the system is hardly out of equilibrium at all, and the bulk viscosity is small. If the relaxation timescale is very long, then it is as if no internal relaxation occurs, and the bulk viscosity is also small. As the two timescales become comparable, bulk viscosity grows larger and could be significant. A textbook discussion of bulk viscosity in fluids is given in Landau & Lifshitz [94].

Bulk viscosity exists in many types of fluids, from gases encountered in everyday life to fluids at the most extreme densities and temperatures in nature. For example, in diatomic gases, a density change leads to a change in kinetic energy of the molecules, which at some finite rate, turns into vibrational and rotational energy [43]. In cosmology, an expanding mixture of matter and radiation is pushed out of thermal equilibrium and exchange energy at some finite rate, leading to energy dissipation [142, 150]. In the expanding quark-gluon plasma formed in relativistic heavy-ion collisions, the matter departs from thermal equilibrium and is reequilibrated by strong interactions [28]. A neutron Fermi liquid, devoid of flavor-changing interactions, has a bulk viscosity stemming from binary collisions which attempt to restore thermal equilibrium [32, 92, 133].

In neutron stars, as the rest of the chapter will demonstrate, a density change pushes the matter out of chemical equilibrium, and flavor-changing interactions (for example, Urca processes) turn on to restore equilibrium, generating entropy as they proceed. A fluid element in the neutron star undergoing a small-amplitude sinusoidal density oscillation will traverse a path in the pressure-volume (PV) plane. If the chemical equilibration occurs very fast or very slow compared to the oscillation frequency, the PV curve will be a line that the fluid element follows forward and then backward over the course of the oscillation. If the chemical equilibration occurs at a similar timescale as the density oscillation, the fluid element will follow a closed curve in the PV plane. The area enclosed by that curve indicates the amount of work done on the fluid element and is maximal when the rate of equilibration matches the density oscillation frequency, a resonance-like behavior. This mechanism of energy dissipation will be explored in much greater detail in the rest of the chapter.

8.1.3 History of bulk viscosity in dense matter

8.1.3.1 The early years: 1965–1980

The study of bulk viscosity in neutron stars started in the mid-1960s, following a resurgence of neutron star research in the late 1950s and early 1960s, where midcentury developments in nuclear interactions and hypernuclear physics were first applied to neutron star structure calculations. For broader context, detailed histories of nuclear physics and its application in neutron stars are presented in the textbooks [67, 80, 126, 130]. In the mid-1960s, researchers were interested in the radial oscillation modes of neutron stars presumed to arise in their initial formation in supernovae. Curiosity about the duration of these oscillations after the birth of the neutron star, as well as the possibility of the conversion of vibrational energy into heat, leading to enhanced radiation from the neutron star surface (see the contemporary review articles [40, 143]), lead to preliminary calculations (Finzi [58], Meltzer & Thorne [104], and Hansen & Tsuruta [85]) of vibrational energy dissipation arising from chemical equilibration. In these early works, which considered neutron stars built of neutron-proton-electron (npe) matter (see Sec. 8.3), modified Urca processes (Sec. 8.2.3) were the equilibration mechanism.

The calculations of Finzi & Wolf [57] and Langer & Cameron [95] in the late '60s set the stage for almost all subsequent neutron star bulk viscosity work. Finzi & Wolf calculated the time evolution of both the temperature and radial pulsation amplitude of a model neutron star, accounting for the chemical equilibration (via modified Urca) throughout a cycle of the oscillation, as well as the associated increase in the neutrino emissivity. They calculated the PdV work done on the fluid element through the oscillation. Langer & Cameron considered hyperonic matter undergoing a sinusoidal density oscillation and strangeness-changing interactions that attempt to reequilibrate the strangeness. This work was the first to spell out the resonant structure of this type of energy dissipation (see Sec. 8.2.3.) Jones was the first to attribute the vibrational damping to a hydrodynamic bulk viscosity coefficient, in his study of hyperonic reactions in pulsating dense matter [89].

The 1970s were very quiet in terms of neutron star bulk viscosity research. However, during that decade the understanding of neutrino-nucleon interactions expanded dramatically, especially with the experimental confirmation of the neutral current interaction in 1973, and it became clear that neutrinos are trapped in core-collapse supernovae and therefore neutrino transport schemes [96] would have to be developed[2]. After a calculation of neutrino mean free paths (MFPs) in dense matter [125], in a 1980 paper [122] Sawyer calculated the bulk viscosity of proto-neutron star matter where neutrinos are trapped.

8.1.3.2 Expansion of applications in cold dense matter: 1980-2000

The 1980s saw the first calculations of bulk viscosity in (non-interacting) quark matter, by Wang & Lu [138, 139], and later, Sawyer [124]. Up until the end of the 1980s, essentially all dense-matter bulk viscosity calculations had in mind a neutron star undergoing radial oscillations that would be damped by bulk viscosity. In 1990, Cutler, Lindblom, & Splinter [47], armed with an improved calculation of bulk viscosity in npe matter by Sawyer [123], calculated the contribution of the bulk viscosity to the damping time of non-radial oscillations of cold neutron stars, and delineated the temperature regimes in which shear and bulk viscosity are each the dominant microscopic dissipation mechanisms (this switchover between the two dissipation mechanisms became more famous later in the context of the r-mode instability window.)

In a 1991 paper, Lattimer *et al.* [97] pointed out that the direct Urca process, long thought (except by Boguta [37]) to be forbidden in cold neutron stars because they were too neutron-rich, may occur in the densest regions of a neutron star, where the proton fraction $x_p \equiv n_p/n_B \gtrsim 0.11{-}0.15$. This result upended prior understandings of neutron star cooling, where it had been thought that if cooling significantly faster than predicted by modified Urca was observed, it was likely due to an exotic phase of matter, like quark matter or a pion condensate. In cold systems, where the beta equilibration rate is much slower than the timescale of the density change, the bulk viscosity is proportional to the beta equilibration rate (see Eq. 8.35), and therefore is strongly enhanced when direct Urca is present (Haensel & Schaeffer [81].)

[2] At present, it seems most likely that the implementation of bulk-viscous effects in neutron star merger simulations will be done through the neutrino transport scheme (see Sec. 8.5.)

In 1992, Madsen [100] studied the bulk viscosity in strange quark matter, extending it to the *suprathermal* regime[3]. Bulk viscosity in the suprathermal regime is enhanced by potentially orders of magnitude compared to the subthermal case. Madsen's suprathermal bulk viscosity work was extended by Goyal *et al.* [70] and then by Gupta *et al.* [71], who calculated the suprathermal bulk viscosity of nuclear matter equilibrating via direct Urca. The calculation of Gupta *et al.* went essentially unnoticed in the literature and was recalculated by Alford, Mahmoodifar, & Schwenzer [20], who account for the more likely possibility of npe matter equilibrating via modified Urca.

Another important development in the 1990s was start of a movement away from the bulk viscosity coefficient, and back to tracking the chemical reactions themselves, at least, in certain situations. Gourgoulhon & Haensel [68] studied the collapse of a neutron star to a black hole, and the ensuing weak interactions that occur due to the compression of the dense matter. Here it is clearly impossible to use the small-amplitude, oscillation-averaged bulk viscosity formalism discussed in this chapter. This study was extended to neutron stars near their maximum allowed mass, where the particle content was evolved along with the hydrodynamic equations during a stellar oscillation [69]. Another physical situation where chemical reactions themselves were tracked is rotochemical heating, studied by Reisenegger [116]. As a neutron star spins down, the centrifugal force decreases and the star contracts, pushing the matter out of chemical equilibrium. The ensuing chemical reactions heat the matter (though, accounting for neutrino cooling, there may or may not be *net* heating.) Any temperature change feeds back on the equilibration rates, and the deviation from chemical equilibrium $\delta\mu$ and the temperature T evolve in nontrivial ways. This analysis was extended to strange stars by Cheng & Dai [46].

At the end of the 1990s, Andersson [25] and Friedman & Morsink [62] conducted the first general-relativistic study of the r-mode, a nonradial mode present in rotating stars. They found that it (in the absence of viscosity) is unstable for all rotation rates, a consequence of the CFS instability [26]. As viscosity can stabilize the r-mode, Andersson, Kokkotas, & Schutz [27] used shear and bulk viscosity calculations to map out the spin rates and core temperatures for which the r-mode is unstable (the *r-mode instability window*). The r-mode and instability window literature is so vast that I will not attempt to detail it here (Andersson's textbook [26] is a nice reference), but after these calculations, the r-mode became a favorite context in which to study bulk viscosity.

8.1.3.3 Bulk viscosity pre-GW170817: 2000-2017

Until about 2000, even though most bulk viscosity calculations were done with cold neutron star matter in mind, no calculation had included the nucleon Cooper pairing that is expected to occur for temperatures less than a few times 10^9 K. This *critical temperature* is a function of density and is still largely uncertain today [129]. Pairing produces a gap in the excitation spectrum at the Fermi surface, exponentially suppressing the rate of

[3]Most bulk viscosity calculations up to this point considered *subthermal* density oscillations, where the matter is pushed only slightly out of chemical equilibrium $\delta\mu \ll T$. *Suprathermal* oscillations push the matter far out of chemical equilibrium $\delta\mu \gg T$, though the formalism developed in this chapter is only sensible when the amplitude of the density oscillation remains small. The first use of the terms *subthermal* and *suprathermal* I could find is by Haensel, Levenfish, & Yakovlev [78].

beta equilibration in degenerate matter [145, 146]. In a series of papers in the early 2000s, Haensel, Levenfish, & Yakovlev calculated the bulk viscosity in superfluid nuclear matter due to the direct Urca process [76], the modified Urca process [77], and in hyperonic matter [79]. Because at low temperatures the subthermal bulk viscosity is proportional to the beta equilibration rate (Eq. 8.35), it is strongly suppressed by superfluidity. However, this suppression can be overcome in the suprathermal regime due to "gap-bridging" [21, 22]. Finally, Gusakov [75] pointed out that in superfluid hydrodynamics, there are actually multiple bulk viscosity coefficients, not just one, and they all contribute significantly to the energy dissipation in superfluid nuclear matter. Gusakov & Kantor [72] extended this analysis to matter with superfluid nucleons and hyperons.

The next decade saw a proliferation of bulk viscosity calculations in ever more exotic forms of matter. Bulk viscosity in hyperonic matter was revisited by Jones [90], Lindblom & Owen [98], van Dalen & Dieperink [137], and Chatterjee & Bandyopadhyay [44]. Bulk viscosity in quark/hadron mixed phases was calculated by Drago, Lavagno, & Pagliara [50] and Pan, Zheng, & Li [111]. A myriad of quark matter phases was studied, including unpaired phases of strange quark matter by Sa'd, Shovkovy, & Rischke [121], Alford, Mahmoodifar, & Schwenzer [20] in the suprathermal regime, and Shovkovy & Wang [131] for anharmonic density oscillations. The bulk viscosity for paired quark matter phases like 2SC (Alford & Schmitt [23]) and other phases that have at least one unpaired quark flavor (Sa'd, Shovkovy, & Rischke [120], Wang & Shovkovy [141] and Berdermann *et al.* [34]), as well as the color-flavor-locked (CFL) phase (Alford *et al.* [11], Manuel & Llanes-Estrada [103], and Alford, Braby, & Schmitt [12]) were calculated. Mannarelli & Manuel [102] and Bierkandt & Manuel [36] calculated all three of the bulk viscosity coefficients (c.f. Gusakov [75]) present in superfluid CFL quark matter. Chatterjee & Bandyopadhyay [45] calculated the bulk viscosity of nuclear matter with a kaon condensate. Bulk viscosity from muon-electron conversion, potentially dominant in superfluid nuclear matter, was calculated by Alford & Good [13]. Finally, Huang *et al.* [87] studied viscosity in strange quark matter under a strong magnetic field, where there are now two bulk viscosity coefficients, one parallel and one perpendicular to the local magnetic field direction. Out of these papers were several [23, 49, 88, 121, 131, 141] that considered multiple equilibration channels, where the bulk-viscous behavior can become quite intricate. Dense $npe\mu$ matter has multiple equilibration channels, and is explored in Sec. 8.3.

A few papers considered the effect of chemical reactions on both dynamical and thermal evolution of neutron stars. Rotochemical heating was extended to a general-relativistic formalism by Fernandez & Reisenegger [56] and to superfluid nuclear matter by Petrovich & Reisenegger [113]. Gusakov, Yakovlev, & Gnedin [74] evolved an oscillating neutron star in general relativity, tracking the oscillation energy dissipation due to bulk viscosity but also the heating from the chemical reactions, essentially expanding on the Finzi & Wolf [57] calculation. Kantor & Gusakov [91] and Gusakov *et al.* [73] did similar analyses, but in superfluid neutron stars.

8.1.3.4 Bulk viscosity after GW170817

In August 2017, the LIGO-VIRGO collaboration detected a gravitational wave signal [1] from the inspiral of two merging neutron stars. Just under two seconds later, a gamma-ray burst was detected by the Fermi Gamma-ray Burst Monitor and soon after, an optical counterpart was identified [2], starting a new chapter in multimessenger astrophysics.

Just prior to this event, Alford *et al.* [10] considered transport properties in nuclear matter under conditions expected to be obtained in neutron star mergers, where the cold matter in the original neutron stars ($T \lesssim 10$ keV [29]) is heated to temperatures of tens of MeV. Alford *et al.* determined that bulk viscosity due to Urca processes might be relevant for neutron star mergers[4], because the timescale on which a sizeable fraction of the energy of an oscillation can be dissipated was found to be milliseconds, well within the (estimated) lifetime of many neutron star merger remnants (see Fig. 3 in the review article [35].) The 2017 neutron star merger and the Alford *et al.* bulk viscosity analysis initiated new interest in weak interactions at high temperatures (say, $T \gtrsim 1$ MeV) and the bulk viscosity stemming from these weak interactions. With only a few exceptions[5], most bulk viscosity papers written since 2017 have had neutron star mergers in mind.

To study viscous effects in the neutron star inspiral, Arras & Weinberg [29] considered a neutron star undergoing forced oscillations due to its companion neutron star, and found that suprathermal bulk viscosity from Urca processes heats the star up to, at most, tens of keV. Ghosh, Pradhan, & Chatterjee studied the same phenomenon but considered hyperonic interactions [66]. Most *et al.* [107] studied the effect of bulk viscosity during tidal deformation in the inspiral. This analysis was refined recently by Ripley, Hegade, & Yunes [117] and a new coefficient, the dissipative tidal deformability, was introduced. Viscous effects in the inspiral phase are discussed further in Sec. 8.5.1.

Alford & Harris [18] calculated the direct Urca rates in $T = 0.5 - 10$ MeV neutrino-transparent nuclear matter, going beyond the Fermi surface (FS) approximation (or strongly degenerate limit) by doing the full integration over the phase space, and found that for $T \gtrsim 1$ MeV, the direct Urca rates become significant even below the direct Urca threshold density and furthermore, the traditional beta equilibrium condition in neutrino-transparent matter has to be modified. Improvements to this calculation were later made by Alford *et al.* [15].

Improving upon the back-of-the-envelope estimate of Alford *et al.* [10], Alford & Harris [19] used the improved direct Urca rates and corrected beta equilibrium condition in a calculation of the bulk viscosity in neutrino-transparent nuclear matter and showed that the bulk viscosity could damp 1 kHz density oscillations in certain thermodynamic conditions in as little as 5 ms, making the case that bulk viscosity should be included in numerical simulations of neutron star mergers[6]. Alford, Harutyunyan, & Sedrakian [5, 6] calculated the bulk viscosity in nuclear matter hot enough to trap neutrinos ($T \gtrsim 5 - 10$ MeV [18]), and found that it is likely too small to have an effect on mergers, but also found

[4]The possibility of bulk viscosity in neutron star mergers was first considered many years earlier, in the inspiral stage by Lai [93] and in the postmerger stage by Ruffert & Janka [119], but was judged not to be of primary importance.

[5]For example, Yakovlev, Gusakov, & Haensel [144] calculated the bulk viscosity in nuclear pasta phases which, unexpectedly, can equilibrate via direct Urca (though at a suppressed rate) and Ofengeim *et al.* [110] improved past calculations of hyperonic bulk viscosity.

[6]Very recently, Alford, Haber, & Zhang [16] found a correction of this calculation due to the density dependence of the difference between the zero-temperature and finite-temperature neutrino-transparent beta equilibrium conditions, and recalculated the bulk viscosity.

interesting features in the bulk viscosity, such as conformal points. Alford, Harutyunyan, & Sedrakian [7–9] considered neutrino-trapped and neutrino-transparent nuclear matter with muons ($npe\mu$ matter) and calculated its bulk viscosity, including the reactions of the muons as well.

Alford & Haber [14] recalculated the hyperonic bulk viscosity, trying to explore the conditions where merger temperatures are high enough to produce a thermal hyperon population, but found that the hyperon bulk viscosity in merger conditions is small. Instead, the peak bulk viscosity was found to be at keV temperatures more relevant to the inspiral phase.

In parallel with the microphysical calculations of the bulk viscosity, a reconsideration of bulk viscosity and chemical reactions within hydrodynamic frameworks has been ongoing. Gavassino, Antonelli, & Haskell [63] confirmed, for a system like npe matter with one independent particle fraction, the equivalence (for small deviations from equilibrium) between tracking the change of chemical composition as it evolves due to weak interactions and a Muller-Israel-Stewart (MIS) theory where a bulk stress Π is introduced into the stress-energy tensor and evolves according to the MIS equation. Gavassino & Noronha [65] recently generalized this equivalence to large deviations from equilibrium. Gavassino [64] showed that in a two-component system like $npe\mu$ matter, the near-equilibrium viscous dynamics are described by a bulk stress Π that evolves in time not according to the MIS equation, but according to a Burgers equation. In two papers, Camelio *et al.* compared the "chemical composition tracking" approach to the MIS approach by implementing both in a simulation of a radially oscillating neutron star [38, 39].

Finally, in the past couple years, bulk viscosity has begun to be implemented in neutron star merger simulations. Most *et al.* [107] postprocessed an inviscid merger simulation to predict the possible strength of bulk viscosity in the merger remnant. Hammond, Hawke, & Andersson [82] and Celora *et al.* [41] contemplated the issues associated with implementation of bulk viscosity in merger simulations. Hammond, Hawke, & Andersson [83] ran merger simulations with infinitely fast and infinitely slow Urca reactions. A few very recent simulations have included flavor-changing interactions, including Radice *et al.* [115], Zappa *et al.* [149], and Most *et al.* [106], and just prior to submission of this chapter, Chabanov & Rezzolla [42] put bulk viscosity into a merger simulation using the MIS formalism. Bulk viscosity in merger simulations is discussed in Sec. 8.5.2.

8.2 BULK VISCOSITY IN NEUTRON-PROTON-ELECTRON MATTER

8.2.1 Dense matter

The nature of dense matter is not fully known. In the vicinity of nuclear saturation density, the matter is uniform, with neutrons, protons, and electrons as degrees of freedom. At low enough temperature, certainly below $T = 1$ MeV, the nucleons form Cooper pairs and the matter becomes a superfluid [129]. In a neutron star merger, after the stars collide, the vast majority of the matter rises to temperatures above 1 MeV, and is therefore not superfluid (though quark matter critical temperatures are expected to be higher [24].) As the density increases significantly beyond n_0, new degrees of freedom may enter the system, including hyperons, a pion condensate, or the nuclear matter might undergo a phase transition to quark matter [17]. As temperature is increased, thermal populations of particles, for example, pions [60], may appear as well.

A sensible first step for modeling dense matter is to consider uniform npe matter. The electrons can be treated as a free Fermi gas, but strong interactions between the nucleons must be considered. This is usually done with Skyrme energy density functionals [52] or relativistic mean-field theories (RMFs) [51]. I will use the IUF RMF [54] throughout this chapter. Of relevance to the discussion in this chapter, the IUF EoS (when muons are not added) has a direct Urca threshold near $n_B = 4n_0$ [15]. RMFs are motivated by the meson-exchange model of the strong interaction, and nucleons interact by exchanging sigma, omega, and rho mesons. The mean field approximation is then taken, the meson dynamics are frozen, and the nucleons act like free particles but with effective masses and chemical potentials [67].

8.2.2 Thermodynamics and susceptibilities

A bulk viscosity calculation in neutrino-transparent npe matter, will require information about the thermodynamics of such as system, and so some relevant thermodynamic relations are reviewed here. The strong and electromagnetic interactions are powerful enough to force the neutrons, protons, and electrons to move together as one fluid. This matter produces neutrinos, but the neutrinos, participating only in the weak interaction, have a much longer MFP and this section will consider thermodynamic conditions where they escape the system. The baryon number of the system n_B is given by $n_B = n_n + n_p$ and charge neutrality of the nuclear matter demands $n_e = n_p$, leaving just one independent particle species, chosen here to be the proton. If the matter is in beta equilibrium, the proton fraction has a fixed value at a particular baryon density and temperature.

A beta equilibrium condition can be derived from any allowed reaction in the system, provided that all particles in the reaction are in thermal equilibrium. Thus, in neutrino-trapped nuclear matter, the beta equilibrium condition is determined by examining the reaction $e^- + p \leftrightarrow n + \nu$, and through a standard thermodynamic procedure involving stoichiometric coefficients [126], the condition

$$\delta\mu_{\text{trapped}} \equiv \mu_n + \mu_\nu - \mu_p - \mu_e = 0 \tag{8.1}$$

fixes the matter in chemical equilibrium. To get the neutrino-transparent condition, it is common to set the neutrino chemical potential in Eq. 8.1 to zero, since neutrino number is not conserved in neutrino-transparent matter, yielding the beta equilibrium condition

$$\delta\mu \equiv \mu_n - \mu_p - \mu_e = 0. \tag{8.2}$$

This equation will be sufficient for this chapter; however, it does pick up corrections at finite temperature[7].

[7] Free-streaming neutrinos are not equivalent to a neutrino Fermi gas with zero chemical potential, so there is no reason to expect $\mu_n = \mu_p + \mu_e$ to be the proper beta equilibrium condition in neutrino-transparent matter. The proper condition is found by equating the neutron-producing rates with the proton-producing rates [15, 18]. However, it was recently found that using $\mu_n = \mu_p + \mu_e$ instead of the condition as finite temperature does not substantially change the bulk viscosity [16].

The first law of thermodynamics in npe matter is [126, 132]

$$dE = -P\,dV + T\,dS + \mu_n\,dN_n + \mu_p\,dN_p + \mu_e\,dN_e\,. \tag{8.3}$$

Normalizing by the volume, and using the Euler equation

$$\varepsilon + P = sT + \mu_n n_n + \mu_p n_p + \mu_e n_e = sT + \mu_n n_B - n_p \delta\mu, \tag{8.4}$$

(derived in, e.g. [132]) leads to

$$d\varepsilon = T\,ds + \mu_n\,dn_n + \mu_p\,dn_p + \mu_e\,dn_e = T\,ds + \mu_n\,dn_B - \delta\mu\,dn_p\,. \tag{8.5}$$

The Gibbs-Duhem equation, in some sense the compliment to the first law of thermodynamics, can be derived by taking the total derivative of the Euler equation and then replacing $d\varepsilon$ by Eq. 8.5, leading to

$$dP = s\,dT + n_n\,d\mu_n + n_p\,d\mu_p + n_e\,d\mu_e = s\,dT + n_B\,d\mu_n - n_p\,d\delta\mu\,. \tag{8.6}$$

In these derivations, I have made use of the conditions $dn_B = dn_n + dn_p$ and $dn_e = dn_p$.

Instead of normalizing the first law (Eq. 8.3) by volume, it can be normalized by baryon number N_B, which is conserved (that is, $dN_B = 0$.) This leads to the expression

$$d\left(\frac{\varepsilon}{n_B}\right) = \frac{P}{n_B^2}\,dn_B + T\,d\left(\frac{s}{n_B}\right) - \delta\mu\,dx_p\,. \tag{8.7}$$

Since it is more natural to specify temperature than entropy in the EoS RMF calculation, I will work at constant temperature in this chapter. The above thermodynamic relations must be Legendre transformed [132], leading to the equations

$$d\,(\varepsilon - sT) = \mu_n\,dn_B - s\,dT - \delta\mu\,dn_p, \tag{8.8}$$

$$d\left(\frac{\varepsilon - sT}{n_B}\right) = \frac{P}{n_B^2}\,dn_B - \frac{s}{n_B}\,dT - \delta\mu\,dx_p\,. \tag{8.9}$$

Many Maxwell relations can be derived from these relations, but the only important one for our purpose comes from Eq. 8.9

$$\left.\frac{\partial P}{\partial x_p}\right|_{n_B,T} = -n_B^2 \left.\frac{\partial \delta\mu}{\partial n_B}\right|_{x_p,T}. \tag{8.10}$$

In calculating the bulk viscosity resulting from small amplitude density oscillations, it is necessary to consider small perturbations around a state of chemical equilibrium. The coefficients of the Taylor expansions will be written in terms of susceptibilities, which are properties of the nuclear EoS. In degenerate nuclear matter, the susceptibilities are essentially unconstrained at present. As a consequence (c.f. Eq. 8.37), the maximum value of the subthermal bulk viscosity is also essentially unconstrained.

In this chapter, I will consider isothermal density oscillations, so the isothermal susceptibilities

$$A \equiv n_B \left.\frac{\partial \delta\mu}{\partial n_B}\right|_{T,x_p} = -\frac{1}{n_B}\left.\frac{\partial P}{\partial x_p}\right|_{n_B,T} \tag{8.11}$$

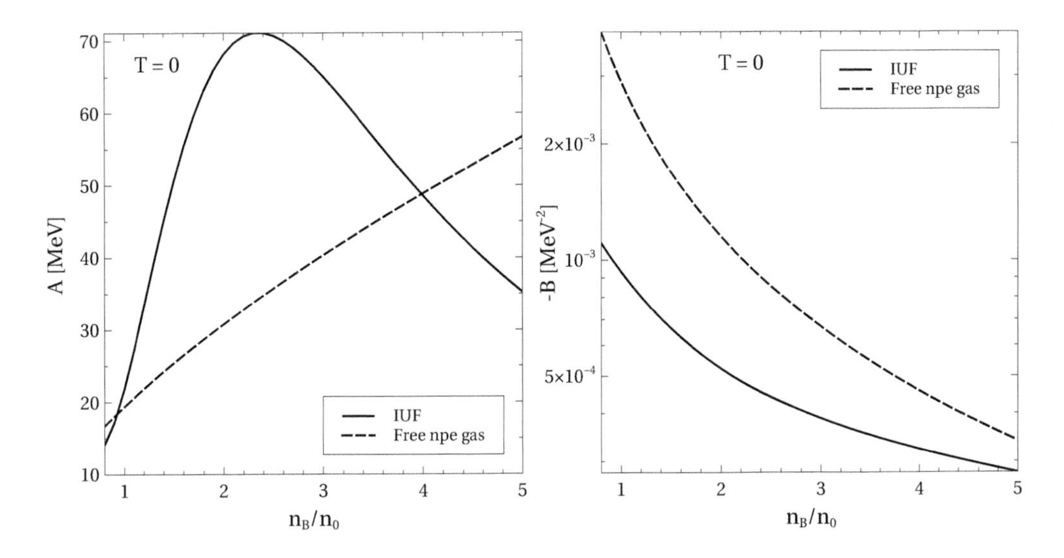

Figure 8.1 Isothermal susceptibilities in npe matter, with the solid lines representing the IUF result and the dashed line representing the free Fermi gas result [20]. The susceptibilities are calculated at zero temperature, but the temperature dependence is small.

and

$$B \equiv \frac{1}{n_B} \frac{\partial \delta\mu}{\partial x_p}\bigg|_{n_B,T} \tag{8.12}$$

are defined. Expressions relating the isothermal and adiabatic (constant entropy per baryon) susceptibilities are provided in the appendix of [19]. Numerically, there is little difference between the two for densities above n_0 and temperatures below 10 MeV [19].

In Fig. 8.1 are plotted the susceptibilities for the IUF EoS, compared with those calculated for a free Fermi gas (from Table 2 in [20].) The susceptibilities do not change significantly with temperature, at least in the neutrino-transparent regime (see, e.g., Fig. 3 in [19]).

8.2.3 Bulk viscosity and beta equilibration

To calculate the bulk viscosity, consider a fluid element undergoing a small amplitude, sinusoidal change in its density

$$n_B(t) = n_B + \Re(\delta n_B e^{i\omega t}) = n_B + \delta n_B \cos{(\omega t)}, \tag{8.13}$$

where $\delta n_B \ll n_B$. Because the beta equilibrium value of the proton fraction (plotted for several typical npe EoSs in Fig. 8.2) is a function of density, when the matter is compressed, the matter departs from beta equilibrium[8]. In response, chemical reactions will try

[8]If the beta-equilibrium value of the proton fraction does not change with density, the system is conformal and the bulk viscosity is zero. Dense matter can be conformal at a particular density where the beta-equilibrium proton fraction is at a minimum or a maximum. This situation is unlikely to occur in uniform npe matter (Fig. 8.2 shows that x_p monotonically increases with density), but can occur in neutrino-trapped matter [5]. Dense npe matter that has zero bulk viscosity from Urca processes at a particular density would still have a (very small) collisional bulk viscosity [92] and thus is not *truly* conformal.

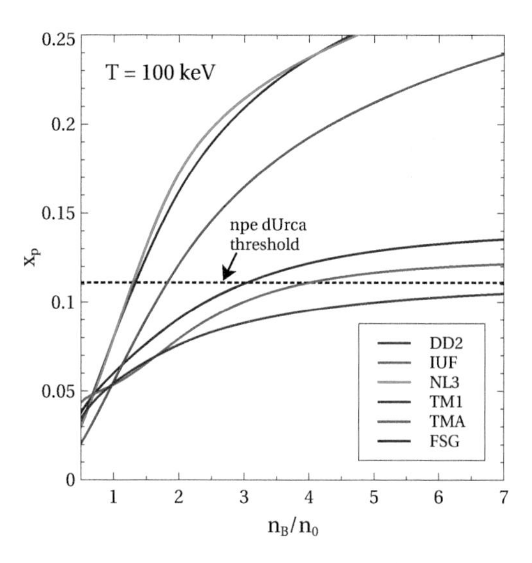

Figure 8.2 Beta-equilibrium proton fraction of several common npe EoSs, as a function of density. At zero temperature, the direct Urca process is kinematically allowed at densities where the proton fraction is greater than 1/9, depicted by the dotted black line. Data obtained from CompOSE [136].

to bring the system to the new value of the beta-equilibrium proton fraction and thus the proton fraction will also be a periodic function

$$x_p(t) = x_p^0 + \Re(\delta x_p e^{i\omega t}) = x_p^0 + \Re(\delta x_p)\cos(\omega t) - \Im(\delta x_p)\sin(\omega t). \tag{8.14}$$

However, the chemical reactions only have a finite rate, and depending on how the reaction rate compares to the density oscillation frequency, there could be a phase lag between the proton fraction and baryon density. The pressure of the nuclear matter is a function of both the density and the proton fraction (and the temperature, but this will be neglected) and will also undergo a harmonic oscillation

$$P(t) = P_0 + \Re(\delta P e^{i\omega t}) = P_0 + \Re(\delta P)\cos(\omega t) - \Im(\delta P)\sin(\omega t) \tag{8.15}$$

with a potential phase lag with respect to the baryon density. Bulk-viscous energy dissipation comes from a phase lag between the pressure and volume of a fluid element throughout an oscillation period. Baryon density is an obvious proxy for volume ($n_B \equiv N_B/V$) and the proton fraction is a proxy for the pressure. The part of the pressure that depends on density will not cause dissipation, but the part of the pressure that depends on x_p will give rise to a term with a phase lag with respect to the baryon density. One should expect the bulk viscosity to be proportional to $\Im(\delta P)$, since that is the coefficient of the part of the pressure (Eq. 8.15) that is out of phase with the density. Finally, when a sinusoidal density oscillation is imposed, the proton fraction, pressure, and beta equilibrium $\delta\mu$ all undergo harmonic oscillations, some potentially with a phase lag. We should expect the temperature to act similarly, but, following Langer & Cameron [95], temperature changes throughout an oscillation will be neglected throughout this chapter.

The oscillation energy (per unit volume) lost due to compression or rarifaction in fluid motion (that is, motion with net divergence) is encapsulated in the bulk viscosity coefficient ζ via the relation

$$\frac{d\varepsilon_{\text{osc}}}{dt} = -\zeta(\nabla \cdot \mathbf{v})^2. \tag{8.16}$$

Using the continuity equation for a comoving fluid element

$$\frac{dn_B}{dt} + n_B \nabla \cdot \mathbf{v} = 0, \tag{8.17}$$

and then averaging over one oscillation period, we find

$$\left\langle \frac{d\varepsilon_{\text{osc}}}{dt} \right\rangle = -\frac{1}{2}\left(\frac{\delta n_B}{n_B}\right)^2 \omega^2 \zeta. \tag{8.18}$$

To finish deriving the bulk viscosity coefficient in terms of $\Im(\delta P)$, we calculate the energy dissipation in terms of the PdV work done in one oscillation

$$\begin{aligned} \left\langle \frac{d\varepsilon_{\text{diss}}}{dt} \right\rangle &= \frac{\omega}{2\pi}\int_0^{2\pi/\omega} dt \, \frac{P(t)}{n_B(t)}\frac{dn_B(t)}{dt} \\ &= \frac{\omega^2}{2\pi}\left(\frac{\delta n_B}{n_B}\right)\Im(\delta P)\int_0^{2\pi/\omega} dt \sin^2(\omega t) \\ &= \frac{\omega}{2}\left(\frac{\delta n_B}{n_B}\right)\Im(\delta P). \end{aligned} \tag{8.19}$$

Since $\langle \dot{\varepsilon}_{\text{osc}} \rangle = -\langle \dot{\varepsilon}_{\text{diss}} \rangle$ (dot denotes time derivative), the bulk viscosity is given by

$$\zeta = \left(\frac{n_B}{\delta n_B}\right)\frac{\Im(\delta P)}{\omega}. \tag{8.20}$$

This expression is general, even for more complex situations where there are multiple equilibrating quantities.

The pressure $P(n_B, T, x_p)$ can be expanded around its equilibrium value P_0

$$\begin{aligned} P &= P_0 + \left.\frac{\partial P}{\partial n_B}\right|_{T,x_p}\Re(\delta n_B e^{i\omega t}) + \left.\frac{\partial P}{\partial T}\right|_{n_B,x_p}\Re(\delta T e^{i\omega t}) + \left.\frac{\partial P}{\partial x_p}\right|_{T,n_B}\Re(\delta x_p e^{i\omega t}), \\ &= P_0 + \frac{1}{\kappa_T}\frac{\delta n_B}{n_B}\cos(\omega t) - n_B A\Re(\delta x_p)\cos(\omega t) + n_B A\Im(\delta x_p)\sin(\omega t), \end{aligned} \tag{8.21}$$

where the definitions of the susceptibility (Eq. 8.11) and the isothermal compressibility [136]

$$\kappa_T = \left(n_B \left.\frac{\partial P}{\partial n_B}\right|_{T,x_p}\right)^{-1}, \tag{8.22}$$

were invoked. Hereafter, I drop the temperature dependence in expansions around beta equilibrium. Matching sine and cosine terms between the two equations for the pressure (Eq. 8.15 and 8.21) yields

$$\Re(\delta P) = \frac{1}{\kappa_T}\frac{\delta n_B}{n_B} - n_B A\Re(\delta x_p) \tag{8.23a}$$

$$\Im(\delta P) = -n_B A\Im(\delta x_p). \tag{8.23b}$$

In order to calculate the bulk viscosity, the imaginary part of the proton fraction oscillation must be found.

The proton fraction evolves through flavor-changing interactions. In neutrino transparent npe matter, the direct Urca (dUrca) processes

$$n \rightarrow p + e^- + \bar{\nu}_e \quad \text{dUrca neutron decay} \tag{8.24a}$$
$$e^- + p \rightarrow n + \nu_e \quad \text{dUrca electron capture} \tag{8.24b}$$

and the modified Urca (mUrca) processes

$$N + n \rightarrow N + p + e^- + \bar{\nu}_e \quad \text{mUrca neutron decay} \tag{8.25a}$$
$$N + e^- + p \rightarrow N + n + \nu_e \quad \text{mUrca electron capture} \tag{8.25b}$$

act to establish beta equilibrium. Because the matter is assumed to be neutrino transparent (due to the long neutrino MFP at low temperatures), neutrinos and antineutrinos are not allowed to be on the left-hand side of any reaction.

At temperatures below about 1 MeV, the npe matter is so strongly degenerate that the Urca rates above can be calculated in the Fermi surface (FS) approximation, where only particles within energy T of the Fermi energy are allowed to participate in the reaction. This approximation illuminates the strong temperature dependence of rates in degenerate nuclear matter. In this approximation, the direct Urca process is only kinematically allowed for densities such that $p_{Fn} < p_{Fp} + p_{Fe}$, which corresponds to proton fractions larger than $1/9$ in npe matter [127]. For the IUF EoS, the direct Urca threshold is near $n_B = 4n_0$ (c.f. Fig. 8.2), and thus below $4n_0$, only the modified Urca process is there to help establish beta equilibrium. As the temperature rises above about 1 MeV, the Boltzmann suppression of reaction participants that lie far from the Fermi surface decreases, causing the direct Urca threshold density to blur (c.f. Fig. 1 in [19]) as the reaction phase space opens up. This effect is important for temperatures above 1 MeV [18], but I will neglect it in this chapter so that I can focus on other aspects of the bulk viscosity calculation.

The proton fraction evolves according to the equation

$$\begin{aligned} n_B \frac{dx_p}{dt} &= \Gamma_{nn\rightarrow npe\nu} - \Gamma_{nep\rightarrow nn\nu} \\ &= \frac{1}{5760\pi^9} G^2 g_A^2 f^4 \frac{m_n^4}{m_\pi^4} \frac{p_{Fn}^4 p_{Fp}}{(p_{Fn}^2 + m_\pi^2)^2} \theta_n \delta\mu \\ &\qquad \times (1835\pi^6 T^6 + 945\pi^4 \delta\mu^2 T^4 + 105\pi^2 \delta\mu^4 T^2 + 3\delta\mu^6). \end{aligned} \tag{8.26}$$

$$\approx \lambda_{npe} \delta\mu \quad \text{(subthermal limit, } \delta\mu \ll T\text{)}, \tag{8.27}$$

with

$$\theta_n = \begin{cases} 1 & p_{Fn} > p_{Fp} + p_{Fe} \\ 1 - \dfrac{3}{8}\dfrac{(p_{Fp} + p_{Fe} - p_{Fn})^2}{p_{Fp} p_{Fe}} & p_{Fn} < p_{Fp} + p_{Fe} \end{cases} \tag{8.28}$$

and[9]

$$\lambda_{npe} = \frac{367}{1152\pi^3} G^2 g_A^2 f^4 \frac{m_n^4}{m_\pi^4} \frac{p_{Fn}^4 p_{Fp}}{(p_{Fn}^2 + m_\pi^2)^2} T^6 \theta_n. \tag{8.29}$$

In this chapter, all λ coefficients are positive. In the above formulas, $G^2 \equiv G_F^2 \cos^2\theta_c = 1.29 \times 10^{-22}$ MeV^{-4}, where G_F is the Fermi coupling constant and θ_C is the Cabibbo angle. The axial vector coupling constant $g_A = 1.26$ and $f \approx 1$. Now, $\delta\mu$ can be expanded around chemical equilibrium

$$\begin{aligned} \delta\mu &= \left.\frac{\partial \delta\mu}{\partial n_B}\right|_{T,x_p} \Re(\delta n_B e^{i\omega t}) + \left.\frac{\partial \delta\mu}{\partial x_p}\right|_{n_B,T} \Re(\delta x_p e^{i\omega t}) \\ &\approx A\frac{\delta n_B}{n_B}\cos(\omega t) + n_B B\left(\Re(\delta x_p)\cos(\omega t) - \Im(\delta x_p)\sin(\omega t)\right), \end{aligned} \tag{8.30}$$

Plugging Eqs. 8.14 and 8.30 into Eq. 8.27, and then matching the coefficients of the sine and cosine terms, a system of two equations for $\Re(\delta x_p)$ and $\Im(\delta x_p)$ is obtained, the solution of which is

$$\Re(\delta x_p) = -\left(\frac{\delta n_B}{n_B}\right)\frac{AB\lambda^2}{n_B(\omega^2 + B^2\lambda^2)} \tag{8.31a}$$

$$\Im(\delta x_p) = -\left(\frac{\delta n_B}{n_B}\right)\frac{A\lambda\omega}{n_B(\omega^2 + B^2\lambda^2)}. \tag{8.31b}$$

Combining Eqs. 8.20, 8.23b, and 8.31b, we obtain the subthermal bulk viscosity in *npe* matter

$$\zeta_{npe} = \frac{A^2\lambda}{\omega^2 + B^2\lambda^2} = \frac{A^2}{|B|}\frac{\gamma}{\omega^2+\gamma^2}, \tag{8.32}$$

where the equilibration rate γ is defined as

$$\gamma \equiv |B|\lambda. \tag{8.33}$$

The expressions for the real and imaginary parts of $\delta x_p(t)$ and $\delta P(t)$, solved for in pursuit of the bulk viscosity, can be used to obtain[10] $x_p(t)$, $P(t)$, and $\delta\mu(t)$

$$x_p(t) = x_p^0 - \left(\frac{\delta n_B}{n_B}\right)\frac{A}{Bn_B}\frac{\gamma}{\omega^2+\gamma^2}\left[\gamma\cos(\omega t) + \omega\sin(\omega t)\right], \tag{8.34a}$$

$$P(t) = P_0 + \left(\frac{\delta n_B}{n_B}\right)\left\{\frac{1}{\kappa T}\cos(\omega t) + \frac{A^2}{B}\frac{\gamma}{\omega^2+\gamma^2}\left[\gamma\cos(\omega t) + \omega\sin(\omega t)\right]\right\}, \tag{8.34b}$$

$$\delta\mu(t) = A\left(\frac{\delta n_B}{n_B}\right)\frac{\omega}{\omega^2+\gamma^2}\left[\omega\cos(\omega t) - \gamma\sin(\omega t)\right]. \tag{8.34c}$$

---

[9] In reality, this is an older version of the modified Urca rate (used in, e.g. Alford & Harris [19].) The factor of m_n^4 should be replaced by $(E_{Fn}^*)^3 E_{Fp}^*$ where $E_{Fi}^* = \sqrt{p_{Fi}^2 + m_*^2}$ [15]. I use the older expression here because while it overestimates the modified Urca rate, it gets the bulk-viscous resonance to be at the same temperature as more realistic calculations [16].

[10] B is assumed to be negative in this calculation, as it is (right panel of Fig. 8.1.)

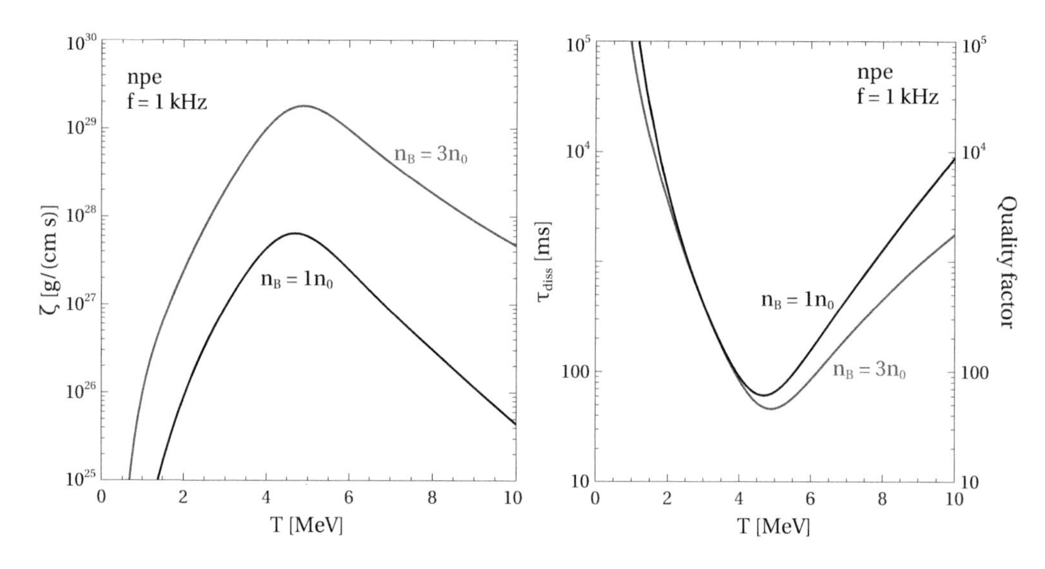

Figure 8.3 Left: Bulk viscosity of neutrino-transparent npe matter undergoing a density oscillation with angular frequency $\omega = 2\pi \times 1$ kHz. Right: Timescale over which bulk viscosity dissipates kinetic energy from the oscillation (left axis), and the quality factor of the oscillation (right axis).

The bulk viscosity of npe matter is plotted as a function of temperature in the left panel of Fig. 8.3. The density oscillation frequency is chosen to be 1 kHz ($\omega = 2\pi \times 1$ kHz), typical of neutron stars and neutron star mergers[11]. The key feature of the bulk viscosity as a function of temperature is the presence of one resonant maximum, occurring at the temperature where the beta equilibration rate γ matches the density oscillation frequency ω. The beta equilibration rate in npe matter is monotonic with temperature ($\lambda_{npe} \sim T^6$, Eq. 8.29), and there are no additional parallel beta equilibration channels (as there will be in $npe\mu$ matter) and thus there is only one resonance.

At low temperatures, the beta equilibration rate is very slow compared to the density oscillation frequency, and thus the matter barely equilibrates upon compression, and therefore the system changes very little throughout the oscillation (beyond the non-dissipative increase and decrease of the pressure due to the density change alone). In this regime, the bulk viscosity is approximately

$$\zeta_{npe}^{\gamma \ll \omega} \approx \frac{A^2}{|B|} \frac{\gamma}{\omega^2}, \tag{8.35}$$

which is small because $\gamma \ll \omega$. This is the form of the bulk viscosity used in studying cold neutron stars, as described in Sec. 8.1.3 (e.g. [76, 77, 79, 98, 144]). The bulk viscosity at fixed frequency ω is proportional to the equilibration rate[12].

[11] The frequency of a sound wave in a neutron star can be estimated by dividing the distance scale (10 km) by the speed of sound in dense matter (0.1) [135], which yields a timescale of 0.33 ms, or a linear frequency of 3 kHz. Sec. 8.5.2 shows evidence for this oscillation frequency from numerical simulations.

[12] Because the bulk viscosity in this limit is proportional to the beta equilibration rate, an early goal in bulk viscosity research was to find "the fastest" equilibration process, as it would lead to the largest bulk viscosity.

At high temperatures, the Urca rate is very fast, and when the density changes, the matter essentially never departs from chemical equilibrium and there is little dissipation. In the high-temperature limit, the bulk viscosity is approximately

$$\zeta_{npe}^{\gamma\gg\omega} \approx \frac{A^2}{|B|}\frac{1}{\gamma}. \tag{8.36}$$

It is independent of the oscillation frequency, and as the rate γ grows faster with increasing temperature, the bulk viscosity decreases. The high-temperature limit is, of course, the low-frequency limit, and this expression for the bulk viscosity describes its behavior with respect to arbitrarily slow density oscillations. It is also common to refer to this "zero-frequency" bulk viscosity coefficient as *the* bulk viscosity, from which a frequency-dependent "effective bulk viscosity" $\zeta_{\text{eff}}(\omega)$ (that is, what I call *the* bulk viscosity) can be obtained [64].

Assuming that the susceptibilities do not depend on temperature, the maximum bulk viscosity occurs at the temperature where $\gamma = \omega$ and is

$$\zeta_{npe,\,\text{max}} = \frac{1}{2\omega}\frac{A^2}{|B|}. \tag{8.37}$$

Thus, in the simple case where there is only one chemical equilibration channel, the location of the maximum bulk viscosity is controlled by the Urca rate, and the strength of the bulk viscosity at resonance is controlled by the susceptibilities, which are properties of the EoS. As we see from the left panel of Fig. 8.3, the bulk viscosity increases with density, which is due to the fact that A increases from n_0 to $3n_0$ as $|B|$ decreases in the same interval (see Fig. 8.1.)

The resonance behavior of bulk viscosity can be further explored by looking at the path of a fluid element in the $x_p n_B$ (left panel of Fig. 8.4) or PV (Fig. 8.5) plane, an analysis first put forward by Langer & Cameron [95], but also discussed by Muto *et al.* [108] and Camelio *et al.* [38]. First, note that the pressure $P = P(n_B, T, x_p)$ is a function of the three thermodynamic variables. (the T-dependence is neglected, as discussed previously.) The density dependence of the pressure is very strong, and while the x_p dependence is relatively weak, it is the source of the bulk viscosity effect. It is natural to separate the pressure Eq. 8.15 throughout the density oscillation into two pieces

$$P(t) = P_0 + \left(\frac{\delta n_B}{n_B}\right)\frac{1}{\kappa_T}\cos{(\omega t)} + P_{xp}, \tag{8.38}$$

which, comparing with Eq. 8.15, serves as a definition of P_{xp}, the part of the pressure that varies due to chemical equilibration.

At one extreme, infinitely slow reactions, $P = P_0 + (\delta n_B/n_B)\kappa_T^{-1}\cos{(\omega t)}$. At the other extreme, with infinitely fast equilibration, $P(t) = P_0 + (\delta n_B/n_B)(\kappa_T^{-1} + A^2/B)\cos{(\omega t)}$. In both cases, the pressure is in phase with the density oscillation. Examining these two limits, an interpretation for A^2/B, the prefactor of the bulk viscosity,

---

However, it was known that some processes were "too fast", that is, faster than the millisecond oscillation timescales and thus on the other side of their resonant peak (e.g. $n + n \leftrightarrow p + \Delta^-$ [95].)

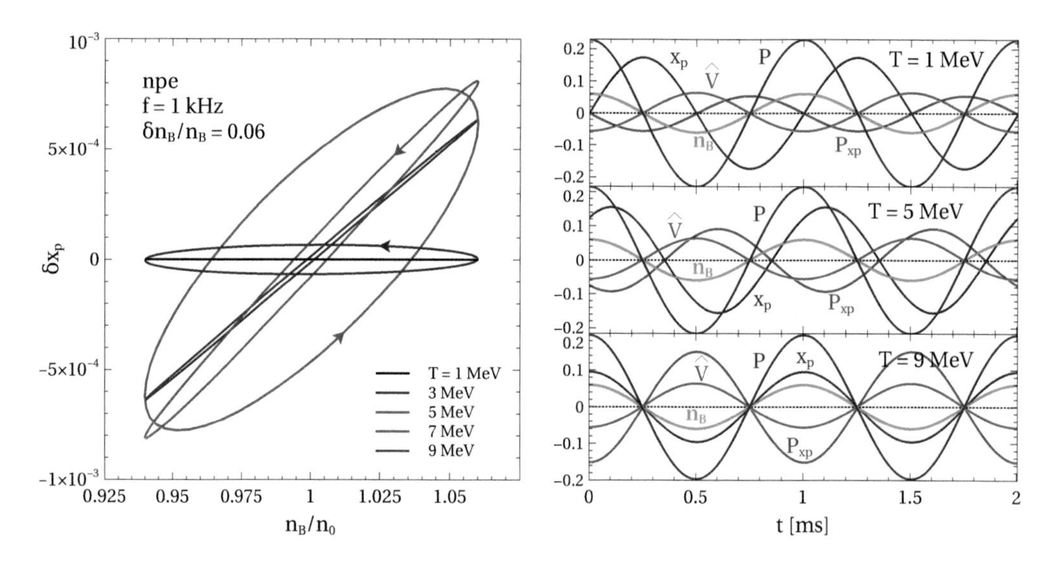

Figure 8.4 Left: Route through the $x_p n_B$ plane of a fluid element in one oscillation period. Right: $P(t)$, $n_B(t)$, $\hat{V}(t) \equiv 1/n_B(t)$, $x_p(t)$, and P_{xp} throughout two cycles, for three different temperatures. The normalization in the right panel is arbitrary—the focus is the phase shifts between the various curves.

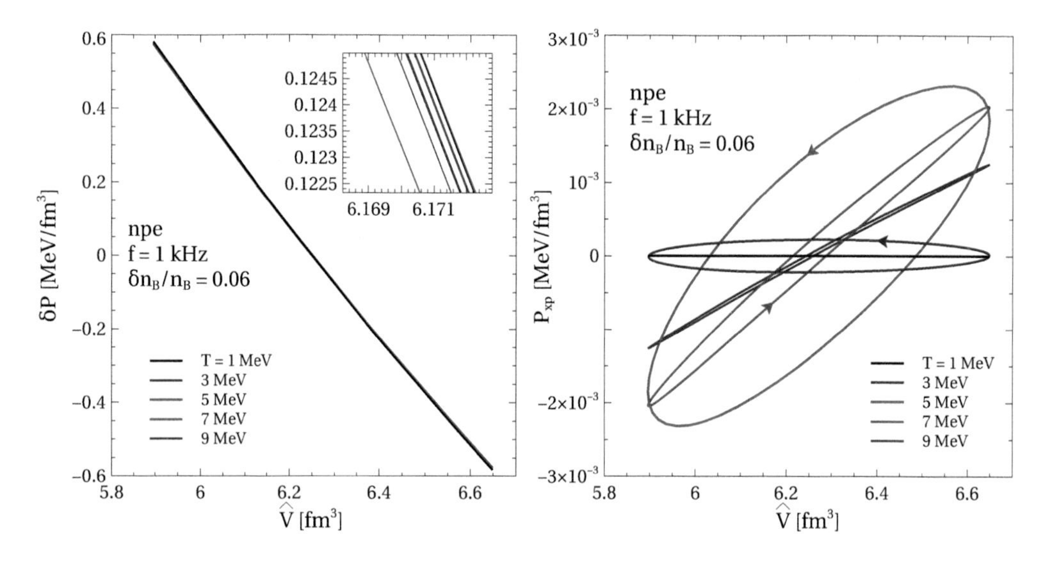

Figure 8.5 Left: Path through the PV plane of a fluid element during one cycle of the density oscillation. All five curves are overlapping due to the fact that the pressure depends much more on the density than the proton fraction, so the inset panel zooms in on a section of the curve to show the five individual curves. Right: Path through the $P_{xp}V$ plane of a fluid element during one cycle of the density oscillation, therefore showing only the part of the pressure that depends on x_p, removing the part of the pressure that depends on n_B as it does not contribute to the PdV work.

becomes clear. The inverse compressibility κ_T^{-1} is the pressure increase caused by a density increase, with no chemical equilibration. With infinitely fast equilibration, the inverse compressibility is corrected by an amount A^2/B. Expanding out the expression, the incompressibility in the $\gamma \to \infty$ limit is

$$\kappa_T^{-1} + \frac{A^2}{B} = n_B \left(\left.\frac{\partial P}{\partial n_B}\right|_{T,x_p} - \left.\frac{\partial P}{\partial x_p}\right|_{T,n_B} \left.\frac{\partial x_p}{\partial \delta\mu}\right|_{T,n_B} \left.\frac{\partial \delta\mu}{\partial n_B}\right|_{T,x_p} \right). \tag{8.39}$$

This means that the prefactor of the bulk viscosity, the factor that determines the value it attains at resonance, is related to the difference between the incompressibility of the matter with infinitely fast chemical equilibration and the incompressibility with infinitely slow chemical equilibration. This result makes clear the connection between the formalism described here and that of, for example, Lindblom & Owen [98], where they find the bulk viscosity is proportional to the difference of the adiabatic indices with infinitely fast or slow equilibration (itself a relativistic generalization of the Landau & Lifshitz [94] and Jones [89] results involving the difference of sound speeds $c_\infty^2 - c_0^2$.)

The area enclosed by the path of a fluid element in the PV plane (left panel of Fig. 8.5) indicates the work done in one cycle, and the direction in the path is traversed indicates if the work is done on the system by the environment, or on the environment by the system [55, 95]. Because dense matter is highly incompressible, the pressure depends strongly on the density and thus the PV curve is close to a line. A better representation of the PdV work is seen by plotting the path of a fluid element in the $P_{xp}V$ plane (right panel of Fig. 8.5), which subtracts off the part of the pressure that does not contribute to the PdV work done - that is, the area of a curve in the PV plane is the same as the area in the $P_{xp}V$ plane. The x_pn_B plane (left panel of Fig. 8.4) gives similar information, as x_p is essentially a proxy of P_{xp}, as n_B is of V.

First consider a fluid element at $T = 1$ MeV. As the density is increased, the matter is pushed out of beta equilibrium and wants to increase the proton fraction to reestablish the equilibrium. However, the Urca processes are much too slow to do so within the duration of the oscillation, and thus the proton fraction remains very close to its original value. In the density-proton fraction plane, the trajectory of one oscillation is very close to a straight line - the density increases and then decreases again, retracing its path. The path of the fluid element through the $P_{xp}V$ plane is similar, enclosing very little area and thus dissipating little energy.

A fluid element with temperature 5 MeV undergoing a 1 kHz density oscillation is very close to experiencing the resonant behavior of bulk viscosity. As the fluid element is compressed, the Urca process slowly turns on and adjusts the proton fraction, though lagging behind the density oscillation. The two are nearly completely out of phase. This leads the fluid element to traverse an ellipse with large area in the x_pn_B plane, as well as the $P_{xp}V$ plane, and much energy is dissipated. It is clear that the curve in the PV plane is traversed counterclockwise, and therefore, the environment does work on the system. That is, the mechanism that causes the density oscillation does work on the fluid element, causing the oscillation to lose energy and the fluid element to heat up (ignoring the resulting enhancement of the neutrino emission which, in the subthermal regime, results in net cooling.)

At higher temperatures, like $T = 9$ MeV, upon compression, the fluid element almost instantaneously adjusts the proton fraction to reestablish beta equilibrium at the new density. Pressure, proton fraction, and density are all in phase with each other and the energy dissipation is very small.

Plotted in the right panel of Fig. 8.3 is the dissipation timescale of a density oscillation due to bulk viscosity. The energy density of a density oscillation in nuclear matter is [19]

$$\varepsilon = \frac{1}{2}\left(\delta n_B\right)^2 \left.\frac{\partial^2 \varepsilon}{\partial n_B^2}\right|_{x_p,T} = \frac{1}{2\kappa_T}\left(\frac{\delta n_B}{n_B}\right)^2. \tag{8.40}$$

The rate of energy dissipation is directly related to the bulk viscosity, through Eq. 8.18. Thus a dissipation timescale can be obtained

$$\tau_{\text{diss}} \equiv \frac{\varepsilon}{d\varepsilon\,/\,dt} = \frac{1}{\kappa_T \omega^2 \zeta}. \tag{8.41}$$

The quality factor of the oscillation, or the number of oscillations before the oscillation energy drops by an e-fold [134], is [98]

$$Q \equiv \frac{\omega}{2\pi}\tau_{\text{diss}}. \tag{8.42}$$

Fig. 8.3 indicates that density oscillations in npe matter can be damped by bulk viscosity in as little as 30 ms. Such a 1 kHz oscillation would have a quality factor of 30. In more extensive scans of the thermodynamic conditions, Alford & Harris [19] and Alford, Haber, & Zhang [16] found dissipation times as little as 5 ms.

8.2.4 Validity of subthermal approximation

In truncating the expression for the beta equilibration rate $\Gamma_{nn\to npe\nu} - \Gamma_{nep\to nn\nu}$ to first order in the departure from beta equilibrium, $\delta\mu$, we gain the ability to solve for the bulk viscosity exactly. But, when using the subthermal bulk viscosity formalism to consider a density oscillation of a specific amplitude δn_B, one must be careful that $\delta\mu$ remains small compared to T.

The maximum value of $\delta\mu$ in an oscillation cycle can be found from Eq. 8.34c, and is given by

$$\delta\mu_{\text{max}} = A\left(\frac{\delta n_B}{n_B}\right)\frac{\omega}{\sqrt{\omega^2+\gamma^2}}. \tag{8.43}$$

In a cold system, where $\gamma \ll \omega$, this reduces to $\delta\mu_{\text{max}} = A(\delta n_B/n_B)$ and thus it is "easy" to deviate strongly from the subthermal regime if $\delta n_B/n_B$ is too large. In a hot system, where $\gamma \gg \omega$, $\delta\mu_{\text{max}} = A(\delta n_B/n_B)(\omega/\gamma)$, which remains small because the system is efficiently beta equilibrated and thus $\delta\mu$ is unable to depart significantly from zero. The ratio $\delta\mu_{\text{max}}/T$ is plotted in the $T(\delta n_B/n_B)$ plane in Fig. 8.6. All points $\{\delta n_B/n_B, T\}$ where $\delta\mu_{\text{max}}/T \ll 1$ lie in the subthermal regime.

In matter with a temperature of 1 MeV, one leaves the subthermal regime if the density oscillation amplitude exceeds a few percent of the background density! Arras & Weinberg observed the suprathermal nature of the bulk viscosity in cold neutron stars in their study of beta reactions in the neutron star inspiral [29]. Alternatively, at a higher temperature like 8 MeV, the subthermal approximation is valid for δn_B of the same order as n_B. The

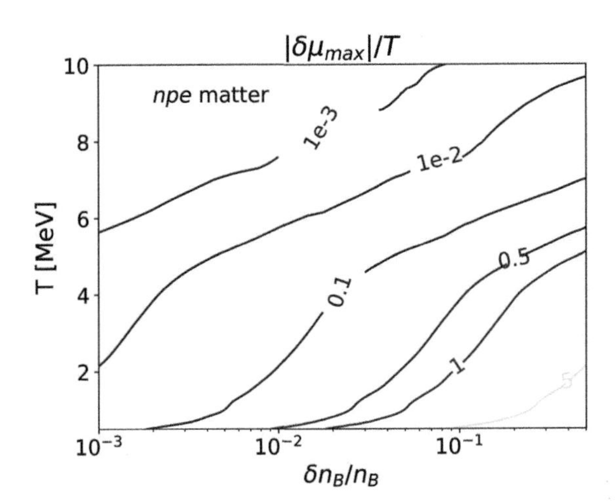

Figure 8.6 Contour plot of $\delta\mu(t)/T$ as a function of temperature and density oscillation amplitude. If $|\delta\mu|/T \ll 1$, the oscillation is subthermal and the formalism developed in this chapter is applicable. For larger density oscillations where $\delta\mu$ nears or exceeds T, the curve given in this figure and Eq. (8.27) no longer accurately describe the oscillation, and one should instead use the full expression Eq. (8.26).

bulk viscosity calculation as formulated here requires $\delta n_B \ll n_B$, or else the notion of expanding around an equilibrium state falls apart. If the density oscillations are no longer small, which does seem to be the case in certain situations in neutron star mergers, then the reaction rates must be directly implemented in the simulation and evolved with the hydrodynamics, instead of ignored and encapsulated in a pre-calculated bulk-viscous contribution to the pressure.

8.3 BULK VISCOSITY IN DENSE MATTER CONTAINING MUONS

8.3.1 Thermodynamics and susceptibilities

Adding muons to the standard npe matter modifies properties of the EoS like the pressure and particle fractions, but also introduces new chemical equilibration channels. In $npe\mu$ matter, the baryon density is still given by $n_B = n_n + n_p$ and the charge neutrality condition is now $n_p = n_e + n_\mu$. Therefore, $npe\mu$ matter has two independent particle species, chosen here to be the proton and muon. There are now two beta equilibrium conditions (with all of the same issues observed in footnote 7, as all flavor-changing reactions involve neutrinos or antineutrinos, which are free-streaming),

$$\delta\mu_1 \equiv \mu_n - \mu_p - \mu_e = 0 \tag{8.44a}$$

$$\delta\mu_2 \equiv \mu_n - \mu_p - \mu_\mu = 0. \tag{8.44b}$$

Another natural choice would be $\mu_e = \mu_\mu$, but this condition is already implied by the two chosen conditions. The thermodynamic formulas from Sec. 8.2.2 pick up terms with muons. I will present them quickly here. The first law of thermodynamics is

$$dE = -P\,dV + T\,dS + \mu_n\,dN_n + \mu_p\,dN_p + \mu_e\,dN_e + \mu_\mu\,dN_\mu\,. \tag{8.45}$$

The Euler equation is

$$\varepsilon + P = sT + \mu_n n_n + \mu_p n_p + \mu_e n_e + \mu_\mu n_\mu = sT + \mu_n n_B - n_p \delta\mu_1 + n_\mu \left(\delta\mu_1 - \delta\mu_2\right), \quad (8.46)$$

The first law, in terms of intensive quantities, is

$$d\varepsilon = T\, ds + \mu_n\, dn_B - \delta\mu_1\, dn_p + (\delta\mu_1 - \delta\mu_2)\, dn_\mu\,. \quad (8.47)$$

The Gibbs-Duhem equation is

$$dP = s\, dT + n_B\, d\mu_n + (n_\mu - n_p)\, d\delta\mu_1 - n_\mu\, d\delta\mu_2\,. \quad (8.48)$$

The first law, in terms of quantities normalized by baryon number N_B, is

$$d\left(\frac{\varepsilon}{n_B}\right) = \frac{P}{n_B^2}\, dn_B + T\, d\left(\frac{s}{n_B}\right) - \delta\mu_1\, dx_p + (\delta\mu_1 - \delta\mu_2)\, dx_\mu\,. \quad (8.49)$$

With a Legendre transformation to make temperature, not entropy per baryon, the control parameter, Eq. 8.47 and 8.49 become

$$d\left(\varepsilon - sT\right) = \mu_n\, dn_B - s\, dT - \delta\mu_1\, dn_p + (\delta\mu_1 - \delta\mu_2)\, dn_\mu \quad (8.50)$$

$$d\left(\frac{\varepsilon - sT}{n_B}\right) = \frac{P}{n_B^2}\, dn_B - \frac{s}{n_B}\, dT - \delta\mu_1\, dx_p + (\delta\mu_1 - \delta\mu_2)\, dx_\mu\,. \quad (8.51)$$

In $npe\mu$ matter, there are many more susceptibilities. They are defined as

$$A_i \equiv n_B \frac{\partial \delta\mu_i}{\partial n_B}\bigg|_{T, x_p, x_\mu} \quad (8.52a)$$

$$B_i \equiv \frac{1}{n_B}\frac{\partial \delta\mu_i}{\partial x_p}\bigg|_{T, n_B, x_\mu} \quad (8.52b)$$

$$C_i \equiv \frac{1}{n_B}\frac{\partial \delta\mu_i}{\partial x_\mu}\bigg|_{T, n_B, x_p}, \quad (8.52c)$$

where $i = 1, 2$. At constant temperature, three Maxwell relations can be derived from Eq. 8.51

$$A_1 = -\frac{1}{n_B}\frac{\partial P}{\partial x_p}\bigg|_{n_B, T, x_\mu} \quad (8.53a)$$

$$A_1 - A_2 = \frac{1}{n_B}\frac{\partial P}{\partial x_\mu}\bigg|_{n_B, T, x_p} \quad (8.53b)$$

$$C_1 = B_2 - B_1. \quad (8.53c)$$

Eq. 8.53c tells us that only five of the susceptibilities, say A_1, A_2, B_1, B_2, and C_2 are independent. The susceptibilities in $npe\mu$ matter are plotted in Fig. 8.7. Note that indeed $C_1 = B_2 - B_1$

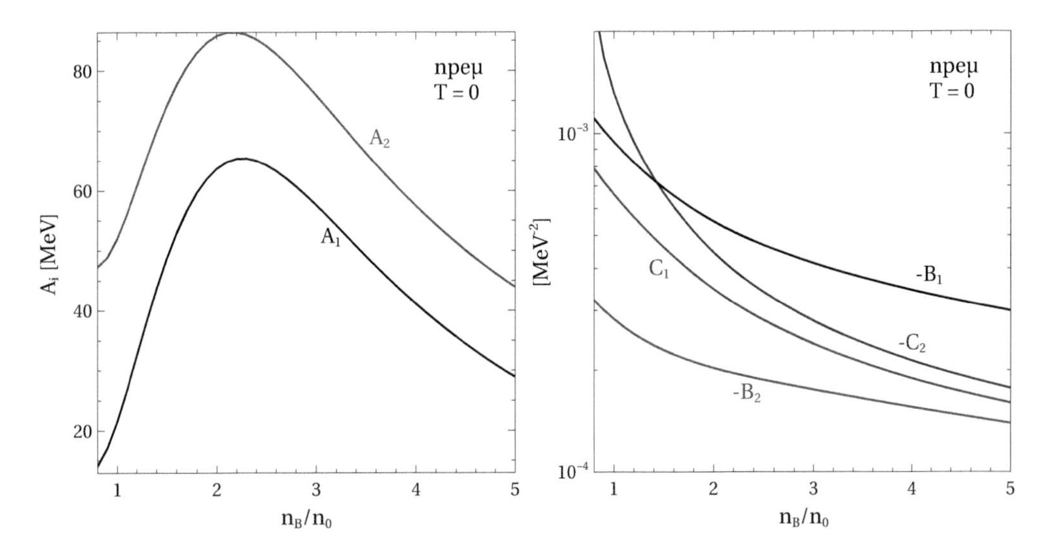

Figure 8.7 Isothermal susceptibilities in $npe\mu$ matter, calculated at zero temperature. The temperature dependence is small.

8.3.2 Bulk viscosity and beta equilibration

The bulk viscosity of $npe\mu$ matter has been calculated in detail in neutrino-transparent matter [8, 9] as well as in neutrino-trapped matter [7, 8]. I present here a simple calculation of the $npe\mu$ bulk viscosity, where, as in the npe bulk viscosity calculation, I leave out complications coming from the full phase space integration of the rates and the modification of the beta equilibrium conditions due to neutrino transparency (as described in footnote 7.) Any change in the temperature throughout a density oscillation is neglected, just as in the npe case.

The derivation of bulk viscosity in $npe\mu$ matter proceeds in a very similar manner to that in npe matter, and the expression for the bulk viscosity (Eq. 8.20)

$$\zeta = \left(\frac{n_B}{\delta n_B}\right)\frac{\Im(\delta P)}{\omega} \tag{8.54}$$

still holds. The baryon density, pressure, and two independent particle fractions can be expanded

$$n_B(t) = n_B + \Re(\delta n_B e^{i\omega t}) = n_B + \delta n_B \cos(\omega t), \tag{8.55a}$$
$$P(t) = P_0 + \Re(\delta P e^{i\omega t}) = P_0 + \Re(\delta P)\cos(\omega t) - \Im(\delta P)\sin(\omega t), \tag{8.55b}$$
$$x_p(t) = x_p^0 + \Re(\delta x_p e^{i\omega t}) = x_p^0 + \Re(\delta x_p)\cos(\omega t) - \Im(\delta x_p)\sin(\omega t), \tag{8.55c}$$
$$x_\mu(t) = x_\mu^0 + \Re(\delta x_\mu e^{i\omega t}) = x_\mu^0 + \Re(\delta x_\mu)\cos(\omega t) - \Im(\delta x_\mu)\sin(\omega t). \tag{8.55d}$$

TABLE 8.1 Chemical equilibration mechanisms at work in $npe\mu$ matter. While the "direct" versions of the chemical processes are written in the middle column, the modified versions of the processes (i.e., those with spectator particles) operate too, and in the conditions discussed in this chapter, are the principle operating processes because the direct processes are kinematically forbidden (though see footnote 13.)

Equilibrating $\delta\mu$	Reactions	$\overrightarrow{\Gamma} - \overleftarrow{\Gamma}$ (subthermal)
1) $\delta\mu_1 \equiv \mu_n - \mu_p - \mu_e$	$n \rightarrow p + e^- + \bar{\nu}_e$ $e^- + p \rightarrow n + \nu_e$	$\lambda_1 \delta\mu_1$
2) $\delta\mu_2 \equiv \mu_n - \mu_p - \mu_\mu$	$n \rightarrow p + \mu^- + \bar{\nu}_\mu$ $\mu^- + p \rightarrow n + \nu_\mu$	$\lambda_2 \delta\mu_2$
3) $\delta\mu_3 = \mu_\mu - \mu_e = \delta\mu_1 - \delta\mu_2$	$\mu^- \rightarrow e^- + \nu_\mu + \bar{\nu}_e$ $e^- \rightarrow \mu^- + \nu_e + \bar{\nu}_\mu$	$\lambda_{\mu e}(\delta\mu_1 - \delta\mu_2)$

The pressure $P(n_B, T, x_p, x_\mu)$ can be expanded around its equilibrium value P_0

$$\begin{aligned} P &= P_0 + \left.\frac{\partial P}{\partial n_B}\right|_{T,x_p,x_\mu} \Re(\delta n_B e^{i\omega t}) + \left.\frac{\partial P}{\partial x_p}\right|_{T,n_B,x_\mu} \Re(\delta x_p e^{i\omega t}) + \left.\frac{\partial P}{\partial x_\mu}\right|_{T,n_B,x_p} \Re(\delta x_\mu e^{i\omega t}), \\ &= P_0 + \frac{1}{\kappa_T}\frac{\delta n_B}{n_B}\cos(\omega t) - n_B A_1 \left[\Re(\delta x_p)\cos(\omega t) - \Im(\delta x_p)\sin(\omega t)\right] \\ &+ n_B(A_1 - A_2)\left[\Re(\delta x_\mu)\cos(\omega t) - \Im(\delta x_\mu)\sin(\omega t)\right]. \end{aligned} \tag{8.56}$$

Matching sine and cosine terms of Eqs. 8.55b and 8.56 yields

$$\Re(\delta P) = \frac{1}{\kappa_T}\frac{\delta n_B}{n_B} - n_B A_1 \Re(\delta x_p) + n_B(A_1 - A_2)\Re(\delta x_\mu) \tag{8.57a}$$

$$\Im(\delta P) = -n_B A_1 \Im(\delta x_p) + n_B(A_1 - A_2)\Im(\delta x_\mu). \tag{8.57b}$$

To solve for the particle fractions, we need to study the flavor-changing interactions in the $npe\mu$ system. The possible reactions are summarized in Table 8.1. The two independent particle fractions x_p and x_μ evolve according to the equations (jumping immediately to the subthermal limit)

$$n_B \frac{dx_p}{dt} = \lambda_1 \delta\mu_1 + \lambda_2 \delta\mu_2, \tag{8.58a}$$

$$n_B \frac{dx_\mu}{dt} = -\lambda_{\mu e}\delta\mu_1 + (\lambda_2 + \lambda_{\mu e})\delta\mu_2. \tag{8.58b}$$

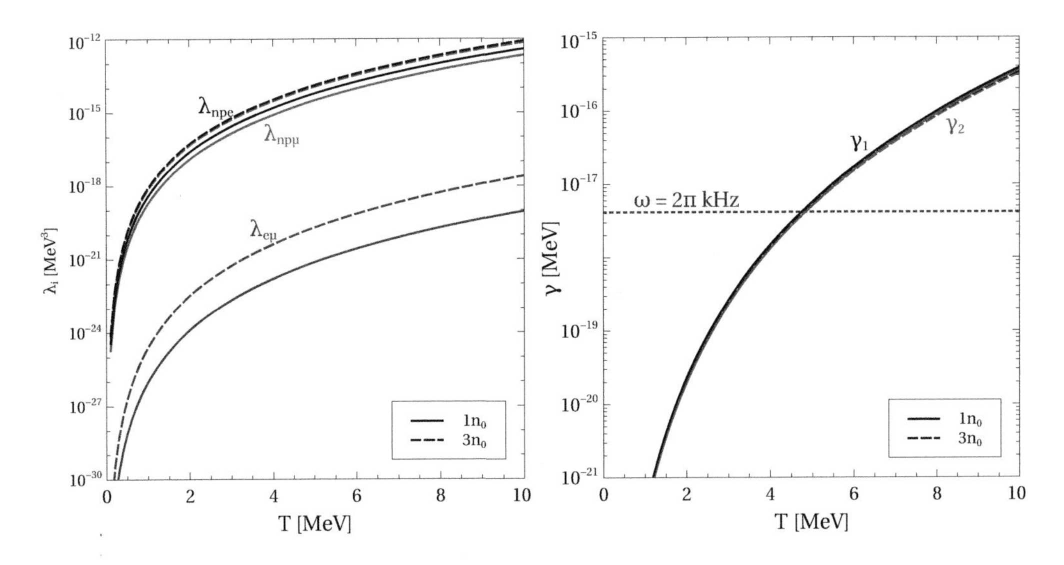

Figure 8.8 Left: Beta reaction rates λ versus temperature (λ_{npe} is λ_1 and $\lambda_{np\mu}$ is λ_2.) Right: Beta equilibration rates γ of the partial bulk viscosities. They are defined as $\gamma_1 \equiv |B_1|\lambda_1$ and $\gamma_2 \equiv |B_2 + C_2|\lambda_2$ and should be compared to the density oscillation frequency, which in neutron star mergers is typically around 1 kHz ($\omega = 2\pi$ kHz).

The subthermal rates λ_i are

$$\lambda_1 = \frac{367}{1152\pi^3} G^2 g_A^2 f^4 \frac{m_n^4}{m_\pi^4} \frac{p_{Fn}^4 p_{Fp}}{(p_{Fn}^2 + m_\pi^2)^2} T^6 \theta_n. \tag{8.59a}$$

$$\lambda_2 = \frac{367}{1152\pi^3} G^2 g_A^2 f^4 \frac{m_n^4}{m_\pi^4} \frac{p_{Fn}^4 p_{Fp}}{(p_{Fn}^2 + m_\pi^2)^2} \frac{p_{F\mu}}{\sqrt{p_{F\mu}^2 + m_\mu^2}} T^6 \theta_n. \tag{8.59b}$$

The expression[13] for $\lambda_{\mu e}$ is given in terms of a convenient curve fit in Eqs. 32-35 in Alford & Good [13]. In the expression for λ_2, the term θ_n is defined in Eq. 8.28, but with p_{Fe} replaced by $p_{F\mu}$. The functions λ_i are plotted in the left panel of Fig. 8.8 in matter described by the IUF EoS.

[13]In neutrino-transparent matter, the process $\mu^- \to e^- + \nu_\mu + \bar{\nu}_e$ is actually allowed, but the reverse process $e^- \to \mu^- + \bar{\nu}_\mu + \nu_e$ is not, which could complicate the analysis here [9]. I neglect both of these "direct" processes in this chapter.

The deviations from chemical equilibrium can be expanded around equilibrium and written in terms of the susceptibilities

$$\delta\mu_1 = A_1 \frac{\delta n_B}{n_B} \cos(\omega t) + n_B B_1 \left[\Re(\delta x_p)\cos(\omega t) - \Im(\delta x_p)\sin(\omega t)\right], \tag{8.60a}$$

$$+ n_B C_1 \left[\Re(\delta x_\mu)\cos(\omega t) - \Im(\delta x_\mu)\sin(\omega t)\right]$$

$$\delta\mu_2 = A_2 \frac{\delta n_B}{n_B} \cos(\omega t) + n_B B_2 \left[\Re(\delta x_p)\cos(\omega t) - \Im(\delta x_p)\sin(\omega t)\right], \tag{8.60b}$$

$$+ n_B C_2 \left[\Re(\delta x_\mu)\cos(\omega t) - \Im(\delta x_\mu)\sin(\omega t)\right].$$

In the same manner as in the previous section, Eqs. 8.58a and 8.58b lead to a system of equations for the real and imaginary parts of δx_p and δx_μ. The solutions are proportional to $\delta n_B/n_B$ and are functions of the five independent susceptibilities, the three rates λ_i, and the density oscillation frequency ω, but are too complicated to be shown here. The bulk viscosity is obtained with Eqs. 8.54 and 8.57b and is given by

$$\zeta_{npe\mu} = \frac{F + G\omega^2}{H + J\omega^2 + \omega^4}, \tag{8.61}$$

with

$$F = \left[\lambda_2\lambda_{\mu e} + \lambda_1(\lambda_2 + \lambda_{\mu e})\right] \tag{8.62a}$$
$$\times \left\{(A_1B_2 - A_2B_1)^2\lambda_1 + \left[(A_1 - A_2)B_2 + A_1C_2\right]^2\lambda_2 + \left[A_2(B_1 - B_2) + A_1C_2\right]^2\lambda_{\mu e}\right\},$$
$$G = A_1^2\lambda_1 + A_2^2\lambda_2 + (A_1 - A_2)^2\lambda_{\mu e}, \tag{8.62b}$$
$$H = \left[\lambda_2\lambda_{\mu e} + \lambda_1(\lambda_2 + \lambda_{\mu e})\right]^2\left[B_2^2 - B_1(B_2 + C_2)\right]^2, \tag{8.62c}$$
$$J = B_1^2\lambda_1^2 + 2B_2^2\lambda_1\lambda_2 + (B_2 + C_2)^2\lambda_2^2 + 2(B_1 - B_2)^2\lambda_1\lambda_{\mu e} + 2C_2^2\lambda_2\lambda_{\mu e}$$
$$+ (B_2 - B_1 - C_2)^2\lambda_{\mu e}^2. \tag{8.62d}$$

This form of the bulk viscosity, $\zeta_{npe\mu} = (F + G\omega^2)/(H + J\omega^2 + \omega^4)$, is present in other systems with two equilibrating quantities, such as neutrino-trapped $npe\mu$ matter [7, 8] and strange quark matter phases [23, 121, 141].

Figure 8.8 indicates that the rate $\lambda_{\mu e}$ is much slower than the other rates. In fact, setting $\lambda_{\mu e} = 0$ in Eq. 8.61 does not affect the bulk viscosity at all in the conditions studied here. The bulk viscosity with $\lambda_{\mu e} = 0$ is

$$\zeta_{npe\mu} = \frac{\lambda_1\lambda_2\left\{(A_1B_2 - A_2B_1)^2\lambda_1 + \left[(A_1 - A_2)B_2 + A_1C_2\right]^2\lambda_2\right\} + (A_1^2\lambda_1 + A_2^2\lambda_2)\omega^2}{\lambda_1^2\lambda_2^2\left[B_2^2 - B_1(B_2 + C_2)\right]^2 + \left[B_1^2\lambda_1^2 + 2B_2^2\lambda_1\lambda_2 + (B_2 + C_2)^2\lambda_2^2\right]\omega^2 + \omega^4}, \tag{8.63}$$

which of course still has its $\zeta_{npe\mu} = (F + G\omega^2)/(H + J\omega^2 + \omega^4)$ form (with different F, G, H, J) because there are still two equilibrating quantities (c.f. Eq. 8.58b with $\lambda_{\mu e} = 0$.) In interpreting Eq. 8.63, it is useful to define the "partial" bulk viscosities — the bulk viscosity due to each reaction λ_i, with $\lambda_{j\neq i} = 0$. We have

$$\zeta_1 \equiv \frac{A_1^2\lambda_1}{B_1^2\lambda_1^2 + \omega^2} = \frac{A_1^2}{|B_1|}\frac{\gamma_1}{\gamma_1^2 + \omega^2} \tag{8.64a}$$

$$\zeta_2 \equiv \frac{A_2^2\lambda_2}{(B_2 + C_2)^2\lambda_2^2 + \omega^2} = \frac{A_2^2}{|B_2 + C_2|}\frac{\gamma_2}{\gamma_2^2 + \omega^2}, \tag{8.64b}$$

where in analogy with Eq. 8.33, the beta equilibration rates γ_i in each partial bulk viscosity are defined

$$\gamma_1 \equiv |B_1|\lambda_1, \tag{8.65a}$$

$$\gamma_2 \equiv |B_2 + C_2|\lambda_2. \tag{8.65b}$$

These equilibration rates γ_i are plotted in the right panel of Fig. 8.8. The two rates γ_1 and γ_2 are nearly equal across the range of temperatures and densities shown in the plot, and they match the 1 kHz frequency of the density oscillation at $T \approx 4 - 5$ MeV.

Eq. 8.64a is just the bulk viscosity in npe matter equilibrating by the Urca process (with electrons) described by Eq. 8.32. In the high-frequency limit the bulk viscosity (Eq. 8.63) expression decouples

$$\zeta_{npe\mu}^{\gamma_i \ll \omega} = \frac{A_1^2\lambda_1 + A_2^2\lambda_2}{\omega^2} = \zeta_1^{\gamma \ll \omega} + \zeta_2^{\gamma \ll \omega} \tag{8.66}$$

into a sum of two individual contributions. The high-frequency limit is applicable in cold neutron stars, which explains why many of the early bulk viscosity papers, which focused on cold neutron stars, considered the total bulk viscosity of a multicomponent system to be a sum of all of the individual contributions (e.g. [76, 77, 79].)

The bulk viscosity of $npe\mu$ matter undergoing a 1 kHz density oscillation is plotted in the left panel of Fig. 8.9. The dashed lines show the results without muons, and the solid lines with muons included (both in the EoS and in equilibration processes). There is still essentially one resonance, which at first thought is surprising, since there are two equilibrating quantities, but this behavior will be explained shortly. The presence of muons is found to increase the bulk viscosity by about a factor of 2-5 at the resonant maximum. The resonance occurs roughly at the same temperature, because adding muons to the system does not significantly change the equilibration rates in the npe sector, and the equilibration rates in the $np\mu$ sector are very close to the npe rates (see the right panel of Fig. 8.8.) The dissipation timescale is shown in the right panel of Fig. 8.9. Matter with muons has a strong enough bulk viscosity to damp density oscillations in less than 10 ms. The quality factor of kHz oscillations in $npe\mu$ matter can be as low as 10.

To further understand the behavior of the bulk viscosity curve, Fig. 8.10 shows the bulk viscosity in $npe\mu$ matter and also the partial bulk viscosities ζ_1 and ζ_2 ($\zeta_{\mu e}$, if defined in an analogous way to the other partial bulk viscosities, is much smaller than ζ_1 and ζ_2.) From this plot, it is clear that the total bulk viscosity is *not* the sum of the partial bulk viscosities

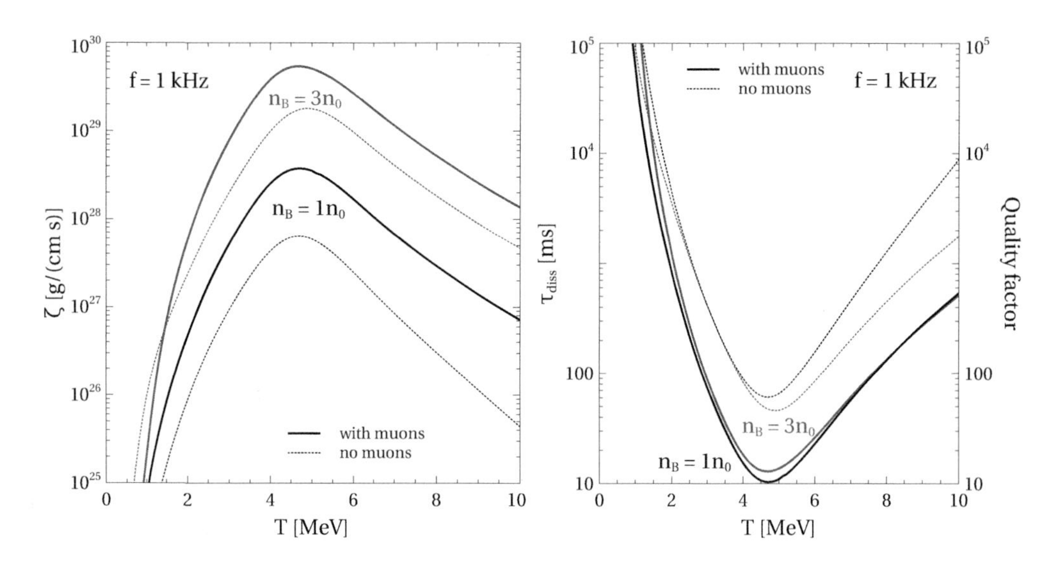

Figure 8.9 Left: Bulk viscosity in $npe\mu$ matter subjected to a harmonic, small-amplitude density oscillation of angular frequency $\omega = 2\pi$ kHz. Right: The timescale of energy dissipation of the oscillation due to the bulk viscosity of $npe\mu$ matter, as well as the quality factor of the oscillation.

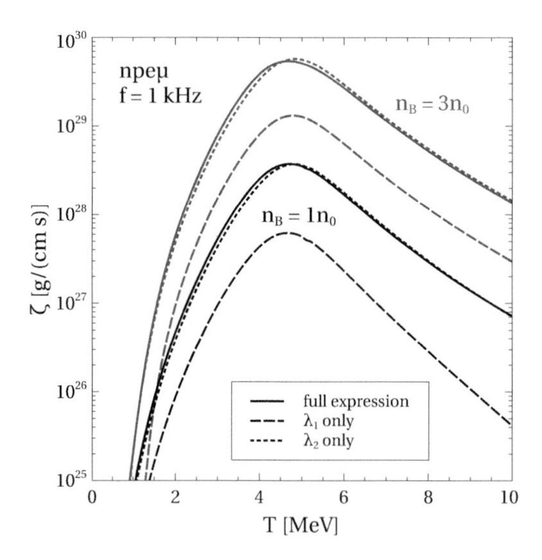

Figure 8.10 Bulk viscosity in $npe\mu$ matter, compared with the partial bulk viscosities ζ_1 and ζ_2.

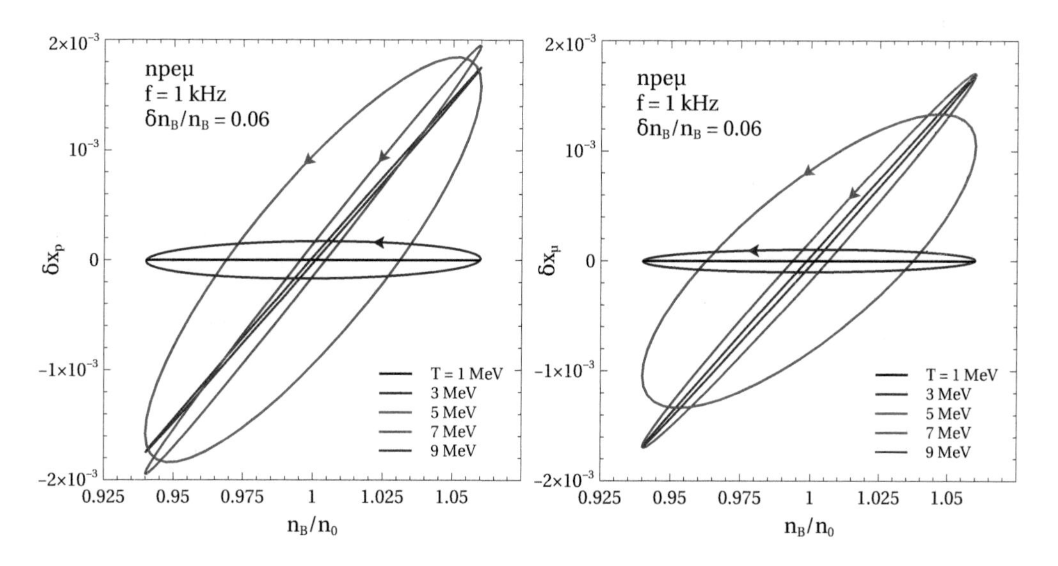

Figure 8.11 Path of a fluid element in the $x_p n_B$ plane (left) and the $x_\mu n_B$ plane (right), over the course of a complete period of the density oscillation. Five trajectories, each at a different temperature (leading to different beta equilibration rates), are shown.

(except in the high-frequency limit, which corresponds to temperatures below ≈ 4 MeV on this plot. The right panel of Fig. 8.8 indicates that the resonances of the partial bulk viscosities ζ_1 and ζ_2 both lie near $T = 4 - 5$ MeV. In the FS approximation used here for the rates, the electron mUrca rate is nearly identical to the muon mUrca rate, because the mUrca rates depend on the lepton type by just a slowly varying function of the lepton density (Eq. 8.59a and 8.59b.) In reality, the below-threshold contribution of the direct Urca rates, which is significant for $T \gtrsim 1$ MeV [15, 18], would split the near degeneracy between the rates [9].

The behavior of the total bulk viscosity near the two overlapping resonances of ζ_1 and ζ_2 is difficult to predict because the resonances interact in a complicated way. The height of the resonance of ζ_2 is larger than ζ_1 because the susceptibility combination $A_2^2/|B_2 + C_2|$ is several times larger than $A_1^2/|B_1|$. Because of this dominance of ζ_2, the total bulk viscosity follows ζ_2 quite closely, and the peak of the curve shifts only slightly to the left, representing the influence of the resonance of ζ_1.

In Fig. 8.11 is shown the path of a fluid element in the $x_p n_B$ (left panel) and $x_\mu n_B$ (right panel) planes over the course of a full oscillation period. The behavior is very much like that observed in the *npe* system. At low temperature, the equilibration rates are very slow and the particle fractions remain close to unchanged throughout the oscillation. As the temperature increases above 1 MeV, the Urca reactions are fast enough to make progress toward establishing beta equilibrium, but the particle fractions still lag behind the density change, creating a curve with finite area in the $x_i n_B$ plane. The area, and thus the dissipation, is maximal at temperatures near 5 MeV, and when the temperature exceeds that, the Urca processes are fast enough to keep the system close to beta equilibrium throughout the oscillation.

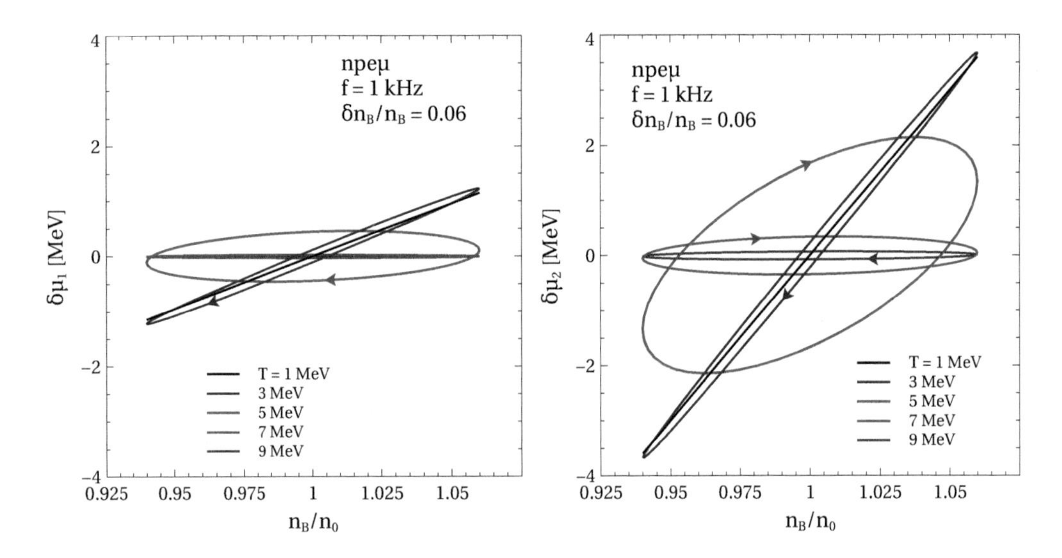

Figure 8.12 Path of a fluid element in the $\delta\mu_1 n_B$ plane (left) and the $\delta\mu_2 n_B$ plane (right) over the course of a complete period of the density oscillation. Five trajectories, each at a different temperature (leading to different beta equilibration rates), are shown.

Fig. 8.12, which plots the path of a fluid element in the $\delta\mu_1 n_B$ (left panel) and $\delta\mu_2 n_B$ plans, shows very similar information. When the system is at high temperature and is efficiently chemically equilibrated, $\delta\mu$ deviates little from zero. When the temperature is low and the system struggles to chemically equilibrate, $\delta\mu$ is able to grow large. In fact, for temperatures of 1 and 3 MeV, at least one of the $\delta\mu$ values exceeds the temperature, and thus the system is no longer in the subthermal limit and the results in the plot cannot be trusted.

8.4 NEUTRINO-TRAPPED NUCLEAR MATTER

The neutrino MFP in dense matter decreases significantly as the temperature increases. At temperatures above about 5 MeV, the MFP[14] falls below a kilometer [18, 118] and neutrinos can be considered trapped inside a ten-kilometer-scale object like a neutron star. In this case, the nuclear matter picks up a new conserved quantity, the total lepton number $Y_L \equiv (n_e + n_\nu)/n_B$. The value of the conserved lepton fraction Y_L is determined by the history of the system, in the same way that the conserved baryon number N_B is. A neutron star in isolation produces neutrinos, but they escape easily due to the low temperature. When two neutron stars merge, much of the matter is heated to temperatures of tens of MeV, trapping any neutrinos produced afterwords. Whatever the lepton number is in a fluid element at the point when it begins to trap, neutrinos becomes conserved.

[14]The neutrino MFP is actually a function of neutrino energy. It is common practice to, at a particular density and temperature, calculate the MFP of a "typical" neutrino with energy $E_\nu \approx 3T$ (the average energy of massless fermions with a Fermi-Dirac distribution at zero chemical potential is $\langle E \rangle = 7\pi^4 T/[180\zeta(3)] \approx 3.15T$) and call that *the* neutrino MFP.

Matter with trapped neutrinos only has one independent particle fraction, say x_p, because the addition of neutrinos comes with the addition of a constraint (conservation of lepton number). Neutrino-trapped nuclear matter, let us say with a net electron neutrino population (as opposed to antineutrinos), achieves chemical equilibrium through the direct Urca electron capture processes

$$n + \nu \leftrightarrow e^- + p. \tag{8.67}$$

Neutron decay and its inverse, as well as modified Urca processes are slower and can be neglected in the case of matter with a net neutrino population. In the limit of strongly degenerate matter (note that this means a substantial neutrino population with Fermi momentum $p_{F\nu}$), the rate of the direct Urca electron capture process in the subthermal limit is [6]

$$\lambda_{\text{dUrca, trapped}} = \frac{1}{12\pi^3} G_F^2 \cos^2\theta_C (1 + 3g_A^2) E^*_{Fn} E^*_{Fp} p_{Fe} p_{F\nu} (p_{Fp} + p_{Fe} + p_{F\nu} - p_{Fn}) T^2. \tag{8.68}$$

This rate is very fast and has no kinematic threshold. For comparison, the subthermal direct Urca rate in neutrino-transparent nuclear matter is [6, 19]

$$\lambda_{\text{dUrca, transparent}} = \frac{17}{240\pi} G_F^2 \cos^2\theta_C (1 + 3g_A^2) E^*_{Fn} E^*_{Fp} p_{Fe} T^4 \theta_{\text{dUrca}}, \tag{8.69}$$

where θ_{dUrca} is one if $p_{Fn} < p_{Fp} + p_{Fe}$ and zero otherwise. The ratio of these two rates (when $\theta_{\text{dUrca}} = 1$) is

$$\frac{\lambda_{\text{dUrca, trapped}}}{\lambda_{\text{dUrca, transparent}}} = \frac{20}{17\pi^2} \frac{p_{F\nu}(p_{Fp} + p_{Fe} + p_{F\nu} - p_{Fn})}{T^2}, \tag{8.70}$$

which is typically much larger than one. Therefore, neutrino-trapped matter equilibrates much faster than even dUrca-equilibrated neutrino-transparent matter and therefore assuming the susceptibilities are not changed dramatically with the addition of neutrinos[15], the bulk-viscous resonance likely occurs at a temperature well below 1 MeV. In fact, calculations by Alford, Harutyunyan, & Sedrakian [7] confirm that this is true. Their calculation includes all relevant neutrino and antineutrino processes, and the rate calculations go beyond the FS approximation by doing the full phase space integration. The neutrino-trapped regime is $T \gtrsim 5$ MeV, and so the bulk viscosity is very far from the resonant peak. The calculations in [7] indicate that the bulk viscosity in neutrino-trapped matter is only strong enough to damp density oscillations on the timescale of seconds or longer.

8.5 EFFECTS OF BULK VISCOSITY IN NEUTRON STAR MERGERS

Neutron star mergers, both in the inspiral and postmerger phase, offer an interesting environment to search for signatures of bulk-viscous dissipation because not only is the matter likely to experience changes in density (ranging from small amplitude oscillations to catastrophic changes in density during a potential collapse to a black hole), but because we can see both gravitational and electromagnetic signals from neutron star mergers, giving the community many tools with which to search for dissipative effects.

[15]Independent of the issue of neutrino-trapped versus neutrino-transparent susceptibilities, at temperatures of tens of MeV, the isothermal and adiabatic susceptibilities differ significantly from each other [8, 33].

8.5.1 Neutron star inspiral

In a typical neutron star binary, the stars orbit around their common center of mass for millions of years, emitting gravitational radiation that takes energy away from their orbit, causing them to grow closer together over time [26]. Eventually, the two stars get close enough to each other that they no longer are effectively point objects, and tidal forces become significant. By the time, the neutron stars are close enough to tidally interact, they (typically) have been cooling via neutrino and photon emission for long enough to have core temperatures around $10^6 - 10^7$ K [114].

Arras & Weinberg [29] studied the role that weak interactions play in dissipating energy in tidally interacting, inspiraling neutron stars. Tidal interactions excite oscillation modes within a star, pushing it out of beta equilibrium. The matter can deviate significantly from beta equilibrium because the Urca rates in $T \ll 1\,\mathrm{MeV}$ matter are much slower than the dynamical timescale of milliseconds. Nevertheless, in response, the Urca processes turn on, attempting to reestablish beta equilibrium, and the neutrino emission is enhanced. In these cold neutron stars, even a very small amplitude density oscillation (say, $\delta n_B/n_B \sim 0.01$) is not subthermal (this should seem plausible from examining Fig. 8.6, though the smallest temperature considered in the plot is around 10^9 K), and therefore one must use the full expression for the Urca rates out of beta equilibrium (e.g. Eq. 8.26 or see the appendix of [106].) As described in Sec. 8.3 and 8.3 in this chapter, the chemical equilibration processes result in PdV work done on the fluid element, leading to heating. In the subthermal limit, the extra cooling from the enhanced neutrino emission wins over the heating, and the fluid element still cools, though more slowly than if it were not vibrating. Suprathermal oscillations can actually lead to so much PdV heating that the fluid element undergoes net heating due to the vibration. This issue is discussed in [56, 59, 140]. Arras & Weinberg find that tidal heating from direct Urca processes can heat an inspiraling neutron star to an average core temperature of, at most, 10 keV. They find little alteration of the inspiral from the tidal excitations.

Hyperonic reactions like $n + p \leftrightarrow \Lambda + p$ have an even faster rate than direct Urca, so they may be able to enhance the tidal heating. Ghosh, Pradhan, & Chatterjee [66] found, with the assumption of tidal oscillations that remain subthermal[16], that hyperonic reactions can lead to tidal heating of the star to temperatures greater than 100 keV and potentially measurable phase shifts in the inspiral gravitational wave signal.

8.5.2 Neutron star merger and remnant

What we know about neutron star merger remnant dynamics at this point comes almost entirely from numerical simulations [31], as no gravitational wave signal from this phase of the merger has yet been observed due to its high frequency, where the LIGO detector is less sensitive [30]. The numerical simulations evolve Einstein's equations coupled to relativistic hydrodynamics. The EoS describing the neutron star matter [109] could be a

[16]The relationship between the density oscillation magnitude δn_B and the deviation from chemical equilibrium $\delta\mu$ depends on susceptibilities of the EoS (see Eq. 8.43 for the relationship in npe matter, which involves the susceptibility A) and the speed of the strangeness equilibration γ with respect to the density oscillation frequency ω. I do not know if tidal oscillations would lead to subthermal or suprathermal deviations from strangeness equilibrium.

simple analytic parameterization, like a polytrope, or it could be the tabulated result of a nuclear theory calculation (see the CompOSE repository [136].) Unlike the neutrons, protons, and electrons, which are considered to be a multicomponent fluid whose chemical composition is described by the proton fraction x_p, in much of the merger remnant the neutrino MFP is not short enough for the neutrinos to be considered part of the npe fluid. Intricate transport schemes [61] are devised to handle the neutrinos.

Merger simulations mostly agree upon the following physical picture following the end of the inspiral phase [30, 31]. When two neutron stars merge, the matter at the interface undergoes dramatic heating due to the compression and from shocks [112]. Its temperature may reach many tens of MeV. The two dense cores retain their structural integrity for some time, bouncing off of each other while gravitational radiation, and perhaps other dissipative mechanisms, damp the energy of the oscillation, bringing them together within ten milliseconds [42, 107]. As the cores lose their structural integrity, what remains is a differentially rotating mass of dense matter, with a density profile that decreases from core to edge, but with a temperature profile that is peaked at a few kilometers away from the center. The densest region of the remnant remains cool, likely with a temperature at or below 10 MeV (see Fig. 4 in [84].) Neutrinos are continuously being produced by weak interactions in the dense matter, and in areas where the temperature increases beyond several MeV, they become trapped and move with the fluid element.

Tracer particles put into inviscid simulations of the merger remnant indicate the presence of density oscillations on millisecond timescales. The fluid elements in Fig. 8.13 start out at $2n_0$ and then experience wild density oscillations when the stars collide, some changing density by over 100% in the first few milliseconds. After five ms or so, the remnant has

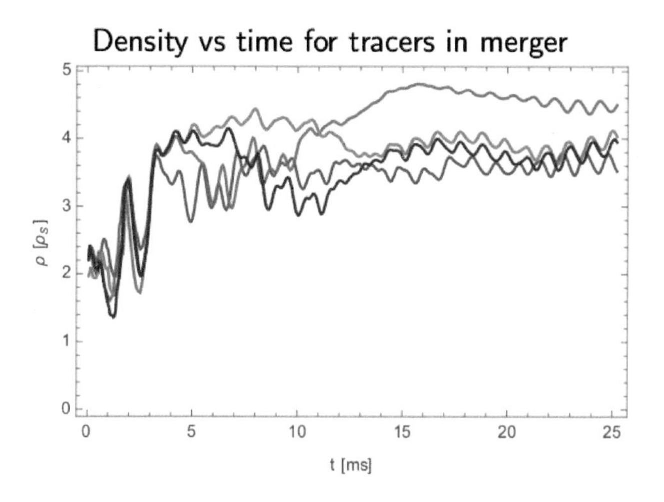

Figure 8.13 Time evolution of the density of several comoving fluid elements during a neutron star merger. The tracer particles initially experience large density oscillations right when the two neutron stars touch (t = 0), but after several milliseconds the oscillation amplitude decreases. The underlying simulation does not include viscosity. Figure courtesy of M. Hanauske and the Rezzolla group and originally displayed in [86]. A similar figure tracking density *and temperature* of tracer particles is given in Fig. 4 of Ref. [10].

settled down and the fluid element density oscillations become a lot closer to small amplitude, periodic oscillations around an equilibrium density. The timescale of the density changes matches the collision frequency of the two cores immediately after merger (see Fig. 3 in [107]) and the back-of-the-envelope estimate in footnote 11.

Because a neutrino transport scheme already exists in neutron star merger simulations, it makes sense to account for bulk viscosity through the reaction network that gives rise to it, if possible, instead of treating it with a Muller-Israel-Stewart formalism. These two methods were compared by Camelio *et al.* [38, 39] in the case of oscillating neutron stars. Considering the reaction network approach avoids the need to pre-calculate the transport coefficient $\zeta(n_B, T, \omega)$, which necessarily assumes small amplitude density oscillations (even for the suprathermal case, the amplitude $\delta n_B \ll n_B$), which Fig. 8.13 indicates are only found after the first few ms post-collision.

The data presented in Fig. 8.14 are from an inviscid neutron star merger simulation [107]. The top row shows the temperatures encountered in the merger remnant at different time slices, as well as selected density contours. In the third panel, the two cores are distinct and are separated by lower density (and hotter) matter. The middle row depicts

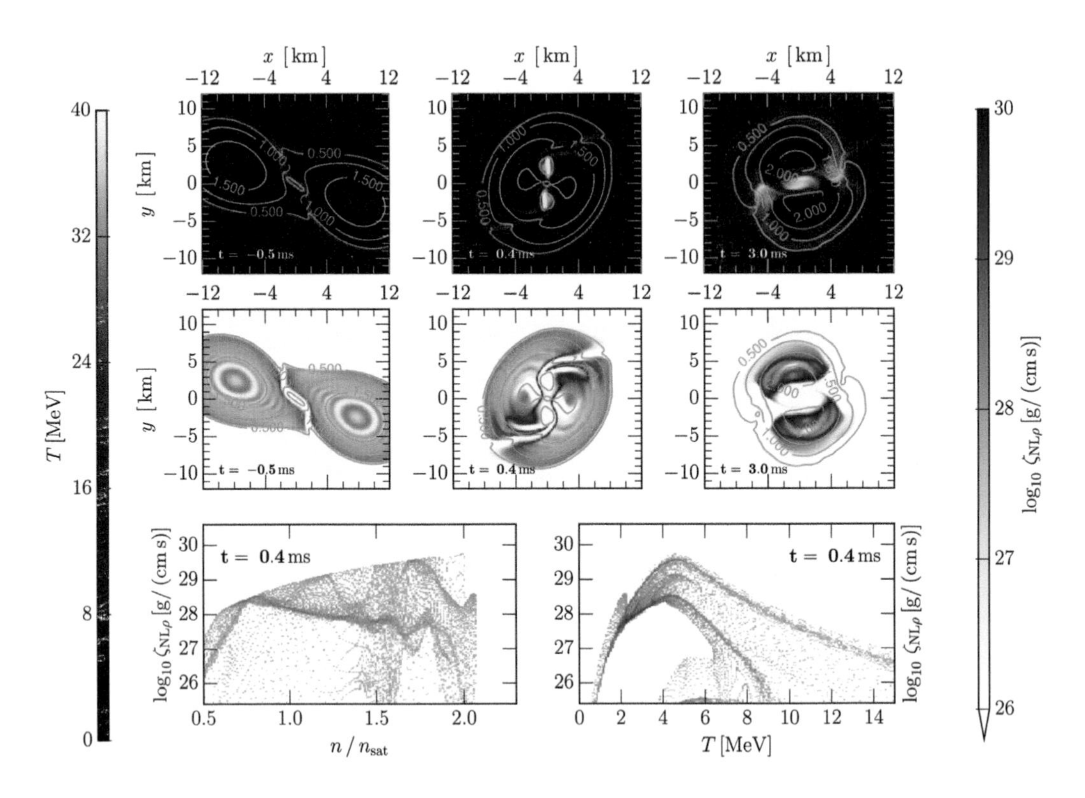

Figure 8.14 Post-processing of an inviscid neutron star merger simulation to predict the strength of the bulk viscosity. Top row: The temperature (left color scale) obtained in the equatorial plane of the merger remnant at three time slices. Center row: The subthermal bulk viscosity (right color scale), assuming a 1kHz density oscillation. Bottom row: Subthermal bulk viscosity of fluid elements in a merger at one time slice, plotted according to their density (left) and temperature (right). Figure reprinted from Most *et al.* [107].

the subthermal bulk viscosity calculated according[17] to Eq. 8.32, using the density and temperature of each fluid element in the equatorial slice. The oscillation angular frequency is assumed to be $\omega = 2\pi$ kHz, and the matter is assumed to be neutrino transparent at all temperatures. The bottom two panels plot the expected bulk viscosity of each fluid element in one time slice according to the density (left panel) and temperature (right panel) of the fluid element. The resonant behaviour of the subthermal, neutrino-transparent bulk viscosity as a function of temperature is apparent in the bottom right panel. This post-processed analysis shows that, neglecting the back reaction of the bulk viscosity and if the matter remains neutrino transparent, large values of the bulk viscosity are attainable in neutron star mergers. This figure represents, in some sense, the best-case scenario for bulk viscosity in merger remnants, because neutrino-trapping will reduce the bulk viscosity in the high-temperature regions of the remnant.

The implementation of advanced neutrino transport schemes (those that can accommodate out-of-beta-equilibrium physics) in merger simulations is still in its early stages, and simulations are not at the point of being able to identify the impact of weak interactions on energy dissipation. To study the edge cases, Hammond, Hawke, & Andersson [83] simulated two neutron star mergers, one with infinitely slow beta equilibration and one with infinitely fast beta equilibration. They found that due to the x_p-dependence of the pressure, the gravitational wave signals of these extreme cases are different, thus highlighting the need for simulations to properly take into account weak interactions. Along the same lines, Most *et al.* [106] implemented Urca reactions, calculated in the FS approximation, in a neutrino-transparent simulation and investigated the impact of the weak interactions on the gravitational wave signal. They found indications that weak interactions alter the gravitational wave signal and generate some entropy. These simulations are optimistic, as neutrino-trapping would set in at high temperatures and significantly decrease the bulk viscosity in these regions (see Sec. 8.4.)

The simulations searching for bulk-viscous effects with the most sophisticated neutrino transport scheme are those of Radice *et al.* [115] and Zappa *et al.* [149]. They compare the gravitational wave signals obtained in simulations with advanced neutrino transport schemes that consider chemical equilibration effects with simulations with simpler neutrino transport schemes that do not. They find that the two schemes produce different gravitational wave signals, but the origin of the difference is still under investigation. In all of these simulations, the resolution may not be high enough to resolve bulk-viscous effects. As the bulk viscosity originating from Urca processes peaks at $T \approx 5$ MeV [16], which is also around the same temperature at which neutrino trapping becomes important, diminishing the bulk viscosity, it seems clear that further investigation into neutrino transport schemes and out-of-equilibrium effects will be needed to truly understand the role of bulk viscosity in neutron star mergers.

A preview of the potential effects of bulk viscosity on postmerger dynamics was provided in a very recent work by Chabanov & Rezzolla [42]. They implemented bulk viscosity with the MIS formalism and assumed that bulk viscosity took a uniform value,

---

[17]The calculation of the equilibration rate γ in Ref. [107] is slightly different than in the npe section of this chapter. Direct and modified Urca are both included, as is some artificial blurring of the direct Urca threshold. See Sec. 2.1 in [107] for full details.

independent of density, temperature, and neutrino-trapping. If the bulk viscosity is larger[18] than a few times 10^{29} g/(cm s), then it can significantly modify the behavior of the merger remnant. They find that bulk viscosity damps the bouncing of two neutron star cores, as predicted in [107]. In addition, the bulk viscosity reduces the bar deformation of the remnant, suppressing the gravitational wave emission and increases the rotational energy of the remnant, raising the f_2 peak of the gravitational wave signal to higher frequencies.

8.6 CONCLUSIONS

Bulk viscosity dissipates energy from systems that experience changes in density, provided that they have an internal degree of freedom that is slow to equilibrate. In dense npe and $npe\mu$ matter, as we have seen, the particle fractions are that degree of freedom, equilibrating through various flavor-changing interactions. Bulk viscosity in a fluid element undergoing small amplitude oscillations is a function of the oscillation frequency and of the equilibration rates and when the two are comparable, maximal bulk-viscous dissipation occurs. This stems from the trajectory of the fluid element in the PV plane during one complete oscillation, where if the two rates are equal, then the pressure and volume are out of phase and the PdV work done on the fluid element is maximal.

Matter in neutron stars undergoes changes in density in many different situations. A newly born neutron star is likely endowed with oscillations resulting from its violent formation, and the radial oscillations are best damped by bulk viscosity. Rotating neutron stars generate r-mode oscillations which, at higher temperatures, are kept stable by bulk viscosity. Inspiraling neutron stars are tidally excited by their companion star and in the course of damping the ensuing oscillations, bulk viscosity can lead to heating of the stars. Neutron star mergers lead to a violently oscillating merger remnant, and bulk viscosity may be able to dissipate enough energy to alter the gravitational wave signal.

The use of neutron star mergers to understand transport properties in dense matter, including bulk viscosity, and from a different angle, the need to include effects like bulk viscosity into our simulations of mergers to better predict observables from neutron star mergers, has been the subject of intensive research over the past few years. If mergers can help us learn about bulk viscosity, this would provide insight into both the weak interaction rates in dense matter and also more obscure properties of the equation of state, like the susceptibilities. The community realizes the potential for this direction of research, and bulk viscosity in merger environments has been included in recent nuclear theory white papers [3, 99] and is part of the science case for Cosmic Explorer [53].

On the theory side of bulk viscosity, there is much work to be done. While it seems likely that bulk viscosity will be included in merger simulations through tracking of the particle degrees of freedom directly (e.g. Urca reactions in the neutrino transport scheme), calculations of the subthermal and suprathermal bulk viscosity are useful for obtaining an idea of the strength of bulk viscosity as a function of density and temperature. If the case looks promising, then the effort can be expended to implement the rates in a numerical simulation. Therefore, it is imperative to calculate the bulk viscosity in other phases of matter, including matter with thermal pions, a pion condensate, quarkyonic matter, and perhaps

[18]For context, this condition on the bulk viscosity is only met (in neutrino-transparent npe matter) in a limited temperature window of 3-6 MeV, though across a wide density range (c.f. [16].)

to revisit quark matter bulk viscosity at higher temperatures. Even in npe matter there are uncertainties. The direct Urca threshold is poorly constrained, and the susceptibilities are almost entirely unconstrained. The peak of bulk viscosity seems to lie at temperatures of around 5 MeV, but at this temperature the assumption of complete neutrino-transparency is likely invalid. It seems inevitable that improvements in neutrino-transport schemes (especially geared to degenerate matter above densities of n_0) will shed light on how bulk-viscous damping occurs as the neutrino MFP shrinks.

Finally, a wide variety of neutron star merger scenarios should be explored. Maybe bulk viscosity is most impactful in mergers with certain EoSs, certain mass ratios, or certain orbital eccentricities. And perhaps other astrophysical events entirely can be additional environments in which bulk viscosity plays a role. After all, the past few decades have seen the evolution of the focus of bulk viscosity research from the radial oscillations of newly born neutron stars, to the r-modes generated in rotating neutron stars, to the inspiral and merger of two neutron stars. Maybe bulk viscosity in other scenarios, like white dwarf mergers, white dwarf collapse to a neutron star, or the collapse of neutron stars to twin hybrid star configurations will be the focus of our future attention.

ACKNOWLEDGEMENTS

I thank Mark Alford, Alex Haber, and Elias Most for helpful comments on the chapter. I also want to thank my collaborators on bulk viscosity research: Mark Alford, Bryce Fore, Alex Haber, Elias Most, Jorge Noronha, Sanjay Reddy, and Ziyuan Zhang. My work is supported by the U.S. Department of Energy grant DE-FG02-00ER41132.

Bibliography

[1] B. P. Abbott et al. GW170817: Observation of Gravitational Waves from a Binary Neutron Star Inspiral. *Phys. Rev. Lett.*, 119(16):161101, 2017.

[2] B. P. Abbott et al. Multi-messenger Observations of a Binary Neutron Star Merger. *Astrophys. J. Lett.*, 848(2):L12, 2017.

[3] P. Achenbach et al. The Present and Future of QCD. 3 2023.

[4] Mark Alford, Matt Braby, M. W. Paris, and Sanjay Reddy. Hybrid stars that masquerade as neutron stars. *Astrophys. J.*, 629:969–978, 2005.

[5] Mark Alford, Arus Harutyunyan, and Armen Sedrakian. Bulk viscosity of baryonic matter with trapped neutrinos. *Phys. Rev. D*, 100(10):103021, 2019.

[6] Mark Alford, Arus Harutyunyan, and Armen Sedrakian. Bulk Viscous Damping of Density Oscillations in Neutron Star Mergers. *Particles*, 3(2):500–517, 2020.

[7] Mark Alford, Arus Harutyunyan, and Armen Sedrakian. Bulk viscosity from Urca processes: npeμ matter in the neutrino-trapped regime. *Phys. Rev. D*, 104(10):103027, 2021.

[8] Mark Alford, Arus Harutyunyan, and Armen Sedrakian. Bulk Viscosity of Relativistic npeμ Matter in Neutron-Star Mergers. *Particles*, 5(3):361–376, 2022.

[9] Mark Alford, Arus Harutyunyan, and Armen Sedrakian. Bulk viscosity from Urca processes: $npe\mu$ matter in the neutrino-transparent regime. 6 2023.

[10] Mark G. Alford, Luke Bovard, Matthias Hanauske, Luciano Rezzolla, and Kai Schwenzer. Viscous Dissipation and Heat Conduction in Binary Neutron-Star Mergers. *Phys. Rev. Lett.*, 120(4):041101, 2018.

[11] Mark G. Alford, Matt Braby, Sanjay Reddy, and Thomas Schafer. Bulk viscosity due to kaons in color-flavor-locked quark matter. *Phys. Rev. C*, 75:055209, 2007.

[12] Mark G. Alford, Matt Braby, and Andreas Schmitt. Bulk viscosity in kaon-condensed color-flavor locked quark matter. *J. Phys. G*, 35:115007, 2008.

[13] Mark G. Alford and Gerald Good. Leptonic contribution to the bulk viscosity of nuclear matter. *Phys. Rev. C*, 82:055805, 2010.

[14] Mark G. Alford and Alexander Haber. Strangeness-changing Rates and Hyperonic Bulk Viscosity in Neutron Star Mergers. *Phys. Rev. C*, 103(4):045810, 2021.

[15] Mark G. Alford, Alexander Haber, Steven P. Harris, and Ziyuan Zhang. Beta Equilibrium Under Neutron Star Merger Conditions. *Universe*, 7(11):399, 2021.

[16] Mark G. Alford, Alexander Haber, and Ziyuan Zhang. Isospin Equilibration in Neutron Star Mergers. 6 2023.

[17] Mark G. Alford, Sophia Han, and Kai Schwenzer. Signatures for quark matter from multi-messenger observations. *J. Phys. G*, 46(11):114001, 2019.

[18] Mark G. Alford and Steven P. Harris. Beta equilibrium in neutron star mergers. *Phys. Rev. C*, 98(6):065806, 2018.

[19] Mark G. Alford and Steven P. Harris. Damping of density oscillations in neutrino-transparent nuclear matter. *Phys. Rev. C*, 100(3):035803, 2019.

[20] Mark G. Alford, Simin Mahmoodifar, and Kai Schwenzer. Large amplitude behavior of the bulk viscosity of dense matter. *J. Phys. G*, 37:125202, 2010.

[21] Mark G. Alford and Kamal Pangeni. Gap-bridging enhancement of modified Urca processes in nuclear matter. *Phys. Rev. C*, 95(1):015802, 2017.

[22] Mark G. Alford, Sanjay Reddy, and Kai Schwenzer. Bridging the Gap by Squeezing Superfluid Matter. *Phys. Rev. Lett.*, 108:111102, 2012.

[23] Mark G. Alford and Andreas Schmitt. Bulk viscosity in 2SC quark matter. *J. Phys. G*, 34:67–102, 2007.

[24] Mark G. Alford, Andreas Schmitt, Krishna Rajagopal, and Thomas Schäfer. Color superconductivity in dense quark matter. *Rev. Mod. Phys.*, 80:1455–1515, 2008.

[25] Nils Andersson. A New class of unstable modes of rotating relativistic stars. *Astrophys. J.*, 502:708–713, 1998.

[26] Nils Andersson. *Gravitational-Wave Astronomy*. Oxford Graduate Texts. Oxford University Press, 11 2019.

[27] Nils Andersson, Kostas D. Kokkotas, and Bernard F. Schutz. Gravitational radiation limit on the spin of young neutron stars. *Astrophys. J.*, 510:846, 1999.

[28] Peter Brockway Arnold, Caglar Dogan, and Guy D. Moore. The Bulk Viscosity of High-Temperature QCD. *Phys. Rev. D*, 74:085021, 2006.

[29] Phil Arras and Nevin N. Weinberg. Urca reactions during neutron star inspiral. *Mon. Not. Roy. Astron. Soc.*, 486(1):1424–1436, 2019.

[30] Luca Baiotti. Gravitational waves from neutron star mergers and their relation to the nuclear equation of state. *Prog. Part. Nucl. Phys.*, 109:103714, 2019.

[31] Luca Baiotti and Luciano Rezzolla. Binary neutron star mergers: a review of Einstein's richest laboratory. *Rept. Prog. Phys.*, 80(9):096901, 2017.

[32] G. Baym and C. Pethick. *Landau Fermi-Liquid Theory: Concepts and Applications*. Wiley, 2008.

[33] Omar Benhar, Alessandro Lovato, and Lucas Tonetto. Properties of Hot Nuclear Matter. 6 2023.

[34] J. Berdermann, D. Blaschke, T. Fischer, and A. Kachanovich. Neutrino emissivities and bulk viscosity in neutral two-flavor quark matter. *Phys. Rev. D*, 94(12):123010, 2016.

[35] Sebastiano Bernuzzi. Neutron Star Merger Remnants. *Gen. Rel. Grav.*, 52(11):108, 2020.

[36] Robert Bierkandt and Cristina Manuel. Bulk viscosity coefficients due to phonons and kaons in superfluid color-flavor locked quark matter. *Phys. Rev. D*, 84:023004, 2011.

[37] J. Boguta. Remarks on the Beta Stability in Neutron Stars. *Phys. Lett. B*, 106:255–258, 1981.

[38] Giovanni Camelio, Lorenzo Gavassino, Marco Antonelli, Sebastiano Bernuzzi, and Brynmor Haskell. Simulating bulk viscosity in neutron stars. I. Formalism. *Phys. Rev. D*, 107(10):103031, 2023.

[39] Giovanni Camelio, Lorenzo Gavassino, Marco Antonelli, Sebastiano Bernuzzi, and Brynmor Haskell. Simulating bulk viscosity in neutron stars. II. Evolution in spherical symmetry. *Phys. Rev. D*, 107(10):103032, 2023.

[40] A. G. W Cameron. Neutron stars. *Ann. Rev. Astron. Astrophys.*, 8:179–208, 1970.

[41] T. Celora, I. Hawke, P. C. Hammond, N. Andersson, and G. L. Comer. Formulating bulk viscosity for neutron star simulations. *Phys. Rev. D*, 105(10):103016, 2022.

[42] Michail Chabanov and Luciano Rezzolla. Impact of bulk viscosity on the post-merger gravitational-wave signal from merging neutron stars. 7 2023.

[43] Sydney Chapman and T. G. Cowling. *The Mathematical Theory of Non-uniform Gases*. 1991.

[44] Debarati Chatterjee and Debades Bandyopadhyay. Effect of hyperon-hyperon interaction on bulk viscosity and r-mode instability in neutron stars. *Phys. Rev. D*, 74:023003, 2006.

[45] Debarati Chatterjee and Debades Bandyopadhyay. Bulk viscosity in kaon condensed matter. *Phys. Rev. D*, 75:123006, 2007.

[46] K. S. Cheng and Z. G. Dai. Chemical heating in strange stars. *Astrophys. J.*, 468:819–822, 1996.

[47] Curt Cutler, Lee Lindblom, and Randall J. Splinter. Damping Times for Neutron Star Oscillations. *Astrophys. J.* , 363:603, November 1990.

[48] Veronica Dexheimer, Jorge Noronha, Jacquelyn Noronha-Hostler, Claudia Ratti, and Nicolás Yunes. Future physics perspectives on the equation of state from heavy ion collisions to neutron stars. *J. Phys. G*, 48(7):073001, 2021.

[49] Hui Dong, Nan Su, and Qun Wang. Baryon number conservation and enforced electric charge neutrality for bulk viscosity in quark matter. *Phys. Rev. D*, 75:074016, 2007.

[50] Alessandro Drago, A. Lavagno, and G. Pagliara. Bulk viscosity in hybrid stars. *Phys. Rev. D*, 71:103004, 2005.

[51] M. Dutra, O. Lourenço, S. S. Avancini, B. V. Carlson, A. Delfino, D. P. Menezes, C. Providência, S. Typel, and J. R. Stone. Relativistic Mean-Field Hadronic Models under Nuclear Matter Constraints. *Phys. Rev. C*, 90(5):055203, 2014.

[52] M. Dutra, O. Lourenco, J. S. Sa Martins, A. Delfino, J. R. Stone, and P. D. Stevenson. Skyrme Interaction and Nuclear Matter Constraints. *Phys. Rev. C*, 85:035201, 2012.

[53] Matthew Evans et al. Cosmic Explorer: A Submission to the NSF MPSAC ngGW Subcommittee. 6 2023.

[54] F. J. Fattoyev, C. J. Horowitz, J. Piekarewicz, and G. Shen. Relativistic effective interaction for nuclei, giant resonances, and neutron stars. *Phys. Rev. C*, 82:055803, 2010.

[55] Enrico Fermi. *Thermodynamics*. 1956.

[56] Rodrigo Fernandez and Andreas Reisenegger. Rotochemical heating in millisecond pulsars. Formalism and non-superfluid case. *Astrophys. J.*, 625:291–306, 2005.

[57] A. Finzi and R. A. Wolf. Hot, Vibrating Neutron Stars. *Astrophys. J.* , 153:835, September 1968.

[58] Arrigo Finzi. Vibrational Energy of Neutron Stars and the Exponential Light Curves of Type-I Supernovae. *Phys. Rev. Lett.* , 15(15):599–601, October 1965.

[59] Sergio Flores-Tulian and Andreas Reisenegger. Non-Equilibrium Beta Processes in Neutron Stars: A Relationship between the Net Reaction Rate and the Total Emissivity of Neutrinos. *Mon. Not. Roy. Astron. Soc.*, 372:276–278, 2006.

[60] Bryce Fore and Sanjay Reddy. Pions in hot dense matter and their astrophysical implications. *Phys. Rev. C*, 101(3):035809, 2020.

[61] Francois Foucart. Neutrino transport in general relativistic neutron star merger simulations. 9 2022.

[62] John L. Friedman and Sharon M. Morsink. Axial instability of rotating relativistic stars. *Astrophys. J.*, 502:714–720, 1998.

[63] L. Gavassino, M. Antonelli, and B. Haskell. Bulk viscosity in relativistic fluids: from thermodynamics to hydrodynamics. *Class. Quant. Grav.*, 38(7):075001, 2021.

[64] Lorenzo Gavassino. Relativistic bulk viscous fluids of Burgers type and their presence in neutron stars. *Class. Quant. Grav.*, 40(16):165008, 2023.

[65] Lorenzo Gavassino and Jorge Noronha. Relativistic bulk-viscous dynamics far from equilibrium. 5 2023.

[66] Suprovo Ghosh, Bikram Keshari Pradhan, and Debarati Chatterjee. Tidal heating as a direct probe of Strangeness inside Neutron stars. 6 2023.

[67] N. K. Glendenning. *Compact stars: Nuclear physics, particle physics, and general relativity*. Springer, 1997.

[68] E. Gourgoulhon and P. Haensel. Upper bounds on the neutrino burst from collapse of a neutron star into a black hole. *Astron. Astrophys.* , 271:187, April 1993.

[69] E. Gourgoulhon, P. Haensel, and D. Gondek. Maximum mass instability of neutron stars and weak interaction processes in dense matter. *Astron. Astrophys.* , 294: 747–756, February 1995.

[70] Ashok Goyal, V. K. Gupta, Pragya, and J. D. Anand. Bulk viscosity of strange quark matter. *Z. Phys. A*, 349:93–98, 1994.

[71] V. K. Gupta, S. Singh, J. D. Anand, and A. Wadhwa. Bulk viscosity of neutron stars. *Pramana*, 49:443–453, 1997.

[72] M. E. Gusakov and E. M. Kantor. Bulk viscosity of superfluid hyperon stars. *Phys. Rev. D*, 78:083006, 2008.

[73] M. E. Gusakov, E. M. Kantor, A. I. Chugunov, and L. Gualtieri. Dissipation in relativistic superfluid neutron stars. *Mon. Not. Roy. Astron. Soc.*, 428:1518–1536, 2013.

[74] M. E. Gusakov, Dima G. Yakovlev, and Oleg Y. Gnedin. Thermal evolution of a pulsating neutron star. *Mon. Not. Roy. Astron. Soc.*, 361:1415, 2005.

[75] Mikhail E. Gusakov. Bulk viscosity of superfluid neutron stars. *Phys. Rev. D*, 76:083001, 2007.

[76] P. Haensel, K. P. Levenfish, and D. G. Yakovlev. Bulk viscosity in superfluid neutron star cores. I. direct urca processes in npe mu matter. *Astron. Astrophys.*, 357:1157–1169, 2000.

[77] P. Haensel, K. P. Levenfish, and D. G. Yakovlev. Bulk viscosity in superfluid neutron star cores. 2. Modified Urca processes in npe mu matter. *Astron. Astrophys.*, 327:130–137, 2001.

[78] P. Haensel, K. P. Levenfish, and D. G. Yakovlev. Adiabatic index of dense matter and damping of neutron star pulsations. *Astron. Astrophys.*, 394:213–218, 2002.

[79] P. Haensel, K. P. Levenfish, and D. G. Yakovlev. Bulk viscosity in superfluid neutron star cores. 3. Effects of sigma- hyperons. *Astron. Astrophys.*, 381:1080–1089, 2002.

[80] P. Haensel, A. Y. Potekhin, and D. G. Yakovlev. *Neutron stars 1: Equation of state and structure*, volume 326. Springer, New York, USA, 2007.

[81] P. Haensel and R. Schaeffer. Bulk viscosity of hot-neutron-star matter from direct URCA processes. *Phys. Rev. D*, 45:4708–4712, 1992.

[82] Peter Hammond, Ian Hawke, and Nils Andersson. Thermal aspects of neutron star mergers. *Phys. Rev. D*, 104(10):103006, 2021.

[83] Peter Hammond, Ian Hawke, and Nils Andersson. Impact of nuclear reactions on gravitational waves from neutron star mergers. *Phys. Rev. D*, 107(4):043023, 2023.

[84] Matthias Hanauske, Jan Steinheimer, Anton Motornenko, Volodymyr Vovchenko, Luke Bovard, Elias R. Most, L. Jens Papenfort, Stefan Schramm, and Horst Stöcker. Neutron star mergers: Probing the eos of hot, dense matter by gravitational waves. *Particles*, 2(1):44–56, 2019.

[85] Carl J. Hansen and Sachiko Tsuruta. Vibrating neutron stars. *Canadian Journal of Physics*, 45:2823, January 1967.

[86] Steven Patrick Harris. *Transport in Neutron Star Mergers*. PhD thesis, Washington U., St. Louis, 2020.

[87] Xu-Guang Huang, Mei Huang, Dirk H. Rischke, and Armen Sedrakian. Anisotropic Hydrodynamics, Bulk Viscosities and R-Modes of Strange Quark Stars with Strong Magnetic Fields. *Phys. Rev. D*, 81:045015, 2010.

[88] Prashanth Jaikumar and Stou Sandalski. Neutrino cooling and spin-down of rapidly rotating compact stars. *Phys. Rev. D*, 82:103013, 2010.

[89] P. B. Jones. Pulsation Damping by the Nonleptonic Weak Interaction in Hyperon Stars. *Astrophys. Lett.* , 5:33, January 1970.

[90] P. B. Jones. Bulk viscosity of neutron star matter. *Phys. Rev. D*, 64:084003, 2001.

[91] Elena M. Kantor and Mikhail E. Gusakov. Temperature effects in pulsating superfluid neutron stars. *Phys. Rev. D*, 83:103008, 2011.

[92] E. E. Kolomeitsev and D. N. Voskresensky. Viscosity of neutron star matter and r-modes in rotating pulsars. *Phys. Rev. C*, 91(2):025805, 2015.

[93] Dong Lai. Resonant oscillations and tidal heating in coalescing binary neutron stars. *Mon. Not. Roy. Astron. Soc.*, 270:611, 1994.

[94] L. D. Landau and E. M. Lifshitz. *Fluid Mechanics*. 1987.

[95] William D. Langer and A. G. W. Cameron. Effects of Hyperons on the Vibrations of Neutron Stars. *Astrophys. Space. Sci.* , 5(2):213–253, October 1969.

[96] J. M. Lattimer. The equation of state of hot dense matter and supernovae. *Ann. Rev. Nucl. Part. Sci.*, 31:337–374, 1981.

[97] J. M. Lattimer, M. Prakash, C. J. Pethick, and P. Haensel. Direct URCA process in neutron stars. *Phys. Rev. Lett.*, 66:2701–2704, 1991.

[98] Lee Lindblom and Benjamin J. Owen. Effect of hyperon bulk viscosity on neutron star r modes. *Phys. Rev. D*, 65:063006, 2002.

[99] Alessandro Lovato et al. Long Range Plan: Dense matter theory for heavy-ion collisions and neutron stars. 11 2022.

[100] J. Madsen. Bulk viscosity of strange quark matter, damping of quark star vibration, and the maximum rotation rate of pulsars. *Phys. Rev. D*, 46:3290–3295, 1992.

[101] Gerald D. Mahan. *Condensed Matter in a Nutshell*. 2011.

[102] Massimo Mannarelli and Cristina Manuel. Bulk viscosities of a cold relativistic superfluid: Color-flavor locked quark matter. *Phys. Rev. D*, 81:043002, 2010.

[103] Cristina Manuel and Felipe J. Llanes-Estrada. Bulk viscosity in a cold CFL superfluid. *JCAP*, 08:001, 2007.

[104] David W. Meltzer and Kip S. Thorne. Normal Modes of Radial Pulsation of Stars at the End Point of Thermonuclear Evolution. *Astrophys. J.* , 145:514, August 1966.

[105] Anthony Mezzacappa, Eirik Endeve, O. E. Bronson Messer, and Stephen W. Bruenn. Physical, numerical, and computational challenges of modeling neutrino transport in core-collapse supernovae. *Liv. Rev. Comput. Astrophys.*, 6:4, 2020.

[106] Elias R. Most, Alexander Haber, Steven P. Harris, Ziyuan Zhang, Mark G. Alford, and Jorge Noronha. Emergence of microphysical viscosity in binary neutron star post-merger dynamics. 7 2022.

[107] Elias R. Most, Steven P. Harris, Christopher Plumberg, Mark G. Alford, Jorge Noronha, Jacquelyn Noronha-Hostler, Frans Pretorius, Helvi Witek, and Nicolás Yunes. Projecting the likely importance of weak-interaction-driven bulk viscosity in neutron star mergers. *Mon. Not. Roy. Astron. Soc.*, 509(1):1096–1108, 2021.

[108] T. Muto, T. Takatsuka, R. Tamagaki, and T. Tatsumi. Chapter VIII. Implications of Various Hadron Phases to Neutron Star Phenomena. *Prog. Theor. Phys.*, 112:221–275, January 1993.

[109] M. Oertel, M. Hempel, T. Klähn, and S. Typel. Equations of state for supernovae and compact stars. *Rev. Mod. Phys.*, 89(1):015007, 2017.

[110] D. D. Ofengeim, M. E. Gusakov, P. Haensel, and M. Fortin. Bulk viscosity in neutron stars with hyperon cores. *Phys. Rev. D*, 100(10):103017, 2019.

[111] Na-Na Pan, Xiao-Ping Zheng, and Jia-Rong Li. Bulk viscosity of Mixed nucleon-hyperon-quark Matter in Neutron stars. *Mon. Not. Roy. Astron. Soc.*, 371:1359–1366, 2006.

[112] Albino Perego, Sebastiano Bernuzzi, and David Radice. Thermodynamics conditions of matter in neutron star mergers. *Eur. Phys. J. A*, 55(8):124, 2019.

[113] Cristobal Petrovich and Andreas Reisenegger. Rotochemical heating in millisecond pulsars: modified Urca reactions with uniform Cooper pairing gaps. *Astron. Astrophys.*, 521:A77, 2010.

[114] A. Y. Potekhin and G. Chabrier. Magnetic neutron star cooling and microphysics. *Astron. Astrophys.*, 609:A74, 2018.

[115] David Radice, Sebastiano Bernuzzi, Albino Perego, and Roland Haas. A new moment-based general-relativistic neutrino-radiation transport code: Methods and first applications to neutron star mergers. *Mon. Not. Roy. Astron. Soc.*, 512(1): 1499–1521, 2022.

[116] Andreas Reisenegger. Deviations from chemical equilibrium due to spindown as an internal heat source in neutron stars. *Astrophys. J.*, 442:749, 1995.

[117] Justin L. Ripley, Abhishek Hegade K. R., and Nicolas Yunes. Probing internal dissipative processes of neutron stars with gravitational waves during the inspiral of neutron star binaries. 6 2023.

[118] Luke F. Roberts and Sanjay Reddy. Charged current neutrino interactions in hot and dense matter. *Phys. Rev. C*, 95(4):045807, 2017.

[119] M. Ruffert, H. Th. Ruffert, and H. Th. Janka. Coalescing neutron stars - a step towards physical models. 3. Improved numerics and different neutron star masses and spins. *Astron. Astrophys.*, 380:544, 2001.

[120] Basil A. Sa'd, Igor A. Shovkovy, and Dirk H. Rischke. Bulk viscosity of spin-one color superconductors with two quark flavors. *Phys. Rev. D*, 75:065016, 2007.

[121] Basil A. Sa'd, Igor A. Shovkovy, and Dirk H. Rischke. Bulk viscosity of strange quark matter: Urca versus non-leptonic processes. *Phys. Rev. D*, 75:125004, 2007.

[122] R. F. Sawyer. Damping of neutron star pulsations by weak interaction processes. *Astrophys. J.*, 237:187–197, 1980.

[123] R. F. Sawyer. Bulk viscosity of hot neutron-star matter and the maximum rotation rates of neutron stars. *Phys. Rev. D*, 39:3804–3806, 1989.

[124] R. F. Sawyer. Damping of Vibrations and of the Secular Instability in Quark Stars. *Phys. Lett. B*, 233:412–416, 1989. [Erratum: Phys.Lett.B 237, 605 (1990), Erratum: Phys.Lett.B 347, 467–467 (1995)].

[125] R. F. Sawyer and A. Soni. Transport of neutrinos in hot neutron star matter. *Astrophys. J.*, 230:859–869, 1979.

[126] Jurgen Schaffner-Bielich. *Compact Star Physics*. Cambridge University Press, 8 2020.

[127] Andreas Schmitt. *Dense matter in compact stars: A pedagogical introduction*, volume 811. 2010.

[128] Andreas Schmitt and Peter Shternin. Reaction rates and transport in neutron stars. *Astrophys. Space Sci. Libr.*, 457:455–574, 2018.

[129] Armen Sedrakian and John W. Clark. Superfluidity in nuclear systems and neutron stars. *Eur. Phys. J. A*, 55(9):167, 2019.

[130] S. L. Shapiro and S. A. Teukolsky. *Black holes, white dwarfs, and neutron stars: The physics of compact objects*. 1983.

[131] Igor A. Shovkovy and Xinyang Wang. Bulk viscosity in the nonlinear and anharmonic regime of strange quark matter. *New J. Phys.*, 13:045018, 2011.

[132] R. Swendsen. *An Introduction to Statistical Mechanics and Thermodynamics: Second Edition*. Oxford Graduate Texts. Oxford University Press, 2020.

[133] J. Sykes and G. A. Brooker. The transport coefficients of a fermi liquid. *Annals of Physics*, 56(1):1–39, January 1970.

[134] John Robert Taylor. *Classical mechanics*, volume 1. Springer, 2005.

[135] Ingo Tews, Joseph Carlson, Stefano Gandolfi, and Sanjay Reddy. Constraining the speed of sound inside neutron stars with chiral effective field theory interactions and observations. *Astrophys. J.*, 860(2):149, 2018.

[136] S. Typel et al. CompOSE Reference Manual. *Eur. Phys. J. A*, 58(11):221, 2022.

[137] E. N. E. van Dalen and A. E. L. Dieperink. Bulk viscosity in neutron stars from hyperons. *Phys. Rev. C*, 69:025802, 2004.

[138] Q. D. Wang and T. Lu. THE DAMPING EFFECTS OF THE VIBRATIONS IN THE CORE OF A NEUTRON STAR. *Phys. Lett. B*, 148:211–214, 1984.

[139] Q. D. Wang and T. Lu. Vibrational damping by quark matter inside neutron stars. *Acta Astrophysica Sinica*, 5(1):59–66, June 1985.

[140] Wei-Hua Wang, Xiao-Ping Zheng, Xi Huang, and Kun Tian. Recursion relations connecting the net reaction rate with the total emissivity of neutrinos in nonequilibrium β processes. *Phys. Rev. C*, 98(1):015801, 2018.

[141] Xinyang Wang and Igor A. Shovkovy. Bulk viscosity of spin-one color superconducting strange quark matter. *Phys. Rev. D*, 82:085007, 2010.

[142] Steven Weinberg. *Gravitation and Cosmology: Principles and Applications of the General Theory of Relativity*. John Wiley and Sons, New York, 1972.

[143] J. A. Wheeler. Superdense stars. *Ann. Rev. Astron. Astrophys.*, 4:393–432, 1966.

[144] D. G. Yakovlev, M. E. Gusakov, and P. Haensel. Bulk viscosity in a neutron star mantle. *Mon. Not. Roy. Astron. Soc.*, 481(4):4924–4930, 2018.

[145] D. G. Yakovlev, A. D. Kaminker, Oleg Y. Gnedin, and P. Haensel. Neutrino emission from neutron stars. *Phys. Rept.*, 354:1, 2001.

[146] D. G. Yakovlev, K. P. Levenfish, and Yu. A. Shibanov. Cooling neutron stars and superfluidity in their interiors. *Phys. Usp.*, 42:737–778, 1999.

[147] Dima G. Yakovlev and C. J. Pethick. Neutron star cooling. *Ann. Rev. Astron. Astrophys.*, 42:169–210, 2004.

[148] Dmitry G. Yakovlev, Pawel Haensel, Gordon Baym, and Christopher J. Pethick. Lev Landau and the concept of neutron stars. *Phys. Usp.*, 56:289–295, 2013.

[149] Francesco Zappa, Sebastiano Bernuzzi, David Radice, and Albino Perego. Binary neutron star merger simulations with neutrino transport and turbulent viscosity: impact of different schemes and grid resolution. 10 2022.

[150] Winfried Zimdahl. 'Understanding' cosmological bulk viscosity. *Mon. Not. Roy. Astron. Soc.*, 280:1239, 1996.

CHAPTER 9

Neutron Star Asteroseismology: Beyond the Mass-Radius Curve

Nils Andersson, Brynmor Haskell

WE discuss how different aspects of extreme density nuclear physics impact on neutron star dynamics. Focussing on seismology aspects, which is natural if we want to explore the full range of connections with equation of state physics, we develop the theory step by step and consider relevant applications and current results along the way. The aim is to provide a gentle pedagogical introduction rather than an exhaustively detailed review.

9.1 INTRODUCTION

Close your eyes and imagine. You are sitting on a beautiful beach. The sun may—or may not, depending on what mood you are in—be setting. The tide is rolling in from the ocean. Gentle waves lap the shore. The sound of the surf is soothing. Breathe in. Breathe out. The rhythm of the sea helps you relax.

The rhythm of the sea. Waves of a vast range of scales, from the surface waves that hit a beach to the large-scale circulation that is intimately linked to our climate. The tides raised by the Moon and the Sun. Some aspects of these waves we understand. Others not. It is a question of obvious importance. Our understanding of ocean waves is closely linked to the development of civilization and the desire to safely navigate the seas to establish new trade routes.

A reliable understanding of tides is needed to avoid maritime disaster. This problem has inspired many great thinkers, dating back to the philosophers of ancient Greece (and very likely before, but the recorded evidence only goes so far). Plato apparently believed the Earth was a large animal and that tides were a result of oscillations of the fluid inside this animal. Instead, Aristotle proposed that the tide was due to wind from the Sun and the Moon striking the water[1] while Timaeus guessed that the Moon pushed down on the atmosphere, thus compressing the sea. Each of these early ideas has some merit. This

[1]Legend has it that he committed suicide in frustration over his failure to understand the problem of tides.

DOI: 10.1201/9781003306580-9

is evident from the simple fact that they lasted—in one form or another—for nearly two thousand years. Our quantitative understanding is much more recent. In order to understand the tides, we need, first of all, Newton's universal law of gravity. We also need the clever mathematical formulation of Laplace, focussing on the harmonic (wave) content of the ocean dynamics. In essence, we need an operational understanding of forced fluid dynamics. But the problem is complicated. The precise shape of coast lines, the varying depth of the ocean, winds and weather patterns all come into play.

Close your eyes and imagine. You are sitting in a grand concert hall. The lights are dimming and there is a hush of anticipation in the audience. The conductor taps his music stand to get the attention of her orchestra. She raises her baton. And so it begins. The overture starts out gently, slowly growing in a long crescendo. And then the percussion explodes into action...

Can you hear the shape of the drums? Can you tell if the violins were made by the masterful hands of Stradivarius or mass produced in a modern factory? Does it make a difference? The short answer is, of course; yes. The sound of an instrument depends on its size and shape—the length and tension of a string or the form of the resonance cavity. But there is more to it than that. The problem is subtle. The precise material and fine details of the craftsmanship distinguishes the instrument of the top professional from the—much less expensive—beginner's version. It may not be easy to explain the difference with mathematics, but we can all hear the difference.

What do we learn from these two flights of fancy? Evidently, the harmonic content of a body's dynamics—its spectrum of oscillations, if you will—may be used to infer the shape and composition of the body, along with the detailed forces at play. This is the principal idea of asteroseismology. Inspired by information gained from the natural oscillations of everyday objects—be it the string of a violin, the taught skin of a drum or some complicated machinery—terrestrial seismology and lessons learned from geophysical fluid dynamics and the close study of earthquakes [14, 34], we may turn our attention to the seismology of stars.

Stars may—throughout their main sequence lifetime and beyond—support a range of oscillations [6]. The nature of these waves depend on the local speed of sound, the matter composition, thermal gradients, magnetic fields and so on. Solar oscillations were first detected in the 1960s, but it was only in 1975 that the origin of the observed variations were associated with the Sun's normal modes. The first observations involved modes with short horizontal wavelengths, but large-scale modes associated with timescales in the 5-minute range were discovered soon after. The study of solar oscillations culminated with the ground-based GONG network of observing stations and instruments on the SOHO (SOlar and Heliospheric Observatory) mission [56], both starting operations in the middle of the 1990s. By now, inferences from helioseismology data for gravity g-modes and low-multipole pressure p-modes have led to measures of the sound speed at different depths and the differential rotation throughout much of the Sun's interior.

Asteroseismology observations of more distant stars are obviously more challenging. The expected oscillations have low amplitude, and astronomical distances are vast. Yet, there has been astonishing progress in the last few decades. Definite detections of solar-like oscillations in other stars were first made in the late 1990s. Follow-up observing campaigns, often using the largest telescopes in the world pointing at the brightest stars, led to analysis of the oscillations in a few main-sequence and red-giant stars. The breakthrough had to wait, however, for the space-based photometry facilitated by the CoRoT and Kepler missions, launched in 2006 and 2009, respectively [71, 92]. These missions combined asteroseismology searches—focussing on oscillations that are stochastically excited by surface convection—with a hunt for exoplanets. We now have access to high-quality asteroseismology data for hundreds of main-sequence and subgiant stars, and tens of thousands of red giants, leading to conclusions about bulk stellar parameters, like mass, radius and age. The NASA TESS mission (launched in 2018) and the ESA PLATO mission (expected to launch in 2026) add a tremendous amount of high-quality data.

Asteroseismology has become a powerful method to characterize host stars in exoplanet systems. Most exoplanet discoveries are indirect—either using transits, Doppler shifts or microlensing. These different observations all measure exoplanet properties relative to the host star. Asteroseismology data for these hosts play a key role in establishing the nature of the smaller bodies orbiting around them. Ultimately, the seismology data may allow us to establish the existence of Earth-like planets, which may in turn lead to evidence that we are not alone in the Universe.

9.2 THE NEUTRON STAR PROBLEM

Before considering the detailed mathematics/physics, we highlight aspects which complicate the analysis of neutron star seismology and motivate why it is natural to focus on the gravitational-wave aspects of the problem.

With the broader asteroseismology context in mind, let us move on to the main topic of this chapter. We want of understand what we can hope to learn from neutron star seismology and to what extent we should expect to make progress on the required observations. The first is, essentially, a theory problem and the second clearly depends on the sensitivity of current and future instruments. Focussing on the modelling, the problem is challenging because neutron stars are known to represent many extremes of physics (in density, pressure, temperature—through the early formation stages and during binary mergers—and magnetic field). In practice, this means that our models will involve some level of ignorance. We need to consider aspects that may not yet be fully (or perhaps even partially) understood. In fact, the composition and state of matter of a neutron star core introduce concepts from across modern physics—fluid dynamics and elasticity, thermodynamics, superfluidity and superconductivity, electromagnetism, nuclear physics—while the models have to be developed in the curved spacetime framework of general relativity. Without giving up this goal, we have to concede that we may not—at least not now—be able to account for all aspects we know we ought to consider. Still, we want to make progress.

Fundamentally, we need an effective description of the nuclear interactions in dense matter. This is typically represented by the equation of state; the pressure–density–temperature relation for bulk matter. Different theoretical models generate different mass–radius relations, which then provide a characteristic radius for a range of masses and a maximum mass above which a neutron star must collapse to a black hole. They also predict quantities like the maximum spin frequency and the moment of inertia. Unfortunately, first principle calculations of the interactions for many-body QCD systems are held back by the fermion sign problem. This becomes particularly problematic at the high densities we encounter in a neutron star core. As we do not have direct predictions for strongly interacting quark matter, we have to (at least to some extent) resort to phenomenology. Experiments and observations help test the theory and support progress [138]. In this sense, two-body interactions are fairly well constrained by laboratory experiments while three-body forces represent the frontier of nuclear physics. At low energies, effective field theories based on QCD symmetries provide a systematic expansion of the nuclear forces, which in turn predict two- and many-nucleon interactions. In addition, there are complementary efforts using lattice approaches to the nuclear forces to provide few-body nucleon-nucleon and more generally baryon-baryon interactions, but this approach remains hampered by uncertainties. Predictions need to match current nuclear data [48], or in situations where data are not yet available (at least) be internally consistent.

The obvious question to ask is if we can use observations to constrain the uncertain theory aspects. This is tricky because neutron stars are not hands-on laboratories. They may be extreme but they are small and distant. Nevertheless, there has been clear progress. We have precise mass estimates for many systems from radio pulsar timing (telling us that the equation of state must allow stars with a mass above $2M_\odot$ [54]). The recent results from NICER also help constrain the neutron star radius (very roughly to the range 11-14 km) [107, 123]. These constraints on bulk properties—mass and radius—should become increasingly precise as more data become available. Future instruments, like the SKA in the radio and the planned eXTP mission for x-ray timing [138], will ensure that this area of exploration remains healthy. Whether this progress will allow us to probe aspects associated with the neutron star interior, e.g. the state and composition of matter, is less clear.

Before we consider neutron star dynamics, let us take a first look at how aspects of the matter equation of state enter the discussion. Simply counting the variables and the available equations from fluid dynamics it is easy to see that we need more information in order to proceed. We have to specify the equation of state for matter. This issue is either easy or, as we have indicated, intricate and complex. The easy option is to "invent" some relation between the fluid pressure (p) and the baryon number density (n), $p = p(n)$ (perhaps inspired by "real" physics, like the ideal gas law or a system of interacting fermions) and then go ahead and solve the equations. Commonly, this strategy would introduce a polytropic relationship (for cold stars)

$$p = K\rho^\Gamma \tag{9.1}$$

with constant values of K and Γ. This may seem quite ad hoc but we nevertheless learn a lot from this exercise. In many ways, the polytropic model provides a convenient parameterisation of our ignorance. Of course, if we aim for realism—and ultimately, we do!—then the equation of state needs to be based on the relevant microphysics. This is where it gets more difficult. Still, any model we choose to work with has to be thermodynamically consistent and it is useful to illustrate what this implies.

Suppose we consider the simplest reasonable case—a star composed primarily of neutrons with a few protons and electrons thrown in the mix—and ignore (for the moment) thermal effects. Then we have the thermodynamical relation

$$p + \varepsilon = n_\mathrm{n}\mu_\mathrm{n} + n_\mathrm{p}\mu_\mathrm{p} + n_\mathrm{e}\mu_\mathrm{e} \, , \tag{9.2}$$

where ε is the energy density and the chemical potentials are defined as

$$\mu_{\rm x} = \left(\frac{\partial \varepsilon}{\partial n_{\rm x}}\right)_{n_{\rm y}} , \tag{9.3}$$

with x=n,p,e (and $\mathrm{y} \neq \mathrm{x}$) for neutrons, protons and electrons, respectively. Evidently, the thermodynamics involves information not explicitly represented in the fluid equations. How do we square this? The answer is instructive, so let us spell it out. First, as long as we are ignoring electromagnetic aspects, it is natural to assume that the matter is locally charge neutral[2]. Then we have $n_{\rm p} = n_{\rm e}$, so we can eliminate one of the number densities from the discussion. This leads to

$$p + \varepsilon = n_{\rm n}\mu_{\rm n} + n_{\rm e}\left(\mu_{\rm p} + \mu_{\rm e}\right) . \tag{9.4}$$

Next, let us make contact with baryon number conservation by changing variables to $n = n_{\rm n} + n_{\rm p}$ and the lepton fraction $Y_{\rm e} = n_{\rm e}/n$. This takes us to

$$p + \varepsilon = n\mu_{\rm n} + nY_{\rm e}\left(\mu_{\rm p} + \mu_{\rm e} - \mu_{\rm n}\right) . \tag{9.5}$$

The final step involves further physics assumptions. In the first instance, let us assume that the matter remains in chemical equilibrium even though the fluid may be sloshing about. This should be a reasonable approximation if the timescales associated with the relevant nuclear reactions are fast compared to the dynamics involved (we will come back to this issue later). With this assumption, given that we are dealing with the Urca reactions (and we assume the matter is transparent to neutrinos, so they freely escape the system), we have the equilibrium condition

$$\mu_{\rm p} + \mu_{\rm e} = \mu_{\rm n} . \tag{9.6}$$

For any given number density n, we can solve this equation for the equilibrium lepton fraction $Y_{\rm e} = Y_{\rm e}(n)$.Once we do this, we are left with a single parameter problem. That is, we know that the equation of state can be expressed in terms of the energy $\varepsilon = \varepsilon(n)$ and the pressure then follows from the thermodynamics. We have simplified the problem to that of a barotropic fluid.

The arguments we have made are consistent, but it is worth adding a caveat at this point. We have—somewhat sneakily—used variable names that connect the fluid model to the thermodynamics. The fluid equations involve the (isotropic) fluid pressure p and we seem to have used the same variable in the thermodynamical relation. Are these two quantities necessarily the same? The answer is not obvious. The association of the two pressures may seem natural—after all, everyone does it!—but it is an assumption, nevertheless. The importance of this becomes apparent when one considers the problem of averaging the physics on the microscale to reach the macroscopic scale of the fluid equations. This comes to the fore in models of turbulence [45], a problem we will not worry too much about here. Nevertheless, it is important to keep in mind that the fluid model is phenomenological. The standard logic is, well... logical, but we need to be aware that we are making assumptions which may impact on the results.

[2]This assumption is hardwired into most nuclear equation of state calculations, anyway...

As already indicated, it is natural to (try to) formulate a seismology strategy in order to explore aspects associated with the dense neutron star interior. Work in this direction has established that the complex interior physics is reflected in a rich spectrum of oscillation modes [24] and one may hope to be able to use observations of associated features to gain insight. Of course, when it comes to neutron stars we are not likely to ever be able to "resolve" surface features. Instead, it is natural to consider the gravitational-wave aspects of the problem. In essence, any deformation/acceleration of the matter in the star will generate gravitational waves and one would expect to be able to express these waves in terms of the star's oscillation modes. The promise of such gravitational-wave asteroseismology [27, 28, 33, 93] relies on the answer to two questions. First, are the mode features robust enough that we can use observations to constrain the physics? Second, are there realistic scenarios where specific oscillation modes are excited to a level where the gravitational-wave signal can be detected by current (or, indeed, future) instruments? Our aim is to argue in favour of affirmative answers to both questions. This is not to suggest that the venture will be in any way straightforward, but fortune favours the brave and the effort may pay off—perhaps even handsomely—in the end.

9.3 FORMULATING THE PROBLEM: THE FUNDAMENTAL MODE

Moving on, we formulate the seismology problem and discuss the main ingredients of a stellar perturbation calculation. We outline the derivation of the fundamental mode of a star and contemplate astrophysical scenarios where this mode may be relevant. The main aim is to gain insight into the relevant phenomenology.

9.3.1 Fluid dynamics

The various oscillation modes of a star are given by solutions to the equations of fluid dynamics, coupled to gravity and subject to suitable boundary/regularity conditions. As we want to emphasize the phenomenology, rather than get bogged down in the mathematical details, we will sketch the problem in Newtonian gravity and then return to point out what changes when we move on to the fully relativistic description. Given that neutron stars have a compactness of order $GM/Rc^2 \sim 0.2$ we expect a relativistic calculation to be an absolute requirement if we want precise answers. Yet, we know from experience that some of the calculations we have to carry out are technically challenging. Given this, it is natural to try to gain intuition before embarking on the detailed calculation. Newtonian estimates are useful in this respect. It is also worth keeping in mind that, for several aspects of the problem the relativistic calculation remains on the to-do list.

Let us start with the oscillations of a Newtonian star. The fluid motion is then described by—ignoring viscosity, crust elasticity, superfluidity, electromagnetism; much of the physics that make neutron stars fascinating—the Euler equation

$$\frac{dv_i}{dt} = \frac{\partial v_i}{\partial t} + v^j \nabla_j v_i = -\frac{1}{\rho} \nabla_i p - \nabla_i \Phi \,. \tag{9.7}$$

This is the equation for momentum conservation, with $v_i = g_{ij} v^j$ the velocity components in a coordinate basis (g_{ij} here is the flat three-metric and the indices $i = 1-3$ and $j = 1-3$

are spatial), p is the isotropic pressure, ρ is the mass density and Φ is the gravitational potential. We are using the tensor language from the start—even though it may seem like overkill at this stage—because it makes the connection to the relativistic problem more transparent. In addition to the momentum equation, we need the continuity equation, which can be be expressed in terms of the baryon number density n:

$$\frac{\partial n}{\partial t} + \nabla_i \left(n v^i \right) = 0 \,. \tag{9.8}$$

For Newtonian problems, one would usually work with mass density $\rho = mn$, with m the baryon mass, but when we consider the relativistic problem later the relevant conservation law is for baryon number so we may as well highlight this from the beginning. We also need the Poisson equation for the gravitational potential

$$\nabla^2 \Phi = 4\pi G \rho \,, \tag{9.9}$$

where g is Newton's gravitational constant.

Moving on to the perturbation problem, the first step involves separating low-amplitude (we wish to linearise the equations) perturbations from a comparatively slowly varying background configuration. The simplest situation is represented by a static star for which the background velocity vanishes and we only need to solve

$$\nabla_i p = -\rho \nabla_i \Phi \,, \tag{9.10}$$

along with the Poisson equation to determine(say) the density profile for the prescribed equation of state. For the perturbations, we generally have a choice. We may either consider the problem at a fixed point in space or in a frame that moves along with the background fluid elements: We may take either the Eulerian or the Lagrangian perspective. As it turns out, we will use a bit of both.

In the covariant framework, the Lagrangian perturbation ΔQ of a quantity Q, be it a scalar, a vector component or a tensor, is related to the Eulerian variation δQ via

$$\Delta Q = \delta Q + \mathcal{L}_\xi Q \,, \tag{9.11}$$

where $\mathcal{L}_\xi$ is the Lie derivative along the displacement ξ^i. This vector is, in turn, defined in terms of the Lagrangian change in the fluid velocity, and we have

$$\Delta v^i = \partial_t \xi^i . \tag{9.12}$$

Noting that

$$\Delta g_{ij} = \nabla_i \xi_j + \nabla_j \xi_i \,, \tag{9.13}$$

it follows that

$$\Delta v_i = \partial_t \xi_i + v^j \nabla_i \xi_j + v^j \nabla_j \xi_i \,. \tag{9.14}$$

Let us now consider the simplest case; a perturbed barotropic fluid with $\varepsilon = \varepsilon(n)$. We want to perturb the continuity equation (9.8) and the Euler equation (9.7). In doing this, it is useful to note that

$$\Delta\left(\frac{\partial}{\partial t} + \mathcal{L}_v\right) = \left(\frac{\partial}{\partial t} + \mathcal{L}_v\right)\Delta\,. \tag{9.15}$$

Cranking through the algebra, we find that

$$\Delta n = -n\nabla_i\xi^i \Longrightarrow \delta n = -\nabla_i(n\xi^i)\,, \tag{9.16}$$

and the perturbed gravitational potential then follows from

$$\nabla^2\delta\Phi = 4\pi G\delta\rho = 4\pi Gm\,\delta n = -4\pi Gm\nabla_i(n\xi^i)\,. \tag{9.17}$$

In order to perturb the Euler equations we first rewrite (9.7) as

$$(\partial_t + \mathcal{L}_v)v_i + \nabla_i\left(h + \Phi - \frac{1}{2}v^2\right) = 0\,, \tag{9.18}$$

where the enthalpy h is defined through

$$\nabla_i h = \frac{1}{\rho}\nabla_i p\,. \tag{9.19}$$

For a barotropic fluid, it is worth noting that

$$\nabla_i p = n\nabla_i\mu = \rho\nabla_i\left(\mu_{\rm n}/m\right)\,, \tag{9.20}$$

so $h = \mu_{\rm n}/m$, which connects to the previous discussion of the thermodynamics. Perturbing (9.18) we have

$$(\partial_t + \mathcal{L}_v)\Delta v_i + \nabla_i\left[\Delta h + \Delta\Phi - \frac{1}{2}\Delta(v^2)\right] = 0, \tag{9.21}$$

which we can rewrite in terms of the displacement vector:

$$\partial_t^2\xi_i + 2v^j\nabla_j\partial_t\xi_i + (v^j\nabla_j)^2\xi_i + \nabla_i\delta\Phi + \xi^j\nabla_i\nabla_j\Phi \\ - (\nabla_i\xi^j)\nabla_j h + \nabla_i\Delta h = 0\,. \tag{9.22}$$

Finally, we need

$$\Delta h = \delta h + \xi^i\nabla_i h = \left(\frac{dh}{dn}\right)\delta n + \xi^i\nabla_i h = -\left(\frac{dh}{dn}\right)\nabla_i(n\xi^i) + \xi^i\nabla_i h \tag{9.23}$$

to arrive at the final version of the perturbed Euler equation:

$$\partial_t^2\xi_i + 2v^j\nabla_j\partial_t\xi_i + (v^j\nabla_j)^2\xi_i + \nabla_i\delta\Phi + \xi^j\nabla_i\nabla_j\left(\Phi + h\right) \\ - \nabla_i\left[\left(\frac{dh}{dn}\right)\nabla_j(n\xi^j)\right] = 0\,. \tag{9.24}$$

Schematically, we may write this as [63]

$$\partial_t^2 \xi_i + B\partial_t \xi_i + C_{ij}\xi^j = 0 \; . \tag{9.25}$$

This will be useful later.

We may consider the final equation (9.24) from two perspectives. First, we can convince ourselves that the problem has the anticipated nature. We have a second time derivative and second space derivatives, which makes sense given that we expect the fluid perturbations to be described by waves. Of course, a closer look tells us that the problem might not—even though we have simplified life by linearising—be all that straightforward to solve. We will come back to some of the complicating features later. At this point, let us highlight a feature that will be important as we make the neutron star model more realistic (and inevitably more complicated). We have to work with (a number of) thermodynamical derivatives and it is helpful to gain some level of intuition of their meaning. The way we have chosen to express the single (barotropic) fluid problem, the equation evidently involves

$$\frac{dh}{dn} = \frac{1}{\rho}\frac{dp}{dn} = \frac{1}{n}\frac{dp}{d\rho} = \frac{1}{n}c_s^2 \; . \tag{9.26}$$

That is, the main parameter that controls the fluid behaviour is the speed of sound, c_s^2. This seems quite natural.

9.3.2 The fundamental mode

As a first illustration of the stellar oscillation problem let us work through the simplest case; a non-rotating uniform density star. This problem, which involves taking the density to be constant and hence assuming the sound speed is infinite, is clearly somewhat artificial but we are not going to worry too much about this. Suffice it to say that, we can make the argument formally more appealing by demonstrating that any dynamics which is safely subsonic ($v^2 \ll c_s^2$) is effectively incompressible. Second, it turns out that an often used model for strange quark stars (the MIT bag model) is fairly well described by the constant density assumption. With these remarks in mind, let us proceed.

First, it is easy to see that, assuming that the background star is non-rotating and taking the density to be constant, the perturbed continuity equation simplifies to

$$\nabla_i \delta v^i = 0 \; . \tag{9.27}$$

Second, in terms of the Eulerian pertubations, the perturbed version of the momentum equation (9.7) takes the form

$$\partial_t \delta v_i + \frac{1}{\rho}\nabla_i \delta p + \nabla_i \delta\Phi = 0 \; , \tag{9.28}$$

where $\delta\Phi$ is the variation in the gravitational potential, which is governed by[3]

$$\nabla^2 \delta\Phi = 0 \; . \tag{9.29}$$

[3]Although... we have to be a little bit careful here because there will be a discontinuity in the background density at the star's surface.

From (9.28), it is easy to see (by taking the curl of the equation) that the perturbed velocity must be irrotational. This means that we may introduce a velocity potential χ such that

$$\delta v_i = \nabla_i \chi \,, \tag{9.30}$$

and write the perturbed Euler equation as

$$\partial_t \chi + \frac{1}{\rho}\delta p + \delta\Phi = \text{ constant} \,. \tag{9.31}$$

This is, basically, the standard Bernoulli argument for incompressible flows. Moreover, we see that χ, δp and $\delta\Phi$ must all solve the homogeneous version of Laplace's equation. This, in turn, tells us that all the variables are naturally expanded in spherical harmonics and, further, it is easy to show that if we insist that the solutions are regular at the centre of the star then each quantity must be proportional to $r^l Y_l^m$ for a given (l, m) multipole.

Finally, at the surface of the star, an oscillation mode must satisfy two conditions. First, the Lagrangian pressure variation must vanish[4]

$$\Delta p = \delta p + \xi^r \partial_r p = 0 \,, \qquad \text{at } r = R \,. \tag{9.32}$$

Second, the continuity of the gravitational potential and its derivative across the surface leads to

$$\partial_r \delta\Phi + \frac{l+1}{R}\delta\Phi = -4\pi G\rho\xi^r \,, \qquad \text{at } r = R \,, \tag{9.33}$$

where the right-hand side does not vanish for the constant density model or, indeed, whenever the star's surface is assumed to be at a finite density.

Now we are ready to solve the problem, and we will do so for three different assumptions. The aim is to illustrate aspects which will be important later. First, let us simplify the problem by making what is known as the Cowling approximation [52]. This means that we ignore the perturbations in the gravitational potential. The motivation for this assumption comes from the analysis of surface waves. The argument is simple: Waves in the low-density surface region will not induce significant variations in the overall gravitational potential of the body. Hence, these variations may—without much loss of precision—be ignored. The Cowling approximation is commonly used, at least in work that mainly aims at gaining intuition, but we have to be mindful that its applicability may be debatable (especially in general relativity!). Anyway, in the Cowling approximation, we only need to solve for the velocity potential χ and the perturbed pressure δp. Working out the algebra (in the Fourier-domain, with a harmonic time dependence $e^{i\omega t}$ for all quantities), it is easy to show that the surface boundary condition leads to the mode frequency

$$\omega^2 = \frac{4\pi G\rho l}{3} = l\frac{GM}{R^3} \,. \tag{9.34}$$

These are the fundamental modes of the star, the only mode-solution that exists for a non-rotating incompressible fluid, and we can draw two important conclusions. First, we note

[4]This is essentially just a statement that a fluid element at the surface remains at the surface when the star is perturbed.

that the problem is degenerate in the azimuthal m-harmonics. For a given l multipole, the mode frequencies are the same regardless of the value of m (in the range $-l \leq m \leq l$). This is due to the spherical symmetry of the background configuration. Second, we see that the mode frequency scales with the (average) density. In effect, an observation of the fundamental mode can be "inverted" to provide insight into the density of the star. This would be a step towards constraining the neutron star equation of state, but we need more information if we want to extract both mass and radius.

Let us try to understand the nature of the fundamental model a bit better. We can do this by introducing an artificial inner boundary at (say) $r = R_c$ and imposing a no-penetration condition at that point, letting $\xi_r(R_c) = 0$. This obviously changes the mode solution and we now arrive at

$$\omega^2 = l\frac{GM}{R^3}\left[1 - \left(\frac{R_c}{R}\right)^{2l+1}\right]\left[1 + \frac{l}{l+1}\left(\frac{R_c}{R}\right)^{2l+1}\right]^{-1} . \tag{9.35}$$

This may not look very instructive, apart from that it is easy to see that we retain the previous solution in the limit $R_c \to 0$. However, if we rewrite the expression in terms of a depth h, such that

$$R_c = R\left(1 - \frac{h}{R}\right) , \tag{9.36}$$

and assume that $h \ll R$ (effectively, consider a shallow ocean) then we get

$$\omega^2 \approx l(l+1)\frac{GM}{R^3}\frac{h}{R} . \tag{9.37}$$

This is more transparent. Noting that the gravitational acceleration near the surface is $g = GM/R^2$ and comparing to the classic (planar) shallow water wave dispersion relation

$$\omega^2 = ghk^2 , \tag{9.38}$$

we see that our result corresponds to a surface wave with wave number $k^2 \approx l^2/R^2$. This, in turn, suggests that the fundamental mode arises due to the presence of the stellar surface, which seems quite intuitive.

The third calculation brings the perturbed gravitational potential back into play. The mode frequency then changes to

$$\omega^2 = \frac{GM}{R^3}\frac{2l(l-1)}{2l+1} . \tag{9.39}$$

This confirm the warning flag we raised about the precision of the Cowling approximation. Evidently, the mode frequencies differs by a factor

$$\sqrt{\frac{2(l-1)}{2l+1}} \to 1 \text{ as } l \to \infty . \tag{9.40}$$

As one might have expected, the Cowling approximation is accurate for short wavelength modes but not very precise for the quadrupole ($l = 2$) modes which may be the most interesting from the gravitational-wave perspective. The main lesson is that, if we simplify the calculation we pay a price in terms of accuracy.

Before we move on, let us make one final point. For each value of l we have two f-modes (associated with the two signs of the square root). Given that the mode solutions behave as $e^{i(\omega t+m\varphi)}$, we can introduce the pattern speed of the wave (tracking the constant phase) as

$$\sigma_p = -\omega/m \, . \tag{9.41}$$

One of the two f-modes moves "backwards" on the star while the other moves "forwards". This observation will be important later.

9.3.3 The perspective of general relativity

In order to describe real neutron star, built from a realistic nuclear physics equation of state, we need to solve the problem in general relativity. The Newtonian problem serves as a useful guide and provides insight into the qualitative behaviour, but if we want precision then we need to use Einstein's theory. For a static star, this means that we start from a metric

$$ds^2 = -e^{\nu} c^2 dt^2 + g_{ij} dx^i dx^j \, . \tag{9.42}$$

where g_{ij} is diagonal. It follows that the static four-velocity is given by

$$u^a = e^{-\nu/2} c \delta_0^a \, . \tag{9.43}$$

Moreover, as the body has to be spherically symmetric we have $p = p(r)$ and a little bit of algebra leads to the familiar equation for hydrostatic equilibrium

$$\partial_r p = -\frac{p+\varepsilon}{2} \partial_r \nu \, . \tag{9.44}$$

In addition, we need to solve the Einstein field equations

$$R_{ab} - \frac{1}{2} g_{ab} R = \frac{8\pi G}{c^4} T_{ab} \, . \tag{9.45}$$

which can be rewritten using

$$g^{ab}\left(R_{ab} - \frac{1}{2} g_{ab} R\right) = -R = \frac{8\pi G}{c^4} g^{ab} T_{ab} = \frac{8\pi G}{c^4} T \, , \tag{9.46}$$

leading to

$$R_{ab} = \frac{8\pi G}{c^4} \left(T_{ab} - \frac{1}{2} g_{ab} T\right) \, . \tag{9.47}$$

The main difference from the Newtonian problem is that, when we turn to the perturbation problem we need to account for the live spacetime in the form of the perturbed metric δg_{ab}. Some of the isses that arise are are easy and others less so. For example, the normalisation of the four-velocity leads to

$$\delta u^0 = \frac{c}{2} e^{-3\nu/2} \delta g_{00} \, . \tag{9.48}$$

This condition fixes the time component so that the perturbed four-velocity has three components, just like in classical fluid dynamics.

Some further calculation tells us how the perturbed metric enters the fluid equations. For example, the energy equation takes the form

$$\partial_t \delta\varepsilon + e^{\nu/2}\delta u^i \partial_i \varepsilon + (p+\varepsilon) D_i \left[e^{\nu/2} \delta u^i \right] = -\frac{1}{2}(p+\varepsilon) e^{-\nu} \partial_t \delta g_{00} \tag{9.49}$$

where D_i is the covariant derivative associated with g_{ij}. Meanwhile, the momentum equation becomes

$$\frac{p+\varepsilon}{c^2} e^{-\nu/2} \partial_t \delta u^i + \frac{1}{2}(\delta p + \delta\varepsilon) g^{ij} \partial_j \nu + g^{ij} \partial_j \delta p$$
$$= -\frac{p+\varepsilon}{2} e^{-\nu} g^{ij} \partial_j \nu \, \delta g_{00} - \frac{p+\varepsilon}{2} g^{ij} e^{-\nu} \left(\frac{2}{c} \partial_t \delta g_{0j} - \partial_j \delta g_{00} \right) . \tag{9.50}$$

Pragmatically, the fluid equations are not significantly more complicated than their Newtonian counterparts. The perturbed metric is, however, more involved. We need

$$\delta R_{ab} = \frac{8\pi G}{c^4} \left(\delta T_{ab} - \frac{1}{2} \delta g_{ab} T - \frac{1}{2} g_{ab} \delta T \right) , \tag{9.51}$$

where the perturbed Ricci tensor involves $\delta\Gamma^a_{bc}$ and its derivatives. This gets messy and if we want to make progress we need to make clever use of the gauge freedom of the problem. Most work on perturbed static stars take a lead from work on black holes and use the Regge-Wheeler gauge [55, 131].

We will not dig for further details here. Instead, we focus on the main changes to the mode problem. The most significant aspect is that the presence of gravitational waves change the boundary conditions. A mode solution must be such that the waves are purely outgoing at spatial infinity—otherwise we have a scattering situation. This means that all modes will be damped and we are dealing with quasinormal modes [110].

As an indication of the impact of the gravitational waves, we may consider the induced damping of the f-mode. Combining the Newtonian mode results for uniform density stars with the standard post-Newtonian multipole formulas [24], we find the damping timescale to be

$$t_{\rm gw} \sim \left(\frac{c^2 R}{GM} \right)^{l+1} \frac{R}{c} . \tag{9.52}$$

For a typical neutron star, the f-mode damps in a fraction of a section. Other modes, in particular the g-modes we will discuss later, couple less strongly to gravity and hence their damping times are much longer. Of course, if we want to detect these gravitational waves we need the modes to be efficient emitters. This is why the f-mode is favoured in many detection scenarios.

At this point, we have enough information to state the asteroseismology argument. The uniform density model is described by two parameters, which we can take to be the density and the radius. An observation of the f-mode frequency would provide the former and then the damping rate would help us infer the radius—as the scaling with the parameters is different. This is the key idea [27, 28, 33, 93], although a better strategy may be to use other modes. For example, a g-mode detection would give some insight into the interior matter composition. We will soon see why this is the case.

9.3.4 Core collapse, newly born neutron stars

Let us consider a practical example of when f-modes may be excited in a NS, and what their astrophysical signature may be. To be specific, we will consider the very first instants of a neutron star's life, immediately after birth. Core collapse supernovae are the main formation channel for these systems (we will consider the possibility of forming massive neutron stars in binary mergers later). Modelling the full core collapse process is technically challenging and relies on computationally expensive 3D simulations. However, once a proto neutron star forms in the core region, it can be modelled with good precision as quasistationary [60]. The discussion from the previous sections apply, with the obvious caveat that we are now dealing with a hot, convective, system, in which neutrino transport plays an important role [41].

In principle, incorporating thermal effects is straightforward. Adding temperature to the previous thermodynamical argument, we have

$$p + \varepsilon = n_\mathrm{n}\mu_\mathrm{n} + n_\mathrm{p}\mu_\mathrm{p} + n_\mathrm{e}\mu_\mathrm{e} + sT \ , \tag{9.53}$$

where the temperature is the chemical potential associated with the entropy density s

$$T = \left(\frac{\partial \varepsilon}{\partial s}\right)_{n_\mathrm{x}} \ . \tag{9.54}$$

However, it is often is more intuitive to work with the temperature rather than the entropy. This change is easily effected via a Legendre transform to the (Helmholtz) free energy. Introducing

$$f = \varepsilon - sT \ , \tag{9.55}$$

we get

$$df = \sum_\mathrm{x} \mu_\mathrm{x} dn_\mathrm{x} - sdT \ , \tag{9.56}$$

which shows that $f = f(n_\mathrm{x}, T)$. We also see that

$$\mu_\mathrm{x} = \left(\frac{\partial f}{\partial n_\mathrm{x}}\right)_{n_\mathrm{y}, T} \ , \tag{9.57}$$

and

$$s = -\left(\frac{\partial f}{\partial T}\right)_{n_\mathrm{x}} \ , \tag{9.58}$$

while the pressure is given by

$$p = p(n, Y_\mathrm{e}, T) = n_\mathrm{n}\mu_\mathrm{n} + n_\mathrm{p}\mu_\mathrm{p} + n_\mathrm{e}\mu_\mathrm{e} - f \ . \tag{9.59}$$

We are now dealing with a multi-parameter equation of state. For a given temperature, we need to keep track of both baryon number and lepton fraction.

In practice, it is not quite so straightforward to incorporate the thermal effects in the equation of state. The calculation involves so-called Fermi-integrals [126], which are complicated to solve. For low temperatures, a perturbative expansion may suffice but newly born neutron stars—or, indeed, merger remnants—may reach high enough temperatures that we need a full solution. We also need to keep in mind that the kind of objects we try to simulate may have vastly different temperature at different densities. The likely upshot of this is that we will always need a robust representation of the thermal effects. Fairly recent work has added a twist to this, highlighting that the notion of beta-equilibrium changes for non-zero temperatures [9]. Thermal effects lead to an imbalance in the forwards and backwards Urca reaction that shift the equilibrium. This, in turn, softens the equation of state [78]. As this may have observational implications, it is important to incorporate this new notion of "warm" equilibrium in models of neutron star dynamics.

Turning to a sample of results from simulations, 1D modelling of proto-neutron stars [76] shows that in general there are two regions in which modes can propagate, a core region where trapped g modes can exist, and a surface region where g and p-modes can be present, separated by a convective region in which modes cannot propagate (we will discuss the nature of these modes in the following section). These conclusions are consistent with the results of 3D simulations, in which most of the energy is radiated away as gravitational waves by lower order g-modes and by the $l = 2$ f-mode [136]. The f-mode holds the best prospects for detection, given that it is the easiest to excite (and has a higher predicted amplitude).

More recently, simulations have tracked the evolution of the f-mode frequency for several seconds following core collapse [136]. The results show that during the collapse, as the core becomes more compact and the average density increases, the frequency of the f-mode sweeps through the most sensitive region for ground based detectors, from several hundreds of Hz up to ≈ 1 kHz [7, 60]. The late time frequency can be, quite independently of the specific equation of state, fitted quite well in terms of the average density of the proto neutron star, as we have seen in (9.39) [127, 132, 137]. Modelling the full spectrum, including g and p-modes, accurately obviously requires an understanding of complex physical mechanisms at work during the supernova, but the above examples nevertheless show how the detection of an f-mode signal could be used to estimate the average density of the proto neutron star, and therefore help constrain the mass and radius of the newly born object. The frequency evolution of the mode frequency during the first few seconds also sheds light on thermal effects (including the rate of neutrino driven cooling).

9.4 ADDING PHYSICS: STRATIFIED AND COMPRESSIBLE FLUIDS

We now add more physics to the problem by accounting for a varying matter composition and possible thermal gradients. We demonstrate how the associated buoyancy leads to the presence of gravity g-modes. We also account for compressibility, which leads to the pressure p-modes.

9.4.1 Buoyancy

Moving on beyond simple barotropic models for neutron star oscillations, we want to account for composition gradients and thermal stratification. These variations lead to buoyancy, which supports a family of low-frequency gravity g-modes.

From a perturbative perspective, any fluid motion tends to drive the system away from chemical and thermal equilibrium. As a result, displaced fluid elements will be distinct from their surroundings. This leads to buoyancy. The problem now becomes difficult because we have to consider the relevant timescales associated with heat flow (and cooling) as well as the rate at which nuclear reactions work to reinstate chemical balance. Both effects involve transport coefficients, specifically the thermal conductivity and bulk viscosity. There are, however, two limits in which the problem simplifies. We may, for example, consider the transport timescales to be much faster than the dynamics we are interested in (the mode oscillations). In this limit, the fluid remains in equilibrium at all times. The opposite limit involves much slower relaxation times. In this case the fluid element does not have time to react to its surroundings and so effectively retains its identity. The matter composition is frozen.

In the fast timescale setting we do not have to worry about buoyancy. The matter remains in local equilibrium and hence the equation of state can be taken to be barotropic. In the slow-timescale limit we have to be more careful. Taking the composition to be frozen, we have $dY_e = 0$ and the perturbed pressure follows as

$$dp = \left(\frac{\partial p}{\partial n}\right)_{T,Y_e} dn + \left(\frac{\partial p}{\partial T}\right)_{n,Y_e} dT \ . \tag{9.60}$$

Now, we typically also impose that the variation is adiabatic. Thus, we need, first of all the specific entropy $S = s/n$ and then

$$dS = \left(\frac{\partial S}{\partial n}\right)_{T,Y_e} dn + \left(\frac{\partial S}{\partial T}\right)_{n,Y_e} dT = 0 \tag{9.61}$$

leads to

$$dp = \left[\left(\frac{\partial p}{\partial n}\right)_{T,Y_e} - \left(\frac{\partial S}{\partial T}\right)^{-1}_{n,Y_e}\left(\frac{\partial S}{\partial n}\right)_{T,Y_e}\left(\frac{\partial p}{\partial T}\right)_{n,Y_e}\right] dn. \tag{9.62}$$

Hence, for Lagrangian perturbations, we have

$$\Delta p = \left[\left(\frac{\partial p}{\partial n}\right)_{T,Y_e} - \left(\frac{\partial S}{\partial T}\right)^{-1}_{n,Y_e}\left(\frac{\partial S}{\partial n}\right)_{T,Y_e}\left(\frac{\partial p}{\partial T}\right)_{n,Y_e}\right] \Delta n \ . \tag{9.63}$$

In the mode calculation, this relation of commonly written as

$$\Delta p = \frac{p\Gamma_1}{n} \Delta n \ . \tag{9.64}$$

The key point here is that the adiabatic index of the perturbations, Γ_1, is different from that of the background, Γ. Given our relative ignorance of the equation of state it is often assumed that Γ_1 is constant. This is not a realistic assumption, but it is convenient as it allows the mode calculation to proceed without pausing to worry about the precise microphysics and the various thermodynamical derivatives (see [25] for an example of a more realistic model). Of course, if we strive for realism then this attitude will not take us very far.

9.4.2 Return to the mode problem

Let us return to the perturbation equations for non-rotating stars. Generalising (9.30), as required for compressible fluids, we assume that the oscillation modes—with label n and frequency ω_n—are associated with a displacement vector (expressed in the coordinate basis associated with the spherical polar coordinates $[r, \theta, \varphi]$) of form

$$\xi^i_n(t, r, \theta, \varphi) = \xi^i(r, \theta, \varphi)e^{i\omega_n t} . \tag{9.65}$$

We then introduce a basis built from $\hat{e}_r$, ∇Y^l_m—the components we need for a general polar mode like the f-mode—and $\hat{e}_r \times \nabla Y^m_l$, which introduces the axial perturbations. This leads to

$$\xi^i = \sum_l \left[\frac{1}{r} W_l Y^m_l \delta^i_r + \left(\frac{1}{r^2} V_l \partial_\theta Y^m_l + \frac{m}{r^2 \sin\theta} U_l Y^m_l \right) \delta^i_\theta \right.$$
$$\left. + \frac{i}{r^2 \sin^2\theta} \left(m V_l Y^m_l + U_l \sin\theta \partial_\theta Y^m_l \right) \delta^i_\varphi \right] , \tag{9.66}$$

where W_l and V_l correspond to polar perturbations, while U_l is axial. It it easy to show that for non-rotating fluid stars we cannot have a purely axial mode with a non-zero frequency, so we will (in the first instance) consider purely polar modes, with $U_l = 0$. Along with this displacement, all scalar perturbations are expanded in spherical harmonics. For example, for the perturbed mass density we have

$$\delta\rho_n = \sum_l \delta\rho_l Y^m_l e^{i\omega_n t} \tag{9.67}$$

and similar for all other scalar quantities. In the following, whenever p, ρ, Φ are used without δ they refer to the value of the pressure, density and gravitational potential, respectively, of the background equilibrium star. As we are ignoring rotation, the unperturbed star is spherical so all background quantities are functions of r. In addition, the spherical symmetry means that our equations are degenerate in m and hence we may effectively ignore the dependence on the azimuthal angle.

Turning to the perturbed Euler equations, we now have

$$-\omega^2_n \xi_i + \frac{1}{\rho} \nabla_i \delta p - \frac{1}{\rho^2} \delta\rho \nabla_i p + \nabla_i \delta\Phi = 0 . \tag{9.68}$$

This leads to the radial Euler component

$$\partial_r \delta p_l - \left(\frac{\partial_r p}{\rho} \right) \delta\rho_l = \frac{\omega^2_n \rho}{r} W_l - \rho \partial_r \delta\Phi_l , \tag{9.69}$$

and from the φ-component of the Euler equation, we get

$$\omega^2_n V_l = \delta\Phi_l + \frac{\delta p_l}{\rho} . \tag{9.70}$$

We also need the perturbed continuity equation, which (after using (9.70)) can be written

$$\partial_r (r W_l) + \frac{r W_l}{\rho} \partial_r \rho = -\frac{r^2}{\rho} \delta\rho_l + \frac{l(l+1)}{\omega^2_n} \left(\delta\Phi_l + \frac{\delta p_l}{\rho} \right) . \tag{9.71}$$

Finally, we have the perturbed Poisson equation

$$\nabla^2 \delta\Phi_l = 4\pi G \delta\rho_l \, . \tag{9.72}$$

The system of equations is closed by the equation of state for the perturbations (9.64). This leads to

$$\delta p_l = \frac{p\Gamma_1}{\rho}\delta\rho_l + \frac{W_l}{r}\left(\frac{p\Gamma_1}{\rho}\frac{d\rho}{dr} - \frac{dp}{dr}\right) \, . \tag{9.73}$$

Introducing the background sound speed

$$c_s^2 = \left(\frac{\partial p}{\partial \rho}\right)_{\beta=0} \equiv \frac{p\Gamma}{\rho} \, , \tag{9.74}$$

where Γ (obviously) does not have to be constant, we get

$$\delta p_l = \frac{p\Gamma_1}{\rho}\delta\rho_l + \frac{pW_l}{r\rho}\left(\Gamma_1 - \Gamma\right)\frac{d\rho}{dr} \, , \tag{9.75}$$

or, since we want to use this relation to remove the perturbed density from the discussion

$$\delta\rho_l = \frac{\rho}{p\Gamma_1}\delta p_l + \left(\frac{\Gamma}{\Gamma_1} - 1\right)\frac{d\rho}{dr}\frac{W_l}{r} \, . \tag{9.76}$$

For later convenience, we introduce the density scale height (which is useful as we want to avoid involving an actual stellar background model in the plane-wave analysis that follows)

$$\frac{1}{H} = -\frac{1}{\rho}\frac{d\rho}{dr} \, , \tag{9.77}$$

(where the sign ensures that H is positive). This takes us to

$$\delta\rho_l = \frac{\rho}{p\Gamma_1}\delta p_l - \left(\frac{\Gamma}{\Gamma_1} - 1\right)\frac{\rho W_l}{rH} \, . \tag{9.78}$$

Finally, we define the Brunt-Väisälä frequency as

$$\mathcal{N}^2 = \frac{\rho g^2}{p}\left(\frac{1}{\Gamma} - \frac{1}{\Gamma_1}\right) = \frac{c_s^2}{H^2}\left(1 - \frac{\Gamma}{\Gamma_1}\right) \, , \tag{9.79}$$

where we have used the local gravitational acceleration (the surface value of which we used earlier)

$$g = \frac{d\Phi}{dr} = -\frac{1}{\rho}\frac{dp}{dr} = \frac{c_s^2}{H} \, . \tag{9.80}$$

We now have

$$\delta\rho_l = \frac{\rho}{p\Gamma_1}\delta p_l + \frac{\mathcal{N}^2 H}{c_s^2}\bar{W}_l \, , \tag{9.81}$$

where $\bar{W}_l = \rho W_l / r$.

Finally, rewriting (9.69) and (9.71) we get

$$\partial_r \delta p_l + \frac{c_s^2}{H}\delta\rho_l = \omega_n^2 \bar{W}_l - \rho\partial_r\delta\Phi_l \, , \tag{9.82}$$

and

$$\partial_r\left(r^2\bar{W}_l\right) = -r^2\delta\rho_l + \frac{l(l+1)}{\omega_n^2}\left(\rho\delta\Phi_l + \delta p_l\right) \, . \tag{9.83}$$

These are the equations we need to solve in order to determine the modes of a non-rotating star.

9.4.3 A local analysis

So far, we have not made any simplifications. We now have a choice. Either we solve these equations numerically (following, for example, the steps laid out in [135]) or we try to gain intuition by introducing approximations. Focussing on qualitative aspects, let us make the Cowling approximation. That is, we set $\delta\Phi_l = 0$. Using (9.83) to remove $\delta\rho_l$ from the problem we are then left to solve

$$\partial_r \delta p_l + \frac{\Gamma}{\Gamma_1}\frac{\delta p_l}{H} = \left(\omega_n^2 - \mathcal{N}^2\right)\bar{W}_l \,, \tag{9.84}$$

and

$$\frac{1}{r^2}\partial_r(r^2\bar{W}_l) + \frac{\mathcal{N}^2 H}{c_s^2}\bar{W}_l = \left(\frac{\mathcal{L}_l^2}{\omega_n^2} - \frac{\Gamma}{\Gamma_1}\right)\frac{\delta p_l}{c_s^2} \,, \tag{9.85}$$

where the Lamb frequency is given by

$$\mathcal{L}_l^2 = \frac{l(l+1)c_s^2}{r^2} \,. \tag{9.86}$$

In order to explore the nature of the waves we are interested in, we adopt the plane-wave approach—assuming that the perturbations behave as e^{ikr}—with

$$\hat{p} = \delta p_l \,, \qquad \hat{W} = r^2\bar{W}_l \,, \qquad \partial_r \to ik \,, \tag{9.87}$$

leading to

$$\left(ik + \frac{\Gamma}{\Gamma_1 H}\right)\hat{p} = \left(\omega_n^2 - \mathcal{N}^2\right)\frac{\hat{W}}{r^2} \,, \tag{9.88}$$

and

$$\left(ik + \frac{\mathcal{N}^2 H}{c_s^2}\right)\frac{\hat{W}}{r^2} = \left(\frac{\mathcal{L}_l^2}{\omega_n^2} - \frac{\Gamma}{\Gamma_1}\right)\frac{\hat{p}}{c_s^2} \,. \tag{9.89}$$

What can we learn from this? First consider the barotopic case, where $\Gamma_1 = \Gamma \implies \mathcal{N} = 0$. Then we have (for short wavelengths, so $kH \gg 1$)

$$\left(ik + \frac{1}{H}\right)\hat{p} \approx ik\hat{p} = \omega_n^2\frac{\hat{W}}{r^2} \,, \tag{9.90}$$

$$ik\frac{\hat{W}}{r^2} = -\left(1 - \frac{\mathcal{L}_l^2}{\omega_n^2}\right)\frac{\hat{p}}{c_s^2} \,, \tag{9.91}$$

and the dispersion relation

$$\omega_n^2 = c_s^2 k^2 + \mathcal{L}_l^2 = \left[\frac{l(l+1)}{r^2} + k^2\right]c_s^2 \,. \tag{9.92}$$

This solution represents the sounds waves in the problem. These are the pressure p-modes, which evidently depend on the speed of sound. The higher overtone p-modes have shorter scales (=larger k) so the frequency increases. In a neutron star, we expect to find an infinite set of p-modes with frequencies higher than that of the f-mode.

In the non-barotropic case, we may still assume that Γ_1 is close to Γ, in which case we have

$$ik\hat{p} \approx \left(\omega_n^2 - \mathcal{N}^2\right) \frac{\hat{W}}{r^2}, \tag{9.93}$$

and

$$ik\frac{\hat{W}}{r^2} \approx -\left(1 - \frac{\mathcal{L}_l^2}{\omega_n^2}\right) \frac{\hat{p}}{c_s^2}, \tag{9.94}$$

which lead to

$$k^2 c_s^2 = \left(\omega_n^2 - \mathcal{N}^2\right) \left(1 - \frac{\mathcal{L}_l^2}{\omega_n^2}\right). \tag{9.95}$$

This dispersion relation leads to two sets of roots—we now have a quadratic equation of ω_n^2. We need to have ω_n^2 either greater than both $\mathcal{N}^2$ and $\mathcal{L}_l^2$ or smaller than both of them (this leads to the notion of a propagation diagram for the modes, see for example [135]). In the first case, for high frequencies, it is easy to see that we retain the p-modes from before. In the low-frequency case, we instead get

$$\omega_n^2 \approx \mathcal{N}^2 \frac{\mathcal{L}_l^2}{k^2 c_s^2} = \frac{l(l+1)}{k^2 r^2} \mathcal{N}^2. \tag{9.96}$$

In essence, the introduction of the stratification has added a set of low-frequency modes to the spectrum. These are the gravity g-modes. It is easy to see that, as the wavelength decreases (=larger k) the frequency decreases. In a neutron star, with stable stratification, we expect to find an infinite set of undamped low-frequency g-modes.

As the frequency of the neutron star g-modes—typically of the order of 100 Hz— lies in the sensitive region for ground-based gravitational wave detectors, they are of particular interest. In particular, their detection could provide information on the interior composition of the star. For example, sharp phase transitions, due to, e.g. the appearance of hyperons or quarks could lead to large changes in the Brunt-Väisälä frequency and thus of the mode frequency in (9.96) [133, 142, 143], although careful modelling of the reactions and timescales involved is necessary to accurately model the spectrum and constrain the equation of state [20, 95]. Furthermore, the frequencies of the g-modes are low enough that they may be resonantly excited during a binary inspiral and merger, impacting on the orbital evolution, as we will see shortly.

9.4.4 Dynamical tides

With a basic understanding of the nature of the modes that may exist in a stratified star, let us consider an application: The tidal excitation of stellar oscillations during binary inspiral. The tide raised by a binary companion may impact on the orbital evolution and could leave an observable imprint on the emerging gravitational-wave signal. This problem has two aspects. The static tide—which is typically expressed in terms of the so-called Love numbers [38, 53, 62, 84], encoding the response of the dense matter in the neutron star interior to tidal potential of the secondary—has already been constrained by the data from the GW170817 event, providing insight into the neutron star radius and hence the (cold) matter equation of state [3]. The second aspect, the dynamical tide—associated with specific modes being in resonance with the evolving tidal potential [94, 97, 122]—has not yet been limited by observations. This is expected to change in the future [22, 85, 118] and hence it worth understanding this contribution better. The possibility that future observations may be able to catch resonances associated with the low-frequency g-modes [86, 96] is tantalising as this would be giving insight into the composition and state of matter.

The main strategy for modelling tidal dynamics is to make use of an expansion in terms of the star's oscillation modes. Going back to (9.25) and focussing on non-rotating stars (so $B = 0$) we know that the oscillation modes are harmonic solutions to

$$\partial_t^2 \xi_i + C_{ij}\xi^j = 0, \tag{9.97}$$

with mode frequency ω_n. Now, one can show that with the inner product ($*$ indicating complex conjugation)

$$\langle \eta, \xi \rangle = \int \eta^{i*} \xi_i \, \rho dV \,, \tag{9.98}$$

for generic fluid displacements ξ^i and η^i representing solutions to the perturbation equations, the operator C_{ij} is Hermitian. This means that we have

$$\langle \eta, C\xi \rangle = \langle \xi, C\eta \rangle^* \,. \tag{9.99}$$

It follows that the symplectic structure

$$W(\eta, \xi) = \langle \eta, \partial_t \xi \rangle - \langle \partial_t \eta, \xi \rangle \,, \tag{9.100}$$

is such that $\partial_t W(\eta, \xi) = 0$. This is easy to show. We get

$$\begin{aligned} \partial_t W &= \langle \partial_t \eta, \partial_t \xi \rangle + \langle \eta, \partial_t^2 \xi \rangle - \langle \partial_t^2 \eta, \xi \rangle - \langle \partial_t \eta, \partial_t \xi \rangle \\ &= -\langle \eta, C\xi \rangle + \langle C\eta, \xi \rangle = -\langle \eta, C\xi \rangle + \langle \xi, C\eta \rangle^* = 0 \,. \end{aligned} \tag{9.101}$$

Now consider two modes, identified by labels $n = \alpha$ and β, corresponding to solutions $\xi^i = \xi^i_\alpha e^{i\omega_\alpha t}$ and $\eta^i = \xi^i_\beta e^{i\omega_\beta t}$ and assume that the frequencies are real (as appropriate for Newtonian normal modes, ignoring dissipation), to get

$$\partial_t W(\eta, \xi) = \langle \eta, \partial_t^2 \xi \rangle - \langle \partial_t^2 \eta, \xi \rangle = -(\omega_\alpha^2 - \omega_\beta^2) \langle \xi_\beta, \xi_\alpha \rangle e^{i(\omega_\alpha - \omega_\beta)t}. \tag{9.102}$$

Assuming no degeneracy in the frequencies, $\omega_\alpha \neq \omega_\beta$, this shows that the two mode solutions are distinct only if they are orthogonal to one another. In effect, we have

$$\langle \xi_\beta, \xi_\alpha \rangle = \mathcal{A}_\alpha^2 \delta_{\alpha\beta}, \tag{9.103}$$

where $\mathcal{A}_\alpha^2$ is a normalisation constant. This result allows us to build a mode-sum representation for the star's response to an external tidal field.

The tidal deformation is a solution to the inhomogeneous equation

$$\partial_t^2 \xi_i + C_{ij}\xi^j = -\nabla_i \chi, \tag{9.104}$$

where the tidal potential χ due to the secondary (with mass M') is

$$\chi = -\frac{GM'}{|x^i - D^i(t)|} = -\sum_{l\geq 2}\sum_{m=-l}^{l} v_{lm} r^l Y_{lm} e^{-im\Phi(t)}, \tag{9.105}$$

with D the binary separation, Φ the orbital phase and

$$v_{lm} = \frac{GM'W_{lm}}{D^{l+1}(t)}. \tag{9.106}$$

For the leading order contribution to the gravitational-wave signal ($l = 2$) the coefficients W_{lm} are $W_{20} = -\sqrt{\pi/5}$, $W_{2\pm2} = \sqrt{3\pi/10}$ while $W_{2\pm1} = 0$.

Inserting the mode-sum decomposition

$$\xi^i(t, x^i) = \sum_n a_n(t)\xi_n^i(x^i), \tag{9.107}$$

into the Euler equation (9.104) and noting that $C_{ij}\xi_n^j = \omega_n^2\xi_i^n$ (keeping the mode label out of the way of the spatial index) we get

$$\left(\ddot{a}_n + \omega_n^2 a_n\right)\xi_i^n = -\nabla_i\chi. \tag{9.108}$$

Now making use of the orthogonality relation (9.103) we arrive at an equation of motion for the mode amplitudes

$$\ddot{a}_n + \omega_n^2 a_n = -\frac{1}{A_n^2}\langle\xi_n, \nabla\chi\rangle. \tag{9.109}$$

The problem is, effectively, that of a driven harmonic oscillator.

At this point, it is natural to define the overlap integral

$$Q_n \equiv -\langle\xi_n, \nabla\chi\rangle = -\int \delta\rho_n^*\chi dV, \tag{9.110}$$

where $\delta\rho_n$ is the density perturbation associated with the mode (and we have integrated by parts). Through Q_n the tidal perturbation of the star inherits the time dependence from χ, i.e. links to the orbital evolution.

If we now assume that the orbital evolution is adiabatic (so that we can take D to be "constant") and expand the perturbations in spherical harmonics, then we have

$$\delta\rho_n(x^i) = \delta\rho_{nl}(r)Y_l^m(\theta, \phi). \tag{9.111}$$

(as we know that the modes for different multipoles decouple in non-rotating stars and the modes do not depend on m). If we also introduce the orbital frequency such that $\Phi = \Omega t$ then the overlap integral is simply written as

$$Q_n = v_{lm} I_n^* e^{-im\Omega t}, \tag{9.112}$$

where the mass-multipole moment for each mode is given by

$$I_n \equiv \int_0^R \delta\rho_{nl}(r) r^{l+2} dr. \tag{9.113}$$

The final steps are easy. With our assumptions, each mode is driven by the tidal field at frequency $m\Omega$. As long as the orbital motion is not at resonance with the mode, the amplitude equation leads to

$$a_n(t) = \frac{v_{lm} I_n^*}{\mathcal{A}_n^2[\omega_n^2 - (m\Omega)^2]} e^{-im\Omega t}. \tag{9.114}$$

Basically, each mode (n) carries an (l, m) dependence and will be excited by the corresponding multipole component of the tidal field χ. However, the tidal forcing is only non-zero for even $l - m$, so modes with odd $l - m$ will not be relevant.

Now we are pretty much done. The final step is to express the tidal response in terms of the Love number k_{lm} which is defined by the relation (quantifying the extent to which the star is deformed by the tidal potential)

$$\delta\Phi_{lm} = 2k_{lm}\chi_{lm}, \tag{9.115}$$

evaluated at the surface the star. This effective Love number (for a time-varying tidal field) encodes the dependence on the orbital motion and represents both the static and the dynamical tide.

Recall the (l, m) component of the tidal potential from (9.105) and note that

$$\delta\Phi = \sum_n a_n \delta\Phi_n , \tag{9.116}$$

where a given mode sources

$$\delta\Phi_n(R) = -\frac{4\pi G}{(2l+1)R^{l+1}} I_n , \tag{9.117}$$

(which follows after using the Green's function for the Laplacian) at the stellar surface. This leads to

$$\delta\Phi(t, R, \theta, \phi) = -4\pi G \sum_n \frac{1}{(2l+1)R^{l+1}} \frac{I_n^2}{\mathcal{A}_n^2[\omega_n^2 - (m\Omega)^2]} v_{lm} e^{-im\Omega t} Y_l^m . \tag{9.118}$$

The sum in this expression includes all the modes. Labelling the subset of modes (f,p g, ...) for a given harmonic (l, m) by n' we have

$$\delta\Phi_{lm}(t, R) = -\frac{4\pi G}{(2l+1)R^{l+1}} v_{lm} e^{-im\Omega t} \sum_{n'} \frac{I_{n'}^2}{\mathcal{A}_{n'}^2[\omega_{n'}^2 - (m\Omega)^2]} , \tag{9.119}$$

and it follows from (9.115) that the mode-sum expression for the tidal response is

$$k_{lm} = \frac{2\pi G}{(2l+1)R^{2l+1}} \sum_{n'} \frac{I_{n'}^2}{\mathcal{A}_{n'}^2[\omega_{n'}^2 - (m\Omega)^2]}. \tag{9.120}$$

This provides an effective (frequency dependent) Love number. Essentially, because we are using (9.114) which holds only away from resonance, the results only encodes what is often called the "equilibrium" tide. If we want an expression valid through resonance then we have to do a little bit more work on the solution for the mode amplitude, but the general framework remains unchanged.

Finally, the static Love number follows in the limit $\Omega \to 0$ and is simply given by

$$k_l = \frac{2\pi G}{(2l+1)R^{2l+1}} \sum_{n'} \frac{I_{n'}^2}{\mathcal{A}_{n'}^2 \omega_{n'}^2}, \tag{9.121}$$

providing a direct link between the static tide and asteroseismology. The relative importance of the modes in (9.121) is demonstrated by the results from [21, 114]. The clear consensus is that the f-mode provides (by far) the dominant contribution to the tidal response, but other modes may nevertheless be within the reach of future observations.

The expression for the equilibrium tide (9.120) includes the effect of resonances (which have so far only been included in gravitational-wave phase modelling through somewhat ad-hoc terms [85]). If we add the solution for the mode amplitude through each resonance [97] then we have the general frequency dependent evolution of the tidal response close to merger (including all modes that may come into play) while at the same time converging to the standard static Love number in the low-frequency limit. The picture that emerges is complex, as it depends on the specific nature of the set of modes used to evaluate the sum, which in turn depends on the composition and state of matter [114, 115, 141]. However, the main contribution to the expansion of the tidal deformation comes from the f-mode. This is as expected because this mode is essentially involves a "shape deformation" of the star, and therefore this mode most closely resembles the deformation due to tides.

Phenomenological models have been constructed based only on an f-mode expansion [22], and it has been demonstrated that models for the evolution of the gravitational-wave signal that include an f-mode do a better job of recovering the full non-linear dynamics from numerical relativity simulations. The effect is expected to be mainly due to the pre-resonance part of the tide, as the f-mode may not be resonant (for a typical equation of state) before the stars touch. There is good evidence for this from numerical simulations which do not show the actual f-mode resonance [68]. However, the effect is still significant as the pre-resonance evolution enhances the tide. This feature is expected to be detectable [119]. We also need to explore the possibility that pre-merger resonances of low-frequency modes may (just about) become detectable with future instruments. The possibility of detecting g-mode features [86] is particularly exciting as it would give insight into the matter composition.

Before moving on let us also note that the gravitational-wave signal may also be affected by a non-linear instability that leads to the growth of a high frequency p-mode coupled to a low frequency g-mode; the so called pg-instability [140]. The amplitude of the supposedly unstable modes was, however, constrained to be small by data from GW170817 and the instability is therefore not expected to play a dominant role in the orbital evolution [86].

9.5 MORE PHYSICS: ELASTICITY AND SUPERFLUIDITY

We consider features associated with the state of matter in a neutron stars. In particular, we outline how the seismology problem changes when we consider aspects like the crust elasticity and core superfluidity.

Neutron stars are born hot—at a temperature reaching well above 1 MeV, or 10^{10} K in everyday units. However, the star rapidly cools through the emission of neutrinos generated (mainly) from the Urca reactions. As a result, mature neutron stars are cold. Very cold, in fact. With a typical core temperature of order $10^6 - 10^7$ K, these systems are many orders of magnitude below the natural nuclear physics reference temperature; the Fermi temperature of the baryons, which is of order 10^{12} K. In effect, neutron star cores are firmly in the realm of low-temperature physics.

We know from both theory and laboratory experiments that matter may either freeze to a solid or form a superfluid as the temperature drops. The superfluid phase is preferred if quantum fluctuations dominate on the short inter-atomic scale, preventing the formation of an ordered lattice. So, for example, Helium forms a superfluid condensate below about 2.7 K, and there have been many exciting experiments on that system, as well as on its (more recently discovered) atomic Bose-Einstein condensate cousins. A solid phase is, of course, more familiar to us. Just think of rivers and lakes freezing over during a cold winter. Or perhaps ice cubes cooling your drink on a hot summer day.

Neutron stars manifest both solid and superfluid phases (in a somewhat bewildering landscape of combinations depending on the matter composition). The outer kilometer or so of a mature neutron star is generally cold enough that ions can form a crystalline structure, and one has an elastic crust (for a review see [49]). This can be seen by analysing the ratio Γ of thermal to Coulomb energies of the ions, i.e.

$$\Gamma = \frac{Z^2 e^2}{a_i k T}, \tag{9.122}$$

with Z the atomic number of the nuclei involved, e the elementary electric charge, k Boltzmann's constant, T the temperature and a_i the inter-ion spacing. If $\Gamma > 1$ we have a liquid, if $\Gamma > 175$ a solid (treating the ions as point particles), a condition which is met at densities of around

$$\rho_s = 100 \left(\frac{T}{10^6 \text{ K}} \right)^3 \text{ g/cm}^3 . \tag{9.123}$$

It is clear from the estimate above in (9.123) that for mature stars with internal temperatures below $\approx 10^9$ K, the outer layers below $\rho \approx 10^{14}$ g/cm^3 form a solid crust. We can also see, however, that for any temperature, at sufficiently low density (generally below 10^7 g/cm^3) there will always be a liquid region, a shallow ocean which can impact on the dynamics of modes [72].

Meanwhile, the outer core of the star is expected to contain a mixture of superfluid neutrons—forming a condensate due to an analogue of Cooper pairing, below a density-dependent critical temperature—alongside a charge neutral conglomerate of protons and electrons (with muons also entering as the density increases and more exotic superfluid phases—hyperons and/or quark colour superconductors anticipated in the deep neutron star core). Ignoring the more complex options, we need to consider two features. First, beyond a density of about 4×10^{11} g/cm^3, neutrons start to drip out of crust nuclei, leading to a superfluid gas coexisting with the crust lattice. Second, the protons in the outer core are expected to form a superconductor. The nature of the superconducting phase is not clear. It may, in fact, vary with depth. At some densities, we may have a type II superconductor, in which the magnetic field is carried by quantised fluxtubes. At other densities, we may be dealing with the type I version, in which the Meissner effect likely leads to the formation of macroscopic non-superconducting regions. We are not going to discuss the issue of superconductivity in any detail here. Suffice it to say that there are a number of important questions left to answer in that direction.

9.5.1 The elastic crust

Dealing first with the crust, we need to add the elastic restoring force to the Euler equation. Assuming the background model is relaxed (so that there are no elastic strains in absence of perturbations), we have

$$-\omega_n^2 \xi_i + \frac{1}{\rho}\nabla_i \delta p - \frac{\delta \rho}{\rho^2}\nabla_i p + \nabla_i \delta \Phi - \frac{1}{\rho}\nabla^j \sigma_{ij} = 0 \,, \tag{9.124}$$

where the elastic stress tensor σ_{ij} is given by

$$\sigma_{ij} = \check{\mu}\left(\nabla_i \xi_j + \nabla_j \xi_i\right) - \frac{2}{3}\check{\mu} g_{ij}\left(\nabla_k \xi^k\right) \,, \tag{9.125}$$

with $\check{\mu}$ denoting the shear modulus. The perturbations also need to satisfy junction conditions (continuity of the components of the traction) at both the top and the bottom of the elastic region.

The fluid modes we already considered will be altered by elasticity, and new modes come into play. Simplistically, the problem allows for a new set of standing waves in the elastic region. We can see this by considering an incompressible model for which $\delta\rho = 0$. In this case, it is somewhat easier to solve equation (9.124), and we find the standard sound waves modified by the presence of the elastic medium, but also a new kind of transverse wave, such that $k_i\xi^i = 0$ (where k_i is the wave vector), which only exists in the presence of elasticity. The dispersion relation of these shear modes is

$$\omega^2 = \frac{\check{\mu}}{\rho}k^2 \,, \tag{9.126}$$

so their frequencies scale with the shear modulus $\check{\mu}$. The fact that the elastic degrees of freedom allow for transverse waves is important. Translating into a full spherical analysis, the elasticity allows for the presence of axial (toroidal) modes with $U_l \neq 0$ already in a non-rotating star. These may lead to resonances with other toroidal modes, like the r-modes in rotating stars [73, 101].

9.5.2 Magnetar QPOs

The physics of the elastic crust is thought to play an important role in strongly magnetised neutron stars, the magnetars. These are young systems for which the observed highly energetic emission in X-rays and gamma rays is thought to be powered by a strong magnetic field, B$\approx 10^{15}$ G. Some of the observed systems exhibit short bursts of hard X-rays and soft gamma-rays, known as "giant flares". These energetic events provide interesting insights. In the hard X-ray tail of the 2004 giant flare of SGR 1806-20, quasi-periodic oscillations (QPOs) were discovered at frequencies of roughly 18, 30, 93, 150, 625 and

1840Hz [91, 130, 139]. Subsequently, QPOs with frequencies of 28, 54, 84 and 155Hz were detected in the tail of the 1998 giant flare of SGR 1900+14 [129]. Additional evidence for QPOs in the giant flares of other objects, and also in some short bursts, has also been found (see [134] for a review). The leading explanation for the observed oscillations, dating back to predictions from [58], is that they are linked to torsional oscillations of the crust.

We can see that this explanation is reasonable by considering the shear mode dispersion relation from (9.126). For typical values for the shear modulus in the crust [128], we can estimate the fundamental shear mode frequency $\nu = \omega/2\pi$ as

$$\nu = 18 \left(\frac{\check{\mu}/\rho}{10^8 \, \mathrm{cm}^2 \, \mathrm{s}^{-2}} \right)^{1/2} \left(\frac{10 \, \mathrm{km}}{R} \right) \, \mathrm{Hz}. \tag{9.127}$$

It is thus natural to associate the lower frequencies that have been observed with nodeless modes for different angular harmonics, while the two highest frequencies in the giant flare of SGR 1806-20 may represent radial overtones. In fact, given that the full modelling in spherical geometry introduces additional dependence on mass and radius—different for different sets of modes—detecting additional overtones can break degeneracies and potentially constrain the equation of state of dense matter [124].

In practice, however, this procedure is not straightforward, as precise modelling of crustal modes requires an understanding of the magnetic field in the star. The shear modes can couple to Alfven modes in the core and potentially be strongly damped as energy is transferred to them [75, 100]—and a detailed understanding of both the magnetosphere and the state of matter in the core of a neutron star (in particular, if the protons in the core are superconducting) is needed [19].

Despite the complications, coupled crust-core oscillations remain the most favoured explanation for the QPO observations. We only have to accept that there is work to be done and that additional observations will be needed to break the many degeneracies present in the mode calculation and truly constrain the properties of dense matter.

9.5.3 The superfluid core

Superfluidity further complicates the problem. In essence, we need to model the neutron star interior (beyond neutron drip) as two (coupled) fluid components. There are different way to write the relevant equations, involving different combinations of the two dynamical degrees of freedom, but an intuitive formulation distinguishes (eventually, we may need more than two components, but the extension is fairly straightforward): the superfluid neutrons, with mass density ρ_n and velocity v_n^i, from the "protons" (really, a conglomerate of all charged components: everything that is not a neutron), represented by ρ_p and v_p^i. In absence of dissipation, each component is represented by a continuity equation (mass conservation)

$$\partial_t \rho_\mathrm{x} + \nabla_i \left(\rho_\mathrm{x} v_\mathrm{x}^i \right) = 0 \, , \tag{9.128}$$

where the "chemical label" is either $\mathrm{x} = \mathrm{n}$ or p. The main point of this description is that the equations are neatly "symmetric". Next, momentum conservation leads to

$$\left(\partial_t + v_\mathrm{x}^j \nabla_j \right) \left(v_i^\mathrm{x} + \varepsilon_\mathrm{x} w_i^\mathrm{yx} \right) + \nabla_i (\Phi + \tilde{\mu}_\mathrm{x}) + \varepsilon_\mathrm{x} w_j^\mathrm{yx} \nabla_i v_\mathrm{x}^j = 0 \, . \tag{9.129}$$

where $y \neq x$ and the relative velocity between the components—a key part of the story—is defined as

$$w_i^{yx} = v_i^y - v_i^x \,. \tag{9.130}$$

In this description, the matter equation of state is represented by an internal energy $\mathcal{E} = \mathcal{E}(n_n, n_p, w_{np}^2)$, notably depending on the relative velocity [120]. The chemical potential for each particle species is now given by

$$\mu_x = m_x \tilde{\mu}_x = \left(\frac{\partial \mathcal{E}}{\partial n_x} \right)_{n_y, w_{np}^2} , \tag{9.131}$$

with m_x the mass of each particle. We also have to consider the entrainment effect, associated with the coefficients

$$\varepsilon_x = \frac{2\alpha}{\rho_x} \quad \text{where} \quad \alpha = \left(\frac{\partial \mathcal{E}}{\partial w_{np}^2} \right)_{n_x} . \tag{9.132}$$

As should be apparent from (9.129), the main novelties brought by superfludity are the two dynamical degrees of freedom (two velocities) and the entrainment coupling. The first of these has a very intuitive implication: The number of fluid oscillation modes will "double". As discussed by, for example, [16] we may describe the superfluid oscillation in terms of two displacements:

$$\rho \xi^i = \rho_n \xi_n^i + \rho_p \xi_p^i \quad \text{and} \quad \psi^i = \xi_p^i - \xi_n^i, \tag{9.133}$$

The first represents the total momentum and the second the difference. The modes of the system then tend to be dominated by the two fluids either moving together in lock-step (dominated by ξ^i) or out of sync (dominated by ψ^i) [16].

An intuitive understanding of the entrainment follows if we note that the fluid momenta are given by

$$p_i^x = m_x \left(v_i^x + \varepsilon_x w_i^{yx} \right) , \tag{9.134}$$

This illustrates the fact that the momentum of each fluid need not be parallel with that fluid's velocity. Furthermore, we can introduce the effective proton mass (say) in a frame moving with the neutrons (setting $v_n^i = 0$):

$$m_p (1 - \varepsilon_p) v_p^i \equiv m_p^* v_p^i , \tag{9.135}$$

with a similar relation for the neutrons defining m_n^*. We then have

$$2\alpha = \rho_p \varepsilon_p = n_p (m_p - m_p^*) \,. \tag{9.136}$$

This notion of a "dynamical effective mass" encodes an important expectation from nuclear physics [13, 44, 106]. In a neutron star core, the strong force endows each neutron with a virtual cloud of protons (and vice versa). This impacts on the effective mass of the particle and, when it moves, alters the particle's momentum. In effect, the mass of a neutron appears different from what it would be in isolation. The entrainment is known to be important for quantitative models of radio pulsar glitches [18].

9.5.4 The quantised vortices

Up to now we have discussed a non-rotating superfluid. Neutron stars, however, rotate, and it is well known from laboratory experiments on superfluid Helium and cold atomic gases that a superfluid in a rotating container is threaded by an array of quantum vortices. These vortices are topological defects that carry the vorticity of the system and "break" the superfluidity locally, allowing for the (irrotational) superfluid to mimic solid body rotation. They are particularly important as they mediate a dissipative interaction between the superfluid and the normal fluid, the so-called "mutual friction", which can have a strong impact on the oscillation spectrum of the star (see [24] for a review).

Essentially, this involves adding a force f_j to the right-hand side of the equation (9.129), which therefore takes the form

$$\left(\partial_t + v^j_{\rm x}\nabla_j\right)\left(v^{\rm x}_i + \varepsilon_{\rm x} w^{\rm yx}_i\right) + \nabla_i(\Phi + \tilde{\mu}_{\rm x}) + \varepsilon_{\rm x} w^{\rm yx}_j \nabla_i v^j_{\rm x} = f^{\rm x}_i/\rho_{\rm x} \,. \tag{9.137}$$

For laminar flows (we will not discuss turbulent flows, or other non-linear forms of mutual friction here), the mutual friction takes the form—well known from the study of superfluid Helium:

$$f^{\rm x}_j = 2\Omega_{\rm n}\kappa\rho_{\rm n}\left(\mathcal{B}'\epsilon_{ijk}\hat{\kappa}^i w^k_{\rm np} + \mathcal{B}\epsilon_{ijk}\hat{\kappa}^i\epsilon^{klm}\hat{\kappa}_l w^{\rm np}_m\right), \tag{9.138}$$

where $\kappa = h/2m_{\rm n}$ is the quantum of circulation, and the vector κ^i is directed along the vortex axis (in the direction of the superfluid angular velocity vector), such that, given a vortex density $\mathcal{N}_{\rm n}$, one has $2\Omega^i_{\rm n} = \mathcal{N}_{\rm n}\kappa^i$. The strength of the mutual friction depends on the dimensionless parameters $\mathcal{B}'$ and $\mathcal{B}$, which encode the microphysical interactions at play. In general, it is expected that $\mathcal{B}' \ll \mathcal{B}$, and $\mathcal{B} \approx 10^{-4}$ in the core of a mature neutron star (see [83] for a review of possible mechanisms giving rise to mutual friction). As we shall see, this means that once the star is old enough to cool below the superfluid transition temperature, mutual friction will play an important role in mode damping, especially for dominantly polar modes like the f-mode.

9.6 NON-EQUILIBRIUM ASPECTS/TRANSPORT

Having added physics that enter already at the level of the unperturbed background, we now consider aspects that come into play when the matter is in motion. Any expansion/compression or shearing of the fluid will introduce non-equilibrium aspects—the most commonly considered being bulk- and shear viscosity—which tend to add damping to the fluid motion. These features, in turn, depend on transport coefficients which need to be calculated at the microphysics level, beyond the equilibrium equation of state.

The gradual evolution of a neutron star is dominated by physics that act on a timescale much longer than the frequency of a typical oscillation mode. For example, the Urca reactions that dictate the rate at which the star cool through its lifetime act on a timescale of months to years in a mature neutron star. Because the associated timescale tends to be long, the effects can be subtle. Especially if we focus on the dramatic neutron star merger dynamics. Having said that, there is recent evidence that the bulk viscosity associated with nuclear reactions in a merger may leave an imprint detectable with next-generation gravitational-wave instruments [79, 109], so the problem is worth our attention. Similarly, it is well known that modes that become unstable—for example, due to the CFS mechanism, we discuss later—may be suppressed by viscosity, often making the outcome sensitive to temperature. In essence, more physics come into play. For now, let us focus on the bulk viscosity (for an in-depth review, see [125]).

9.6.1 Bulk viscosity

In order to understand the emergence of bulk viscosity, we need to relax the assumption we made in the discussion of stratification due to internal composition/temperature gradients. More generally, once we account for out-of-equilibrium nuclear reactions, the perturbed lepton fraction Y_e evolves according to (for small deviations from equilibrium—i.e. in what is known as the sub-thermal limit [8, 10])

$$(\partial_t + v^j \nabla_j)\Delta Y_\mathrm{e} = \frac{\gamma}{n}\Delta\beta \tag{9.139}$$

Here $\Delta\beta$ represents the deviation from (cold) beta-equilibrium and γ encodes the (dominant) reaction rate. Considering β as a function of ρ and Y_e, and assuming that the star is non-rotating (so that $v^i = 0$), we have

$$\partial_t \Delta\beta = \left(\frac{\partial\beta}{\partial\rho}\right)_{Y_\mathrm{e}} \partial_t \Delta\rho + \left(\frac{\partial\beta}{\partial Y_\mathrm{e}}\right)_{\rho} \partial_t \Delta Y_\mathrm{e} \tag{9.140}$$

which, once we consider (9.139), becomes

$$\partial_t \Delta\beta = \left(\frac{\partial\beta}{\partial\rho}\right)_{Y_\mathrm{e}} \partial_t \Delta\rho + \left(\frac{\partial\beta}{\partial Y_\mathrm{e}}\right)_{\rho} \frac{\gamma}{n}\Delta\beta \tag{9.141}$$

That is,

$$\partial_t \Delta\beta - \mathcal{A}\Delta\beta = \mathcal{B}\partial_t\Delta\rho \tag{9.142}$$

with

$$\mathcal{A} = \left(\frac{\partial\beta}{\partial Y_\mathrm{e}}\right)_{\rho} \frac{\gamma}{n}, \qquad \mathcal{B} = \left(\frac{\partial\beta}{\partial\rho}\right)_{Y_\mathrm{e}} \tag{9.143}$$

The coefficients $\mathcal{A}$ and $\mathcal{B}$ are time independent, as they are evaluated for the equilibrium background, so if we work in the frequency domain (as we typically do when we consider stellar oscillations), then we have a harmonic time dependence $e^{i\omega t}$ and it follows that

$$\Delta\beta = \frac{\mathcal{B}}{1 + i\mathcal{A}/\omega}\Delta\rho \tag{9.144}$$

Now consider the timescales involved. Noting that $\mathcal{A}$ needs to be negative in order for the system to relax towards equilibrium, we introduce a characteristic reaction time as

$$t_R = -\frac{1}{\mathcal{A}} \tag{9.145}$$

We then see that, if the reactions are fast compared to the dynamics (associated with a timescale $\sim 1/\omega$) then $|t_R\omega| \ll 1$ and we have

$$\Delta\beta \approx 0\,. \tag{9.146}$$

In this limit, the fluid elements reach equilibrium before executing an oscillation. The fluid remains in beta-equilibrium and hence the perturbations are (effectively) barotropic.

As a rough estimate of the relevant timescale, we may use [77]

$$t_R \sim 10^{13}\left(\frac{10^8\mathrm{K}}{T}\right)^6 \mathrm{s} \tag{9.147}$$

for the modified Urca reactions. This means that—for all modes/rotation rates we may conceivably be interested in—the matter will not be in the fast-reaction regime. We need to consider the slow-reaction problem.

In the limit of slow reactions, we have $|t_R\omega| \gg 1$ and we can Taylor expand (9.144) to get

$$\Delta\beta \approx \mathcal{B}\left(1 - i\mathcal{A}/\omega\right)\Delta\rho \approx \mathcal{B}\Delta\rho \tag{9.148}$$

Using this result in

$$\Delta p = \left(\frac{\partial p}{\partial \rho}\right)_\beta \Delta\rho + \left(\frac{\partial p}{\partial \beta}\right)_\rho \Delta\beta \tag{9.149}$$

we have

$$\Delta p = \left[\left(\frac{\partial p}{\partial \rho}\right)_\beta + \left(\frac{\partial p}{\partial \beta}\right)_\rho \left(\frac{\partial \beta}{\partial \rho}\right)_{x_\mathrm{p}}\right]\Delta\rho \equiv \frac{p\Gamma_1}{\rho}\Delta\rho \tag{9.150}$$

recovering the result from before (although expressed in terms of different thermodynamical derivatives).

The challenge is, however, not the fast/slow-reaction limits. In many situations, we need to consider the intermediate regime. That is, we have to solve (9.144). This leads to the emergence of bulk viscosity. It is easy to see from (9.144) that this is a resonant effect (as the magnitude of $\Delta\beta$ diverges when $\omega^2 \sim \mathcal{A}^2$). This resonant feature—which strongly depends on temperature and the matter involved—will enter the discussion later when we consider unstable modes. For numerical simulations, the problem is more complicated as we can then not appeal to frequency domain arguments. Instead, one would typically introduce the effect in terms of a bulk-viscous contribution to the pressure. The implementation, especially in General Relativity, of this presents a technical technical challenge that has only recently begun to be addressed [42, 43, 46, 69, 70, 109].

9.6.2 Post-merger oscillations

A binary neutron star merger can have four possible outcomes, depending on the mass of two stars and the equation of state of dense matter. The first possibility is that the binary is massive enough that the merger product collapses promptly to a black hole. For lower masses, the merger product may be a hypermassive neutron star, supported by differential rotation, but subsequently collapsing—on a timescale of seconds—as the star spins down [31]. For even lower masses, a supramassive neutron star may be formed, which still has a higher mass that the maximum mass of a static star, but is supported by rigid rotation. Again, it will collapse once the star has spin down enough, but on the much longer timescale of timescale of $10-10^4$ s [121]. Finally, if the merging neutron stars are light enough and the equation of dense matter is stiff enough, the merger may result in a stable, lasting neutron star.

We therefore expect to see, except for the case in which the merger leads to prompt collapse to a black hole, a ringdown phase in which the quasinormal modes of the stable, or metastable, neutron star are excited. In particular, we expect—and numerical simulations confirm [36]—that most of the energy will be emitted through the $l = 2$ f-mode of the hot remnant, with a frequency around 1-4 kHz. Other peaks are present in simulated spectra, due to the presence of other modes and also non-linear couplings to higher order modes [35]. The overall picture is complicated but a measurement of the f-mode frequency from a merger remnant—a feature which appears extremely robust—would give us a handle on the mass and equation of state of the system [32].

If the star is supramassive (or even stable), there is also the possibility of longer-lived emission due to f-modes and possibly r-modes. As we will see in the next section, it is possible for these modes to undergo an instability that allows them to grow to large amplitude and emit gravitational waves for relatively long periods of time, until the secular evolution of the neutron star spin and temperature evolves the system out of the unstable region. For merger remnants, the f-mode is expected to be the most efficient at emitting gravitational waves, but the r-mode may remain unstable for longer, and the relative importance therefore depends mainly on the unknown—or at least, not very precisely known...—saturation amplitude of the modes [57, 117].

Searches were carried out for post-merger signals after the GW170817 binary merger event, but they were not sensitive enough to reach the theoretically expected values (the upper limits were approximately an order of magnitude above the theoretical expectations), and in fact no signal was found [1, 2]. Nevertheless, future searches, especially with next-generation detectors, with increased sensitivity in the kHz range, and more refined data analysis techniques, may allow for the detection of signals from the oscillations of a merger remnant. This is an exciting prospect as it would turn neutron star mergers into a cosmic laboratory for extreme density physics.

9.7 ADDING ROTATION

Recognizing that stars tend to rotate, we discuss how the seismology problem changes when we account for rotation. We explain how rotating stars host new families of modes associated with the Coriolis force and how different classes of waves may be driven unstable due to the emission of gravitational waves.

Stars and planets rotate. Some fast, some slow, but they all tend to have some degree of spin. This impacts on (read: complicates) the seismology problem. The main changes follow from the introduction of the fictitious forces associated with a rotating frame. First, and perhaps most intuitively, the centrifugal force deforms the star. Rapidly spinning objects have a significant centrifugal bulge. This breaks the spherical symmetry of the background on top of which we are trying to work out perturbations and waves. As a result, the oscillation modes are no longer represented by individual spherical harmonics. For faster-spinning stars, we need to include many harmonics and the mode calculations become unwieldy[5]. Essentially, the shape change associated with rotation alters all the modes that are present already in a non-rotating star. In addition, the introduction of the Coriolis force brings new modes into existence. The frequency of these so-called inertial modes is proportional to the rotation rate. Their presence makes the rotating star problem richer. Another feature that comes into play for rotating stars is the possibility of different instabilities. Oscillation modes that are stable in a non-rotating star may become unstable due to the rotation.

9.7.1 The CFS argument

Let us start by outlining the formal instability argument. In the rotating case, the perturbed Euler equation can be written (recall (9.25))

$$\partial_t^2 \xi + B\partial_t \xi + C\xi = 0 \, . \tag{9.151}$$

Moreover, with the inner product (9.98) one can show that

$$\langle \eta, B\xi \rangle = - \langle \xi, B\eta \rangle^* \, , \tag{9.152}$$

and (as before)

$$\langle \eta, C\xi \rangle = \langle \xi, C\eta \rangle^* \, . \tag{9.153}$$

Given these relations and assuming that η and ξ both solve the perturbed Euler equation (9.151)—as before—it is relatively straightforward to show that

$$W(\eta, \xi) = \left\langle \eta, \partial_t \xi + \frac{1}{2} B\xi \right\rangle - \left\langle \partial_t \eta + \frac{1}{2} B\eta, \xi \right\rangle \, , \tag{9.154}$$

is a conserved quantity

The fact that W is conserved motivates the definition of the canonical energy

$$E_c = \frac{1}{2} W(\partial_t \xi, \xi) = \frac{1}{2} \left[\langle \partial_t \xi, \partial_t \xi \rangle + \langle \xi, C\xi \rangle \right] \, , \tag{9.155}$$

and, for axisymmetric systems like rotating stars, the canonical angular momentum

$$\begin{aligned} J_c &= \frac{1}{2} W(\partial_\varphi \xi, \xi) \\ &= \frac{1}{2} \left\{ \left\langle \partial_\varphi \xi, \partial_t \xi + \frac{1}{2} B\xi \right\rangle - \left\langle \partial_{t\varphi}^2 \xi + \frac{1}{2} B\partial_\varphi \xi, \xi \right\rangle \right\} \\ &= - \, \mathrm{Re} \left\langle \partial_\varphi \xi, \partial_t \xi + \frac{1}{2} B\xi \right\rangle \, . \end{aligned} \tag{9.156}$$

These two quantities are also conserved. This fact is key to the seminal Friedman and Schutz proof [64] that the emission of gravitational waves may drive stellar oscillation

[5]One option would be to avoid some of these issues by carrying out time evolutions of the perturbation equations, see for example [67]. This strategy has advantages but also some drawbacks. In particular, for more realistic neutron star models, it is quite difficult to extract specific modes from the numerical results.

modes unstable. In essence, if the oscillating system is coupled to radiation (gravitational waves) which carries away positive energy (in the sense that $\partial_t E_c < 0$), then any initial data for which $E_c < 0$ will lead to an instability.

For real frequency modes, we have

$$E_c = -\frac{\omega}{m} J_c = \sigma_p J_c \,, \tag{9.157}$$

where σ_p is the pattern speed of the mode (as before). Moreover, defining mode solutions (also, as before)

$$\xi^j = \xi^j_n e^{i\omega_n t} \,, \tag{9.158}$$

one can show that (for uniform rotation)

$$\sigma_p - \Omega\left(1 + \frac{1}{m}\right) \leq \frac{J_c/m^2}{\langle \xi_n, \xi_n \rangle} \leq \sigma_p - \Omega\left(1 - \frac{1}{m}\right) \,. \tag{9.159}$$

Given this result, consider modes with finite frequency in the $\Omega \to 0$ limit. Then (9.159) implies that co-rotating modes (with $\sigma_p > 0$) must have $J_c > 0$, while counter-rotating ones (for which $\sigma_p < 0$) will have $J_c < 0$. In both cases, it follows from (9.157) that $E_c > 0$, which means that the two sets of modes are stable. Now focus on a small neighbourhood of a point where $\sigma_p = 0$ (at a finite rotation rate). This would correspond to a point—at this point fiducial—where an initially counter-rotating mode becomes co-rotating. In this region $J_c < 0$. However, because of (9.157), E_c will change sign at the point where σ_p (or, equivalently, the frequency ω) vanishes. Since the mode was stable in the non-rotating limit the change of sign indicates the onset of instability [64].

Let us apply this logic to an oscillating star. Start in the non-rotating limit. As we have seen, the mode-problem leads to eigenvalues for ω^2, which, in turn, yields equal values $\pm|\omega|$ for the forwards and backwards propagating modes (corresponding to $m = \pm|m|$, respectively). The two sets modes are affected by rotation in different ways. A backwards moving mode will be dragged forwards by the rotation, and if the star spins sufficiently fast the mode will move forwards with respect to the inertial frame. However, the mode is still moving backwards in the rotating frame. The gravitational waves from such a mode carry positive angular momentum away from the star but, since the perturbed fluid actually rotates slower than it would in absence of the perturbation, the angular momentum of the mode is negative. The emitted gravitational waves make the angular momentum increasingly negative and lead to what has become known as the CFS (Chandrasekhar-Friedman-Schutz) instability.

The CFS instability is generic in rotating stars. To see this, we only need to note that one can always find a non-axisymmetric f-mode with a low pattern speed (fixed m and a large l). Formally, this means that, when the star is rotating, there must be an unstable f-mode no matter how slow the rotation rate is. However, in reality the short-wavelength modes that become unstable first are not efficient gravitational-wave emitters. The short lenghtscale modes are also efficiently damped by shear viscosity so the actual instability analysis is a bit more involved. We will return to this issue later.

9.7.2 f-mode astrophysics

Detailed analysis of the f-mode instability in a realistic neutron star reveals that the backwards propagating mode can only be dragged forward by rotation and become CFS unstable for very high rotation rates, so that large scale, lower multipole, f-modes can only become unstable close to the Keplerian breakup frequency. In fact the $l = m = 2$ mode generally does not go unstable before breakup in Newtonian gravity, but can be unstable in general relativity if the equation of state is stiff. Generally, the $l = m = 4$ mode provides the strongest contribution to the gravitational-wave signal [24]

Furthermore, the mode can only grow unstable as long as viscosity does not damp it on a shorter timescale, and in the case of the f-mode, the superfluid mutual friction (see section 9.5) is very efficient at damping the mode at all rotation rates [17, 66]. The upshot is that the f-mode instability may not operate in mature (cold) neutron stars.

There is, however, an astrophysical situation in which the f-mode instability may be relevant, namely the case of a newborn neutron star, which is hot enough to not be superfluid (so there is no mutual friction damping to worry about) and may be rotating close to its Keplerian frequency f_k. In this case, the f-mode may indeed be CFS unstable and grow to large amplitudes. In fact, while in Newtonian gravity, the f-mode is only unstable for frequencies above $f \approx 0.9 f_k$, in general relativistic simulations the unstable region has been shown to extend down to $f \approx 0.8 f_k$, depending on the details of the composition and the equation of state [57]. The duration and amplitude of the associated gravitational-wave signal then depend mainly on the details of the (presently unknown) saturation mechanism, which will determine how long it will take the source to spin down and exit the instability window [117]. The situation is more promising for the r-mode instability, which may play a leading role in the spin evolution of young neutron stars, as we shall see in the next section.

9.7.3 The r-mode instability

We have already referred to the inertial modes, which owe their existence to the Coriolis force. Let us consider these modes in a bit more detail, starting from the velocity perturbations (δv^i) in the rotating frame of the star. We then have

$$\partial_t \delta v_i + 2\epsilon_{ijk}\Omega^j \delta v^k + \frac{1}{\rho}\nabla_i \delta p - \frac{1}{\rho^2}\delta\rho \nabla_i p + \nabla_i \delta\Phi = 0\,, \tag{9.160}$$

while the continuity equation remains as before.

The nature of the inertial modes stems from the static limit of the problem (where time variations vanish and the rotation frequency $\Omega^i \to 0$). In that case, the equations decouple into two sets [103]. We have either

$$\nabla_i \delta\Phi + \frac{1}{\rho}\nabla_i \delta p - \frac{1}{\rho^2}\delta\rho \nabla_i p = 0\,, \tag{9.161}$$

along with (9.72), or

$$\nabla_i(\rho \delta v^i) = 0\,. \tag{9.162}$$

The first set of equations represents perturbations that take us to a neighbouring equilibrium star. In a slowly rotating star, this solution picks up rotational corrections at order Ω, which suggests a solution such that

$$[\delta\rho, \delta p, \delta\Phi] = \mathcal{O}(1) \quad \text{and} \quad \delta v^i = \mathcal{O}(\Omega) \,. \tag{9.163}$$

This problem is equivalent to considering the dynamics of the original configuration (albeit for a slight different central density). In essence, there is nothing new. The new dynamical aspects are, instead, associated with the second set of perturbations, for which we have

$$\delta v^i = \mathcal{O}(1) \quad \text{and} \quad [\delta\rho, \delta p, \delta\Phi] = \mathcal{O}(\Omega) \,. \tag{9.164}$$

This leads to the inertial modes. Moreover, as we can always multiply the perturbation equations by a constant, we can normalise the Lagrangian displacement associated with the perturbation, in the rotation frame simply given by

$$\delta v^i = \partial_t \xi^i \,, \tag{9.165}$$

such that

$$\xi^i = \mathcal{O}(1) \Longrightarrow \delta v^i = \mathcal{O}(\Omega) \quad \text{and} \quad [\delta\rho, \delta p, \delta\Phi] = \mathcal{O}(\Omega^2) \,. \tag{9.166}$$

Evidently, we cannot (completely) determine the density perturbations etcetera without accounting for the change in shape due to the centrifugal force (which enters at order Ω^2).

In general, the problem is quite complicated. There is, however, an exception. For barotropic stars we can find a purely toroidal mode solution. That is, a solution such that $U_l \neq 0$ while $W_l = V_l = 0$. This solution represents the r-modes. Analogous to the Rossby waves in the Earth's oceans, they were first discussed for stars in the late 1970s [113]. Working out the—linear in Ω— mode problem for a purely toroidal solution (for simplicity assuming a constant density star) we have, first of all,

$$[l(l+1)\omega_r - 2m\Omega]U_l = 0 \tag{9.167}$$

where ω_r is the rotating-frame mode frequency. This shows that the only way to avoid a trivial solution is to have the mode frequency

$$\omega_r = \frac{2m\Omega}{l(l+1)} \,. \tag{9.168}$$

Moreover, we see that only a single U_l can be nonzero. In addition, we have the two equations

$$[(l\omega_r - 2m\Omega)r\partial_r U_l + 2ml\Omega U_l]\, Q_{l+1} = 0 \,, \tag{9.169}$$

$$\{[(l+1)\omega_r + 2m\Omega]r\partial_r U_l + 2m(l+1)\Omega U_l\}\, Q_l = 0 \,, \tag{9.170}$$

where

$$Q_l = \left[\frac{(l-m)(l+m)}{(2l-1)(2l+1)}\right]^{1/2} \,. \tag{9.171}$$

These equations are not compatible—the problem is overdetermined—unless $l = m$, in which case we have $Q_l = 0$ and the second equation is automatically satisfied. In this case, we find that the eigenfunction is

$$U_m = r^{m+1} . \tag{9.172}$$

This represents the single r-mode that exists (for $l = m$) in a slowly rotating barotropic star.

From these results, it is easy to see that the pattern speed for a typical r-mode is

$$\sigma_r = -\frac{2\Omega}{l(l+1)} < 0 , \tag{9.173}$$

according to an observer rotating with the star. On the other hand, an inertial observer would find that

$$\sigma_i = \Omega\frac{(l-1)(l+2)}{l(l+1)} > 0 . \tag{9.174}$$

In essence, the modes appear retrograde in the rotating system but an inertial observer always finds them to be prograde. Connecting this conclusion with the CFS argument, we see that the r-modes should be driven unstable by gravitational-wave emission in all rotating stars [23, 65]. This may seem a somewhat drastic statement given the implication that gravitational waves would ultimately prevent stars from rotating. Of course, the reality is a little bit different. In particular, we need to consider different dissipation mechanisms (=viscosity) that may dominate the gravitational-wave emission and prevent the instability.

9.7.4 The instability window

The r-mode problem is intricate. The balance between gravitational-wave emission—which drives the instability—and various damping agents—which serve to suppress it—depends on the detailed physics of the star's interior. We need to consider a range of transport properties for supranuclear matter. In addition, it is interesting to ask how astrophysical observations—e.g. in X-rays, radio or, indeed, gravitational waves—constrain the instability scenario and how this, in turn, influences the theory.

To make progress, we need to understand the impact of different aspects of neutron star physics—dissipation channels like shear and bulk viscosity, the state of matter (superfluid mutual friction and the interfaces with the elastic crust) and the role of the star's magnetic field.

Newtonian gravity r-mode calculations show that the instability growth time is [26, 102]

$$t_{\rm gw} \approx 50 \left(\frac{M}{1.4M_\odot}\right)^{-1} \left(\frac{R}{10\,\rm km}\right)^{-4} \left(\frac{P}{10^{-3}}\right)^{6} \rm s . \tag{9.175}$$

This estimate shows that the r-mode instability is significantly stronger than that of the f-mode [90]. This is a little bit surprising because we know the f-mode is an efficient gravitational-wave emitter, while the r-mode mainly radiates through the gravitomagnetic coupling (the current multipoles). The result arises as the r-mode is unstable already for slowly rotating stars while the f-mode instability only comes into play in fast-spinning stars. If both modes were to be unstable at the same time, the f-mode would dominate [57].

The growth time estimate suggests that the r-mode may grow fast enough to be of significance for rapidly spinning neutron stars. However, in order to understand the relevance of the instability, we should also consider possible damping effects. An unstable mode must grow fast enough that it is not completely killed by viscosity. In the first instance, one typically considers the effects of bulk- and shear viscosity. At relatively low temperatures (below a few times 10^9 K), the main viscous dissipation mechanism in a fluid star arises from internal friction (momentum transport due to particle scattering), modelled in terms of a macroscopic shear viscosity. In a normal fluid star, neutron-neutron scattering provides the most important contribution. The effect of the corresponding shear viscosity is usually estimated using the viscosity coefficient

$$\eta = 2 \times 10^{18} \left(\frac{\rho}{10^{15}\,\mathrm{g/cm^3}} \right)^{9/4} \left(\frac{T}{10^9\mathrm{K}} \right)^{-2} \mathrm{g/cm\,s}\,. \tag{9.176}$$

For the r-modes, this leads to a dissipation time-scale

$$t_{\mathrm{sv}} \approx 7 \times 10^7 \left(\frac{M}{1.4M_\odot} \right)^{-5/4} \left(\frac{R}{10\,\mathrm{km}} \right)^{23/4} \left(\frac{T}{10^9\mathrm{K}} \right)^2 \mathrm{s} \tag{9.177}$$

Clearly, the gravitational-wave driven instability growth dominates the shear viscous damping at moderate and high neutron star core temperatures.

However, at high temperatures (above a few times 10^9 K), bulk viscosity is the dominant dissipation mechanism in a neutron star core doiminated by npe-matter. We have already seen how bulk viscosity represents an estimate of the energy is dissipated from the fluid motion as weak interactions try to re-establish equilibrium. The mode energy lost through bulk viscosity is carried away by neutrinos, so the story changes if the temperature is high enough that the neutrinos are trapped in the star (which then again supresses the viscous damping, see figure 9.1). An estimate of the bulk viscosity damping requires the Lagrangian density perturbation. For r-modes, this quantity vanishes at leading order so the calculation must be carried (at least) to order Ω^2 in the slow-rotation expansion. In the standard case, where β-equilibrium is regulated by the modified Urca reactions, the relevant bulk viscosity coefficient is

$$\zeta = 6 \times 10^{25} \left(\frac{\rho}{10^{15}\,\mathrm{g/cm^3}} \right)^2 \left(\frac{T}{10^9\,\mathrm{K}} \right)^6 \left(\frac{\omega_r}{1\,\mathrm{Hz}} \right)^{-2} \mathrm{g/cms}\,. \tag{9.178}$$

This leads to an estimated bulk viscosity timescale

$$t_{\mathrm{bv}} \approx 3 \times 10^{11} \left(\frac{M}{1.4M_\odot} \right) \left(\frac{R}{10\,\mathrm{km}} \right)^{-1} \left(\frac{P}{10^{-3}} \right)^2 \left(\frac{T}{10^9\,\mathrm{K}} \right)^{-6} \mathrm{s} \tag{9.179}$$

Evidently, the bulk viscosity is to weak to prevent the instability at moderate and low temperatures

From the above estimates, we see that a gravitational-wave driven instability will only be active in a certain temperature range. In order to have an instability, we need t_{gw} to be smaller in magnitude than both t_{sv} and t_{bv}. In the schematic illustration of Figure 9.1,

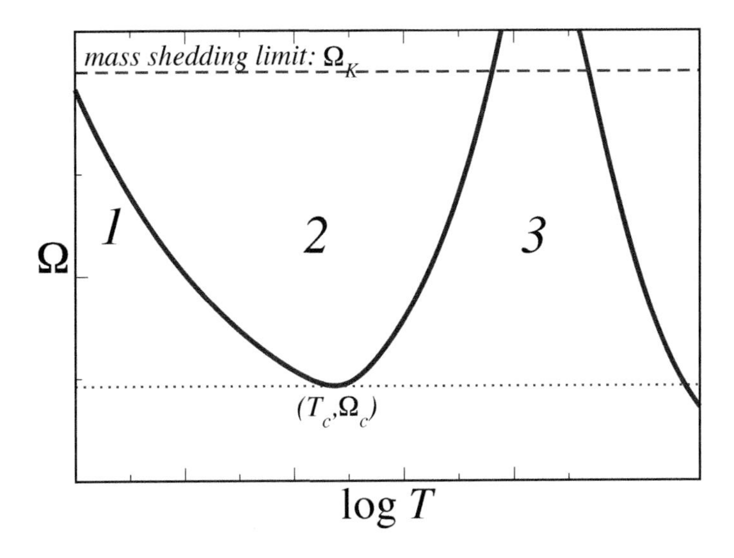

Figure 9.1 Schematic illustration of the CFS instability window: At low temperatures (region 1) dissipation due to shear viscosity prevents the instability. At high temperatures (region 3), of the order of 10^{10} K, bulk viscosity suppresses the instability. The main instability window is expected at temperatures near $T_c \approx 10^9$ K (region 2). The instability window may change considerably if we consider other aspects of physics, like superfluidity or the presence of hyperons. Specifically, the temperature at which the bulk viscosity "resonance" occurs may shift significantly for exotic matter.

region 2 indicates a temperature window where the growth time due to gravitational radiation is short enough to overcome the viscous damping and drive the mode unstable. The critical rotation rate, above which the mode is unstable, follows from

$$\frac{1}{2E}\frac{dE}{dt} = -\frac{1}{t_{\rm gw}} + \frac{1}{t_{\rm bv}} + \frac{1}{t_{\rm sv}} = 0 \tag{9.180}$$

Detailed calculations show that viscosity stabilizes the r-modes below $\Omega_c \approx 0.04\Omega_K$ [26, 102]. For a typical neutron star, this suggests that the r-modes may be unstable at rotation periods shorter than 25 ms. While this estimate is interestingly close to the birth spin period of the Crab pulsar inferred from observational data, $P_o \approx 19$ ms, we know that recycled neutron stars spin much faster than this, suggesting that the simple picture we have painted so far is missing some key colour.

9.7.5 r-mode astrophysics

9.7.5.1 *Young neutron stars*

Let us first of all consider how the r-mode instability may act in a young neutron star. In this case, if the star is born hot and rapidly rotating, it will start its evolution in the top right part of region 3 in figure 9.1. As it cools down due to neutrino emission, it will evolve to the left, towards lower temperatures, and enter the unstable region 2. Once it enters the unstable region, the r-mode will grow to a large enough amplitude that nonlinear interactions with other modes come into play [30], and the mode saturates. Subsequently, viscous heating will balance the cooling and the neutron stars will spin down rapidly as energy is emitted in gravitational waves [111]. Eventually the system will exit the instability window and the mode will be damped. In fact, it has been suggested that this could be the mechanism that spins down all newly born neutron stars and leads to the, relatively low, birth spin inferred for the observed pulsar population [111].

The exact point in parameter space at which a system exits the instability window depends on how large the saturation amplitude is and the precise viscosity coefficients, therefore on details of the equation of state (where we need to consider non-equilibrium transport properties). Nevertheless it has been shown that, as the exit point will be close to the minimum of the instability curve, it is relatively insensitive to the exact details of the equation of state [11]. This allows us to estimate, within the uncertainties associated with the equation of state, whether any currently observed pulsars may still be at the end of their r-mode-driven evolution.

Interestingly there may be such a candidate; the young X-ray pulsar PSR J0537-6910. Not only is this the most rapidly rotating young neutron star currently observed, but it also undergoes large "glitches" (i.e. sudden spin-up events thought to be due to superfluid components in the interior (see [29] for a review). Regular timing of the pulsar in X-rays has allowed monitoring of the evolution of the spin frequency ν and revealed that after glitches the braking index $n = \ddot{\nu}\nu/\dot{\nu}^2$ decreases, with observations suggesting that it may tend asymptotically to a value $n \approx 7$ [15, 59, 88]. This is interesting, as the braking index gives an indication of the mechanisms driving the spin-down. If we write the spin-down torque as a power-law in the spin frequency, i.e.

$$\dot{\nu} = -k\nu^n , \tag{9.181}$$

then one can verify that the index of the power-law n is exactly the braking index that can be measured observationally as $n = \ddot{\nu}\nu/\dot{\nu}^2$. If electromagnetic dipole radiation is driving the spin-down, then $n \approx 3$, while $n \approx 7$ is exactly the value expected for a constant amplitude r-mode spinning down the star.

Given the above theoretical and observational motivations, searches for an r-mode signal were carried out in data from both O2 and O3 runs of the ground-based detector network [4, 61]. No signal was observed, but the most recent search in O3 data (supported by X-ray ephemeris provided by NICER), has constrained the parameter space for the r-mode-driven spin-down scenario [4]. Observations in the next observing runs, accompanied by X-ray observations, therefore have the potential to detect a signal or, in the event of a non-detection, rule out an r-mode driven spin-down for PSR J0537-6910.

Let us look at this scenario in a little more detail, as it is an interesting example of how multimessenger observations of oscillations may allow us to constrain the equation of state of dense matter. In the event of a detection of a signal loud enough to confirm a gravitational-wave driven spin-down for the pulsar, the measured (in X-ray) spin-down rate $\dot{\nu}$ can be linked to the wave strain h_0 and frequency f (the spin frequency of the pulsar is ν) recorded by the detectors as:

$$h_o = \frac{1}{d}\sqrt{\frac{10G}{c^3} I \frac{\nu|\dot{\nu}|}{f^2}} , \tag{9.182}$$

where I is the moment of inertia of the star and d its distance from Earth. The frequency f of the mode for a relativistic star, on the other hand, is not simply $f = 4/3\nu$ as suggested by the Newtonian result in (9.168), but rather a function of the compactness $\mathcal{C} = GM/Rc^2$ [89]. A measurement of f alone would thus allow for a constraint on the ratio M/R, but in a system such as PSR J0537-6910 where we may have reason to *expect* an r-mode driven spin-down, the relation in (9.182) allows for an additional constraint on I, which breaks the degeneracy between mass M and radius R and can deliver a much tighter constraint on the equation of state.

9.7.5.2 Accreting systems

Let us now focus on the low temperature part of the r-mode instability window, which we expect to be relevant for old neutron stars with core temperatures below $T \lesssim 10^9$ K. Neutron stars in binaries that are accreting from a less evolved companion are an interesting target for gravitational-wave searches; the neutron stars in Low Mass X-ray Binaries (LMXBs), which are thought to be the progenitors of the observed millisecond radio pulsars.

In these systems, angular momentum is transferred from the disk to the star as the neutron star accretes matter, spinning it up. Given enough time, one might expect the neutron star to be spun up to its breakup frequency. This will depend on the details of the equation of state but, based on general considerations of stability and causality, the breakup frequency has to be well above 1 kHz [81]. However, no neutron star has been observed rotating close to this limit, and there appears to be an observational cut-off in the spin distribution at around 700 Hz [47, 116], which cannot be easily explained in terms of observational selection effects. Poorly constrained aspects of the accretion torque, or evolutionary effects, may explain the observations, but it has also been suggested that gravitational-wave torques, due to mountains or modes of oscillation excited during accretion, may lead to spin-equilibrium well below the breakup frequency [37, 112].

Let us consider again the "standard" r-mode instability window from figure (9.1). Once a system has entered the stable region close to the minimum of the curve—as already described—it will continue to cool and spin down due to electromagnetic wave emission, without re-entering the unstable part of parameter space. If, however, an old, cold system, is spun-up by accretion, it can re-enter the unstable region from "below", at (core) temperatures $T \lesssim 10^8$ K. At this point the r-mode can become unstable, and if the saturation amplitude—the amplitude that the mode can reach before the growth is halted by energy being redistributed to higher order modes—is large enough, potentially provide the spin-down torque needed to explain the observed spins of accreting neutron stars in LMXBs.

If the saturation amplitude is high, the system would heat significantly, as the mode grows, then spin down rapidly, exiting the unstable region, leading to a very low duty cycle for gravitational-wave emission. If, on the other hand, the saturation amplitude is low, as some calculations suggest [39, 40], the system essentially "slides" along the instability curve, and emits gravitational waves for longer, albeit at a lower level. Either way, we do not expect to find systems well inside the unstable region. This idea was tested in [82, 87] by populating the instability window using observations of spin frequency and surface temperature of accreting neutron stars in LMXBs. While there is some modelling required to obtain the interior temperature of the star, leading to significant error bars, it is still the case that there are a large number of sources well inside the "minimal" instability window.

It is thus likely the case that additional physics must be at work to damp the r-modes and modify the instability window. In fact, several additional effects, including superfluid mutual friction, hyperon bulk viscosity and coupling to crustal modes may be at play, and lead to a significantly different instability window at low temperatures [80]. An intriguing possibility involves exotic forms of matter, like deconfined quark matter [12], where the resonant enhancement of the bulk viscosity at certain neutron star temperatures—evident in Figure 9.1—could provide the dissipation to completely damp these modes in LMXBs. Another possibility is that the r-mode is indeed unstable in these systems, but the saturation amplitude is so low that there is no significant contribution to the thermal and spin evolution of the neutron star. Limits on thermal X-ray emission set stringent bounds on the r-mode amplitude [104, 105] several orders of magnitude below theoretical predictions for the conventional saturation mechanisms, such as mode coupling [40].

An answer to the conundrum may come from gravitational-wave observations of these systems. In the third observing run of the LIGO-Virgo-KAGRA detectors, O3, observations started probing below the spin-equilibrium limit for Scorpius X-1 [5]. This is an interesting target as it is the most luminous LMXB in the sky, but unfortunately it is not visible as an X-ray pulsar, so the spin frequency of the neutron star is not known. This means that limits were set only in a limited frequency range, where a signal would have been detected had gravitational waves been setting the spin-equilibrium of the system. Future, observations of this system and other LMXBs, with increased sensitivity, have the potential to help us understand the nature of gravitational-wave emission in these systems and map the r-mode instability window. There is unfinished business here.

9.8 CONCLUDING REMARKS: DID YOU FORGET THE MAGNETIC FIELD?

We set out to provide a pedagogical introduction to the different ways that neutron star equation of state physics impacts on observable phenomena, mainly relating to large-scale dynamics. Adding relevant complicating features step-by-step, we have shown that there is much more to the problem than calculating mass-radius curves for equilibrium matter. Things get "more interesting" when we consider non-equilibrium aspects, as we then have to consider the composition and state of matter in the neutron star interior. Any—current or future—observational probe of such features would be of immense value and it seems entirely possible that the next-generation of gravitational-wave instruments will lead to breakthroughs in this respect. Of course, in parallel with building advanced detectors, we need to improve our understanding of the theory...

As we are reaching the conclusions and we are safely running out of space, it is appropriate to comment on issues we have not touched upon. In particular, we should own up to the fact that we (more or less completely) ignored the neutron star magnetic field. There were two reasons for this. The first reason was pragmatic: we had to draw the line somewhere and the magnetic field complicates the discussion to the extent where the calculations we have done could not be laid out as succinctly. The second reason was more based on our current level of "ignorance". We have focussed on neutron star seismology and it is—intuitively—obvious that electromagnetic aspects will enter any analysis of fluid motion and waves. In principle, we know how to deal with this problem. The main technical issue that arises is that we break the (essentially) spherical symmetry we assumed for the unperturbed stellar background. The calculations get more involved, but so be it [51]. There is, however, a snag. Magnetic stars are prone to different instabilities and neutron stars particularly so [50, 99]. The implications of this may seem somewhat drastic: none of the global magnetic field configurations we are currently "able to build" are stable [98]. Evidently, nature disagrees because we obviously observe electromagnetic emission from pulsars and magnetars, so neutron stars certainly have long-term stable magnetic fields. We must be doing something wrong. While we are grappling to find a resolution, likely involving some aspect of physics that has so far been ignored [74, 108], it does not make much sense to invest too much effort into exploring the implications of the seismology problem. Perfectly aware that we will need to revisit this problem at some point in the future, we claim this as a legitimate excuse for ignoring the magnetic field for now.

ACKNOWLEDGEMENTS

NA acknowledges support from STFC via grant number ST/R00045X/1. BH acknowledges support from the National Science Center Poland (NCN) via OPUS grant 2019/33/B/ST9/00942.

Bibliography

[1] B. P. Abbott, R. Abbott, T. D. Abbott, F. Acernese, K. Ackley, and C. et al. Adams. Search for Gravitational Waves from a Long-lived Remnant of the Binary Neutron Star Merger GW170817. *Astrophys. J.*, 875(2):160, April 2019.

[2] B. P. Abbott, R. Abbott, T. D. Abbott, F. Acernese, K. Ackley, and et al. Adams. Search for Post-merger Gravitational Waves from the Remnant of the Binary Neutron Star Merger GW170817. *Astrophys. J. Lett.*, 851(1):L16, December 2017.

[3] B. P. Abbott et al. GW170817: Observation of Gravitational Waves from a Binary Neutron Star Inspiral. *Phys. Rev. Lett.*, 119(16):161101, October 2017.

[4] R. Abbott, T. D. Abbott, S. Abraham, F. Acernese, K. Ackley, and A. et al. Adams. Constraints from LIGO O3 Data on Gravitational-wave Emission Due to R-modes in the Glitching Pulsar PSR J0537-6910. *Astrophys. J.*, 922(1):71, November 2021.

[5] R. Abbott, H. Abe, F. Acernese, K. Ackley, S. Adhicary, and N. et al. Adhikari. Model-based Cross-correlation Search for Gravitational Waves from the Low-mass X-Ray Binary Scorpius X-1 in LIGO O3 Data. *Astrophys. J. Lett.*, 941(2):L30, December 2022.

[6] Conny Aerts, Jørgen Christensen-Dalsgaard, and Donald W. Kurtz. *Asteroseismology*. 2010.

[7] Chaitanya Afle, Suman Kumar Kundu, Jenna Cammerino, Eric R. Coughlin, Duncan A. Brown, David Vartanyan, and Adam Burrows. Measuring the properties of

f -mode oscillations of a protoneutron star by third-generation gravitational-wave detectors. *Phys. Rev. D*, 107(12):123005, June 2023.

[8] Mark Alford, Arus Harutyunyan, and Armen Sedrakian. Bulk Viscous Damping of Density Oscillations in Neutron Star Mergers. *Particles*, 3(2):500–517, June 2020.

[9] Mark G. Alford and Steven P. Harris. β equilibrium in neutron-star mergers. *Phys. Rev. C*, 98(6):065806, December 2018.

[10] Mark G. Alford, Simin Mahmoodifar, and Kai Schwenzer. Large amplitude behavior of the bulk viscosity of dense matter. *Journal of Physics G Nuclear Physics*, 37(12):125202, December 2010.

[11] Mark G. Alford and Kai Schwenzer. Gravitational Wave Emission and Spin-down of Young Pulsars. *Astrophys. J.*, 781(1):26, January 2014.

[12] Mark G. Alford and Kai Schwenzer. What the Timing of Millisecond Pulsars Can Teach us about Their Interior. *Phys.Rev.Lett.*, 113:251102, 2014.

[13] M. A. Alpar, S. A. Langer, and J. A. Sauls. Rapid postglitch spin-up of the superfluid core in pulsars. *Astrophys. J.*, 282:533–541, July 1984.

[14] Z. Alterman, H. Jarosch, and C. L. Pekeris. Oscillations of the Earth. *Proceedings of the Royal Society of London Series A*, 252(1268):80–95, August 1959.

[15] N. Andersson, D. Antonopoulou, C. M. Espinoza, B. Haskell, and W. C. G. Ho. The Enigmatic Spin Evolution of PSR J0537-6910: r-modes, Gravitational Waves, and the Case for Continued Timing. *Astrophys. J.*, 864(2):137, September 2018.

[16] N. Andersson and G. L. Comer. On the dynamics of superfluid neutron star cores. *MNRAS*, 328(4):1129–1143, December 2001.

[17] N. Andersson, K. Glampedakis, and B. Haskell. Oscillations of dissipative superfluid neutron stars. *Phys. Rev. D*, 79(10):103009, May 2009.

[18] N. Andersson, K. Glampedakis, W. C. G. Ho, and C. M. Espinoza. Pulsar Glitches: The Crust is not Enough. *Phys. Rev. Lett.*, 109(24):241103, December 2012.

[19] N. Andersson, K. Glampedakis, and L. Samuelsson. Superfluid signatures in magnetar seismology. *MNRAS*, 396(2):894–899, June 2009.

[20] N. Andersson and P. Pnigouras. The g-mode spectrum of reactive neutron star cores. *MNRAS*, 489(3):4043–4048, November 2019.

[21] N. Andersson and P. Pnigouras. Exploring the effective tidal deformability of neutron stars. *Phys. Rev. D*, 101(8):083001, April 2020.

[22] N. Andersson and P. Pnigouras. The phenomenology of dynamical neutron star tides. *MNRAS*, 503(1):533–539, May 2021.

[23] Nils Andersson. A New Class of Unstable Modes of Rotating Relativistic Stars. *Astrophys. J.*, 502(2):708–713, August 1998.

[24] Nils Andersson. *Gravitational-Wave Astronomy: Exploring the Dark Side of the Universe*. 2019.

[25] Nils Andersson and Fabian Gittins. Formulating the r-mode Problem for Slowly Rotating Neutron Stars. *Astrophys. J.*, 945(2):139, March 2023.

[26] Nils Andersson, Kostas Kokkotas, and Bernard F. Schutz. Gravitational Radiation Limit on the Spin of Young Neutron Stars. *Astrophys. J.*, 510(2):846–853, January 1999.

[27] Nils Andersson and Kostas D. Kokkotas. Gravitational Waves and Pulsating Stars: What Can We Learn from Future Observations? *Phys. Rev. Lett.*, 77(20):4134–4137, November 1996.

[28] Nils Andersson and Kostas D. Kokkotas. Towards gravitational wave asteroseismology. *MNRAS*, 299(4):1059–1068, October 1998.

[29] Danai Antonopoulou, Brynmor Haskell, and Cristóbal M. Espinoza. Pulsar glitches: observations and physical interpretation. *Reports on Progress in Physics*, 85(12):126901, December 2022.

[30] Phil Arras, Eanna E. Flanagan, Sharon M. Morsink, A. Katrin Schenk, Saul A. Teukolsky, and Ira Wasserman. Saturation of the r-Mode Instability. *Astrophys. J.*, 591(2):1129–1151, July 2003.

[31] Thomas W. Baumgarte, Stuart L. Shapiro, and Masaru Shibata. On the Maximum Mass of Differentially Rotating Neutron Stars. *Astrophys. J. Lett.*, 528(1):L29–L32, January 2000.

[32] A. Bauswein and N. Stergioulas. Unified picture of the post-merger dynamics and gravitational wave emission in neutron star mergers. *Phys. Rev. D*, 91(12):124056, June 2015.

[33] Omar Benhar, Valeria Ferrari, and Leonardo Gualtieri. Gravitational wave asteroseismology reexamined. *Phys. Rev. D*, 70(12):124015, December 2004.

[34] Hugo Benioff, Frank Press, and Stewart Smith. Excitation of the Free Oscillations of the Earth by Earthquakes. *Journal of Geophys. Research*, 66(2):605–619, February 1961.

[35] Sebastiano Bernuzzi, Tim Dietrich, and Alessandro Nagar. Modeling the Complete Gravitational Wave Spectrum of Neutron Star Mergers. *Phys. Rev. Lett.*, 115(9):091101, August 2015.

[36] Sebastiano Bernuzzi, Tim Dietrich, Wolfgang Tichy, and Bernd Brügmann. Mergers of binary neutron stars with realistic spin. *Phys. Rev. D*, 89(10):104021, May 2014.

[37] Lars Bildsten. Gravitational Radiation and Rotation of Accreting Neutron Stars. *Astrophys. J. Lett.*, 501(1):L89–L93, July 1998.

[38] T. Binnington and E. Poisson. Relativistic theory of tidal Love numbers. *Phys. Rev. D*, 80(8):084018, October 2009.

[39] Ruxandra Bondarescu, Saul A. Teukolsky, and Ira Wasserman. Spin Evolution of Accreting Neutron Stars: Nonlinear Development of the R-mode Instability. *Phys. Rev.*, D76:064019, 2007.

[40] Ruxandra Bondarescu and Ira Wasserman. Nonlinear Development of the R-Mode Instability and the Maximum Rotation Rate of Neutron Stars. *Astrophys.J.*, 778:9, 2013.

[41] A. Burrows and D. Vartanyan. Core-collapse supernova explosion theory. *Nature*, 589(7840):29–39, January 2021.

[42] Giovanni Camelio, Lorenzo Gavassino, Marco Antonelli, Sebastiano Bernuzzi, and Brynmor Haskell. Simulating bulk viscosity in neutron stars. I. Formalism. *Physical Review D*, 107(10):103031, May 2023.

[43] Giovanni Camelio, Lorenzo Gavassino, Marco Antonelli, Sebastiano Bernuzzi, and Brynmor Haskell. Simulating bulk viscosity in neutron stars. II. Evolution in spherical symmetry. *Physical Review D*, 107(10):103032, May 2023.

[44] Brandon Carter, Nicolas Chamel, and Pawel Haensel. Entrainment coefficient and effective mass for conduction neutrons in neutron star crust: simple microscopic models. *Nucl. Phys. A*, 748(3-4):675–697, February 2005.

[45] T. Celora, N. Andersson, I. Hawke, and G. L. Comer. Covariant approach to relativistic large-eddy simulations: The fibration picture. *Phys. Rev. D*, 104(8):084090, October 2021.

[46] T. Celora, I. Hawke, P. C. Hammond, N. Andersson, and G. L. Comer. Formulating bulk viscosity for neutron star simulations. *Phys. Rev. D*, 105(10):103016, May 2022.

[47] Deepto Chakrabarty, Edward H. Morgan, Michael P. Muno, Duncan K. Galloway, Rudy Wijnands, Michiel van der Klis, and Craig B. Markwardt. Nuclear-powered millisecond pulsars and the maximum spin frequency of neutron stars. *Nature*, 424(6944):42–44, July 2003.

[48] N. Chamel, A. F. Fantina, J. M. Pearson, and S. Goriely. Symmetry energy from nuclear masses and neutron-star observations using generalised Skyrme functionals. In *J. Phys. Conf. Ser.*, volume 665 of *Journal of Physics Conference Series*, page 012066, January 2016.

[49] Nicolas Chamel and Pawel Haensel. Physics of Neutron Star Crusts. *Living rev. relativ.*, 11(1):10, December 2008.

[50] Riccardo Ciolfi, Samuel K. Lander, Gian Mario Manca, and Luciano Rezzolla. Instability-driven Evolution of Poloidal Magnetic Fields in Relativistic Stars. *Astrophys. J. Lett.*, 736(1):L6, July 2011.

[51] A. Colaiuda and K. D. Kokkotas. Coupled polar-axial magnetar oscillations. *MNRAS*, 423(1):811–821, June 2012.

[52] T. G. Cowling. The non-radial oscillations of polytropic stars. *MNRAS*, 101:367, January 1941.

[53] T. Damour and A. Nagar. Relativistic tidal properties of neutron stars. *Phys. Rev. D*, 80(8):084035, October 2009.

[54] P. B. Demorest, T. Pennucci, S. M. Ransom, M. S. E. Roberts, and J. W. T. Hessels. A two-solar-mass neutron star measured using Shapiro delay. *Nature*, 467(7319):1081–1083, October 2010.

[55] S. Detweiler and L. Lindblom. On the nonradial pulsations of general relativistic stellar models. *Astrophys. J.*, 292:12–15, May 1985.

[56] V. Domingo, B. Fleck, and A. I. Poland. The SOHO Mission: an Overview. *Solar Physics*, 162(1-2):1–37, December 1995.

[57] Daniela D. Doneva, Kostas D. Kokkotas, and Pantelis Pnigouras. Gravitational wave afterglow in binary neutron star mergers. *Phys. Rev. D*, 92(10):104040, November 2015.

[58] Robert C. Duncan. Global Seismic Oscillations in Soft Gamma Repeaters. *Astrophys. J. Lett.*, 498(1):L45–L49, May 1998.

[59] R. D. Ferdman, R. F. Archibald, K. N. Gourgouliatos, and V. M. Kaspi. The Glitches and Rotational History of the Highly Energetic Young Pulsar PSR J0537-6910. *Astrophys. J.*, 852(2):123, January 2018.

[60] V. Ferrari, G. Miniutti, and J. A. Pons. Gravitational waves from newly born, hot neutron stars. *MNRAS*, 342(2):629–638, June 2003.

[61] Liudmila Fesik and Maria Alessandra Papa. First Search for r-mode Gravitational Waves from PSR J0537–6910. *Astrophys. J.*, 895(1):11, 2020. [Erratum: Astrophys.J. 897, 185 (2020)].

[62] É É. Flanagan and T. Hinderer. Constraining neutron-star tidal Love numbers with gravitational-wave detectors. *Phys. Rev. D*, 77(2):021502, January 2008.

[63] J. L. Friedman and B. F. Schutz. Lagrangian perturbation theory of nonrelativistic fluids. *Astrophys. J.*, 221:937–957, May 1978.

[64] J. L. Friedman and Bernard F. Schutz. Secular instability of rotating Newtonian stars. *Astrophys. J.*, 222:281, 1978.

[65] John L. Friedman and Sharon M. Morsink. Axial Instability of Rotating Relativistic Stars. *Astrophys. J.*, 502(2):714–720, August 1998.

[66] E. Gaertig, K. Glampedakis, K. D. Kokkotas, and B. Zink. f-Mode Instability in Relativistic Neutron Stars. *Phys. Rev. Lett.*, 107(10):101102, September 2011.

[67] Erich Gaertig and Kostas D. Kokkotas. Gravitational wave asteroseismology with fast rotating neutron stars. *Phys. Rev. D*, 83(6):064031, March 2011.

[68] Rossella Gamba and Sebastiano Bernuzzi. Resonant tides in binary neutron star mergers: Analytical-numerical relativity study. *Phys. Rev. D*, 107(4):044014, February 2023.

[69] L. Gavassino. Can We Make Sense of Dissipation without Causality? *Phys. Rev. X*, 12(4):041001, October 2022.

[70] L. Gavassino, M. Antonelli, and B. Haskell. Bulk viscosity in relativistic fluids: from thermodynamics to hydrodynamics. *Classical and Quantum Gravity*, 38(7):075001, April 2021.

[71] Ronald L. Gilliland, Timothy M. Brown, Jørgen Christensen-Dalsgaard, Hans Kjeldsen, Conny Aerts, Thierry Appourchaux, Sarbani Basu, Timothy R. Bedding, William J. Chaplin, Margarida S. Cunha, Peter De Cat, Joris De Ridder, Joyce A. Guzik, Gerald Handler, Steven Kawaler, László Kiss, Katrien Kolenberg, Donald W. Kurtz, Travis S. Metcalfe, Mario J. P. F. G. Monteiro, Robert Szabó, Torben Arentoft, Luis Balona, Jonas Debosscher, Yvonne P. Elsworth, Pierre-Olivier Quirion, Dennis Stello, Juan Carlos Suárez, William J. Borucki, Jon M. Jenkins, David Koch, Yoji Kondo, David W. Latham, Jason F. Rowe, and Jason H. Steffen. Kepler Asteroseismology Program: Introduction and First Results. *Publ. Astron. Soc. Pacific*, 122(888):131, February 2010.

[72] Fabian Gittins, Thomas Celora, Aru Beri, and Nils Andersson. Modelling Neutron-Star Ocean Dynamics. *Universe*, 9(5):226, May 2023.

[73] Kostas Glampedakis and Nils Andersson. Crust-core coupling in rotating neutron stars. *Phys. Rev. D*, 74(4):044040, August 2006.

[74] Kostas Glampedakis, Nils Andersson, and Lars Samuelsson. Magnetohydrodynamics of superfluid and superconducting neutron star cores. *MNRAS*, 410(2):805–829, January 2011.

[75] Kostas Glampedakis, Lars Samuelsson, and Nils Andersson. Elastic or magnetic? A toy model for global magnetar oscillations with implications for quasi-periodic oscillations during flares. *MNRAS*, 371(1):L74–L77, September 2006.

[76] Sarah E. Gossan, Jim Fuller, and Luke F. Roberts. Wave heating from proto-neutron star convection and the core-collapse supernova explosion mechanism. *MNRAS*, 491(4):5376–5391, February 2020.

[77] P. Haensel, K. P. Levenfish, and D. G. Yakovlev. Adiabatic index of dense matter and damping of neutron star pulsations. *Astron. Astrophys.*, 394:213–217, October 2002.

[78] P. Hammond, I. Hawke, and N. Andersson. Thermal aspects of neutron star mergers. *Phys. Rev. D*, 104(10):103006, November 2021.

[79] P. Hammond, I. Hawke, and N. Andersson. Impact of nuclear reactions on gravitational waves from neutron star mergers. *Phys. Rev. D*, 107(4):043023, February 2023.

[80] B. Haskell. R-modes in neutron stars: Theory and observations. *International Journal of Modern Physics E*, 24(9):1541007, August 2015.

[81] B. Haskell, J. L. Zdunik, M. Fortin, M. Bejger, R. Wijnands, and A. Patruno. Fundamental physics and the absence of sub-millisecond pulsars. *Astron. Astrophys.*, 620:A69, December 2018.

[82] Brynmor Haskell, Nathalie Degenaar, and Wynn C. G. Ho. Constraining the physics of the r-mode instability in neutron stars with X-ray and ultraviolet observations. *Mon. Not. Roy. Astron. Soc.*, 424(1):93–103, July 2012.

[83] Brynmor Haskell and Armen Sedrakian. Superfluidity and Superconductivity in Neutron Stars. In Luciano Rezzolla, Pierre Pizzochero, David Ian Jones, Nanda Rea, and Isaac Vidaña, editors, *Astrophysics and Space Science Library*, volume 457 of *Astrophysics and Space Science Library*, page 401, January 2018.

[84] T. Hinderer. Tidal Love Numbers of Neutron Stars. *Astrophys. J.*, 677(2): 1216–1220, April 2008.

[85] Tanja Hinderer, Andrea Taracchini, Francois Foucart, Alessandra Buonanno, Jan Steinhoff, Matthew Duez, Lawrence E. Kidder, Harald P. Pfeiffer, Mark A. Scheel, Bela Szilagyi, Kenta Hotokezaka, Koutarou Kyutoku, Masaru Shibata, and Cory W. Carpenter. Effects of Neutron-Star Dynamic Tides on Gravitational Waveforms within the Effective-One-Body Approach. *Phys. Rev. Lett.*, 116(18):181101, May 2016.

[86] Wynn C. G. Ho and Nils Andersson. New dynamical tide constraints from current and future gravitational wave detections of inspiralling neutron stars. *Phys. Rev. D*, 108(4):043003, August 2023.

[87] Wynn C. G. Ho, Nils Andersson, and Brynmor Haskell. Revealing the Physics of r Modes in Low-Mass X-Ray Binaries. *Phys. Rev. Lett.*, 107(10):101101, September 2011.

[88] Wynn C. G. Ho, Cristóbal M. Espinoza, Zaven Arzoumanian, Teruaki Enoto, Tsubasa Tamba, Danai Antonopoulou, Michał Bejger, Sebastien Guillot, Brynmor Haskell, and Paul S. Ray. Return of the Big Glitcher: NICER timing and glitches of PSR J0537-6910. *MNRAS*, 498(4):4605–4614, November 2020.

[89] Ashikuzzaman Idrisy, Benjamin J. Owen, and David I. Jones. R -mode frequencies of slowly rotating relativistic neutron stars with realistic equations of state. *Phys. Rev. D*, 91(2):024001, January 2015.

[90] James R. Ipser and Lee Lindblom. The Oscillations of Rapidly Rotating Newtonian Stellar Models. II. Dissipative Effects. *Astrophys. J.*, 373:213, May 1991.

[91] G. L. Israel, T. Belloni, L. Stella, Y. Rephaeli, D. E. Gruber, P. Casella, S. Dall'Osso, N. Rea, M. Persic, and R. E. Rothschild. The Discovery of Rapid X-Ray Oscillations in the Tail of the SGR 1806-20 Hyperflare. *Astrophys. J. Lett.*, 628(1):L53–L56, July 2005.

[92] David G. Koch, William J. Borucki, Gibor Basri, Natalie M. Batalha, Timothy M. Brown, Douglas Caldwell, Jørgen Christensen-Dalsgaard, William D. Cochran, Edna DeVore, Edward W. Dunham, III Gautier, Thomas N., John C. Geary, Ronald L. Gilliland, Alan Gould, Jon Jenkins, Yoji Kondo, David W. Latham, Jack J. Lissauer, Geoffrey Marcy, David Monet, Dimitar Sasselov, Alan Boss, Donald Brownlee, John Caldwell, Andrea K. Dupree, Steve B. Howell, Hans Kjeldsen, Søren Meibom, David Morrison, Tobias Owen, Harold Reitsema, Jill Tarter, Stephen T. Bryson, Jessie L. Dotson, Paul Gazis, Michael R. Haas, Jeffrey Kolodziejczak, Jason F. Rowe, Jeffrey E. Van Cleve, Christopher Allen, Hema Chandrasekaran, Bruce D. Clarke, Jie Li, Elisa V. Quintana, Peter Tenenbaum, Joseph D. Twicken, and Hayley Wu. Kepler Mission Design, Realized Photometric Performance, and Early Science. *Astrophys. J. Lett.*, 713(2):L79–L86, April 2010.

[93] K. D. Kokkotas, T. A. Apostolatos, and N. Andersson. The inverse problem for pulsating neutron stars: a 'fingerprint analysis' for the supranuclear equation of state. *MNRAS*, 320(3):307–315, January 2001.

[94] Kostas D. Kokkotas and Gerhard Schafer. Tidal and tidal-resonant effects in coalescing binaries. *MNRAS*, 275(2):301–308, July 1995.

[95] Hao-Jui Kuan, Christian J. Krüger, Arthur G. Suvorov, and Kostas D. Kokkotas. Constraining equation-of-state groups from g-mode asteroseismology. *MNRAS*, 513(3):4045–4056, July 2022.

[96] Hao-Jui Kuan, Christian J. Krüger, Arthur G. Suvorov, and Kostas D. Kokkotas. Constraining equation-of-state groups from g-mode asteroseismology. *MNRAS*, 513(3):4045–4056, July 2022.

[97] D. Lai. Resonant Oscillations and Tidal Heating in Coalescing Binary Neutron Stars. *MNRAS*, 270:611, October 1994.

[98] S. K. Lander and D. I. Jones. Are there any stable magnetic fields in barotropic stars? *MNRAS*, 424(1):482–494, July 2012.

[99] Paul D. Lasky, Burkhard Zink, Kostas D. Kokkotas, and Kostas Glampedakis. Hydromagnetic Instabilities in Relativistic Neutron Stars. *Astrophys. J. Lett.*, 735(1):L20, July 2011.

[100] Yuri Levin. QPOs during magnetar flares are not driven by mechanical normal modes of the crust. *MNRAS*, 368(1):L35–L38, May 2006.

[101] Yuri Levin and Greg Ushomirsky. Crust-core coupling and r-mode damping in neutron stars: a toy model. *MNRAS*, 324(4):917–922, July 2001.

[102] Lee Lindblom, Benjamin J. Owen, and Sharon M. Morsink. Gravitational Radiation Instability in Hot Young Neutron Stars. *Phys. Rev. Lett.*, 80(22):4843–4846, June 1998.

[103] Keith H. Lockitch and John L. Friedman. Where are the R-Modes of Isentropic Stars? *Astrophys. J.*, 521(2):764–788, August 1999.

[104] Simin Mahmoodifar and Tod Strohmayer. Upper Bounds on r-mode Amplitudes from Observations of Low-mass X-Ray Binary Neutron Stars. *Astrophys. J.*, 773(2):140, August 2013.

[105] Simin Mahmoodifar and Tod Strohmayer. Where Are the r-modes? Chandra Observations of Millisecond Pulsars. *Astrophys. J.*, 840(2):94, May 2017.

[106] Gregory Mendell. Superfluid Hydrodynamics in Rotating Neutron Stars. I. Nondissipative Equations. *Astrophys. J.*, 380:515, October 1991.

[107] M. C. Miller, F. K. Lamb, A. J. Dittmann, S. Bogdanov, Z. Arzoumanian, K. C. Gendreau, S. Guillot, A. K. Harding, W. C. G. Ho, J. M. Lattimer, R. M. Ludlam, S. Mahmoodifar, S. M. Morsink, P. S. Ray, T. E. Strohmayer, K. S. Wood, T. Enoto, R. Foster, T. Okajima, G. Prigozhin, and Y. Soong. PSR J0030+0451 Mass and Radius from NICER Data and Implications for the Properties of Neutron Star Matter. *Astrophys. J. Lett.*, 887(1):L24, December 2019.

[108] Nicolás A. Moraga, Francisco Castillo, Andreas Reisenegger, Juan A. Valdivia, and Mikhail E. Gusakov. Magneto-thermal evolution in the cores of adolescent neutron stars: The Grad-Shafranov equilibrium is never reached in the 'strong-coupling' regime. *arXiv e-prints*, page arXiv:2309.14182, September 2023.

[109] Elias R. Most, Alexander Haber, Steven P. Harris, Ziyuan Zhang, Mark G. Alford, and Jorge Noronha. Emergence of microphysical viscosity in binary neutron star post-merger dynamics. *arXiv e-prints*, page arXiv:2207.00442, July 2022.

[110] Hans-Peter Nollert. TOPICAL REVIEW: Quasinormal modes: the characteristic 'sound' of black holes and neutron stars. *Classical and Quantum Gravity*, 16(12):R159–R216, December 1999.

[111] Benjamin J. Owen et al. Gravitational waves from hot young rapidly rotating neutron stars. *Phys. Rev.*, D58:084020, 1998.

[112] J. Papaloizou and J. E. Pringle. Gravitational radiation and the stability of rotating stars. *Mon. Not. Roy. Astron. Soc.*, 184:501–508, August 1978.

[113] J. Papaloizou and J. E. Pringle. Non-radial oscillations of rotating stars and their relevance to the short-period oscillations of cataclysmic variables. *MNRAS*, 182: 423–442, February 1978.

[114] A. Passamonti, N. Andersson, and P. Pnigouras. Dynamical tides in neutron stars: the impact of the crust. *MNRAS*, 504(1):1273–1293, June 2021.

[115] A. Passamonti, N. Andersson, and P. Pnigouras. Dynamical tides in superfluid neutron stars. *MNRAS*, 514(1):1494–1510, July 2022.

[116] A. Patruno, B. Haskell, and N. Andersson. The Spin Distribution of Fast-spinning Neutron Stars in Low-mass X-Ray Binaries: Evidence for Two Subpopulations. *Astrophys. J.*, 850(1):106, November 2017.

[117] Pantelis Pnigouras and Kostas D. Kokkotas. Saturation of the f -mode instability in neutron stars: Theoretical framework. *Phys. Rev. D*, 92(8):084018, October 2015.

[118] Geraint Pratten, Patricia Schmidt, and Tanja Hinderer. Gravitational-wave asteroseismology with fundamental modes from compact binary inspirals. *Nat. Commun.*, 11:2553, May 2020.

[119] Geraint Pratten, Patricia Schmidt, and Natalie Williams. Impact of Dynamical Tides on the Reconstruction of the Neutron Star Equation of State. *Phys. Rev. Lett.*, 129(8):081102, August 2022.

[120] Reinhard Prix. Variational description of multifluid hydrodynamics: Uncharged fluids. *Phys. Rev. D*, 69(4):043001, February 2004.

[121] Vikram Ravi and Paul D. Lasky. The birth of black holes: neutron star collapse times, gamma-ray bursts and fast radio bursts. *MNRAS*, 441(3):2433–2439, July 2014.

[122] Andreas Reisenegger and Peter Goldreich. Excitation of Neutron Star Normal Modes during Binary Inspiral. *Astrophys. J.*, 426:688, May 1994.

[123] T. E. Riley, A. L. Watts, S. Bogdanov, P. S. Ray, R. M. Ludlam, S. Guillot, Z. Arzoumanian, C. L. Baker, A. V. Bilous, D. Chakrabarty, K. C. Gendreau, A. K. Harding, W. C. G. Ho, J. M. Lattimer, S. M. Morsink, and T. E. Strohmayer. A NICER View of PSR J0030+0451: Millisecond Pulsar Parameter Estimation. *Astrophys. J. Lett.*, 887(1):L21, December 2019.

[124] Lars Samuelsson and Nils Andersson. Neutron star asteroseismology. Axial crust oscillations in the Cowling approximation. *MNRAS*, 374(1):256–268, January 2007.

[125] Andreas Schmitt and Peter Shternin. Reaction Rates and Transport in Neutron Stars. In Luciano Rezzolla, Pierre Pizzochero, David Ian Jones, Nanda Rea, and Isaac Vidaña, editors, *Astrophysics and Space Science Library*, volume 457 of *Astrophysics and Space Science Library*, page 455, January 2018.

[126] A. S. Schneider, C. Constantinou, B. Muccioli, and M. Prakash. Akmal-Pandharipande-Ravenhall equation of state for simulations of supernovae, neutron stars, and binary mergers. *Phys. Rev. C*, 100(2):025803, August 2019.

[127] Hajime Sotani, Tomoya Takiwaki, and Hajime Togashi. Universal relation for supernova gravitational waves. *Phys. Rev. D*, 104(12):123009, December 2021.

[128] T. Strohmayer, S. Ogata, H. Iyetomi, S. Ichimaru, and H. M. van Horn. The Shear Modulus of the Neutron Star Crust and Nonradial Oscillations of Neutron Stars. *Astrophys. J.*, 375:679, July 1991.

[129] Tod E. Strohmayer and Anna L. Watts. Discovery of Fast X-Ray Oscillations during the 1998 Giant Flare from SGR 1900+14. *Astrophys. J. Lett.*, 632(2):L111–L114, October 2005.

[130] Tod E. Strohmayer and Anna L. Watts. The 2004 Hyperflare from SGR 1806-20: Further Evidence for Global Torsional Vibrations. *Astrophys. J.*, 653(1):593–601, December 2006.

[131] Kip S. Thorne and Alfonso Campolattaro. Non-Radial Pulsation of General-Relativistic Stellar Models. I. Analytic Analysis for $l \geq 2$. *Astrophys. J.*, 149:591, September 1967.

[132] Alejandro Torres-Forné, Pablo Cerdá-Durán, Martin Obergaulinger, Bernhard Müller, and José A. Font. Universal Relations for Gravitational-Wave Asteroseismology of Protoneutron Stars. *Phys. Rev. Lett.*, 123(5):051102, August 2019.

[133] Vinh Tran, Suprovo Ghosh, Nicholas Lozano, Debarati Chatterjee, and Prashanth Jaikumar. g -mode oscillations in neutron stars with hyperons. *Phys. Rev. C*, 108(1):015803, July 2023.

[134] R. Turolla, S. Zane, and A. L. Watts. Magnetars: the physics behind observations. A review. *Reports on Progress in Physics*, 78(11):116901, November 2015.

[135] Wasaburo Unno, Yoji Osaki, Hiroyasu Ando, H. Saio, and H. Shibahashi. *Nonradial oscillations of stars*. 1989.

[136] David Vartanyan, Adam Burrows, Tianshu Wang, Matthew S. B. Coleman, and Christopher J. White. Gravitational-wave signature of core-collapse supernovae. *Phys. Rev. D*, 107(10):103015, May 2023.

[137] David Vartanyan, Adam Burrows, Tianshu Wang, Matthew S. B. Coleman, and Christopher J. White. Gravitational-wave signature of core-collapse supernovae. *Phys. Rev. D*, 107(10):103015, May 2023.

[138] Anna L. Watts, Nils Andersson, Deepto Chakrabarty, Marco Feroci, Kai Hebeler, Gianluca Israel, Frederick K. Lamb, M. Coleman Miller, Sharon Morsink, Feryal Özel, Alessandro Patruno, Juri Poutanen, Dimitrios Psaltis, Achim Schwenk, Andrew W. Steiner, Luigi Stella, Laura Tolos, and Michiel van der Klis. Colloquium:

Measuring the neutron star equation of state using x-ray timing. *Rev. Mod. Phys.*, 88(2):021001, April 2016.

[139] Anna L. Watts and Tod E. Strohmayer. Detection with RHESSI of High-Frequency X-Ray Oscillations in the Tailof the 2004 Hyperflare from SGR 1806-20. *Astrophys. J. Lett.*, 637(2):L117–L120, February 2006.

[140] Nevin N. Weinberg, Phil Arras, and Joshua Burkart. An Instability due to the Non-linear Coupling of p-modes to g-modes: Implications for Coalescing Neutron Star Binaries. *Astrophys. J.*, 769(2):121, June 2013.

[141] Hang Yu and Nevin N. Weinberg. Resonant tidal excitation of superfluid neutron stars in coalescing binaries. *MNRAS*, 464(3):2622–2637, January 2017.

[142] Tianqi Zhao, Constantinos Constantinou, Prashanth Jaikumar, and Madappa Prakash. Quasinormal g modes of neutron stars with quarks. *Phys. Rev. D*, 105(10):103025, May 2022.

[143] Tianqi Zhao and James M. Lattimer. Universal relations for neutron star f -mode and g -mode oscillations. *Phys. Rev. D*, 106(12):123002, December 2022.

CHAPTER 10

Probing the Equation of State Through Binary Neutron Star Mergers: Numerical Simulations and Nuclear Uncertainties

Domenico Logoteta, Albino Perego

DURING the merger of two neutron stars (NSs), matter reaches conditions that are never realized in other places or events in the Universe. But, interestingly, the tendency in the evolution of the thermodynamical variables inside such a violent collision is somehow opposite to the one that occurs during the formation of the NSs that compose the merging system. Indeed, the two colliding NSs have previously formed at the heart of two massive stars undergoing successful core-collapse explosions. In these explosions, the plasma in the stellar code is compressed starting from 10^{10-11} $\mathrm{g\ cm^{-3}}$ up to a few times nuclear saturation density ($n_0 \approx 0.16\ \mathrm{fm^{-3}}$), while the temperature increases from a fraction to a few tens of MeV. If the iron core of the star is almost isospin symmetric, electron captures on nuclei and on free protons decrease the electron fraction, efficiently converting protons into neutrons. The resulting proto-neutron star cools down through the emission of neutrinos and photons on a time scale of millions of years [1]. During their entire life and up to a few orbits before the merger, the matter inside NSs can be described as cold ($T = 0$) and β-equilibrated, with densities in the center reaching several times n_0, depending on the NS gravitational mass and on the properties of the matter forming the NS. During the subsequent merger, the matter density reaches the maximum density allowed by nature, possibly forming a black hole (BH) or leaving behind a fast-rotating massive neutron star (MNS). At the same time, the temperature can increase up to $\sim 100\mathrm{MeV}$, see

DOI: 10.1201/9781003306580-10

e.g. [2], well above the Fermi temperature of the system:

$$T_{\rm Fermi} \sim 30 \ {\rm MeV} \left(\frac{\rho}{3 \times 10^{14} \ {\rm g \ cm^{-3}}} \right)^{2/3} , \tag{10.1}$$

while the activation of weak interactions, and specifically of charged current reactions, modifies once again the composition. In particular, the decompression and the heating of very neutron-rich matter can result in the leptonization of the system, i.e. in the increase of the relative amount of protons with respect to neutrons. Differently from BHs (whose behavior is characterized by the scale invariance of General Relativity with respect to the gravitational mass for vacuum solutions), self-gravitating non-singular objects have a maximum mass above which the gravitational collapse is unavoidable. In the case of white dwarves, gravity is balanced by the pressure produced by a degenerate electron gas, which can be described in very good approximation as an ideal, cold, Fermi gas. In this case, considering the rotation to be negligible, the limiting mass is the Chandrasekhar mass. NSs are bound systems in which the scale invariance is broken by the presence of matter, but whose behavior cannot ignore the strong interaction occurring inside baryonic matter around or above saturation density. In other words, NSs are gravitationally bound systems, glued together by the nuclear force. It is now well established that the behavior and the properties of dense matter, possibly at finite temperature and out-of-equilibrium composition, influence both the properties of isolated, cold NSs, as well as the collision dynamics of merging BNSs, and leave an imprint in all the relevant aspects of the merger. At the same time, it is also clear that the observation of gravitational waves (GWs) and electromagnetic radiation, as well as the study of the nucleosynthesis occurring in compact binary mergers involving at least one NS, can shed light and provide unique insights on the behavior of nuclear matter in regimes which will be never tested on Earth. Simulations of BNS mergers, and in particular the ones performed in full General Relativity (often called Numerical Relativity simulations), are a necessary ingredient to predict the outcome of these collisions, as well as to test the dependency of all the merger outcomes (from the GW emission during the late inspiral up to the electromagnetic emission occurring from hours to years after the merger) on the input microphysics, and in particular on the EOS describe the properties of matter at densities around and in excess of saturation density.

In this contribution, we will present our current understanding of the role of the nuclear EOS in BNS mergers based on the outcome of Numerical Relativity merger simulations, and we will provide concrete examples of how future multimessenger detections will be able to constraints nuclear properties. We will start by presenting an overview of the merger dynamics as it has emerged in the last few years from sophisticated and detailed simulations. After that, we will present the properties and the methodologies behind the nuclear EOSs that are currently used in BNS merger modeling. We will then present in detail quantitative results that connect BNS merger observables to EOS properties. Finally, we will show a few concrete examples on how future multimessenger observations will provide unique constraints on the nuclear EOS, and thus on the physics of dense matter.

10.1 OVERVIEW OF BNS MERGER DYNAMICS AND OBSERVABLES

Throughout the GW-driven inspiral, matter inside the two NSs is in cold, neutrinoless, weak equilibrium, and degenerate baryons are the dominant source of pressure inside their cores. For binary separations much larger than the NS radius, the dynamics of a BNS is indistinguishable from the one of a binary BH. For quasi-circular orbits, the dynamics of the system is governed by the binary chirp mass, $\mathcal{M}_c$. The latter is defined as $\mathcal{M}_c = (M_A M_B)^{3/5}/M^{1/5}$, where $M_{A,B}$ are the gravitational masses of the two NSs at infinite separation, with $M_A \geq M_B$, while $M \equiv M_A + M_B$ is the total mass of the system. We further define the mass ratio as $q \equiv M_B/M_A \leq 1$. The GW frequency ($f_{\rm GW}$) is twice the Keplerian orbital frequency. As the emission of GWs removes energy and angular momentum from the binary, both the GW frequency and amplitude ($h_{\rm GW}$) increase with time according to (at leading order)

$$h_{\rm GW} \sim \frac{4\pi^{2/3}}{c^4 d_L}(G\mathcal{M}_c)^{5/3} f_{\rm GW}^{2/3}, \qquad \frac{df_{\rm GW}}{dt} \approx \frac{96\pi^{8/3}}{5}\left(\frac{G\mathcal{M}_c}{c^3}\right)^{5/3} f_{\rm GW}^{11/3},$$

where d_L is the luminosity distance of the source, producing the characteristic chirp signal. During the last few orbits before merger, the mutual tidal interaction becomes relevant and the two NSs get deformed. This deformation can be described as a dissipative process which accelerates the inspiral. In particular, at a given distance, the tides increase the orbital frequency, and, as the motion is accelerated, the BNS merges earlier and at a lower frequency. See, for example, [3–5] and references therein . The degree of deformation is expressed in terms of the so-called Love numbers . In particular, in the linear response regime, at leading order each star develops a quadrupole moment, Q_{ij}, proportional to the tidal field of the companion, $\mathcal{E}_{ij}$: $Q_{ij} = -\lambda_2 \mathcal{E}_{ij}$. Since the only macroscopic length scale of the star is its radius $R_{\rm NS}$, because of dimensional arguments λ_2 can be expressed as $\lambda_2 = (2/3)(k_2 R_{\rm NS}^5/G)$ where G is the gravitational constant and k_2 is the dimensionless quadrupole gravitoelectric coefficient, see e.g. [6]. The calculation of k_2 (and more in general of all Love numbers) reduces to the solution of stationary perturbations of spherical relativistic stars, under the assumption that the external field varies sufficiently slowly compared with the deformation time scale (adiabatic tides) [7–9]. For typical EOS, $k_2 \sim 0.025 - 0.125$ for NSs in the range 1-2 $M_\odot$. Moreover, since the inspiral dynamics introduces also an explicit dependence on the NS masses, it is useful to introduce the dimensionless tidal coefficient $\Lambda = (2/3)k_2[R_{\rm NS}c^2/(GM_{\rm NS})]^5$, where c is the speed of light and $M_{\rm NS}$ is the NS gravitational mass . It is then clear that the tidal coefficients have a strong dependency on the NS compactness, defined as the dimensionless ratio between the NS mass and radius , $C_{\rm NS} \equiv GM_{\rm NS}/(R_{\rm NS}c^2)$. The mass and radius of an isolated, cold NS are intimately related through the EOS of NS matter. Indeed, starting from a certain central density (or enthalpy), the NS structure can be obtained by integrating the stellar structure equation in General Relativity. In the case of non-rotating stellar objects, the latter are called Tolman-Oppenheimer-Volkoff (TOV) equations [10]. In the case of (neutrino-less) β-equilibrated, cold matter, the EOS is just a function of the matter density. Independently from the nuclear interaction and even from the relevant thermodynamics degrees of freedom, NS matter EOSs are often refereed as "soft" or "stiff". An EOS is said to be softer than another one if, for a given density interval (usually, above n_0), the former provides systematically a lower pressure than the latter. Softer EOSs usually provides less massive heaviest NSs and smaller radii, leading to more compact NSs. For more details, see e.g. Section 10.2.4.2. Given the above definitions, we estimate the compactness and the tidal coefficient of a single star as

$$\mathcal{C}_{\rm NS} \approx 0.171 \left(\frac{M_{\rm NS}}{1.4 M_\odot}\right) \left(\frac{R_{\rm NS}}{12~{\rm km}}\right)^{-1} , \tag{10.2}$$

and

$$\Lambda_{\rm NS} \approx 460 \left(\frac{k_2}{0.1}\right) \left(\frac{\mathcal{C}_{\rm NS}}{0.171}\right)^{-5} . \tag{10.3}$$

At leading order, the tides decrease the cumulative GW phase with respect to the equivalent binary BH case by a quantity $\Delta\Phi_{\rm tides}$ which is proportional the so called binary tidal parameter, $\tilde{\Lambda}$, defined as

$$\tilde{\Lambda} \equiv \frac{16}{13} \left(\frac{(M_A + 12 M_B) M_A^4}{M^5} \Lambda_A + (A \leftrightarrow B) \right) . \tag{10.4}$$

In the above formula, $\Lambda_{i=A,B}$ is the tidal coefficient of each star. A larger $\tilde{\Lambda}$ implies a larger NS deformation, typically related to less compact NSs. Since the stellar radius, for a given mass in equilibrium conditions, depends on the EOS, it is clear that the tidal deformation carries the imprint of the EOS on the merger dynamics and GW signal. During the inspiral, tidal deformation and resonances dissipate also energy. However, the increase in temperature is expected to be modest, $\Delta T \lesssim 0.1$ MeV [11]. The GW signal increases in amplitude and frequency, producing the so-called chirp signal. The moment of merger is conventionally defined as the moment when they reach their maximum within the chirp evolution.

At merger, the orbital speed of the two stars can be estimated as $v_{\rm orb} \sim \Omega_{\rm orb} r \approx \sqrt{GM/(R_A + R_B)}$, where $\Omega_{\rm orb}$ is the orbital frequency (which is approximately half of the GW frequency, i.e. $\Omega_{\rm GW} \sim 2\Omega_{\rm orb}$) and $r \sim R_A + R_B$ is the two stars separation. Assuming that $R_A \approx R_B$, we obtain

$$v_{\rm orb}/c \sim \sqrt{\mathcal{C}_A \left(\frac{1+q}{2}\right)} \approx 0.39 \left(\frac{\mathcal{C}_A}{0.15}\right)^{1/2} \left(\frac{1+q}{2}\right)^{1/2} . \tag{10.5}$$

At the same time, the radial infall velocity due to the loss of GWs can be estimated as $v_r \sim 2\Omega_{\rm orb}\dot{\Omega}_{\rm orb} r/(3\Omega_{\rm orb}^2)$ while $\dot{\Omega}_{\rm orb} \approx \dot{\Omega}_{\rm GW}/2 \sim 3456/250 \left(G\mathcal{M}_c/c^3\right)^5 \Omega_{\rm GW}^{11}$, leading to

$$v_{\rm rad}/c \sim \frac{2\pi}{5} \mathcal{C}_A^3 q(1+q) \approx 0.034 \left(\frac{\mathcal{C}_A}{0.15}\right)^3 \frac{q(1+q)}{2} . \tag{10.6}$$

So, even at merger, $v_{\rm rad}/c \ll v_{\rm orb}/c$ and the dynamics is dominated by the orbital motion. The corresponding dynamical timescale is then given by

$$t_{\rm dyn} \sim 2\pi/\Omega_{\rm orb} \approx 1.5 {\rm ms} \left(\frac{M}{2.8 M_\odot}\right)^{-1/2} \left(\frac{\mathcal{C}_A}{0.15}\right)^{-3/2} . \tag{10.7}$$

Equation 10.5 allows to determine the energy scale of the merger, expressed in terms of the kinetic energy of the colliding stars:

$$K_{\rm orb} \sim \frac{1}{2} M v_{\rm orb}^2 \approx 3.79 \times 10^{53} {\rm erg} \left(\frac{M}{2.8 M_\odot} \right) \left(\frac{\mathcal{C}_A}{0.15} \right) \left(\frac{1+q}{2} \right) . \tag{10.8}$$

The fate of the remnant depends in the leading order on the masses of the two NSs and on the strength of the nuclear interaction in contrasting the gravitational pull. If the masses are relatively heavy or the EOS is very soft, the remnant is prone to collapse to a BH, even within one dynamic timescale (prompt collapse). Instead, if the collapse happens within a few dynamic time scales, the remnant is said to be short-lived, otherwise long-lived. If not collapsed, the central object inside the remnant is usually more massive than the heaviest, cold, non-rotating NS. This is due to the fact that (differential) rotation and thermal support can temporally prevent the collapse. Such an object is usually referred as MNS. If the remnant does not collapse to a BH, due to its initial oblateness and far from equilibrium structure, the remnant is initially a strong GW emitter. The dominant mode is the $m = 2$ mode, typical of a rotating bar, whose extremes are the NS cores. The bar retains most of the angular momentum available at merger and numerical simulations show that it rotates with a characteristic period $P_{\rm rmn} \lesssim 1{\rm ms}$, so that the corresponding GW frequency is $f_{\rm GW,rmn} \gtrsim 2{\rm kHz}$. See, for example, [12]. Typical GW luminosities are of the order of 10^{55} erg s^{-1}, while the spectrum is characterized by the presence of discrete features or peaks. The main one, usually called f_2, is a broad peak at frequencies around $2 - 4$ kHz. This peak frequency relates to the nuclear EOS since it depends on the compactness of the forming remnant. GWs carry away energy and angular momentum, producing a back-reaction that quickly damps the bar mode, favoring the fusion of the two cores into a single one characterized by lower energy (i.e., more bound) and angular momentum. This phase is often referred as the GW-dominated post-merger phase. This phase lasts $10-20{\rm ms}$ post-merger, corresponding to several dynamical time scales. This is indeed the timescale over which the NS cores fuse. During this process, the forming remnant bounces several times due to the succession of contraction and expansion episodes, triggered by gravity and nuclear repulsion, respectively. As a result, the maximum density and temperature increase immediately after merger, before starting oscillating because of the bounce dynamics. Nevertheless, an important difference distinguishes the behavior of these two quantities: the maximum density, averaged over the oscillation period, increases monotonically, while the maximum temperature, after having reached its peak value immediately after merger, settles to a lower, almost constant value. Despite the large relative collision speed, the formation of shocks inside the two merging cores is prevented by the high speed of sound of nuclear matter at densities $n_B \gtrsim 2n_0$. Thus, matter inside the cores remains at low entropy ($s \sim 1 k_{\rm B} {\rm baryon}^{-1}$) and degenerate ($T \sim 10$ MeV $\ll T_{\rm Fermi}$). Only when pressure perturbations reaches lower density regions, they steepen into shock waves that propagates outwards, increasing the matter temperature for $n_B \lesssim n_0$.

While the densest part of the cores rotate and fuse, compressed matter at the contact interface is pushed outwards. Compression and shear dissipation increase its temperature forming a pair of co-rotating hot spots displaced by an angle of $\sim \pi/2$ with respect to the densest cores, see e.g. [13]. This structure survives until the cores have completed their fusion (or until BH formation). At that point, the hot spots have evolved into a hot annulus. The core of the remnant remains relatively cold instead. We estimate the fraction of the orbital kinetic energy that goes into thermal energy ($X_{\rm th}$) by comparing the kinetic energy at merger ($K_{\rm orb} \sim Mv_{\rm orb}^2/2$, see Equation 10.8) with the kinetic energy of the remnant ($K_{\rm rmn} \sim I_{\rm rmn}\Omega_{\rm rmn}^2/2$), where $I_{\rm rmn} = k_I M_{\rm rmn} R_{\rm rmn}^2$ is the moment of inertia of the remnant characterized by a mass $M_{\rm rmn} \lesssim M$ and a radius $R_{\rm rmn} \sim R_{\rm NS}$, while $\Omega_{\rm rmn} = 2\pi/P_{\rm rmn}$ is the rotational pulsation associated to a typical rotational period $P_{\rm rmn} \approx 0.8$ ms, see e.g. [14]. In particular, $1 - X_{\rm th} \sim K_{\rm rmn}/K_{\rm orb}$ so that

$$1 - X_{\rm th} \sim 0.4 \left(\frac{P_{\rm rmn}}{0.8{\rm ms}}\right)^{-2} \left(\frac{k_I}{2/5}\right) \left(\frac{R_{\rm rmn}}{15{\rm km}}\right)^2 \left(\frac{\mathcal{C}_A}{0.15}\right)^{-1} \left(\frac{1+q}{2}\right)^{-1} . \tag{10.9}$$

Since a fraction of this energy can also be radiated by GWs, we consider this a safe upper limit for $X_{\rm th}$. This implies that a fraction close to 50% of the orbital speed at merger can be converted into thermal energy inside the remnant. Assuming to distribute this thermal energy to the $N_{\rm rmn} \sim M_{\rm rmn}/m_n$ baryons that form the remnant, i.e., $X_{\rm th}K_{\rm orb} \sim N_{\rm rmn}k_B T_{\rm rmn}$, we can provide an estimate of the temperature inside the hot part of the remnant:

$$k_B T_{\rm rmn} \sim \frac{1}{2} m_n v_{\rm orb}^2 X_{\rm th} \approx 36 \text{ MeV} \left(\frac{X_{\rm th}}{0.5}\right) \left(\frac{v_{\rm orb}/c}{0.39}\right)^2 . \tag{10.10}$$

In the above expressions, m_n is the neutron mass, taken to be the representative baryon mass inside the remnant. The above estimate shows how the EOS can affect the dynamics and the properties of the remnant, especially through the compactness. Indeed, if an EOS produces relatively high compactness NSs (a "softer" EOS), the two NSs collide with a larger orbital speed and the collision is more violent, due to the larger infall velocity and kinetic energy. The resulting dynamics is faster and the remnant is hotter. The average temperature is larger due to the fact that both more kinetic energy is available and the dissipation in internal energy is more efficient.

Starting from the moment of merger, tidal torques and remnant bounces expel matter, decompressing it from nuclear to sub-nuclear densities ($n_B \lesssim 0.1 n_0$). Due to its angular momentum content, this matter settles in a thick ($(H/R)_{\rm disk} \sim 1/2$) accretion disk around the central part of the remnant. The mass of these disks can vary considerably, up to a few tenths of $M_\odot$, and systematically increases for very asymmetric binaries (for which the tidal interaction is more effective) and for less compact NSs, see, for example [15–17]. Their radial extension is of the order of a few hundreds of kilometers, even if most of the mass is contained within a few tens of kilometers. On the other hand, if a BH has formed, the innermost part of the disk is swallowed inside the apparent horizon, reducing the disk mass by $\sim 50\%$.

After the GW-dominated phase, the subsequent evolution of the remnant (formed by a massive NS or a BH surrounded by an accretion disk) is governed by neutrino emission and viscous processes of magnetic origin. The increase in temperature inside high-density matter activates weak reactions, and neutrinos of all flavors are copiously produced during and after the merger [18]. One of the most relevant reaction is the scattering of neutrinos off free nucleons. The corresponding mean free path, comparable to the one of charged current absorption reactions, can be estimated as

$$\lambda_\nu \sim (n_B \sigma_\nu)^{-1} \approx\sim 2.36 \times 10^3 \text{ cm} \left(\frac{\rho}{10^{14} \text{ g cm}^{-3}} \right)^{-1} \left(\frac{E_\nu}{10 \text{ MeV}} \right)^{-2}, \qquad (10.11)$$

where $\sigma_\nu \approx \sigma_0 \left(E_\nu/m_e c^2\right)^2$, and $\sigma_0 = G_F^2 (m_e c^2)^2 (g_V^2 + 3g_A^2)/(\pi(\hbar c)^4) \approx 2.43 \times 10^{-44} \text{cm}^{-2}$ is the typical scattering cross section and E_ν the incoming neutrino energy. In the previous expression, G_F is the Fermi constant, g_V and g_A the vector and axial coupling constants, $\hbar$ the reduced Planck constant. The dominant production channel for electron neutrinos is the electron capture on free protons, $p + e^- \rightarrow n + \nu_e$. The mean energy of the produced neutrinos is $E_{\nu_e} \approx (k_B T) F_5(\mu_e/k_B T)/F_4(\mu_e/k_B T)$, see, e.g. [19–22] . In the case of very degenerate electrons,

$$E_{\nu_e} \sim \frac{5}{6} \mu_e \approx 74.1 \text{ MeV} \left(\frac{Y_e}{0.05} \right)^{1/3} \left(\frac{\rho}{3 \times 10^{14} \text{g cm}^{-3}} \right)^{1/3}.$$

Similarly, positron capture on free neutrons, $n + e^+ \rightarrow p + \bar{\nu}_e$, produces electron antineutrinos whose mean energy is given by $E_{\bar{\nu}_e} = (k_B T) F_5(-\mu_e/T)/F_4(-\mu_e/T)$ and

$$E_{\bar{\nu}_e} \approx 5 k_B T \approx 100 \text{ MeV} \left(\frac{k_B T}{20 \text{ MeV}} \right).$$

By considering the mean free path given by Equation 10.11 and the typical production energy of neutrinos, it is clear that, deep inside the remnant, where temperature and density are high enough, the neutrino optical depth $\tau \sim \ell/\lambda_\nu \gg 1$ (where ℓ is the relevant length scale) and neutrinos are trapped, i.e. their diffusion timescale is larger than the dynamical timescale. The diffusion timescale can be estimated using random-walk arguments as:

$$t_{\nu,\text{diff}} \sim \frac{3 \tau_\nu \ell}{c} \sim \frac{3 \ell^2}{c \lambda_\nu}. \qquad (10.12)$$

In the case of the diffusion from the remnant, $\ell \sim R_{\text{rmn}}$ and

$$t_{\nu,\text{diff,rmn}} \approx 1.88 \text{ s} \left(\frac{R_{\text{rmn}}}{15 \text{ km}} \right)^2 \left(\frac{\rho_{\text{rmn}}}{10^{14} \text{ g cm}^{-3}} \right) \left(\frac{k_B T_{\text{rmn}}}{25 \text{ MeV}} \right)^2, \qquad (10.13)$$

where we have assumed thermal neutrinos such that $E_\nu \sim 3 k_B T$. Similarly, for the diffusion of neutrinos from the innermost part of the disk, $\ell \sim H_{\text{disk}}$ and

$$t_{\nu,\text{diff,disk}} \approx 15.6 \text{ ms} \left(\frac{(H/R)_{\text{disk}}}{1/2} \right)^2 \left(\frac{R_{\text{disk}}}{50 \text{ km}} \right)^2 \left(\frac{\rho_{\text{disk}}}{10^{12} \text{ g cm}^{-3}} \right) \left(\frac{k_B T_{\text{disk}}}{5 \text{ MeV}} \right)^2. \qquad (10.14)$$

Pair processes, as electron-positron annihilation, plasmon decay and nucleon-nucleon bremsstrahlung, produce neutrinos of all flavours. The former two reactions are relevant at very high temperatures, while the latter at large densities. Overall, inside the remnant and in the innermost part of the disk, charged current reactions and pair reactions bring neutrinos into thermal and weak equilibrium with matter. Due to the trapping conditions, neutrinos form an ultra-relativistic gas in thermal equilibrium which behaves as a fluid component. In the case of electron flavor (anti)neutrinos, the equilibrium chemical potentials are $\mu_{\nu_e,\mathrm{eq}} = \mu_e - \mu_n + \mu_p$ and $\mu_{\bar{\nu}_e,\mathrm{eq}} = -\mu_{\nu_e,\mathrm{eq}}$, where μ_e, μ_n and μ_p are the relativistic chemical potentials of electrons, neutrons and protons, respectively. Since in cold, β-equilibrated matter protons are less degenerate than neutrons and electrons, an increase in temperature decrease μ_p more significantly than μ_n and μ_e, so that $\mu_{\nu_e,\mathrm{eq}} < 0$. Thus, electron neutrinos are suppressed by degeneracy, while electron antineutrinos dominate.

The most important effect of neutrino emission is matter cooling. In particular, neutrino emission is powered by the internal energy of the remnant, as well by the conversion of gravitational energy into internal energy inside the accretion disk. The typical neutrino luminosities can be estimated by dividing the thermal internal energy content of the remnant and of the disk by the neutrino diffusion timescale. Using the energy scale of the merger given by the kinetic energy of the colliding stars, Equation 10.8, taking into account the fraction converted in internal energy, Equation 10.9, and the diffusion timescale from the remnant, Equation 10.13, the resulting average luminosity over the diffusion timescale ranges between 10^{52} and 10^{53} $\mathrm{erg\ s^{-1}}$. After being emitted at the last scattering surface (typically located between 10^{11} and 10^{12} $\mathrm{g\ cm^{-3}}$ for the relevant neutrino energies, see e.g. [23]), neutrinos stream freely. Along their path, they irradiate matter inside the disk and expanding matter beyond it, called ejecta (see below), with a non-negligible probability of being re-absorbed. While the neutron-rich conditions favours the production and emission of $\bar{\nu}_e$'s over ν_e's deep inside the remnant, the absorption of electron neutrino on free neutrons, i.e. $\nu_e + n \rightarrow p + e^+$, is favoured in optically thin conditions and increases the electron fraction of matter far enough from the last scattering surface. Given the fact that GWs and ν's are the main sources of energy emission during the first (tens of milliseconds) after merger, we compare the energy radiated in both channels:

$$\Delta E_{\mathrm{GW}} \sim \langle L_{\mathrm{GW}} \rangle \Delta t_{\mathrm{GW}} \approx 0.11 M_{\odot} c^2 \left(\frac{\langle L_{\mathrm{GW}} \rangle}{10^{55} \mathrm{erg\ s^{-1}}} \right) \left(\frac{\Delta t_{\mathrm{GW}}}{20 \mathrm{ms}} \right),$$

and

$$\Delta E_{\nu} \sim \langle L_{\nu} \rangle \Delta t_{\nu} \approx 0.03 M_{\odot} c^2 \left(\frac{\langle L_{\nu} \rangle}{5 \times 10^{52} \mathrm{erg\ s^{-1}}} \right) \left(\frac{\Delta t_{\nu}}{1 \mathrm{s}} \right).$$

Both these energy variations account for the increase in binding energy of the central remnant, as well as the removal of kinetic and internal energy from the remnant.

While most of the matter falls toward the center of the system, forming the MNS (or accreting onto the BH) and the accretion disk around it, a small fraction of it becomes unbound and enriches the interstellar medium [24]. This is the so-called ejecta. Coming from the surface and from the interior of a NS, these ejecta is rich in neutrons and it is a natural site for the production of heavy elements beyond iron through the so-called r-process nucleosynthesis, see e.g. [25, 26] and references therein. Different ejection mechanisms operate on different timescales, leaving an imprint on the ejecta properties. The first ejecta to be emitted is the so-called dynamical ejecta, see, e.g. [15, 27–30]. Its ejection happens on a few dynamical timescales, i.e. within 5–10ms after merger. Two main mechanisms are responsible for it. The tidal interaction between the two merging stars produces tidal tails which are partially detached from the colliding cores. Matter inside these tails can be ballistically expelled, as their speed at merger is comparable or larger than the escape velocity from the forming remnant. This mechanism is especially relevant in the case of stiff

EOSs or for very asymmetric BNSs, leading to more deformed BNSs. The emerging cold ejecta retain low entropy ($s \lesssim 10~\mathrm{k_B~baryon^{-1}}$), and weak interaction has little chances to modify its initial composition, compared with cold, neutrino-less weak equilibrium. In the case of soft EOSs or equal mass BNSs, the merger dynamics is dominated by the core collision dynamics and the mechanism responsible for the ejection is the development of outgoing shock waves inside the remnant. These ejecta have a broader range of electron fraction, since the activation of weak interactions inside them or the effect of neutrino irradiation are stronger, especially in the polar direction above the remnant. On timescales much larger than the dynamical timescale, matter ejection is dominated by the viscous and thermal evolution of the disk. For example, turbulent viscosity of magnetic origin transfers angular momentum outward, expanding the disk. A similar effect can be also achieved by considering the heat produced by the dissipative effect of the resulting stress inside the disk. However, matter inside the expanding disk cools down producing freeze-out from Nuclear Statistical Equilibrium (NSE). The resulting recombination of free neutrons and protons into α particles and heavy nuclei releases $\sim 8.8~\mathrm{MeV~baryon^{-1}}$ of energy, which is enough to unbind matter located further than $\sim 200~\mathrm{km}$ from the central remnant, see e.g. [31–34]. Similar ejection mechanisms can be caused also by neutrino absorption or interaction with a strong, large-scale magnetic field (see, for example, [35–39]). These mechanisms have been proven to be very efficient in unbinding a significant fraction of the accretion disk (between 0.1 and 0.4). The presence of a massive NS or of a BH in the center possibly affects the amount and the composition of these ejecta, due to stronger or weaker neutrino irradiation, as well as to the smaller or larger accretion rate, respectively (see e.g. [40–42]). Since, for a given total mass, the lifetime of the remnant depends on the EOS, the composition of the ejecta ultimately depends on the nuclear EOS. These ejecta are often called viscous or secular ejecta. Such a dependence on the EOS is even stronger in the case of the so-called spiral wave wind ejecta [16, 43]. Indeed, if a massive NS forms, in addition to the GW-loud $m = 2$ bar deformation mode, a $m = 1$ instability develops, see e.g. [44]. Being relatively GW-quiet, this deformation survives the GW-dominated early post-merger phase. Due to the interaction of this mode with the innermost part of the disk, spiral waves propagate inside the accretion disk, transporting angular momentum outward. The net result is the expansion of the outer part of the disk and the formation of an outflow. The existence and the extent of such a wind is related to the nuclear EOS in many ways: EOSs more prone to produce a long-lived MNS after the merger will have less spiral wave wind ejecta, both due to the usually larger disk mass and size, and to the presence of the central engine producing such an ejection mechanism.

The freshly synthesized r-process nuclei in the merger ejecta are however unstable, since they are produced on the neutron-rich side of the nuclear chart, often far from the valley of stability. Starting from a few seconds after merger and continuing for several days, if not weeks, various decay processes (including β decays, but also α decays and fission, depending on the composition) occur, releasing nuclear energy that constantly heats up the expanding material. While matter is initially too dense to irradiate significantly, starting from a few hours after the merger, the expanding matter has reached low enough opacity such that the photon diffusion time scale becomes comparable to the dynamical time scale. Then, the ejecta produce a quasi-thermal emission known as kilonova [45]. Depending on the amount of matter, on the ejecta speed, and on the composition (with ejecta containing lanthanides and actinides have an opacity to photons ten times larger than the one that does not), the duration, peak and color of the kilonova can change dramatically. It is clear that all these properties are ultimately related to the EOS of NS matter through the merger dynamics and the remnant fate, see e.g. [15, 46].

10.2 EQUATIONS OF STATE FOR BNS MERGERS

10.2.1 General properties

The EOS of a physical system is the relationship between matter pressure (P), energy density (ε), baryonic density (n_B) and temperature (T), together with information on the matter composition. An EOS is an essential ingredient not only to study the structure of an isolated NS, but also to understand BNS mergers since it describes the behavior of matter inside the colliding stars. Ultimately, while performing numerical simulations, it is necessary to close the set of hydrodynamical equations governing the system dynamics.

In order to describe NS matter through an adequate EOS, a first necessary step is to discuss the particle content and to define a meaningful set of independent variables. A NS cannot be made by pure neutron matter and a finite proton fraction is indeed necessary. In addition, to ensure charge neutrality, a net fraction of negatively charged, massive leptons must be present. In the limit where thermal effects can be neglected ($T = 0$), the minimal particle content necessary to describe NS matter is a gas of neutron, protons and electrons . As the density increases, electrons become more and more degenerate, until the electron chemical potential reaches the value of the muon rest mass allowing for the net conversion of electrons into muons. This process takes place at a baryonic density around $n_B \gtrsim 2n_0$. Notice that this conversion reduces the electron Fermi energy and turns out to be energetically favourable. The description of electrons and muons in terms of an ideal Fermi gas of massive particles is accurate and widely used. Similarly, for high enough density, hyperons, physical pions or even a phase transition to pure quark matter could occur. The physics governing their appearance is far from being understood and it is a vibrant research field on its own, together with the investigation of their effects on NSs and BNS mergers. However, in this contribution, we will focus to the case in which hyperons, pions or free quarks can be neglected in the description of NS matter. In the case of isolated NS made of neutrons, protons, electrons and muons, not only matter reaches temperature well below the Fermi temperature, Equation 10.1, such that it can be modelled as a completely degenerate system ($T = 0$), but it also reaches neutrino-less, β-equilibrium condition, i.e. the chemical potential of electrons (μ_e), protons (μ_p) and neutrons (μ_n) satisfy $\mu_e + \mu_p + \mu_n = 0$, while the chemical potential of muons (μ_μ) equals the one of electrons, ($\mu_e = \mu_\mu$) . Under these conditions, at a given baryon density, all the abundances are determined and the resulting EOS becomes just a function of one variable, e.g. the baryon density. On the other hand, when matter is heated up and weak reactions produce out-of-equilibrium conditions, the particle content must increase. Antiparticles should be in general always included inside the EOS at finite temperature, as well as photons. Also neutrinos appear, but they are not always in thermal and reaction equilibrium with the rest of the matter inside a BNS merger. Their inclusion in merger simulations requires radiation transport, reproducing the hydrodynamics and thermodynamics limit in which neutrinos behave as an additional species inside the EOS for high enough densities and temperatures. Assuming NSE conditions, a finite temperature, composition-dependent NS EOS can be for example expressed a function of the baryon density, matter temperature and lepton abundances, recalling that the proton abundance can always be obtained by enforcing charge neutrality, i.e. $n_p = n_e + n_\mu$. In the following, we will focus on the description of the EOS that governs the baryons and on its inclusions in BNS merger simulations.

So far, two different approaches have been adopted. In the first one, the starting point is the usage of the $T = 0$, beta-equilibrated EOS also to describe matter in the merger. In this approach, matter composition is considered to be "frozen", in the sense that the system is considered in equilibrium with respect to weak interactions and neutrino abundances are assumed to be negligible. This condition, together with the requirement of charge neutrality of the whole system, fixes matter composition for a given baryonic density. Zero temperature EOSs can be provided in analytic or tabulated form. In the case of analytic EOSs, it is usually assumed that they can be parametrized by a simple polytropic relation between pressure (P) and matter density (ρ):

$$P = K\rho^\gamma \,, \tag{10.15}$$

where γ is a polytropic index and K a constant. The full EOS is finally completed by the relation between matter density and baryonic density: $\rho = n_B m_u$ being m_u the atomic mass unit. Thermal effects are often included in an approximated way, consisting in supplementing the cold EOS with a thermal component added by hand. The full EOS turns out to be in this case:

$$P(\rho, \varepsilon) = P_{\text{cold}}(\rho, \varepsilon) + (\Gamma_{\text{th}} + 1)(\tilde{\varepsilon} - \tilde{\varepsilon}_{\text{cold}})\rho \,, \tag{10.16}$$

where $\tilde{\varepsilon}$ is the specific internal energy.

The second approach includes the effect of finite temperature and arbitrary composition in a more consistent way. In particular, in the case of finite temperature and composition-dependent EOSs, all thermodynamical variables are given out of β-equilibrium and as function of T. It is apparent that the latter provide a more detailed and general description of the system. Some of the most popular methods and models to produce this kind of EOSs will be discussed below in some detail. Finite temperature EOSs are usually provided as function of three independent quantities: baryonic density, temperature and electron fraction (Y_e). The latter coincides with the proton fraction in the case of neutrino and muon free matter. For fixed (n_B, T and Y_e) a typical table contains the following thermodynamical quantities: pressure, energy density, entropy (S), proton chemical potential (μ_p), neutron chemical potential (μ_n), electron chemical potential μ_e, matter composition (x_i). Matter composition includes the relevant degrees of freedom that are expected to play a role in BNS merger simulations. Besides neutrons, protons and electrons, a typical table provides also the abundances of light and heavy nuclei that can be present below saturation density and temperatures below $\sim$ 10 MeV. We should note that more general tables that take into account of the possible formation of additional degrees of freedom besides nucleons and leptons can also be produced. Hyperons and a deconfinement phase transition to quark matter are the first new ingredients expected to play a role in BNS merger simulations. We want finally to point out that the inclusion of muons and neutrinos in a unique table may significantly increase the dimension of the latter. To overcome this problem, it may be convenient to produce different tables, separating the hadronic from the leptonic species.

10.2.2 Specific models for the cold, hardonic EOSs

Despite intense efforts over the last decades, the EOS of NS matter is still uncertain, especially well above saturation density. There are multiple challenges in its calculations, mostly related to i) our lack of detailed knowledge of the nuclear interaction; ii) our difficulties in computing properties of a strongly interacting and highly correlated many-body system. It is then natural that the problem of computing the EOS of NS matter and, in particular, of its hadronic part, has furnished the development of several different models and techniques, possibly resulting in very different EOSs. In this section, we discuss in detail some of the current methods and models used to develop finite temperature and composition-dependent EOSs. There is clearly no unique way to classify such EOSs; however, it looks natural to distinguish between phenomenological and microscopic approaches. Phenomenological approaches are in turn divided into relativistic and non-relativistic.

10.2.2.1 Phenomenological, relativistic approaches

Relativistic approaches are often based on a relativistic Lagrangian density inspired by quantum field theory. Hadron fields are assumed to interact with mesons by a minimal Yukawa coupling. In this approach, all baryon and meson fields are treated as classical fields, and the various operators entering the Lagrangian density are replaced by their expectation values calculated in the nuclear medium. This approach is usually referred to as Relativistic Mean Field (RMF) approach . A typical Walecka-type Lagrangian density can be written as:

$$
\begin{aligned}
\mathcal{L} &= \sum_N \bar{\Psi}_N \left[\gamma_\mu D_N^\mu - m_N^*\right] \Psi_N \\
&+ \sum_l \bar{\psi}_l \left[i\gamma_\mu \partial^\mu - m_l\right] \psi_l \\
&+ \frac{1}{2}\left(\partial_\mu \sigma \partial^\mu \sigma - m_\sigma^2 \sigma^2\right) - \frac{1}{3!} k\sigma^3 - \frac{1}{4!}\lambda\sigma^4 \\
&+ \frac{1}{2} m_\omega^2 \omega_\mu \omega^\mu - \frac{1}{4}\Omega_{\mu\nu}\Omega^{\mu\nu} + \frac{1}{4!}\xi g_\omega^4 \left(\omega_\mu \omega^\mu\right)^2 \\
&+ \frac{1}{2} m_\rho^2 \boldsymbol{b}_\mu \boldsymbol{b}^\mu - \frac{1}{4}\mathbf{B}_{\mu\nu}\mathbf{B}^{\mu\nu} \\
&+ \left[\Lambda_\omega \left(g_\omega^2 \omega_\mu \omega^\mu\right) + \Lambda_N g_\sigma^2 \sigma^2\right] \left(g_\rho^2 \boldsymbol{b}_\mu . \boldsymbol{b}^\mu\right) .
\end{aligned}
\tag{10.17}
$$

where $D_N^\mu = i\partial^\mu - g_{\omega N}\omega^\mu - g_{\rho N}\boldsymbol{\tau}_N.\boldsymbol{b}^\mu$ and $m_N^* = m_N - g_{\sigma N}\sigma$ is the baryon effective mass. Ψ_N and ψ_l are the baryon and lepton Dirac fields, respectively, and σ, ω, and ρ represent the scalar, vector, and isovector meson fields, which describe the nuclear interaction. Lepton masses and bare baryon masses are denoted by m_l and m_N, respectively. The coupling constants of i-meson (where $i = \sigma, \omega, \rho$) with a nucleon N are denoted by $g_{i,N}$, where the index N runs over nucleons: n, p. The above Lagrangian can be extended to include the hyperons of the baryonic octect: Λ, Σ^-, Σ^0, Σ^+, Ξ^- and Ξ^0 or even the Δ quartet. The couplings $g_{i,N}$ set the strength of the interaction between two baryons and a meson. In the following, in order to address the couplings between meson (i) and nucleon (N), we shall use the notation $g_{i,N} = g_i$. Λ_ω is a coupling constant of the mixed $\rho\omega$-term [47], introduced to adjust the behaviour of symmetry energy near saturation density, n_0. The coupling of Λ_N allows us to finally consider a mixing between the σ- and ρ-meson. The constants k and ξ are the weights of the σ and ω self-interaction terms and $\boldsymbol{\tau}_B$ is the isospin operator, namely the vector of the three Pauli matrices. The mesonic field tensors are given by their usual antisymmetric expressions: $\Omega_{\mu\nu} = \partial_\mu \omega_\nu - \partial_\nu \omega_\mu$, $\boldsymbol{B}_{\mu\nu} = \partial_\mu \boldsymbol{b}_\nu - \partial_\nu \boldsymbol{b}_\mu$, and $\phi_{\mu\nu} = \partial_\mu \phi_\nu - \partial_\nu \phi_\mu$. In this notation, the $\boldsymbol{b}$-field is associated to the ρ-meson. So far, various classes of RMF models have been developed according to the terms retained in the Lagrangian density, Equation (10.17). In some models, the couplings g_σ, g_ω, and g_ρ have a density dependence, and terms proportional to k, λ, ξ, Λ_ω, and Λ_N are set to zero, while in other models g_σ, g_ω, and g_ρ are constant, and all terms of the above Lagrangian are kept. In the following, we refer to the first class of models as density-dependent (DD) mean-field models and to the seconds as non-linear (NL) mean-field models. In the Hartree approximation, the equations of motion of the various fields can be easily obtained starting from a typical Lagrangian density, like Equation (10.17), and the Euler-Lagrange equations. In this approximation, all of the fields retain only a density dependence. The total set of equations is therefore reduced to a set of nonlinear ordinary equations, which can be solved numerically with a multidimensional Newton-Raphson method. All the parameters of the model (mainly coupling constants and some meson's masses) are fit to reproduce properties of nuclear matter around saturation density, nuclei and/or neutron star properties like maximum mass and radii. If additional degrees of freedom like hyperons are included, then new couplings and/or parameters are involved. They are usually fixed using the existing data on single- and double-hypernuclei, hypernuclear matter, together with $SU(6)$ relations which relate nucleonic couplings with the hyperonic ones.

10.2.2.2 *Phenomenological, non-relativistic approaches*

Skyrme models are based on a functional of the energy density whose typical form can be written in terms of the nuclear density, n_B, and the proton fraction, $y = n_p/n_B$:

$$\varepsilon(n_B, y) = \varepsilon_{\rm kin}(n_B, y) + V_{\rm pot}(n_B, y) \tag{10.18}$$

where:

$$\varepsilon_{\rm kin}(n_B, y) = \frac{\hbar^2 \tau_n}{2m_n^*} + \frac{\hbar^2 \tau_p}{2m_p^*} \tag{10.19}$$

is the kinetic energy density contribution and

$$V_{\rm pot}(n_B, y) = [a + 4by(1-y)]n_B^2 + [c + 4dy(1-y)]n_B^{(1+\delta)} \tag{10.20}$$

is the term that takes into account the interaction between nucleons. Inside $\varepsilon_{\rm kin}$, $m^*_{n,p}$ are the effective neutron and proton masses defined by:

$$\frac{\hbar^2}{2m_i^*} = \frac{\hbar^2}{2m_i} + \alpha_1 n_t + \alpha_2 n_{-t} \ , \tag{10.21}$$

with m_i being the bare mass of the nucleon i and n_t (n_{-t}) the nucleonic density corresponding to the component t ($-t$) of the isospin. $V_{\rm pot}$ is given by the sum of two terms: the first, proportional to n_B^2 arises from the contribution of two-body nuclear interaction while the second term proportional to n_B^δ is introduced to mimic the effect of many-body nuclear interaction. $a, b, c, d, \alpha_1, \alpha_2, \delta$ are parameters of the model fitted to reproduce the properties of nuclear matter around saturation density and finite nuclei binding energies and radii. We note that at zero temperature the full ε can be written in analytic form while the extension at finite temperature requires the evaluation of Fermi integrals. Skyrme models have been employed mainly in the nucleonic sector and just a few works tried to extend the formalism to include hyperons [48].

10.2.2.3 Microscopic approaches

The Brueckner-Hartree-Fock (BHF) approach represents the lowest order of the Brueckner-Bethe-Goldstone many-body theory [49] which is based on a linked cluster expansion (the hole-line expansion) of the energy per nucleon of nuclear matter. The various terms of the expansion can be represented by Goldstone diagrams, grouped according to the number of independent hole lines (i.e. lines representing empty single-particle states in the Fermi sea). The main idea behind the BHF approach is that the repulsive hard core, present in the bare baryon-baryon interaction, does not allow conventional perturbation theory to calculate the energy per particle of the system. The typical matrix elements turn out indeed very large due to the presence of short-range repulsion. The BHF approach designs a method to obtain the matrix elements of the baryon-baryon interaction of such size that the perturbation theory converges quite quickly. The bare baryon-baryon interaction is replaced by the so-called G-matrix, which describes the interaction between two baryons in the presence of a surrounding medium. The G-matrices are obtained by solving the coupled-channel Bethe–Goldstone equation, written schematically as:

$$\begin{aligned} G(\omega)_{B_1B_2,B_3B_4} &= V_{B_1B_2,B_3B_4} \\ &+ \sum_{B_iB_j} V_{B_1B_2,B_iB_j} \frac{Q_{B_iB_j}}{\omega - E_{B_i} - E_{B_j} + i\eta} G(\omega)_{B_iB_j,B_3B_4} \,, \end{aligned} \tag{10.22}$$

where the first (last) two subindices indicate the initial (final) two-baryon states compatible with a given value S of the strangeness, namely nucleon-nucleon (NN) for $S = 0$ and nucleon-hyperon (NY) for $S = -1$; $V_{B_1B_2,B_3B_4}$ is the bare baryon-baryon interaction; $Q_{B_iB_j}$ is the Pauli operator, that prevents the intermediate baryons B_i and B_j from being scattered to states below their respective Fermi momenta; ω is the starting energy, which corresponds to the sum of the nonrelativistic single-particle energies of the interacting baryons. The single-particle energy of a baryon B_i is given by:

$$E_{B_i}(\vec{k}) = M_{B_i} + \frac{\hbar^2 k^2}{2M_{B_i}} + \mathrm{Re}[\mathrm{U}_{B_i}(\vec{k})] \,. \tag{10.23}$$

where M_{B_i} denotes the rest mass of the baryon, and the single-particle potential U_{B_i} represents the average field "felt" by the baryon owing to its interaction with the other baryons of the medium. In the BHF approximation, U_{B_i} is calculated through the "on-shell energy" G-matrix, and is given by:

$$\mathrm{U}_{B_i}(\vec{k}) = \sum_{B_j} \sum_{\vec{k}'} n_{B_j}(|\vec{k}'|) \langle \vec{k}\vec{k}' | G(E_{B_i}(\vec{k}) + E_{B_j}(\vec{k}'))_{B_iB_j,B_iB_j} | \vec{k}\vec{k}' \rangle_{\mathcal{A}} \,, \tag{10.24}$$

where $n_{B_j}(|\vec{k}|)$ is the occupation number of the species B_j, and the index $\mathcal{A}$ indicates that the matrix elements are properly antisymmetrised when baryons B_i and B_j belong to the same isomultiplet. In order to determine the single-particle potentials when solving the Bethe–Goldstone equation, two possibilities are usually adopted. The first one is the so-called continuous prescription, where the single-particle potential is assumed to be continuous above the Fermi momentum. The second possibility is to adopt the so-called gap choice: in this case the single-particle potential is set to zero above the Fermi momentum. It has been shown by the authors of Ref. [50] that the contribution to the energy per particle from the three-hole line diagrams, which include correlations between three particle, is minimized by employing continuous prescription. The inclusion of three-baryon forces in the framework of the BHF approach follows a strategy implemented in the case of pure nuclear matter calculations where a density-dependent NN interaction is obtained from the original NNN force by averaging over the coordinates of one of the nucleons [51, 52]. In the case of a generic BBB interaction, such an average is performed with respect to one of the baryons. The exact treatment of three-baryon forces in the BHF approach would require the solution of several coupled Bethe-Faddev equations in the medium. This task is currently too difficult to deal with. The Bethe-Goldstone equation is usually solved numerically using the inversion matrix technique [53] after a partial waves expansion.

10.2.3 Finite temperature extensions

As we have stated previously, thermal effects play an important role in BNS mergers, where matter is heated due to shocks and compression. It is therefore necessary to develop finite temperature EOSs that could be used in numerical simulations. The different approaches we have discussed in the previous sections can be extended at finite temperature in a straightforward way. In the case of RMF approaches, starting from the Lagrangian density, Equation (10.17), the mean field equations for the various mesons have a natural extension at finite temperature. The particle densities, which are analytic at zero temperature, should be calculated by standard Fermi integrals. An important point that is worth to stress is that in a finite temperature calculation the contribution of antiparticle must be explicitly taken into account; therefore, for instance, the net particle density should be intended as the difference between particle and antiparticle fractions. Explicit formulas and a review of finite temperature formalism can be found in Refs. [54, 55]. Skyrme models can also be generalized in direct way in order to include thermal contributions. In this case the functional of density, Equation 10.18, has to be generalized adding a temperature dependence. This is done calculating the kinetic energy densities for neutron and proton $\tau_{n,p}$ and the particle densities using again standard Fermi integrals (see Ref. [56] for further discussions and explicit formulas). Finally, concerning the BHF approach, the first step to incorporate temperature effects is to extend the Bethe-Goldstone equation at finite temperature. This requires to rewrite the Pauli operator taking into account that the scattering between particle takes place in a thermal bath. In this way, the G-matrices acquire also a temperature dependence and so the single particle potentials $U_i(k)$ do. Chemical potentials of neutron and protons are obtained inverting the normalization condition for each species. The full EOS is finally derived by calculating the free energy density, $F = \varepsilon - Ts$, being s the total entropy density, that is calculated according to a free Fermi gas formula. In this way, some particle–particle correlations, whose inclusion would require to go beyond the BHF approximation in the particle-hole expansion, are neglected. Once F has been computed, for convenience it could be useful to perform an analytic fit of F as function of n_B and T. In this way, from F it is possible to derive all the relevant thermodynamical quantities according to standard thermodynamics relations and in an analytic way. For instance the total pressure can be obtained by $P = \sum_{i=n,p} \mu_i n_i + Ts - \varepsilon$ where μ_i (n_i) are the chemical potential (particle density) of the i-th nucleon and (see Ref. [57] for further discussions).

10.2.4 Specific examples of EOSs

10.2.4.1 Generic properties and definitions

We now discuss in some detail five specific EOSs based on the three general approaches introduced in the previous section. The first two EOSs we consider are HS(DD2) (hereafter DD2) [58, 59] and SFHo [60]. They have been constructed according to the general Lagrangian density given by Equation 10.17 but retain only some of the terms related to the nuclear interaction. In particular, DD2 uses linear, but density-dependent coupling constants [61], while the RMF parametrization of SFHo employs constant couplings adjusted to reproduce, besides nuclear matter properties, the measurements of NS radius from low-mass X-ray binaries (see [60] and references therein). With reference to Equation 10.17, all the nonlinear terms proportional to k, λ, ξ, Λ_ω and Λ_N are neglected in both these two RMF models. EOSs that include these non-linear contributions are known in the literature as non-linear Walecka models. In the case of the DD2 model, the density dependence of the couplings is taken in the form:

$$g_m(n_B) = g_m(n_0)h_m(x) \ , \tag{10.25}$$

with $x = n_B/n_0$, $m = \sigma, \omega$ and:

$$h_m(x) = a_m \frac{1 + b_m(x + d_m)^2}{1 + c_m(x + d_m)^2} \ , \tag{10.26}$$

for the isoscalar coupling and:

$$h_\rho(x) = \exp(-a_\rho(x - 1)) \ , \tag{10.27}$$

for the isovector couplings. The values of the parameters a_m, b_m, c_m, d_m and the density dependence in the above functions were originally proposed in Ref. [61] in order to reproduce the density behaviour of the σ-, ω-, and ρ-couplings found in microscopic Dirac-Brueckner-Hartree-Fock calculations [62]. The DD2 and the SFHo EOSs contain neutrons, protons, light nuclei such as deuterons, helions, tritons, and alpha particles, and heavy nuclei in NSE ([60]). It is important to note that in the case of RMF-based EOSs, it is possible to describe the full density range covered by the EOS using the same approach, namely the same Lagrangian density. This is particularly relevant in the case of low density and temperature neutron star matter where clusters and finite nuclei appear.

The third and fourth EOS we are going to consider are the LS220 [63] and the SLy4 [56, 64] ones, which are based on a liquid droplet model of Skyrme interaction (see Equation 10.18). The LS220 EOS includes surface effects and models α-particles as an ideal, classical, non-relativistic gas. Heavy nuclei are treated using the single-nucleus approximation (SNA). The energy density functional form for the LS220 EOS is obtained putting $d = 0$, $\alpha_1 = \alpha_2 = 0$ in Equation 10.18. The SLy4 EOS has been originally derived in Ref. [64] at zero temperature. Later on, it was extended to finite temperature and full composition-dependent form in Ref. [56]. The SLy4 model is an improved version of the LS220 one that includes non-local isospin asymmetric terms and a better treatment of nuclear surface properties. In addition, the size of heavy nuclei is described in a more consistent way. The transition between the uniform and non-uniform phase is achieved by a first-order transition, i.e. choosing the phase with lower free energy. Notice that also in the case of Skyrme models it is possible to use the same framework to describe all the required density range.

The last representative EOS we consider is the BLh EOS, derived in the framework of non-relativistic BHF approach. For the homogeneous nuclear phase, this EOS employs a purely microphysical approach based on a specific nuclear interaction. The interactions between nucleons are described through a potential derived perturbatively in the chiral effective field theory [65]. Specifically, the local potential reported in [66] and calculated up to next-to-leading order (N3LO) was used as the two-body interaction. This potential takes into account the possible excitation of a Δ-resonance in the intermediate states of the nucleon–nucleon interaction. The above potential was then supplemented by a three-nucleon force calculated up to N2LO and including again the contributions from the Δ-excitation. The parameters of the three-nucleon force were determined to reproduce the properties of symmetric nuclear matter at saturation density [67]. For the non-homogeneous nuclear phase, there is no straightforward extension of these microphysical methods to subsaturation densities. Thus, the low-density part ($n_B \leq 0.05$ fm^{-3}) of the SFHo EOS was smoothly connected to the high-density BLh EOS. This necessary extension has been tested and compared with different finite-temperature, composition-dependent tabulated EOS. We note indeed that the extension of the BHF approach to finite systems is much more complex than RMF models and a low-density EOS based on this formalism is not yet available.

The saturation properties of the four EOSs we have discussed so far are reported in Table 10.1. The meaning of the five quantities reported in the table is the following: n_0 (saturation density) is obtained as the value of the baryon density where of the minimum of the energy per particle of symmetric nuclear matter occurs, and $(E/A)_0$ is the corresponding value of E/A at n_0; the symmetry energy (E_{sym}) is defined as $E_{\text{sym}} = (1/2)(\partial^2(E/A)/\partial\beta^2))_{(n_0,\, \beta=0)}$ being $\beta = (n_n - n_p)/n_B$ the asymmetry of nuclear matter; the slope of the symmetry energy is given by $L = 3n_B(\partial(E/A)/\partial n_B)_{n_0}$; and finally the incompressibility of nuclear matter is defined as $K = 9n_B^2(\partial^2(E/A)/\partial n_B^2)_{n_0}$. We note that although the values of the various quantities are all referred to the saturation density n_0, they have a well-defined density (and temperature) dependence. Just to mention two particular examples about the relevance of the nuclear matter parameters at large nuclear density, we note that the behaviour with density of E_{sym} determines with large extent the proton fraction of β-stable nuclear matter, while $K(n_B)$ plays a crucial role in the

TABLE 10.1 Saturation properties for the five representative EOSs considered (first five lines). In the last line for each quantity is reported the range values provided by experimental measurements.

EOS	n_0 [fm^{-3}]	$(E/A)_0$ [MeV]	E_{sym} [MeV]	L [MeV]	K [MeV]
DD2	0.149	−16.04	31.67	54.04	242.7
SFHo	0.158	−16.19	31.57	47.10	254.4
LS220	0.155	−16.00	28.61	73.82	220.0
SLy4	0.159	−16.62	32.00	45.96	229.9
BLh	0.171	−15.54	34.22	80.32	202.4
Exp	$(0.15 \div 0.17)$	$(-16 \div -15)$	$(29 \div 37)$	$(40 \div 110)$	$(190 \div 300)$

dynamics of BNS mergers, as we shall see in Section 10.3.1. The four representative EOSs we have considered provide reasonable values for all the reported nuclear matter quantities. We note however that there are substantial difference about the derivation of these parameters in the various approaches. In particular, while in the phenomenological RMF and Skyrme approaches some of the nuclear matter parameters can be actually adjusted varying the coupling constants (or other parameters of the model), for the BHF approach they are output of the calculation which in turn depends only on the nuclear interaction. Clearly the uncertainties on the latter can modify the final result; an efficient way to take these uncertainties under control is to quantify the theoretical error by Chiral Effective Field theory [65].

10.2.4.2 Cold, beta-equilibrated EOSs

The cold, β-equilibrated slice of the EOS plays a relevant role in the merger phenomenology and in its description .

First of all, it sets the properties of the two merging NSs, until the very final orbits before the actual coalescence. In particular, the EOS determines the maximum NS mass, the NS compactness (which encodes both information on the NS mass and on the radius), as well as the gravitational binding energy of each NS. The latter is defined as the difference between the (gravitational) mass and the baryonic mass of the star. In Figure 10.1, we present the most relevant properties of the NS sequences computed by considering the representative EOSs introduced in Section 10.2.4. These sequences were obtained by solving the Tolman-Oppenheimer-Volkoff equations using cold, β-equilibrated EOSs, i.e. the equations that describe the hydrostatic equilibrium in General Relativity. We notice that, within our sample, DD2 is the EOS that produces the largest maximum NS mass. At the same time, for a given NS mass, it produces the largest radius, the smallest compactness and the smallest binding energy. DD2 is indeed a typical example of stiff EOS. On the other hand, EOSs such as SFHo and SLy4 are typical examples of a soft EOS. Indeed, they predict more compact and gravitationally bound NSs, characterized by systematically

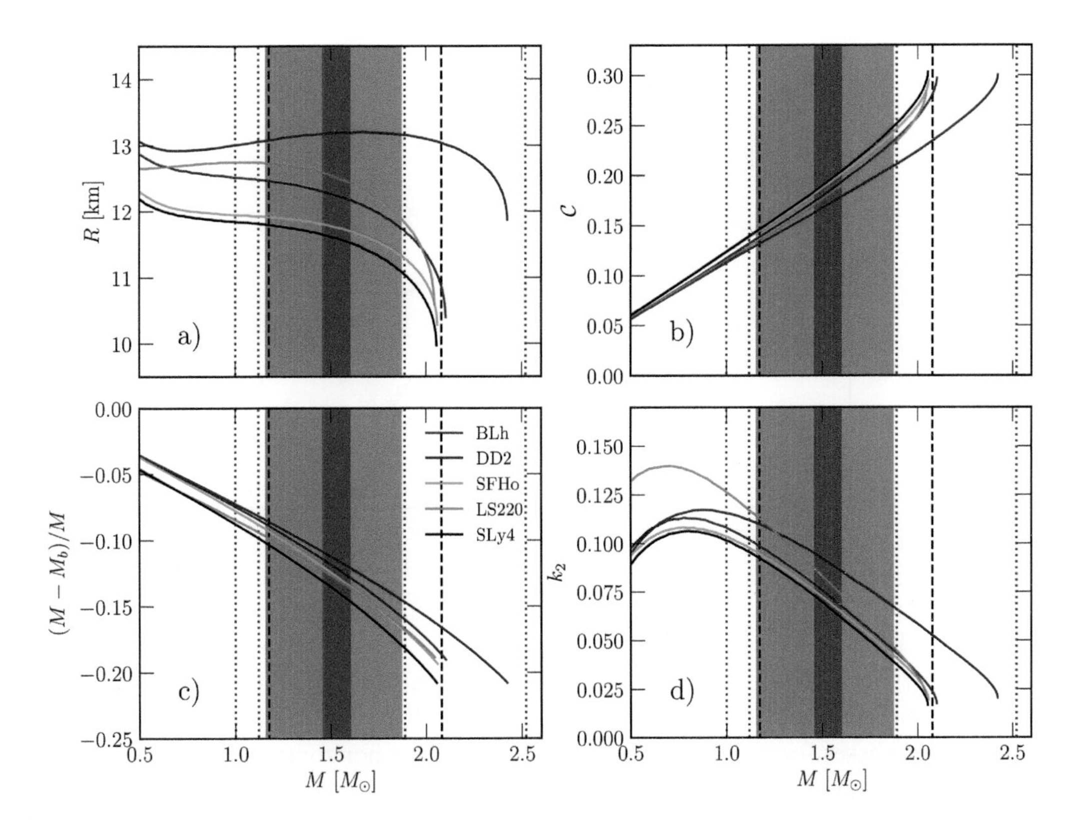

Figure 10.1 Properties of the five different cold EOSs discussed in Section 10.2.4.2, see also Table 10.1. For each EOS, we report (a) the mass-radius diagram, (b) the compactness, (c) the relative NS binding energy, computed as the difference between the gravitational and the baryonic mass, normalized to the latter; (d) the dimensionless (gravitoelectric dipole) Love number k_2. The black, dashed lines correspond to the lightest and heaviest NS observed so far, while the blue and red shaded areas correspond to the inferred NS masses of GW170817 and GW190425, respectively, in the case of low-spin prior analysis. The dotted blue and red lines bracket the mass intervals in the high-spin prior analysis [68, 69].

smaller radii . BLh and LS220 have an intermediate behavior between these two extremes. We notice also that DD2 can reach values of the compactness and of the relative binding energy similar to the ones of SFHo and SLy4, but it requires much larger masses, closer to its maximum mass. Indeed, once normalized to their maximum values and expressed as a function of the mass normalized to its maximum, the curves of the compactness and of the binding energy tend to collapse to a single curve within a few percents. The fourth panel (d) presents the gravitoelectric Love number k_2 . Being more extended and less gravitationally bound, the NSs predicted by DD2 are significantly more prone to be tidally deformed, as testified by values of k_2 larger by $20-25\%$ than the other EOSs. The latter have more comparable values (within 10%), at least in the relevant mass range $1.1M_\odot \lesssim M_{\rm NS} \lesssim 2M_\odot$. The above properties are features that refer to isolated NSs. Once the NSs are in a merging binary, the property that influences the most the GW signal during the inspiral are the chirp

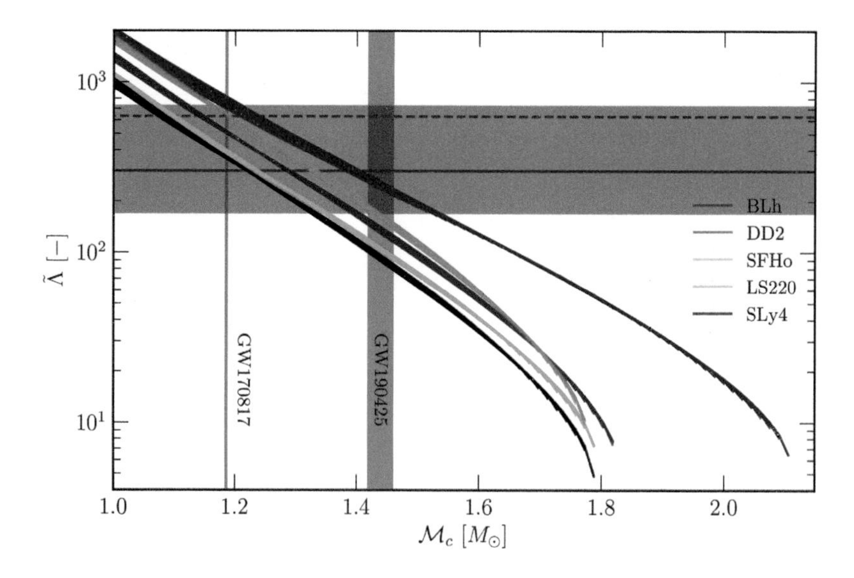

Figure 10.2 Binary tidal coefficient ($\tilde{\Lambda}$) as a function of the chirp mass for the same EOSs presented in Figure 10.1. The bands represent different values of $\tilde{\Lambda}$ obtained by changing the mass ratio between 0.5 and 1. The dashed vertical bands mark the measured intervals of the chirp masses for GW170817 and GW190425. The horizontal blue band marks the 90% confidence level highest posterior density interval from GW analysis [68, 69].

mass and the binary tidal parameter, $\tilde{\Lambda}$. In Figure 10.2, we present the values of $\tilde{\Lambda}$ as a function of the chirp mass. For each $\mathcal{M}_c$, the variability due to the mass ratio is contained within the band. It is evident that the chirp mass is the most relevant binary parameter setting $\tilde{\Lambda}$, followed by the EOS, while the mass ratio has the smallest impact. Since the chirp mass is very well measured from the GW chirp signal, the measurement of $\tilde{\Lambda}$ from a BNS merger with a high enough signal-to-noise ratio can discriminate among different EOSs, especially for a binary whose $\mathcal{M}_c$ is not far from GW170817's one.

However, the cold, β-equilibrated EOS is also relevant to understand the behavior of the remnant in the post-merger phase. Indeed, the forming remnant is characterized by a fast rotating, hot mantle, engulfing a core which rotates at a lower frequency and close to rigid rotation. Inside the core, the lack of shocks guarantees that the matter temperature stays well below the corresponding Fermi temperature. Thus, it is not surprising that a first-order approximation of the core can be described by a massive NS obtained by employing the cold, β equilibrated EOS in computing the stellar structure of a non-rotating NS. Better approximations can be done simply by considering isoentropic and/or slowly rotating NS stationary solutions. A clear evidence of this behavior is represented by the fact that when the massive NS collapses in the center to a BH the central density is usually below, but not far from the central density of the heaviest cold, irrotational NS.

In the left panel of Figure 10.3, we present the relation between the maximum central density and the NS mass for the same EOS presented above. Stiff EOSs, as DD2, produce NS cores and BNS merger remnant characterized by lower densities, with maximum central densities ($n_{B,\max}$) that do not go beyond $6n_0$. Softer EOSs reach larger central densities (up to $8n_0$ in the sample explored here) both in the core of isolated NS close to the

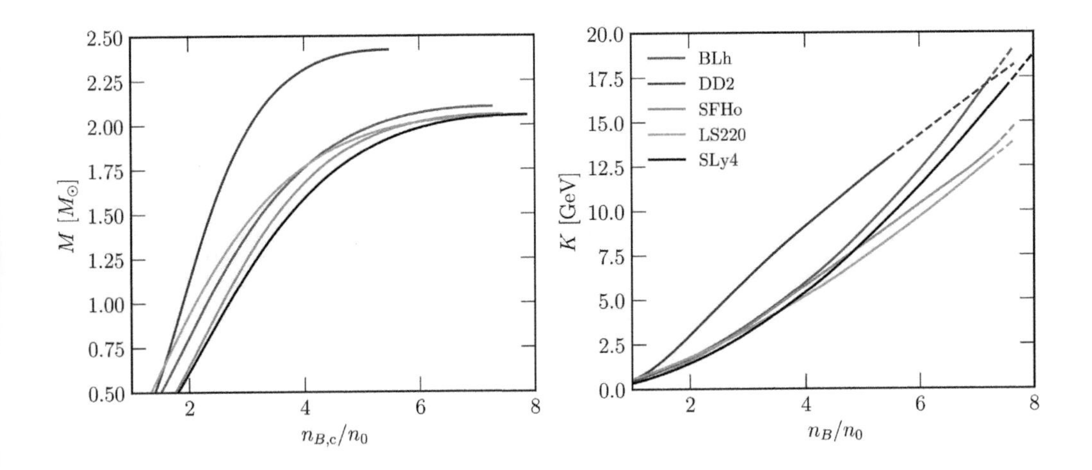

Figure 10.3 Left: Neutron star mass as a function of the central density, $n_{B,c}$, for the same EOSs used in Figure 10.1. Right: Modulus of the nuclear incompressibility, K, defined as $K \equiv \partial P/\partial n_B|_{T=0,Y_{e,\mathrm{eq}}}$, for the same set of EOSs. The transition between the solid and the dashed lines marks the crossing of the maximum allowed central density.

maximum mass and in the center of BNS merger remnants. The remnant of a BNS merger is a very dynamical system, far from equilibrium. The fusion of the two cores is characterized by the alternation of gravitational compression and expansion. The latter is due to the nuclear incompressibility. However, differently from the one that characterizes symmetric nuclear matter around n_0, the one that matters for the dynamics of a BNS merger remnant is the nuclear incompressibility of very asymmetric matter, around $n_{B,\mathrm{max}}$. In the right panel of Figure 10.3, we present $K \equiv \partial P/\partial n_B|_{T=0,Y_{e,\mathrm{eq}}}$, i.e. the gradient of the pressure with respect to the baryon density for cold, β-equilibrated matter. It is clear that, for a given density, stiff EOSs provide a larger incompressibility, at least up to 6 or 7 times n_0. However, in the case of stiff NSs large densities require much larger NS masses to be achieve, as visible in the left panel. Indeed, close to the NS stability limit represented by $n_{B,\mathrm{max}}$ (which can be seen as the largest density achieved by the curves on the left panel or as the transition between the solid and the dashed curves on the right panel), softer EOS can reach comparable or even larger values of K once considering $n_B \lesssim n_{\mathrm{B,max}}$.

Since the core of the remnant is close to a very massive NS, the maximum NS mass sets the mass scale of the remnant in relation to its fate. In particular, it is reasonable to assume that the mass at which a prompt collapse occurs, M_{prompt} , is proportional to the maximum NS mass:

$$M_{\mathrm{prompt}} = k_{\mathrm{prompt}} M^{\mathrm{max}}_{\mathrm{NS,TOV}} \,. \tag{10.28}$$

where k_{prompt} depends on the EOS and on the mass ratio. Additionally, in Ref. [70], a classification of supra massive NS applicable to BNS merger remnants and rooted on on $M^{\mathrm{max}}_{\mathrm{NS,TOV}}$ was introduced. Indeed, considering that:

i) rigid rotation increases the maximum sustainable NS mass by $k_{\mathrm{rot}} \sim 20\%$ at the Keplerian mass-shedding limit, irrespective of the EOS, where we have introduced the maximum rigidly rotating NS mass, $M^{\mathrm{max}}_{\mathrm{NS,Kep}} = k_{\mathrm{rot}} M^{\mathrm{max}}_{\mathrm{NS,TOV}}$;

ii) differential rotation and thermal support can further increase the maximum mass that a NS can support until the prompt collapse mass, $M_{\rm prompt}$, is reached;

the remnant is often said to be one of the following objects:

- a stable MNS, if $M_{\rm rmn} \approx M < M^{\rm max}_{\rm NS,TOV}$;
- a supramassive neutron star (SMNS), if $M^{\rm max}_{\rm NS,TOV} \leq M_{\rm rmn} < M^{\rm max}_{\rm NS,Kep}$. A SMNS is stable also once the differential rotation has been removed from the remnant, which eventually rotates as a rigid body and collapses to a BH on the spindown timescale;
- a hypermassive neutron star (HMNS), if $M^{\rm max}_{\rm NS,Kep} \leq M_{\rm rmn} < M_{\rm prompt}$. In this case, the removal of rotational and thermal support due to angular momentum redistribution and neutrino cooling can induce the collapse on a timescale ranging from a few to several dynamical timescale, comparable with the cooling and the viscous timescale of the remnant.

Despite its apparent robustness, this classification hides a series of flaws which limit its validity and applicability to the remnants of BNS mergers. First of all, it is based on NS structure calculations obtained using a using cold, β-equilibrated, neutrino-less EOS. Despite being a good first-order approximation, the remnant core is expected to have low, but finite entropy, to have trapped neutrinos and to have a composition different from the one of isolated neutron stars (at a given density). Moreover, the mass of the remnant is only in rough approximation the total mass of the system (defined in terms of the gravitational mass of the two stars at infinity). Gravitational wave emission, mass ejection and the disk formation can alter the matter content of the remnant, reducing it also by a few percents, depending on the EOS and on the binary mass ratio. For example, in Ref. [14], it was shown that the remnant often contains a significant amount of angular momentum, larger than the Keplerian limit for uniform rotation. This implies that, in the transition between differential and uniform rotation, an hypothetical massive remnant can significantly alter its mass, irrespectively of the original NS masses.

10.2.4.3 Thermal and composition effects

The calculation of the EOS at finite temperature and arbitrary composition depends sensitively both on the nuclear interaction and on the many-body technique. To show the sensitivity of currently used EOSs to these essential inputs, we compare two quantities for four different EOSs: the gradient of the pressure with respect to the baryon density, $\partial P/\partial n_B|_{T,Y_e}$, and the electron neutrino chemical potential at equilibrium, $\mu_{\nu_e,\rm eq}$. The former is a quantity that generalize the incompressibility coefficient introduced in Section 10.2.4.2. It is dynamically relevant since it sets the capacity of nuclear matter to react to compression. The latter is computed as $\mu_{\nu_e,\rm eq} = \mu_e + \mu_p - \mu_n$. Since the electrons are treated in the same way in all four cases, the differences between the different EOSs are just due to $\mu_p - \mu_n$. Such a quantity is relevant in setting the composition during the merger, as well as the properties of the trapped neutrino gas inside the remnant.

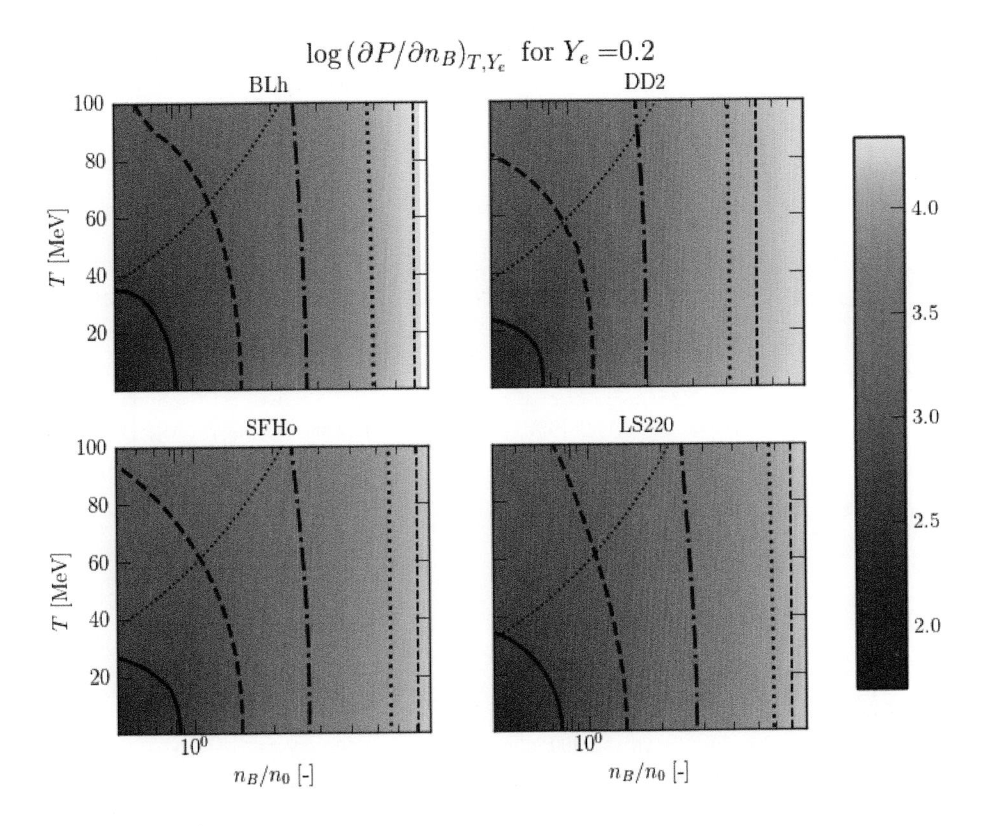

Figure 10.4 Logarithm of the gradient of the pressure with respect to the baryon density, n_B, at $Y_e = 0.2$ and as a function of the baryon density and temperature, T, for four different EOSs (BLh, HS(DD2), SFHo and LS220). The dependence on Y_e is weak. The temperature is evaluated in intervals relevant to the core of BNS merger remnants. The thick solid, dashed, dot-dashed and dotted lines represent contours equal to 2.5, 3.0, 3.5 and 4.0, respectively. The thin dotted lines correspond to the Fermi temperature, T_F, while the thin, vertical, dashed lines mark the maximum baryon density for each EOS.

In Figure 10.4, we present $\partial P/\partial n_B|_{T,Y_e}$ for a broad range of n_B and T, while we fix $Y_e = 0.2$. This is due to the fact that this quantity has a very weak dependence on the composition, while it is very sensitive to the baryon density. Its dependence on the temperature is relevant, at least around and below n_0 and for $T \gtrsim T_F$, while it becomes negligible below it. Deep inside the core, where $n_B > 2n_0$ and $T \lesssim 10$ MeV, thermal effects are negligible, while the differences between the different EOSs reflect the corresponding behaviors already discussed in Section 10.2.4.2. However, in the hot mantle around the massive NS core, where $n_B \sim n_0$ and $T \gtrsim 50$ MeV, the incompressibility of RMF models such as SFHo and DD2 is systematically larger than the non-relativistic BLh and LS220 models. In Figures 10.5 and 10.6 we present $\mu_{\nu_e,\mathrm{eq}}$ and at two different densities, namely $n_B = 2n_0$ and $n_B = 4n_0$. For all EOSs, the qualitative behavior of this quantity is similar and reflects properties that, broadly speaking, do not depend on the details of the nuclear interaction: lower electron fractions imply larger neutron densities and larger neutron chemical

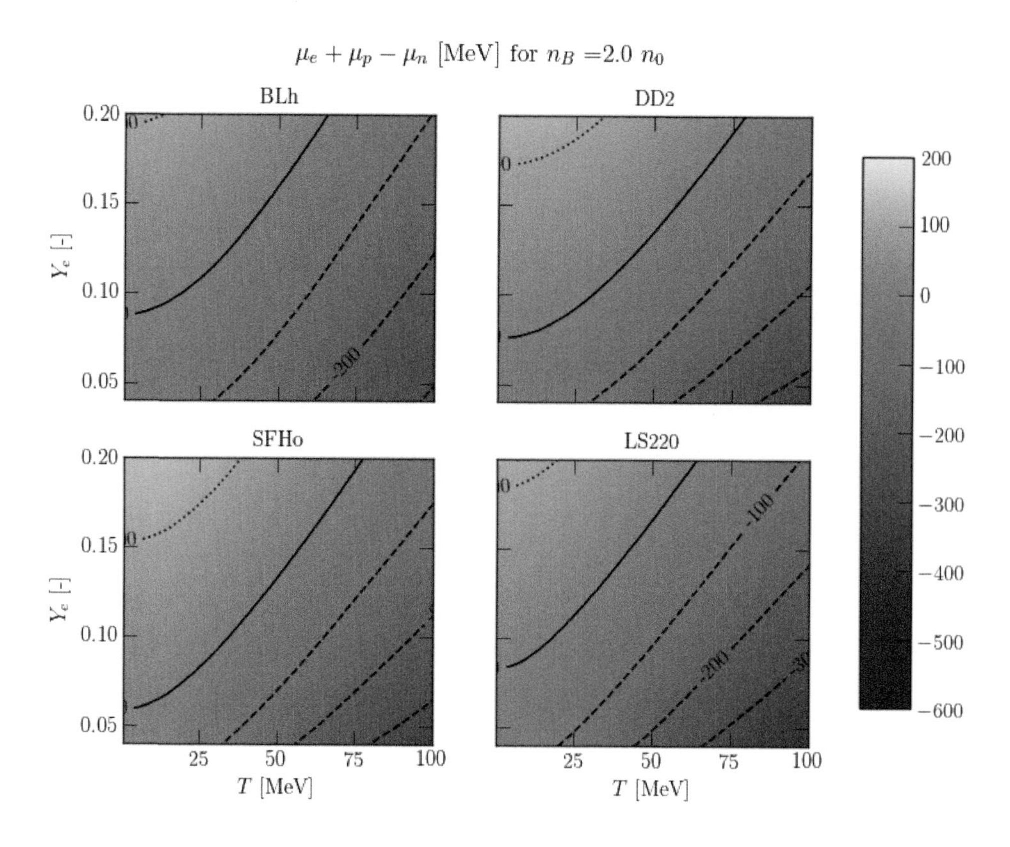

Figure 10.5 Chemical potential of electron neutrinos at equilibrium, computed as $\mu_{\nu_e,\text{eq}} = \mu_e + \mu_p - \mu_n$, at $n_B = 2n_0$ and as a function of temperature, T, and electron fraction, Y_e, and for four different EOSs (BLh, HS(DD2), SFHo and LS220). The temperature and the electron fraction are evaluated in intervals relevant to the core of BNS merger remnants. The solid line represent the $\mu_{\nu_e,\text{eq}} = 0$ contours. The dashed (dotted) lines correspond to negative (positive) values of $\mu_{\nu_e,\text{eq}}$ at intervals of one hundred.

potentials. At the same time, for a given composition, the degree of degeneracy of the less abundant protons is more sensitive to an increase in temperature than the one of the neutrons. So, while the chemical potential of the neutron is large and positive in most of the explored area, the one of the protons decreases faster towards the one of a non-degenerate ideal gases. The $\mu_{\nu_e} = 0$ line separates regions where electron neutrino dominates (top-left corner) from regions where electron anti-neutrino dominates (bottom-right corner). In particular, when $\eta_{\nu_e} \equiv \mu_{\nu_e}/T \ll 0$, i.e. where electron antineutrinos are degenerate, the absolute value of the chemical potential sets, for example, the neutrino density and mean energy. For example, for the more abundant, degenerate electron antineutrinos:

$$n_{\bar{\nu}_e} \sim \frac{4\pi}{3}\left(\frac{-\mu_{\nu_e}}{hc}\right)^3,$$

and

$$\langle E_{\bar{\nu}_e} \rangle \approx -\frac{4}{3}\mu_{\nu_e},$$

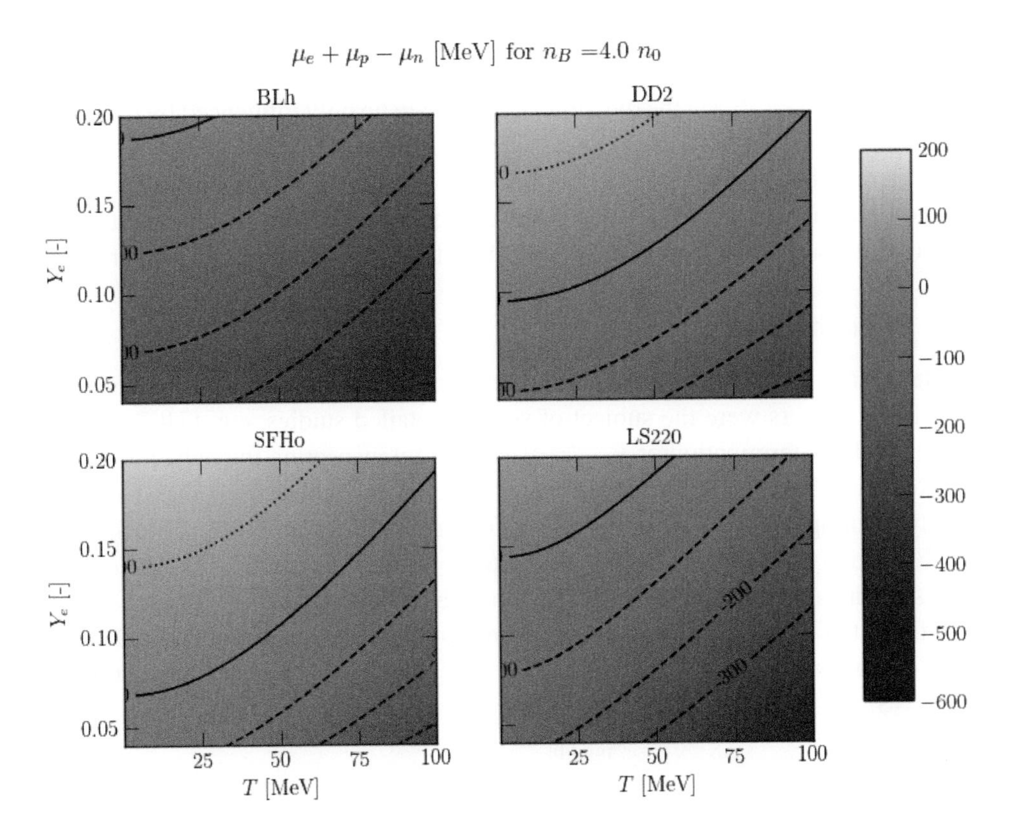

Figure 10.6 Same as in Figure 10.5, but for $n_B = 4n_0$.

while the non-degenerate electron neutrinos form a thermal gas suppressed by degeneracy:

$$n_{\nu_e} \sim 8\pi \left(\frac{k_B T}{hc}\right)^3 \exp\left(\frac{\mu_{\nu_e}}{k_B T}\right),$$

and

$$\langle E_{\nu_e} \rangle \approx 3 k_B T \,.$$

However, quantitative differences between the different EOSs testify the role of the interaction in setting the chemical potentials of the different baryon species, ultimately affecting the neutrino properties. Those differences become more pronounced at large (but not extreme) densities, see e.g. Figure 10.6. In particular, despite the very different maximum mass and compactness, we notice again that the results obtained by the two RMF models are relatively similar and do not depend very sensitively on the baryon density, while the largest differences are obtained by the BLh EOS, which predicts systematically smaller μ_{ν_e}, especially at large densities. The result is that electron antineutrinos are potentially more abundant and more degenerate in remnant described by the BLh EOS, which will consequently affect the composition, favouring the conversion of neutrons into protons. This effect is already visible at $T = 0$, since the β-equilibrated composition predicted by the BLh EOS is less neutron-rich than the one obtained, for example, by the DD2 or the SFHo EOS above saturation density.

10.3 THE ROLE OF THE NUCLEAR EOS IN BNS MERGERS

As we have discussed above, the dynamic of a BNS are heavily influenced by the properties of the hadronic EOS over a broad range of densities between n_0 and $n_{B\max}$. Additionally, also the thermal behaviour can influence in many ways the dynamics and the observables related to BNS mergers. Here, we want to discuss in more detail some quantitative aspects emerging from the detailed simulations and directly connected with the nuclear EOS.

10.3.1 Prompt mass threshold

The determination of the threshold mass for prompt collapse in the case of symmetric (i.e. $q = 1$) BNS mergers were the subject of several detailed studies, e.g. [71–74]. Recently, in Ref. [74] the determination of $k_{\rm prompt}$ was done by inspecting the behavior of binaries at the transition between the formation of a (usually, fast collapsing) massive NS and a promptly forming BH for a sample of 23 EOSs in full General Relativity, using the Numerical Relativity code `WhiskyTHC` at multiple resolutions [75–77]. In agreement with previous findings, it was found that $k_{\rm prompt}$ can be expressed as a linear function of properties of isolated NS sequences obtained by employing cold, β-equilibrated EOSs, such as the maximum compactness, $\mathcal{C}_{\max} \equiv GM^{\max}_{\rm NS,TOV}/(c^2 R^{\max}_{\rm NS,TOV})$ i.e. the compactness of the heaviest NS:

$$k_{\rm prompt} = a\mathcal{C}_{\max} + b\,.$$

The fitting coefficients were determined to be $a = -3.36 \pm 0.20$ and $b = -2.35 \pm 0.06$. Other empirical relations of this kind involve modified versions of the maximum compactness, e.g. $\mathcal{C}^*_{1.4} \equiv GM^{\max}_{\rm NS,TOV}/(c^2 R_{1.4})$ or $\mathcal{C}^*_{1.6} \equiv GM^{\max}_{\rm NS,TOV}/(c^2 R_{1.6})$, where R_X is the radius of a NS with a mass X, instead of $\mathcal{C}_{\max}$. These relations confirm that the maximum NS mass sets the mass scale of the prompt collapse for the central remnant through the coefficient $k_{\rm prompt} \in [1.2, 1.6]$. For example, for the EOSs presented in the previous section, it was found 1.39 ± 0.01 for BLh, 1.35 ± 0.01 for DD2, 1.37 ± 0.01 for SFHo, 1.45 ± 0.01 for LS220, 1.36 ± 0.01 for SLy4.

Such analysis were recently extended to the case of asymmetric binaries, see e.g. [78–80]. In particular, in Ref. [81] it was shown that the threshold mass for prompt collapse in case of generic binaries can be expressed as a the value in the symmetric case times a function that depends both on the EOS and on the mass ratio:

$$M_{\rm prompt}(q) = M_{\rm prompt}(q = 1) f_{\rm EOS}(q)\,. \tag{10.29}$$

The smaller angular momentum (for fixed total mass) that characterizes asymmetric binaries suggests that $f_{\rm EOS}(q) < 1$, due to the smaller rotational support of the remnant. While this seems to be the case for $q \lesssim 0.7$, some EOSs show $f_{\rm EOS}(q) < 1$ also for $q \lesssim 1$, while others have a possibly decreasing behaviour, i.e. $f_{\rm EOS}(q) < 1$ for $q \lesssim 1$. The interesting finding is that the discriminant between the two regimes correlates with the EOS incompressibility around $n_{B,\max}$. In particular, $f_{\rm EOS}$ can be expressed as

$$f_{\rm EOS}(q) = \alpha_{\rm EOS} q + \beta_{\rm EOS} \tag{10.30}$$

separately in the two regimes, i.e. $q \geq q^*$ and $q < q^*$, with $q^* \approx 0.725$. While the $\beta_{\rm EOS}$'s are fixed by requiring the continuity at q^*, the $\alpha_{\rm EOS}$ coefficients show a linear correlation to $K_{\rm eq}$. In particular,

$$\alpha_{{\rm EOS},q<q^*} = -(22 \pm 1){\rm TeV}^{-1} + (0.58 \pm 0.01) \ , \tag{10.31}$$

and

$$\alpha_{{\rm EOS},q>q^*} = -(4.7 \pm 1){\rm TeV}^{-1} + (0.064 \pm 0.017) \ , \tag{10.32}$$

meaning that the slopes appearing in Equation 10.30 decrease as the incompressibility increases. This confirms the tendency of EOS with large incompressibility at $n_B \lesssim n_{B,\rm max}$ to partially compensate the lack of rotational support with a more pronounced nuclear repulsion triggered by the fusion of the two cores in the remnant. The presence of a sharp transition at $q = q^*$ suggests the existence of two merger regimes, one ($q \gtrsim 0.7$) characterized by the fusion of two NS of comparable masses, and the other one by a tidally dominated merger in which the lighter, secondary NS is tidally disrupted and accreted by the heavier, primary NS (see also [17]).

10.3.2 Remnant lifetime

The properties of the EOS do not only affect the prompt collapse threshold. It also influences the stability of the remnant in the case of a (meta-)stable massive NS. In Figure 10.7, we collect data from a large set of Numerical Relativity simulations targeted to GW170817

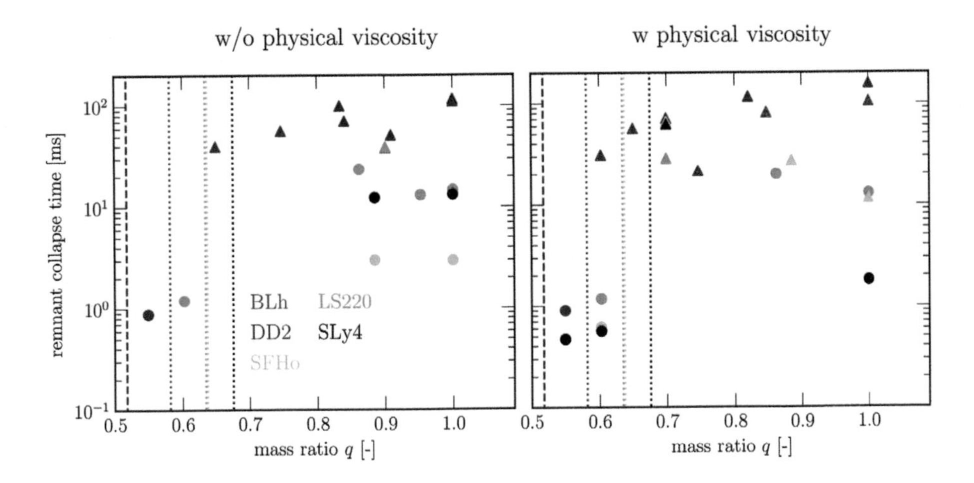

Figure 10.7 Remnant collapse time as a function of the mass ratio in simulations targeted to GW170817 and taken from Refs [16, 17, 82]. On the right(left), we consider models (not) including physical viscosity of turbulent magnetic origin [83, 84]. The different colors refer to different EOSs while the triangle represents lower limits for models not collapsed during the simulated time. Dotted, vertical lines represent the theoretical threshold for prompt collapse in the case of GW170817 chirp mass. Since for DD2 the prompt collapse limit should correspond to much a lower value of q ($q \lesssim 0.4$), we represent the limit for a SMNS case (dashed, blue line).

and published in Refs. [16, 17, 82]. For each BNS merger model, we consider the simulations with the highest available resolution and we consider the BH formation time. If the remnant does not collapse within the end of the simulation, we consider the latter as a lower limit of the collapse time. We distinguish simulations including physical viscosity (right panel) from simulations that do not contain it (left panel). Different colours represent different EOSs. It is clear that EOSs that have large $M^{\rm max}_{\rm NS,TOV}$, as DD2, are more stable and do not show any signs of instability. Indeed, only for $q \lesssim 0.5$ the resulting BNS associated to GW170817 would have enough mass to become a HMNS (dashed, blue vertical line). All the other EOSs produce a prompt collapse for small enough q. We notice that the location of the prompt collapse binaries is in agreement with the relation found in [26] (dotted, vertical lines). However, we also notice that, despite having maximum NS masses that differ just by a few percents (DD2 has $\sim 15\%$ larger $M^{\rm max}_{\rm NS,TOV}$), EOSs characterized by smaller values of $K_{\rm eq}$ around $n_{\rm B,max}$, as SFHo, SLy4 and LS220, are more prone to collapse within the GW-dominated phase (i.e. within the first 15–$20{\rm ms}$ post-merger), while BLh (which has the largest incompressibility among the sample) produces a much more stable remnant, which will collapse on the secular time scale of the latter. This suggests that the incompressibility at the largest allowed density can have an impact on the stability of the remnant. At the same time, a direct comparison between the simulations with and without physical viscosity underlines the stabilizing role of the latter. Indeed, since the core rotates more slowly than the hot envelope around it, the angular momentum transfer operated by viscosity will move angular momentum inward inside the remnant, helping it to be stable on longer timescales.

10.3.3 Disk and ejecta mass

The EOS has an impact also on the mass of the disk that forms after the merger. In Figure 10.8, we present the mass of the disk as a function of the binary tidal parameter $\tilde{\Lambda}$ for a large set of simulations targeted to GW170817 (circles, taken from Refs [16, 17, 82]) and to GW190425 (triangles, taken from Ref. [85]). We distinguish between high and low mass ratios. Since the chirp masses are fixed, for a fixed EOS the mass ratio has a minor impact. Indeed, simulations were performed with the same EOS cluster in the vertical direction. For example, in the case of GW170817 one can distinguish between sequences of simulations using, from the left to the right, SLy4, SFHo, BLh, LS220 and DD2. In the case of GW190425, the sequence is SLy4, SFHo, BLh and DD2. At a first sight, there is no clear dependence describing the behaviour of $M_{\rm disk}$. However, we can recognize some trends. To interpret them, one has to keep in mind the two main mechanisms behind the disk formation: the tidal deformation of the two NSs, producing spiral arms from which the disk is produced; the ejection of matter in the remnant bounces. Looking at the equal and quasi-equal mass cases, there is a clear distinction between the GW170817 and the GW190425. Since all GW190425 cases undergo prompt collapse, the disk has little or no time to form, especially in the equal mass cases. A significant disk mass ($M_{\rm disk} \gtrsim 10^{-2} m_\odot$) can be obtained only for $q \lesssim 0.8$ due to the tidal disruption of the secondary NS. Also for GW170817, it seems that more symmetric BNS mergers tend to produce lighter disks. Looking at the $\tilde{\Lambda}$ dependence, a larger deformability produces on average heavier disks, due to the tendency of the NSs to be deformed and produce the disk from the tidal arms.

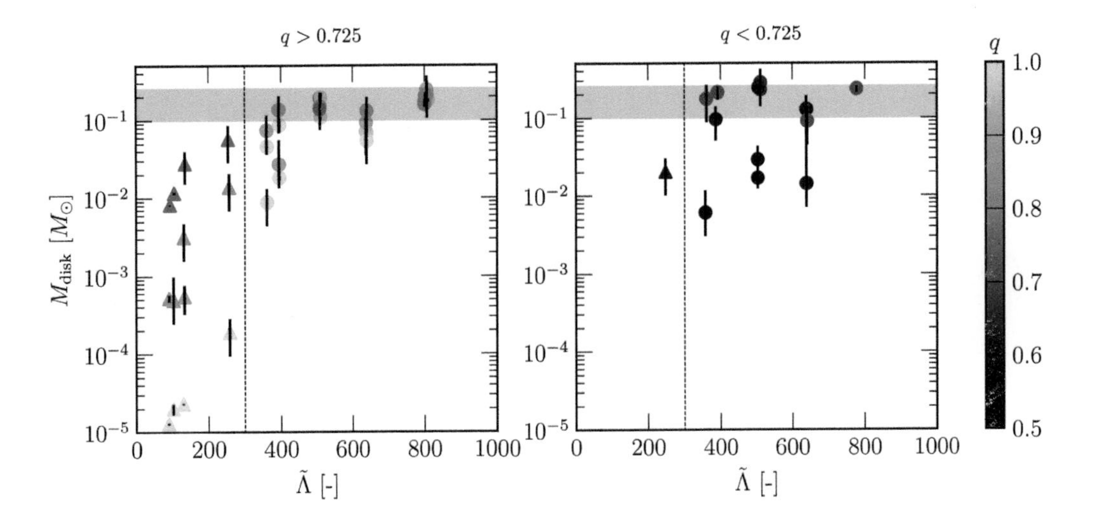

Figure 10.8 Disk mass as a function of the binary tidal parameter, $\tilde{\Lambda}$, from simulations targeted to GW170817 (circles, taken from Refs [16, 17, 82]) and to GW190425 (triangles, taken from Ref. [85]) . In the left panel we consider more symmetric binaries ($q \geq 0.725$), while in the right panel we consider very asymmetric ones ($q \leq 0.725$). The actual value of q is colour coded. The red band represents the disk mass interval obtained by considering $0.04 M_\odot$ of disk ejecta and corresponding to 20–40% of it.

However, the inversion between LS220 and BLh indicates that the tidal deformability is not the only relevant parameter related to the EOS. Indeed, BLh produces heavier disks than LS220 since the bounces that expel matter from the remnant into the disk are stronger due to the larger nuclear incompressibility. Moving to the $q < 0.725$ case, in which the merger is dominated by the tidal interaction, the dependence on the EOS encoded into $\tilde{\Lambda}$ is almost lost. Some similarities can be observed in the case of the dynamical ejecta, presented in Figure 10.9. Looking at the left-hand panel, in the case of collision dominated merger, matter expelled by high chirp mass binaries undergoing prompt collapse (as in the case of GW190425-like simulations) is usually one or two orders of magnitude smaller than the one expelled by GW170817-like events. Moreover, for these binaries, a significant mass ejection happens only in the case of very asymmetric mergers, in which the ballistic ejection of tidally deformed tails partially compensates for the very deep gravitational dwell. The situation changes moving to GW170817-like events. Since in the case of very symmetric mergers, matter ejection comes mostly from shocks, binaries characterized by smaller values of $\tilde{\Lambda}$ and $q \lesssim 1$ produce more violent mergers and eject more mass. In the case of binaries characterized by a large deformability (as in the case of DD2 or LS220), the tidal deformation of the secondary is also relevant in producing tidal ejecta and increasing the ejecta mass. In the case of very asymmetric BNS mergers (right panel), the merger is dominated by the tidal destruction of the secondary and both the asymmetry degree and the NS deformability contribute to increase the dynamical ejecta mass.

By comparing the disk and the ejecta mass, one can conclude that the disk mass is usually larger than the dynamical ejecta mass. More specifically, in the case of more

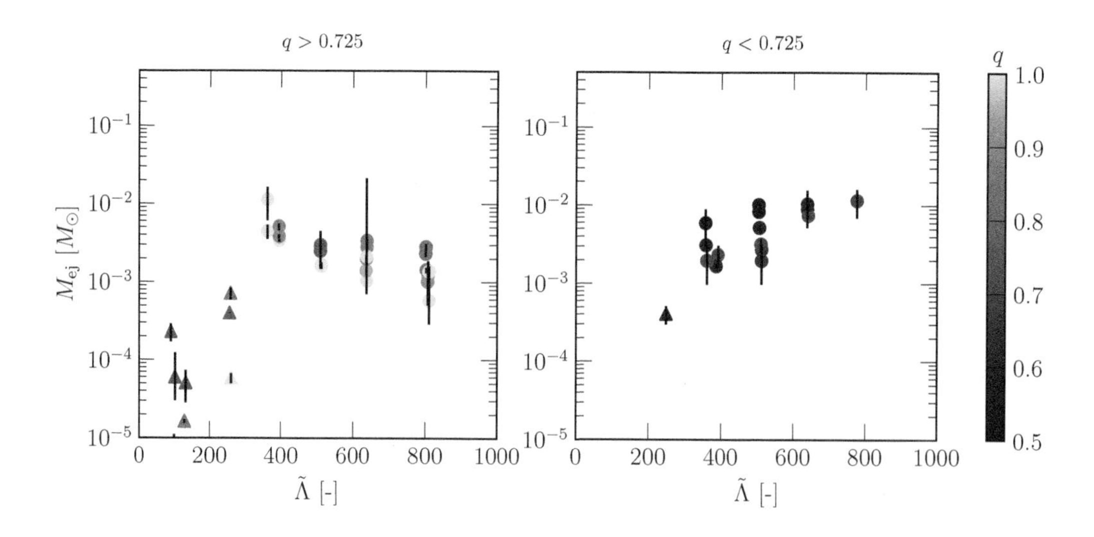

Figure 10.9 Same as in Figure 10.8, but for the dynamical ejecta mass.

symmetric BNSs characterized by larger deformabilities (as in the case of DD2) and/or by the presence of an EOS with a large incompressibility at $n_{B,max}$ (as in the case of BLh), the disk mass significantly dominates over the dynamical ejecta mass. The difference becomes milder in the case for less deformable stars.

10.3.4 Gravitational wave emission

The emission of GWs from the late inspiral, merger and early post-merger phase has been the subject of several recent works. For example, in [86], the authors examined a large sample of BNS merger simulations in Numerical Relativity. They found that the GW emission has its maximum peak at merger and the peak luminosity (once properly normalized by the mass ratio of the binary) depends on the EOS only through the binary tidal deformability almost monotonically. In particular, less deformable binaries, either characterized by large NS masses or very soft EOS, emits more GW radiation than more deformable ones. Promptly collapsing binaries are indeed the most luminous mergers ($L_{\text{peak}} \sim 10^{57}$erg s^{-1}), while binaries producing a long-lived remnant have peak luminosities down to one order of magnitude smaller. The total amount of energy emitted in GWs results from the combined effect of the luminosity strength and of the duration of the GW-loud post-merger phase. The largest amount of energy

$$\Delta E_{\text{GW}} < 0.126 M_\odot c^2 \left(\frac{M}{2.8 M_\odot} \right)$$

is indeed radiated by weakly or moderately deformable, metastable remnants for which the emission lasts at least $10 - 15$ milliseconds post-merger.

When considering the post-merger main peak, f_2, its value (for a fixed total mass of the binary) was found to correlate with properties of cold, irrotational NSs, as for example the radius of a 1.6 $M_\odot$ NS [87, 88]. It was also shown that it correlates with the tidal deformability of the binary [89]. It is clear that both relations are possibly relevant to constrain the nuclear EOS if the postmerger signal is detected [88–92]. These relations may look somehow surprising, since during the post-merger the remnant reaches densities a few times larger than the ones that characterize a $1.6\ M_\odot$ NS or even the two inspiraling stars. The origin of that is related to the fact that the frequency of the post-merger GW emission is practically twice the rotational frequency of the remnant. The latter is initially set by the tidal deformability of the binary, see, e.g. [89]. Due to GW emission, the remnant loses angular momentum and energy. However, the two loss rates are proportional to each other so that the rotational frequency remains approximately constant, and so f_2 does.

10.3.5 Neutrino emission

The increase in the number of available BNS merger simulations including neutrino transport allows us to collect valuable and quantitative information about the features of neutrino luminosities emerging from merger. For example, in Ref. [82] a sample of 55 different BNS models (corresponding to 66 independent simulations) were analysed. The corresponding total mass ranges from 2.6 to $3.44\ M_\odot$ and the mass ratio from 0.55 and 1. Each model includes a finite T, composition-dependent EOS, such that $\tilde{\Lambda}$ in the sample varies between 90 and 1108. The simulation sample is very homogeneous, since all simulations were performed with the `WhiskyTHC` code [75–77], employing the same neutrino radiation treatment, i.e. a leakage+M0 scheme [93]. The simulations allowed to extract the neutrino properties at the edge of the computational domain and, in particular, the integrated neutrino luminosities and mean energies. The nuclear EOS influences the properties of the neutrino emission in several ways. First of all, through the fate of the merger. Indeed, prompt collapsing merger are characterized by a very weak ($L_{\nu,\mathrm{tot}} \lesssim 1 \times 10^{53}\mathrm{erg\ s^{-1}}$) and short ($\Delta t \lesssim 2$ ms) neutrino emissions. This is due to the fact that the prompt collapse removes the main sources of neutrino radiation, namely the MNS and the disk. At the opposite extreme, in case of a long-lived remnant, the emission during the first 10–15ms post-merger is characterized by a series of sharp peaks, corresponding to the oscillatory behavior of the remnant and to the presence of arms inside the disk. At later times, when the central remnant and the disk have become more axisymmetric, luminosities become smooth and they decrease on the cooling timescale. The formation of a BH at any time post-merger clearly reduces the luminosities, by suddenly decreasing them by more than 50%. Despite the lack of the central massive remnant, a partial contribution to the luminosity is provided by the accretion disk in this scenario. Assuming to consider non promptly collapses mergers, the EOS, and in particular the $\tilde{\Lambda}$ parameter associated with the binary, influences the height of the luminosity peaks, as well as the average values of the luminosities. In particular, more compact binaries, corresponding to softer EOSs and/or more massive stars, (thus, lower $\tilde{\Lambda}$) are characterized by larger luminosities. The largest total neutrino luminosities can even exceed 10^{54}ms during the first ms post-merger. This effect is partially mitigated by the mass ratio, with more asymmetric binaries producing lower luminosities than symmetric ones (typically, BNS characterized by $q = 0.7$ have total luminosities around $5 \times 10^{53}\mathrm{erg\ s^{-1}}$ at peak). This can be understood as, in the case of symmetric binaries, the merger is dominated by the collision of the cores, providing more violent and hotter mergers. While, in the case of very asymmetric binaries, the tidal nature of the merger translates in less pronounced remnant bounces and in lower averaged remnant temperatures. Different flavors are characterized by different luminosities. Since the decompression and the heating up of neutron-rich matter favours matter leptonization (i.e. the conversion of the more abundant neutrons into protons), the emission of $\bar{\nu}_e$ is initially favored over the one of ν_e's and heavy flavor neutrinos. Their ratio can be as high as $\sim 3-4$ during the first ms post-merger. Moreover, their is a correlation between the luminosities of different flavors, such that more luminous mergers in $\bar{\nu}_e$'s are also more luminous in the other neutrino species. At later times, when the luminosities start to smoothly decrease, the electron flavor neutrino and antineutrinos luminosities approach each other, i.e. $L_{\nu_e} \approx L_{\bar{\nu}_e}$.

The variety observed in the neutrino luminosities and related to the EOS through the fate or through the tidal deformability seems not to be present in the mean energies. Indeed, for all the explored simulations, the mean neutrino energies, averaged over the first tens of milliseconds after merger, lay in narrow intervals characterized by $\langle E_{\nu_e} \rangle \approx 10-11$ MeV, $\langle E_{\bar{\nu}_e} \rangle \approx 14-15$ MeV and $\langle E_{\nu_{\mu,\tau}} \rangle \approx 20-23$ MeV, with no clear dependence on the EOS. Such a hierarchy translates the different depths at which neutrinos of different species decouple from neutron-rich matter, see, e.g. [23].

10.4 PROBING THE NUCLEAR EOS THROUGH MULTIMESSENGER OBSERVATIONS

The detection of GWs from the inspiral of a BNS in GW170817 disclosed the potential of multimessenger astronomy (and, in particular, of GW astronomy) in setting stringent constraints on the EOS of neutron stars. In particular, the binary tidal deformability extracted from the GW signal was able to constraint the EOS for densities around maximum densities inside the two merging NSs, i.e. $n_B \lesssim 2n_0$. Constraints at larger densities come from the values of the maximum mass of isolated, cold neutron stars, as well as on the combined measurement of mass and radius of heavy neutron stars, see, e.g. [94, 95] In a more indirect way, the combination of information extracted from the GW signal and of the properties of the electromagnetic counterparts associated with GW170817, supported by detailed BNS merger modelling, also allowed to constrain the properties of nuclear matter over a broad interval of densities.

An example of a possible joint analysis containing information from the gravitational wave signal and from the interpretation of the electromagnetic emission, as well as from detailed binary neutron star merger simulations, was discussed in Ref. [96]. Starting from the constraint obtained by the GW signal on the binary tidal deformability parameter ($\tilde{\Lambda} \lesssim 800$ at 90% confidence level [97]), the authors discussed a possible upper limit on $\tilde{\Lambda}$ by considering that the kilonova emission requires $\sim 0.05 M_{\odot}$ of ejecta. Since this is around one order of magnitude larger than the typical amount of dynamical ejecta from BNS mergers, it must come from the disk evolution. However, the disk mass has a correlation with $\tilde{\Lambda}$, as also showed, for example, in Refs. [15, 85, 98]. In particular, small $\tilde{\Lambda}$ corresponds to BNS more prone to collapse, but also to very light disk, since the collapse to BH prevents the formation of very massive disks, unless the mass ratio is very far from 1. The original analysis, based on a set of 29 simulations not targeted to GW170817, but with $M \in [2.6, 3.4]$ and relatively symmetric ($q \gtrsim 0.85$), but employing four different finite T, composition dependent EOSs and neutrino physics, pointed out to an empirical limit of $\tilde{\Lambda} \gtrsim 400$ to ensure a massive enough disk. This translated in a lower limit on $\tilde{\Lambda}$, complementary to in the initial upper limit provided by the GW signal, and it disfavors very soft EOSs. Such a limit is still compatible and constraining, even for more recent $\tilde{\Lambda}$ re-analysis [68]. In Figure 10.8, we have repeated the same analysis as in Ref. [96], but using simulations targeted to GW170817. In the case of core fusion-dominated mergers ($q \gtrsim 0.725$), the $\tilde{\Lambda} \gtrsim 400$ limit still holds. If the binary asymmetry is allowed to decrease below $q < 0.725$, the upper limit is weakened down to $\tilde{\Lambda} \gtrsim 350$.

Stringent constraints on the EOS of matter and on the resulting NS properties can also be obtained by combining detailed modelling of the electromagnetic signals with EOS-dependent relations obtained by BNS merger simulations. For example, in Ref. [99] the authors performed Bayesian inference and model selection on AT2017gfo, the kilonova associated with GW170817. They used semi-analytical, multicomponents and anisotropic models that also account for non-spherical ejecta, see, e.g. [100]. The best-fitting model was an anisotropic three-component composed of dynamical ejecta plus neutrino and viscous winds. Using the dynamical ejecta parameters inferred from the best-fitting model and numerical-relativity relations connecting the ejecta properties to the binary properties [98], the binary mass ratio was constrained to be $q < 1.54$ and the reduced tidal parameter to $120 \lesssim \tilde{\Lambda} \lesssim 1110$. Moreover, using the universal relation presented in Refs. [95, 101], the radius of a $1.4\ M_{\odot}$ NS was constrained to 12.2 ± 0.5 km at $1\ \sigma$ level. To do that, the authors combined the marginalized posterior distribution for the chirp mass coming from the GW170817 measurement [102] and the posterior on the binary tidal parameter obtained with the joint analyses AT2017gfo+GW170817. Similar and consistent constraints were presented also in Refs. [94, 95, 101, 103–108]. Another example of the possible impact of the NS EOS on electromagnetic counterpart observables is provided by the so-called kilonova afterglow. The presence of a high-speed ($v \gtrsim 0.6c$) tail in the dynamical ejecta distribution is a generic feature of BNS merger simulations. These ejecta are usually located at the head of the expanding matter. Their expansion time scale is so small that it can lead to a peculiar, incomplete r-process nucleosynthesis characterized by the presence of a relatively large fraction of unburnt free neutrons, decaying into protons, powering a bright, ultraviolet pre-kilonova transient, see, e.g. [109, 110]. Once these very fast ejecta have swept a portion of space whose interstellar medium mass is comparable to their mass, the dissipation of their kinetic energy can power a non-thermal synchrotron emission, similar to the one that characterizes the afterglow emission of gamma-ray bursts , see, e.g. [111]. The expected properties are, however, different from the latter, and its peak is expected to occur between hundreds of days and several years after the merger, also depending on the interstellar medium density. In Ref. [112], it was shown that the properties of the fast tail ejecta depend on the ejection mechanism, which in turn depends on the binary mass ratio and on the EOS. In particular, in models with moderately soft EOS (e.g. BLh or SFHo) or small mass ratio ($q \lesssim 0.8$), most of the fast ejecta originate at the first bounce and are located across the equator. However, for equal mass binaries with very soft EOSs, e.g. SLy4, additional, more polar, mass ejection occurs at the second bounce. In general, symmetric binaries characterized by soft EOSs carries more kinetic energy to power the afterglow emission. In the case of GW170817, the fading non-thermal emission remained generally consistent with the afterglow powered by synchrotron radiation produced by the interaction of a gamma-ray burst jet with the ambient medium. However, some recent observations found a possible signature of a re-brightening [113] in the temporal and spectral behavior in the X-ray band, compatible with such kilonova afterglow emission. More specifically, the authors of Ref. [112] found that equal mass models with soft EOSs and high-mass ratio models with stiff EOSs could be disfavoured by this interpretation, since they typically predict afterglows that peak too early to explain this possible re-brightening. Moderate stiffness and mass ratio models, instead, could be in better overall agreement with the data. Similar analysis on the nuclear EOS involving both GW and electromagnetic signals can be found, for example, also in Ref. [101, 114, 115].

While future (multimessenger) observations reported by second-generation GW detectors, as well as new mass-radius measurements, can sharpen our understanding of the nuclear EOS in the wake of what was done in the past few years, a qualitative change is expected to occur with third-generation GW observatories, such as the Einstein Telescope and Cosmic Explorer. This is related to different factors: first of all, the increased sensitivity will enlarge the detection horizon and, consequently, the number of detected events; second, the increased sky localization capability will favour more multimessenger events in which GW and electromagnetic information could be coherently combined; third, the detectors will be for the first time sensitive in the kHz band, allowing to detect also the GW emission in the post-merger phase. Differently from the tidal deformation that characterizes the inspiral, the post-merger signal is sensitive to the properties of the EOS below, but not far from $n_{B,\max}$, as well as to finite temperature effects.

For example, in Ref. [116], the authors used data from 289 simulations of BNS mergers in Numerical Relativity, all published by the CoRe collaboration and whose GW data are available in the CoRe catalogue [117]. Most of the simulations were performed using microphysics, neutrino transport and a subgrid model for magnetohydrodynamics turbulence. Being a phenomenological study, these simulations were performed with 14 different EOSs to bracket EOS uncertainties. In particular, they included five finite-temperature, composition-dependent nucleonic models, one finite-temperature hybrid model accounting for deconfined quark matter and one finite-temperature hadronic model with Λ hyperons. Additionally, some simulations including piecewise polytropes were also considered. These large simulation sample allowed to identify a robust quasi-universal relation connecting the postmerger peak gravitational-wave frequency, f_2, and the value of the density at the center of the irrotational, heaviest NS, $n_{B,\max}$:

$$n_{B,\max} = \frac{a_0 c^6}{G^3 m_b M^2}\left[1 + a_1 \left(\frac{c^3}{\pi G M f_2}\right)^{1/6}\right], \tag{10.33}$$

where m_b is the baryon mass. The fitting coefficients were found to be $(a_0, a_1) = (0.135905, -0.59506)$. This relation offers the possibility for precision EOS constraints with next-generation ground-based GW interferometers. By considering the ET design sensitivity, the authors perform full Bayesian analyses of the post-merger signals at different signal-to-noise ratios and employ the Fisher matrix approach to estimate the uncertainties of the parameters measured from the inspiral signals. In particular, they used the `BAJES` pipeline [118] and an updated version of the `NRPM` model [119] for BNS post-merger signals. Their results indicate that post-merger signals can be detected with a signal-to-noise ratio as low as 7.5 in the post-merger, corresponding to a signal-to-noise ratio of ~ 140 in the whole GW signal. For a fiducial BNS merger, the f_2 frequency (at 90% credibility level) can be determined with a 12% uncertainty at the detection threshold, improving down to 3% if the post-merger signal-to-noise ratio increases to ~ 12. The value of $n_{B,\max}$ is then recovered from the posteriors of M (derived from the inspiral signal) and of f_2 with a 15% level of uncertainty. The potential impact of this quasi-universal relation increases when considering another empirical quadratic relation that connects the radius of the heaviest, irrotational NS, $R^{\max}_{\mathrm{NS,TOV}}$, to f_2 and M, as presented in Ref. [119]:

$$R^{\max}_{\mathrm{NS,TOV}} = M\left(b_0 + b_1 f_2 + b_2 (f_2)^2\right) \tag{10.34}$$

The fitting coefficients were found to be $(b_0, b_1, b_2) = (5.81 \pm 0.13, -123.4 \pm 7.2, 1121 \pm 99)$. Thanks to the two above relations, the posteriors on (M, f_2) can be mapped into $(n_{B,\max}, R^{\max}_{\mathrm{NS,TOV}})$. At f_2 detection threshold, $R^{\max}_{\mathrm{NS,TOV}}$ can be recovered with an uncertainty of the order of 30%. By using a large sample (two millions) of cold neutron star EOSs, built just to fulfil all the present astrophysical constraints (but with no explicit information about the underlying nuclear input) [120], the posteriors on $(n_{B,\max}, R^{\max}_{\mathrm{NS,TOV}})$ were able to recover the maximum mass of an irrotational NS with an uncertainty of the order of $\sim 15\%$, as well as to tightly constrain the value of the pressure between $2n_0$ and $n_{B,\max}$, an interval which is very much complementary to present theoretical uncertainties provided by chiral effective field theory calculations.

The impact of thermal effects in the post-merger was the topic of the work presented in Ref. [121]. In this work, the authors performed general relativistic BNS merger simulations including a finite-temperature, composition-dependent EOS and neutrino transport via an two-moment (M1), gray transport scheme [122]. For the EOSs, three different Skyrme models were produced using the framework presented in Ref. [56], parameterized to produce the same cold nuclear matter properties but different specific heat content. In these models, the specific heat of baryonic matter is controlled by the temperature-independent effective masses of neutrons (m_n^*) and protons (m_p^*) and at first order it reads [123]:

$$c_V = \left(\frac{\pi}{3}\right)^{2/3} \frac{T}{n_B}\left(n_p^{1/3} m_p^* + n_n^{1/3} m_n^*\right) , \tag{10.35}$$

such that an increase in the effective masses leads to a larger specific heat capacity for matter in the merger remnant, for fixed conditions of density, temperature and composition. The poorly known values of m_n^* and m_p^* were chosen to reproduce two nuclear matter observables at n_0: the effective mass for symmetric nuclear matter, $m^* = m_n^* \approx m_p^*$, and the neutron–proton effective mass splitting for pure neutron matter, Δm^*. The former were chosen in the range $m^* = (0.55, 0.75, 0.95) m_n$, while the latter as $\Delta m^* = 0.1 m_n$, spanning present theoretical and experimental uncertainties. By analysing the behavior of temperature and density at early time post-merger, the authors found that for higher m^* the remnant's inner core is denser, but cooler. Conversely, below saturation density, models with larger m^* have also a larger temperature. This is an indirect, dynamical effect, since for larger m^* NS surface material becomes hotter as it falls deeper into the gravitational potential before being expelled by the remnant. The GW signal was identical during the inspiral for the different EOS models, while it differed in the frequency evolution, amplitude, damping times and modulations in the post-merger due to finite-temperature effects.

These differences had a visible effect in the GW spectrum, introducing a shift in the f_2 peak toward larger frequency values for larger m^*. Such an effect can be detected if the GW signal has a post-merger signal-to-noise ratio of at least 15. The behavior of the density and temperature, as well as of the GW signal, can be understood in terms of the specific heat. An EOS characterized by a larger m^* (and thus by a larger specific heat) is softer, in the sense that it requires more energy to increase the temperature, leaving less thermal pressure available to support the star. In the merger, this produces a more compact and rapidly rotating remnant with lower temperatures. Despite the simplifications contained by this work (for example, the simple relationship between m^* and c_V and their missing dependence on the temperature), this study is a proof of concept demonstrating that future third-generation detectors can use the post-merger signal to constrain finite-temperature effects in the EOS.

The analysis of the role of nuclear incompressibility in the case of prompt merger BNS merger performed in Ref. [81] and summarized in Section 10.3.1 also revealed that the measurement of the threshold mass for prompt collapse at two different mass ratios (which very likely will require the sensitivity and the detection rates anticipated by third generation GW detectors) can provide a direct measurement of the nuclear incompressibility. In particular, it was shown that, depending on its actual value and on the available mass ratios, the incompressibility at maximum baryon density can be determined with a possible accuracy of $15 - 20\%$. Based on a large sample of available EOSs, the authors also speculated that a measurement of $K_{\max}$ can also provide some indication of the innermost composition of the NSs, with values $\lesssim 12\text{GeV}$ possibly disfavoring the presence of hyperons or quarks.

The presence of muons in cold NSs is well established. However, they have never been included consistently in BNS mergers, considering also their role in the neutrino transport. Preliminary studies showed that they could have an impact, since they change the properties of neutrinos and possibly affect the matter pressure by $\sim 5\%$ [124].

10.5 CONCLUSIONS AND OUTLOOK

The advent of multimessenger astronomy, which coincides with the first, direct detections of GWs from coalescing compact binaries, has provided a unique chance of exploring and studying some of the more extreme events taking place in the universe, like BNS, NS-BH and BH-BH mergers. Such investigations could not be carried on without the recent improved sensitivity of GW detectors. Binary systems involving at least one neutron star, have provided constraints on the features and behavior of dense matter EOS in a regime of densities and temperatures that cannot be reproduced by ground-based experiments. Current GW measurements constrain alone the nuclear matter EOS up to a couple of times n_0. The combination of GW and electromagnetic signals has shown to provide wider and more stringent constraints. This is related to the fact that, while the behaviour of the low density part of the nuclear EOS determines the deformability of the NSs during the inspiral, the high-density part influences the subsequent merger and post-merger dynamics, the fate of the remnant, as well as the amount and the properties of the ejecta. All these aspects are at the very heart of the post-merger GWs emission, as well as of the photon emissions that characterize the electromagnetic counterparts, as, for example the gamma-ray burst and the kilonova. A key aspect related to the possibility of employing multimessenger detections to constraint the nuclear EOS is the availability of detailed, ab initio simulations of BNS mergers. The latter evolve consistently the space-time and the fluid dynamics, and require a nuclear EOS to describe the properties of the involved matter. On the one hand, the interpretation of the detected signals in light of the outcome of these simulations has proved to be a viable and precious method to set constraints on the binary properties and on the EOS at high densities. On the other hand, the availability of detailed models spanning present nuclear uncertainties is key to predict the range and the properties of the multimessenger signals, to guide their searches and even to help in designing future GW detectors and telescopes. In particular, future third-generation GW detectors, that forecast to have a much improved sensitivity in the kilohertz band, may open a direct window on the high density regime of the EOS, shedding light on the possible presence of exotic components expected to be formed in a BNS mergers like hyperons or a decofinement phase transition to quark matter. They will also increase the number and the quality of multimessenger observations, boosting the role of detailed simulations in understanding the observed signals. It is therefore expected that in the next years, there will be a strong improvement in the understanding of the behaviour of matter in extreme conditions of density and temperature and consequently of the astrophysical phenomena governed by the EOS of dense matter.

FURTHER READING

Perego, A., Bernuzzi, S., and Radice, D. (2019). Thermodynamics conditions of matter in neutron star mergers. *The European Physical Journal A*, 55: Issue 8, article id. 124.

Radice, D., Bernuzzi, S. and Perego, A. (2020). The Dynamics of Binary Neutron Star Mergers and GW170817 . *Annual Review of Nuclear and Particle Science,* 70: 95–119.

Shibata, M. and Hotokezaka, K. (2019). Merger and Mass Ejection of Neutron Star Binaries. *Annual Review of Nuclear and Particle Science,* 69: 41–64.

Oertel, M., Hempel M., Klaehn, T. and Typel, S. (2017). Equations of state for supernovae and compact stars. *Reviews of Modern Physics*89, 1, id.015007.

Bibliography

[1] A. Y. Potekhin, J. A. Pons and D. Page, Space Sci. Rev. **191** (2015) no.1-4, 239-291 doi:10.1007/s11214-015-0180-9 [arXiv:1507.06186 [astro-ph.HE]].

[2] A. Perego, S. Bernuzzi and D. Radice, Eur. Phys. J. A **55** (2019) no.8, 124 doi:10.1140/epja/i2019-12810-7 [arXiv:1903.07898 [gr-qc]].

[3] L. Blanchet, Living Rev. Rel. **17** (2014), 2 doi:10.12942/lrr-2014-2 [arXiv: 1310.1528 [gr-qc]].

[4] M. Maggiore, Oxford University Press, 2007, ISBN 978-0-19-171766-6, 978-0-19-852074-0 doi:10.1093/acprof:oso/9780198570745.001.0001.

[5] M. Maggiore, Oxford University Press, 2018, ISBN 978-0-19-857089-9.

[6] T. Damour and A. Nagar, Phys. Rev. D **81** (2010), 084016 doi:10.1103/PhysRevD.81.084016 [arXiv:0911.5041 [gr-qc]].

[7] T. Damour and A. Nagar, Phys. Rev. D **80** (2009), 084035 doi:10.1103/PhysRevD.80.084035 [arXiv:0906.0096 [gr-qc]].

[8] T. Hinderer, Astrophys. J. **677** (2008), 1216-1220 doi:10.1086/533487 [arXiv:0711.2420 [astro-ph]].

[9] T. Binnington and E. Poisson, Phys. Rev. D **80** (2009), 084018 doi:10.1103/PhysRevD.80.084018 [arXiv:0906.1366 [gr-qc]].

[10] J. R. Oppenheimer and G. M. Volkoff, Phys. Rev. **55** (1939), 374–381 doi:10.1103/PhysRev.55.374.

[11] D. Lai, Mon. Not. Roy. Astron. Soc. **270** (1994), 611 doi:10.1093/mnras/270.3.611 [arXiv:astro-ph/9404062 [astro-ph]].

[12] S. Bernuzzi, D. Radice, C. D. Ott, L. F. Roberts, P. Moesta and F. Galeazzi, Phys. Rev. D **94** (2016) no.2, 024023 doi:10.1103/PhysRevD.94.024023 [arXiv:1512.06397 [gr-qc]].

[13] W. Kastaun, R. Ciolfi and B. Giacomazzo, Phys. Rev. D **94** (2016) no.4, 044060 doi:10.1103/PhysRevD.94.044060 [arXiv:1607.02186 [astro-ph.HE]].

[14] D. Radice, A. Perego, S. Bernuzzi and B. Zhang, Mon. Not. Roy. Astron. Soc. **481** (2018) no.3, 3670-3682 doi:10.1093/mnras/sty2531 [arXiv:1803.10865 [astro-ph.HE]].

[15] D. Radice, A. Perego, K. Hotokezaka, S. A. Fromm, S. Bernuzzi and L. F. Roberts, Astrophys. J. **869** (2018) no.2, 130 doi:10.3847/1538-4357/aaf054 [arXiv:1809.11161 [astro-ph.HE]].

[16] V. Nedora, S. Bernuzzi, D. Radice, B. Daszuta, A. Endrizzi, A. Perego, A. Prakash, M. Safarzadeh, F. Schianchi and D. Logoteta, Astrophys. J. **906** (2021) no.2, 98 doi:10.3847/1538-4357/abc9be [arXiv:2008.04333 [astro-ph.HE]].

[17] S. Bernuzzi, M. Breschi, B. Daszuta, A. Endrizzi, D. Logoteta, V. Nedora, A. Perego, F. Schianchi, D. Radice and F. Zappa, *et al.* Mon. Not. Roy. Astron. Soc. **497** (2020) no.2, 1488-1507 doi:10.1093/mnras/staa1860 [arXiv:2003.06015 [astro-ph.HE]].

[18] D. Eichler, M. Livio, T. Piran and D. N. Schramm, Nature **340** (1989), 126-128 doi:10.1038/340126a0.

[19] S. W. Bruenn, Astrophys. J. Suppl. **58** (1985), 771-841 doi:10.1086/191056.

[20] M. H. Ruffert, H. T. Janka and G. Schaefer, Astron. Astrophys. **311** (1996), 532-566 [arXiv:astro-ph/9509006 [astro-ph]].

[21] S. Rosswog and M. Liebendoerfer, Mon. Not. Roy. Astron. Soc. **342** (2003), 673 doi:10.1046/j.1365-8711.2003.06579.x [arXiv:astro-ph/0302301 [astro-ph]].

[22] A. Burrows, S. Reddy and T. A. Thompson, Nucl. Phys. A **777** (2006), 356-394 doi:10.1016/j.nuclphysa.2004.06.012 [arXiv:astro-ph/0404432 [astro-ph]].

[23] A. Endrizzi, A. Perego, F. M. Fabbri, L. Branca, D. Radice, S. Bernuzzi, B. Giacomazzo, F. Pederiva and A. Lovato, Eur. Phys. J. A **56** (2020) no.1, 15 doi:10.1140/epja/s10050-019-00018-6 [arXiv:1908.04952 [astro-ph.HE]].

[24] J. M. Lattimer and D. N. Schramm, Astrophys. J. Lett. **192** (1974), L145 doi:10.1086/181612.

[25] J. J. Cowan, C. Sneden, J. E. Lawler, A. Aprahamian, M. Wiescher, K. Langanke, G. Martínez-Pinedo and F. K. Thielemann, Rev. Mod. Phys. **93** (2021) no.1, 15002 doi:10.1103/RevModPhys.93.015002 [arXiv:1901.01410 [astro-ph.HE]].

[26] A. Perego, F. K. Thielemann and G. Cescutti, doi:10.1007/978-981-15-4702-7_13-1 [arXiv:2109.09162 [astro-ph.HE]].

[27] S. Rosswog, T. Piran and E. Nakar, Mon. Not. Roy. Astron. Soc. **430** (2013), 2585 doi:10.1093/mnras/sts708 [arXiv:1204.6240 [astro-ph.HE]].

[28] A. Bauswein, S. Goriely and H. T. Janka, Astrophys. J. **773** (2013), 78 doi:10.1088/0004-637X/773/1/78 [arXiv:1302.6530 [astro-ph.SR]].

[29] K. Hotokezaka, K. Kiuchi, K. Kyutoku, H. Okawa, Y. i. Sekiguchi, M. Shibata and K. Taniguchi, Phys. Rev. D **87** (2013), 024001 doi:10.1103/PhysRevD.87.024001 [arXiv:1212.0905 [astro-ph.HE]].

[30] Y. Sekiguchi, K. Kiuchi, K. Kyutoku and M. Shibata, Phys. Rev. D **91** (2015) no.6, 064059 doi:10.1103/PhysRevD.91.064059 [arXiv:1502.06660 [astro-ph.HE]].

[31] R. Fernández and B. D. Metzger, Mon. Not. Roy. Astron. Soc. **435** (2013), 502 doi:10.1093/mnras/stt1312 [arXiv:1304.6720 [astro-ph.HE]].

[32] O. Just, A. Bauswein, R. A. Pulpillo, S. Goriely and H. T. Janka, Mon. Not. Roy. Astron. Soc. **448** (2015) no.1, 541-567 doi:10.1093/mnras/stv009 [arXiv:1406.2687 [astro-ph.SR]].

[33] D. M. Siegel and B. D. Metzger, Phys. Rev. Lett. **119** (2017) no.23, 231102 doi:10.1103/PhysRevLett.119.231102 [arXiv:1705.05473 [astro-ph.HE]].

[34] S. Fujibayashi, M. Shibata, S. Wanajo, K. Kiuchi, K. Kyutoku and Y. Sekiguchi, Phys. Rev. D **101** (2020) no.8, 083029 doi:10.1103/PhysRevD.101.083029 [arXiv:2001.04467 [astro-ph.HE]].

[35] A. Perego, S. Rosswog, R. M. Cabezón, O. Korobkin, R. Käppeli, A. Arcones and M. Liebendörfer, Mon. Not. Roy. Astron. Soc. **443** (2014) no.4, 3134-3156 doi:10.1093/mnras/stu1352 [arXiv:1405.6730 [astro-ph.HE]].

[36] D. M. Siegel, R. Ciolfi and L. Rezzolla, Astrophys. J. Lett. **785** (2014), L6 doi:10.1088/2041-8205/785/1/L6 [arXiv:1401.4544 [astro-ph.HE]].

[37] D. Martin, A. Perego, A. Arcones, F. K. Thielemann, O. Korobkin and S. Rosswog, Astrophys. J. **813** (2015) no.1, 2 doi:10.1088/0004-637X/813/1/2 [arXiv:1506.05048 [astro-ph.SR]].

[38] R. Fernández, A. Tchekhovskoy, E. Quataert, F. Foucart and D. Kasen, Mon. Not. Roy. Astron. Soc. **482** (2019) no.3, 3373-3393 doi:10.1093/mnras/sty2932 [arXiv:1808.00461 [astro-ph.HE]].

[39] I. M. Christie, A. Lalakos, A. Tchekhovskoy, R. Fernández, F. Foucart, E. Quataert and D. Kasen, Mon. Not. Roy. Astron. Soc. **490** (2019) no.4, 4811-4825 doi:10.1093/mnras/stz2552 [arXiv:1907.02079 [astro-ph.HE]].

[40] B. D. Metzger, A. L. Piro and E. Quataert, Mon. Not. Roy. Astron. Soc. **390** (2008), 781 doi:10.1111/j.1365-2966.2008.13789.x [arXiv:0805.4415 [astro-ph]].

[41] B. D. Metzger and R. Fernández, Mon. Not. Roy. Astron. Soc. **441** (2014), 3444-3453 doi:10.1093/mnras/stu802 [arXiv:1402.4803 [astro-ph.HE]].

[42] J. Lippuner, R. Fernández, L. F. Roberts, F. Foucart, D. Kasen, B. D. Metzger and C. D. Ott, Mon. Not. Roy. Astron. Soc. **472** (2017) no.1, 904-918 doi:10.1093/mnras/stx1987 [arXiv:1703.06216 [astro-ph.HE]].

[43] V. Nedora, S. Bernuzzi, D. Radice, A. Perego, A. Endrizzi and N. Ortiz, Astrophys. J. Lett. **886** (2019) no.2, L30 doi:10.3847/2041-8213/ab5794 [arXiv:1907.04872 [astro-ph.HE]].

[44] D. Radice, S. Bernuzzi and C. D. Ott, Phys. Rev. D **94** (2016) no.6, 064011 doi:10.1103/PhysRevD.94.064011 [arXiv:1603.05726 [gr-qc]].

[45] L. X. Li and B. Paczynski, Astrophys. J. Lett. **507** (1998), L59 doi:10.1086/311680 [arXiv:astro-ph/9807272 [astro-ph]].

[46] B. D. Metzger, Living Rev. Rel. **23** (2020) no.1, 1 doi:10.1007/s41114-019-0024-0 [arXiv:1910.01617 [astro-ph.HE]].

[47] C. J. Horowitz and J. Piekarewicz, Phys. Rev. C **64** (2001), 062802 doi:10.1103/PhysRevC.64.062802 [arXiv:nucl-th/0108036 [nucl-th]].

[48] S. Balberg and A. Gal, Nucl. Phys. A **625** (1997), 435-472 doi:10.1016/S0375-9474(97)81465-0 [arXiv:nucl-th/9704013 [nucl-th]].

[49] B. D. Day, Rev. Mod. Phys. **39** (1967), 719-744 doi:10.1103/RevModPhys.39.719.

[50] H. Q. Song, M. Baldo, G. Giansiracusa and U. Lombardo, Phys. Rev. Lett. **81** (1998), 1584-1587 doi:10.1103/PhysRevLett.81.1584.

[51] J. W. Holt, N. Kaiser and W. Weise, Phys. Rev. C **81** (2010), 024002 doi:10.1103/PhysRevC.81.024002 [arXiv:0910.1249 [nucl-th]].

[52] D. Logoteta, Phys. Rev. C **100** (2019) no.4, 045803 doi:10.1103/PhysRevC.100.045803.

[53] M. I. Haftel and F. Tabakin, Nucl. Phys. A **158** (1970), 1-42 doi:10.1016/0375-9474(70)90047-3.

[54] D. P. Menezes and C. Providencia, Phys. Rev. C **60** (1999), 024313 doi:10.1103/PhysRevC.60.024313.

[55] S. S. Avancini, M. E. Bracco, M. Chiapparini and D. P. Menezes, Phys. Rev. C **67** (2003), 024301 doi:10.1103/PhysRevC.67.024301 [arXiv:nucl-th/0212080 [nucl-th]].

[56] A. S. Schneider, L. F. Roberts and C. D. Ott, Phys. Rev. C **96** (2017) no.6, 065802 doi:10.1103/PhysRevC.96.065802 [arXiv:1707.01527 [astro-ph.HE]].

[57] D. Logoteta, A. Perego and I. Bombaci, Astron. Astrophys. **646** (2021), A55 doi:10.1051/0004-6361/202039457 [arXiv:2012.03599 [nucl-th]].

[58] M. Hempel and J. Schaffner-Bielich, Nucl. Phys. A **837** (2010), 210-254 doi:10.1016/j.nuclphysa.2010.02.010 [arXiv:0911.4073 [nucl-th]].

[59] M. Hempel, T. Fischer, J. Schaffner-Bielich and M. Liebendorfer, Astrophys. J. **748** (2012), 70 doi:10.1088/0004-637X/748/1/70 [arXiv:1108.0848 [astro-ph.HE]].

[60] A. W. Steiner, J. M. Lattimer and E. F. Brown, Astrophys. J. Lett. **765** (2013), L5 doi:10.1088/2041-8205/765/1/L5 [arXiv:1205.6871 [nucl-th]].

[61] S. Typel, G. Ropke, T. Klahn, D. Blaschke and H. H. Wolter, Phys. Rev. C **81** (2010), 015803 doi:10.1103/PhysRevC.81.015803 [arXiv:0908.2344 [nucl-th]].

[62] B. Ter Haar and R. Malfliet, Phys. Rept. **149** (1987), 207-286 doi:10.1016/0370-1573(87)90085-8

[63] J. M. Lattimer and F. D. Swesty, Nucl. Phys. A **535** (1991), 331-376 doi:10.1016/0375-9474(91)90452-C

[64] F. Douchin and P. Haensel, Astron. Astrophys. **380** (2001), 151 doi:10.1051/0004-6361:20011402 [arXiv:astro-ph/0111092 [astro-ph]].

[65] R. Machleidt and D. R. Entem, Phys. Rept. **503** (2011), 1-75 doi:10.1016/j.physrep.2011.02.001 [arXiv:1105.2919 [nucl-th]].

[66] M. Piarulli, L. Girlanda, R. Schiavilla, A. Kievsky, A. Lovato, L. E. Marcucci, S. C. Pieper, M. Viviani and R. B. Wiringa, Phys. Rev. C **94** (2016) no.5, 054007 doi:10.1103/PhysRevC.94.054007 [arXiv:1606.06335 [nucl-th]].

[67] D. Logoteta, I. Bombaci and A. Kievsky, Phys. Rev. C **94** (2016) no.6, 064001 doi:10.1103/PhysRevC.94.064001 [arXiv:1609.00649 [nucl-th]].

[68] B. P. Abbott *et al.* [LIGO Scientific and Virgo], Phys. Rev. X **9** (2019) no.1, 011001 doi:10.1103/PhysRevX.9.011001 [arXiv:1805.11579 [gr-qc]].

[69] B. P. Abbott *et al.* [LIGO Scientific and Virgo], Astrophys. J. Lett. **892** (2020) no.1, L3 doi:10.3847/2041-8213/ab75f5 [arXiv:2001.01761 [astro-ph.HE]].

[70] T. W. Baumgarte, S. L. Shapiro and M. Shibata, Astrophys. J. Lett. **528** (2000), L29 doi:10.1086/312425 [arXiv:astro-ph/9910565 [astro-ph]].

[71] K. Hotokezaka, K. Kyutoku, H. Okawa, M. Shibata and K. Kiuchi, Phys. Rev. D **83** (2011), 124008 doi:10.1103/PhysRevD.83.124008 [arXiv:1105.4370 [astro-ph.HE]].

[72] A. Bauswein, T. W. Baumgarte and H. T. Janka, Phys. Rev. Lett. **111** (2013) no.13, 131101 doi:10.1103/PhysRevLett.111.131101 [arXiv:1307.5191 [astro-ph.SR]].

[73] S. Köppel, L. Bovard and L. Rezzolla, Astrophys. J. Lett. **872** (2019) no.1, L16 doi:10.3847/2041-8213/ab0210 [arXiv:1901.09977 [gr-qc]].

[74] R. Kashyap, A. Das, D. Radice, S. Padamata, A. Prakash, D. Logoteta, A. Perego, D. A. Godzieba, S. Bernuzzi and I. Bombaci, *et al.* Phys. Rev. D **105** (2022) no.10, 103022 doi:10.1103/PhysRevD.105.103022 [arXiv:2111.05183 [astro-ph.HE]].

[75] D. Radice, L. Rezzolla and F. Galeazzi, ASP Conf. Ser. **498** (2015), 121-126 [arXiv:1502.00551 [gr-qc]].

[76] D. Radice, L. Rezzolla and F. Galeazzi, Class. Quant. Grav. **31** (2014), 075012 doi:10.1088/0264-9381/31/7/075012 [arXiv:1312.5004 [gr-qc]].

[77] D. Radice, L. Rezzolla and F. Galeazzi, Mon. Not. Roy. Astron. Soc. **437** (2014), L46-L50 doi:10.1093/mnrasl/slt137 [arXiv:1306.6052 [gr-qc]].

[78] A. Bauswein, S. Blacker, V. Vijayan, N. Stergioulas, K. Chatziioannou, J. A. Clark, N. U. F. Bastian, D. B. Blaschke, M. Cierniak and T. Fischer, Phys. Rev. Lett. **125** (2020) no.14, 141103 doi:10.1103/PhysRevLett.125.141103 [arXiv:2004.00846 [astro-ph.HE]].

[79] S. D. Tootle, L. J. Papenfort, E. R. Most and L. Rezzolla, Astrophys. J. Lett. **922** (2021) no.1, L19 doi:10.3847/2041-8213/ac350d [arXiv:2109.00940 [gr-qc]].

[80] M. Kölsch, T. Dietrich, M. Ujevic and B. Bruegmann, Phys. Rev. D **106** (2022) no.4, 044026 doi:10.1103/PhysRevD.106.044026 [arXiv:2112.11851 [gr-qc]].

[81] A. Perego, D. Logoteta, D. Radice, S. Bernuzzi, R. Kashyap, A. Das, S. Padamata and A. Prakash, Phys. Rev. Lett. **129** (2022) no.3, 032701 doi:10.1103/PhysRevLett.129.032701 [arXiv:2112.05864 [astro-ph.HE]].

[82] M. Cusinato, F. M. Guercilena, A. Perego, D. Logoteta, D. Radice, S. Bernuzzi and S. Ansoldi, doi:10.1140/epja/s10050-022-00743-5 [arXiv:2111.13005 [astro-ph.HE]].

[83] D. Radice, Astrophys. J. Lett. **838** (2017) no.1, L2 doi:10.3847/2041-8213/aa6483 [arXiv:1703.02046 [astro-ph.HE]].

[84] D. Radice, Symmetry **12** (2020) no.8, 1249 doi:10.3390/sym12081249 [arXiv:2005.09002 [astro-ph.HE]].

[85] A. Camilletti, L. Chiesa, G. Ricigliano, A. Perego, L. C. Lippold, S. Padamata, S. Bernuzzi, D. Radice, D. Logoteta and F. M. Guercilena, doi:10.1093/mnras/stac2333 [arXiv:2204.05336 [astro-ph.HE]].

[86] F. Zappa, S. Bernuzzi, D. Radice, A. Perego and T. Dietrich, Phys. Rev. Lett. **120** (2018) no.11, 111101 doi:10.1103/PhysRevLett.120.111101 [arXiv:1712.04267 [gr-qc]].

[87] A. Bauswein and H. T. Janka, Phys. Rev. Lett. **108** (2012), 011101 doi:10.1103/PhysRevLett.108.011101 [arXiv:1106.1616 [astro-ph.SR]].

[88] K. Hotokezaka, K. Kiuchi, K. Kyutoku, T. Muranushi, Y. i. Sekiguchi, M. Shibata and K. Taniguchi, Phys. Rev. D **88** (2013), 044026 doi:10.1103/PhysRevD.88.044026 [arXiv:1307.5888 [astro-ph.HE]].

[89] S. Bernuzzi, T. Dietrich and A. Nagar, Phys. Rev. Lett. **115** (2015) no.9, 091101 doi:10.1103/PhysRevLett.115.091101 [arXiv:1504.01764 [gr-qc]].

[90] K. Chatziioannou, J. A. Clark, A. Bauswein, M. Millhouse, T. B. Littenberg and N. Cornish, Phys. Rev. D **96** (2017) no.12, 124035 doi:10.1103/PhysRevD.96.124035 [arXiv:1711.00040 [gr-qc]].

[91] P. J. Easter, P. D. Lasky, A. R. Casey, L. Rezzolla and K. Takami, Phys. Rev. D **100** (2019) no.4, 043005 doi:10.1103/PhysRevD.100.043005 [arXiv:1811.11183 [gr-qc]].

[92] K. W. Tsang, T. Dietrich and C. Van Den Broeck, Phys. Rev. D **100** (2019) no.4, 044047 doi:10.1103/PhysRevD.100.044047 [arXiv:1907.02424 [gr-qc]].

[93] D. Radice, F. Galeazzi, J. Lippuner, L. F. Roberts, C. D. Ott and L. Rezzolla, Mon. Not. Roy. Astron. Soc. **460** (2016) no.3, 3255-3271 doi:10.1093/mnras/stw1227 [arXiv:1601.02426 [astro-ph.HE]].

[94] B. P. Abbott *et al.* [LIGO Scientific and Virgo], Phys. Rev. Lett. **121** (2018) no.16, 161101 doi:10.1103/PhysRevLett.121.161101 [arXiv:1805.11581 [gr-qc]].

[95] S. De, D. Finstad, J. M. Lattimer, D. A. Brown, E. Berger and C. M. Biwer, Phys. Rev. Lett. **121** (2018) no.9, 091102 [erratum: Phys. Rev. Lett. **121** (2018) no.25, 259902] doi:10.1103/PhysRevLett.121.091102 [arXiv:1804.08583 [astro-ph.HE]].

[96] D. Radice, A. Perego, F. Zappa and S. Bernuzzi, Astrophys. J. Lett. **852** (2018) no.2, L29 doi:10.3847/2041-8213/aaa402 [arXiv:1711.03647 [astro-ph.HE]].

[97] B. P. Abbott *et al.* [LIGO Scientific and Virgo], Phys. Rev. Lett. **119** (2017) no.16, 161101 doi:10.1103/PhysRevLett.119.161101 [arXiv:1710.05832 [gr-qc]].

[98] V. Nedora, F. Schianchi, S. Bernuzzi, D. Radice, B. Daszuta, A. Endrizzi, A. Perego, A. Prakash and F. Zappa, Class. Quant. Grav. **39** (2022) no.1, 015008 doi:10.1088/1361-6382/ac35a8 [arXiv:2011.11110 [astro-ph.HE]].

[99] M. Breschi, A. Perego, S. Bernuzzi, W. Del Pozzo, V. Nedora, D. Radice and D. Vescovi, Mon. Not. Roy. Astron. Soc. **505** (2021) no.2, 1661-1677 doi:10.1093/mnras/stab1287 [arXiv:2101.01201 [astro-ph.HE]].

[100] A. Perego, D. Radice and S. Bernuzzi, Astrophys. J. Lett. **850** (2017) no.2, L37 doi:10.3847/2041-8213/aa9ab9 [arXiv:1711.03982 [astro-ph.HE]].

[101] D. Radice and L. Dai, Eur. Phys. J. A **55** (2019) no.4, 50 doi:10.1140/epja/i2019-12716-4 [arXiv:1810.12917 [astro-ph.HE]].

[102] R. Gamba, S. Bernuzzi and A. Nagar, Phys. Rev. D **104** (2021) no.8, 084058 doi:10.1103/PhysRevD.104.084058 [arXiv:2012.00027 [gr-qc]].

[103] E. Annala, T. Gorda, A. Kurkela and A. Vuorinen, Phys. Rev. Lett. **120** (2018) no.17, 172703 doi:10.1103/PhysRevLett.120.172703 [arXiv:1711.02644 [astro-ph.HE]].

[104] M. W. Coughlin, T. Dietrich, B. Margalit and B. D. Metzger, Mon. Not. Roy. Astron. Soc. **489** (2019) no.1, L91-L96 doi:10.1093/mnrasl/slz133 [arXiv:1812.04803 [astro-ph.HE]].

[105] G. Raaijmakers, S. K. Greif, T. E. Riley, T. Hinderer, K. Hebeler, A. Schwenk, A. L. Watts, S. Nissanke, S. Guillot and J. M. Lattimer, *et al.* Astrophys. J. Lett. **893** (2020) no.1, L21 doi:10.3847/2041-8213/ab822f [arXiv:1912.11031 [astro-ph.HE]].

[106] C. D. Capano, I. Tews, S. M. Brown, B. Margalit, S. De, S. Kumar, D. A. Brown, B. Krishnan and S. Reddy, Nature Astron. **4** (2020) no.6, 625-632 doi:10.1038/s41550-020-1014-6 [arXiv:1908.10352 [astro-ph.HE]].

[107] R. Essick, I. Tews, P. Landry, S. Reddy and D. E. Holz, Phys. Rev. C **102** (2020) no.5, 055803 doi:10.1103/PhysRevC.102.055803 [arXiv:2004.07744 [astro-ph.HE]].

[108] T. Dietrich, M. W. Coughlin, P. T. H. Pang, M. Bulla, J. Heinzel, L. Issa, I. Tews and S. Antier, Science **370** (2020) no.6523, 1450-1453 doi:10.1126/science.abb4317 [arXiv:2002.11355 [astro-ph.HE]].

[109] B. D. Metzger, A. Bauswein, S. Goriely and D. Kasen, Mon. Not. Roy. Astron. Soc. **446** (2015), 1115-1120 doi:10.1093/mnras/stu2225 [arXiv:1409.0544 [astro-ph.HE]].

[110] A. Perego, D. Vescovi, A. Fiore, L. Chiesa, C. Vogl, S. Benetti, S. Bernuzzi, M. Branchesi, E. Cappellaro and S. Cristallo, *et al.* Astrophys. J. **925** (2022) no.1, 22 doi:10.3847/1538-4357/ac3751 [arXiv:2009.08988 [astro-ph.HE]].

[111] K. Hotokezaka and T. Piran, Mon. Not. Roy. Astron. Soc. **450** (2015), 1430-1440 doi:10.1093/mnras/stv620 [arXiv:1501.01986 [astro-ph.HE]].

[112] V. Nedora, D. Radice, S. Bernuzzi, A. Perego, B. Daszuta, A. Endrizzi, A. Prakash and F. Schianchi, Mon. Not. Roy. Astron. Soc. **506** (2021) no.4, 5908-5915 doi:10.1093/mnras/stab2004 [arXiv:2104.04537 [astro-ph.HE]].

[113] A. Hajela, R. Margutti, J. S. Bright, K. D. Alexander, B. D. Metzger, V. Nedora, A. Kathirgamaraju, B. Margalit, D. Radice and C. Guidorzi, *et al.* Astrophys. J. Lett. **927** (2022) no.1, L17 doi:10.3847/2041-8213/ac504a [arXiv:2104.02070 [astro-ph.HE]].

[114] A. Bauswein, O. Just, H. T. Janka and N. Stergioulas, Astrophys. J. Lett. **850** (2017) no.2, L34 doi:10.3847/2041-8213/aa9994 [arXiv:1710.06843 [astro-ph.HE]].

[115] L. Rezzolla, E. R. Most and L. R. Weih, Astrophys. J. Lett. **852** (2018) no.2, L25 doi:10.3847/2041-8213/aaa401 [arXiv:1711.00314 [astro-ph.HE]].

[116] M. Breschi, S. Bernuzzi, D. Godzieba, A. Perego and D. Radice, Phys. Rev. Lett. **128** (2022) no.16, 161102 doi:10.1103/PhysRevLett.128.161102 [arXiv:2110.06957 [gr-qc]].

[117] A. Gonzalez, F. Zappa, M. Breschi, S. Bernuzzi, D. Radice, A. Adhikari, A. Camilletti, S. V. Chaurasia, G. Doulis and S. Padamata, *et al.* Class. Quant. Grav. **40** (2023) no.8, 085011 doi:10.1088/1361-6382/acc231 [arXiv:2210.16366 [gr-qc]].

[118] M. Breschi, R. Gamba and S. Bernuzzi, Phys. Rev. D **104** (2021) no.4, 042001 doi:10.1103/PhysRevD.104.042001 [arXiv:2102.00017 [gr-qc]].

[119] M. Breschi, S. Bernuzzi, F. Zappa, M. Agathos, A. Perego, D. Radice and A. Nagar, Phys. Rev. D **100** (2019) no.10, 104029 doi:10.1103/PhysRevD.100.104029 [arXiv:1908.11418 [gr-qc]].

[120] D. A. Godzieba, D. Radice and S. Bernuzzi, Astrophys. J. **908** (2021) no.2, 122 doi:10.3847/1538-4357/abd4dd [arXiv:2007.10999 [astro-ph.HE]].

[121] J. Fields, A. Prakash, M. Breschi, D. Radice, S. Bernuzzi and A. da Silva Schneider, Astrophys. J. Lett. **952** (2023) no.2, L36 doi:10.3847/2041-8213/ace5b2 [arXiv:2302.11359 [astro-ph.HE]].

[122] D. Radice, S. Bernuzzi, A. Perego and R. Haas, Mon. Not. Roy. Astron. Soc. **512** (2022) no.1, 1499-1521 doi:10.1093/mnras/stac589 [arXiv:2111.14858 [astro-ph.HE]].

[123] C. Constantinou, B. Muccioli, M. Prakash and J. M. Lattimer, Phys. Rev. C **89** (2014) no.6, 065802 doi:10.1103/PhysRevC.89.065802 [arXiv:1402.6348 [astro-ph.SR]].

[124] E. Loffredo, A. Perego, D. Logoteta and M. Branchesi, Astron. Astrophys. **672** (2023), A124 doi = ”10.1051/0004-6361/202244927 [arXiv: 2209.04458 [astro-ph.HE]]

Index